Contents

1
Introduction

Taxonomy is one of the oldest fields of biology with a long history of development. Man from time immemorial has known and been dependent on the plant world for innumerable needs. This dependency urged even the prehistoric man to identify the plants and also to classify them into different groups, such as food plants, poisonous plants, medicinal plants, etc., which were important for him. This was the beginning of plant taxonomy which included **identification, nomenctature** and **classification** of plants. From the simple methods of recognition of economically useful plants by the earliest man, today it has become a highly complex and all-embracing biological science.

Plant taxonomy or systematics has long been treated as a primitive or orthodox science, in the sense that it is primarily based on morphology. It is also the most controversial, misunderstood, and maligned subject. These adjectives can be applied because of the very nature of the subject. One of the aims of taxonomy is to provide help to non-taxonomists. Its principles and practices are thus more often scrutinised by non-specialists, as compared to other sciences. Much of the distrust and criticism of this subject is due to the lack of understanding of these users.

In fact, the concept of taxonomy or systematics (the two terms interchangaeble) has no obvious, finite, single aim or purpose. It is the science of biological diversity and fulfils three important and interconnected roles in modern biology: (a) Systematics is reponsible for synthesizing knowledge about organisms and for integrating information from all other areas of biology into a single internally consistent and coherent understanding of the diversity of life. (b) It is responsible also for developing a classification system that reflects the patterns of evolutionary interrelationship. Thus, it provides a maximally predictive framework for comparative biology and evolutionary studies. (c) Systematics is involved in developing an explicit, universal,and stable system of names.

The two basic goals of systematics are, therefore, conflicting to each other. Accumulation of data from various fields of biology and synthesis of this data will result in an improved and more stable classification. At the same time, continuous addition of new data will not allow a classification system to be stable. This conflict between the two views i.e. stability and change in systematics can be removed by providing new, improved and more comprehensive synthesis of data.

Alpha taxonomists group organisms at different levels of hierarchy simply on the basis of morphological characteristics. Unfortunately, the delimitation of taxa, circumscription of taxa and phylogenetic relationships amongst various taxa, are controversial particularly in groups where there is no discrete morphological variability pattern. Hence, there is need to look for other details which can be accumulated from various sources—research performed in different herbaria of the world, in laboratories, in field and garden experiments, and also from libraries. Not only that, even studies on fossils have helped taxonomists to learn about the origin, early evolution and phylogeny of angiosperms. Much new information from paleobotanical studies and from comparative studies of gymnospermous groups and ferns have recently been very helpful in preliminary recognition of ancestral angiosperms and their characteristics. Presently another field that is gaining importance is Molecular Systematics. In recent years advances in molecular genetic techniques have revolutionized the study of phylogeny. Molecular investigations—particularly sequence data form ribosomal DNA have been applied to questions ranging from infraspecific variability to the evolution of green plants.

Duties of a taxonomist: The knowledge of plant taxonomy is helpful to the common man in various ways. The name of wild plant may be of little general interest, but to a plant breeder it may prove to be a useful plant in having important genes for disease resistance or better yield. It is a great responsibility of plant taxonomists to provide the correct names and classification of such plants. Similarly, a chemist may find a new source of an important chemical, but the name of this source plant has to be provided by the taxonomist. Extensive use of this plant science has been made in many fields. In forestry, for example, trees must be named and classified because of their innumerable economic importance. Forest lands are often leased out for grazing. A taxonomist would be able to guide as to which plants are palatable to cattle, what methods of propagation must be adopted for the plants, or what should be the maximum grazing limit so that plants do not totally disappear due to overgrazing. In agriculture also the knowledge of taxonomy is useful in various ways. A taxonomist is ready to help a plant breeder with his broad knowledge of which plants can be used as suitable stock for hybridisation experiments. Even the introduction of seeds and plants from one country to another benefits from the expertise of a taxonomist. In range management also, plant taxonomists play an important role, by suggesting which plant species have better soil-binding capacity, or which plant can act as a suitable windbreak so that erosion does not take place.

Ecologists with sound knowledge of plant taxonomy can easily idenify soil type or climatic condition of a region from the plants that are growing there. These are only some of the duties performed by plant taxonomists.

Aims and Objectives. Plant taxonomy is primarily concerned with the identification, nomenclature, classification and evolutionary relationship between diverse plant groups.

Identification. Is the determination of a specimen as being identical to or different from a previously known plant. A previously known specimen will not always be available, and the unknown specimen might prove to be a new one. The process of naming is usually not involved at this stage. For example, of the three specimens A, B and C, A is *Solanum nigrum* and B and C are unknown. Specimen B is similar to A and therefore it is *S. nigrum*; but specimen C is dissimilar and therefore further comparison with additional known specimens is advisable. However, there are also other methods of identification, such as consulting (a) a flora, monograph or revision, (b) a herbarium specimen, or (c) a garden specimen, and the comparison may provide the information required. If, however, a specimen shows no similarity to any known specimen, then it may be counted as a completely new taxon.

Nomenclature. Simple indentification is not enough. These identified specimens must be given correct, valid names—a name by which these plants can be made known to the rest of the world. Nomenclature is an orderly application of names for which certain rules have to be followed. As it is of international importance, the rules for nomenclature are listed in the International Code of Botanical Nomenclature (ICBN). The basic rule is that binomial nomenclature should be followed for naming all specimens that are treated as plants. This means that each plant must have two names—the first is the generic name and the second the specific epithet. Apart from this basic rule, many others are proposed by the International Association for Plant Taxonomy (IAPT). These rules are finalized every 5 years during the International Botanical Congress.

Classification. Classification is an arrangement of plants or plant groups hierarchially. The various major units of classification are divisions, subdivisions, classes, subclasses, orders, suborders, tribes, genera and species. There are infraspecific units too, apart from these nine major ones, and all these together constitute the plant kingdom. Different classification criteria have given rise to different system of classification. As civilized man came to understand the utility of the plants, he classified them as food, fruit, medicine and many such groups.

At different times, different bases have been used for classification. Many classificatory systems of earlier times were based on economic uses of plants or their habits; some of the systems suggested by herbalists were based on medicinal uses of plants. These systems were, however, incomplete in the sense that those plants which did not fit into these systems were ignored. The only system of classification that was complete, at least for the plants known until that time, was the one based on the natural relationship amongst the members of different plant groups. Even this "natural relationship" meant only similarity of plant parts based on the study of comparative morphology.

In fact, a natural classification should depict the natural relationship existing among plants as they evolved from the ancestral to the most advanced forms. All the later classificatory systems, such as those of Cronquist (1968, 1981), Dahlgren (1983a), Thorne (1992b) and Takhtajan (1987), are both natural and phylogenetic.

Molecular systematics has certainly revolutionised plant taxonomy but still it is not the ultimate. Moreover, such sophisticated techniques may not be easily available to every taxonomist. Classical taxonomy still remains the foundation for all taxonomic studies, be it identification, classification or establishing origin and evolution. Molecular taxonomic tools are complementary to the earlier used tools such as herbarium and botanical gardens and libraries. The knowledge synthesized by classical taxonomist is further strengthened by molecular evidences.

A taxonomist is a collector, analyser and synthesizer of information from all fields of evidence for use in characterization, identification and classification of organisms.

2

Taxonomic Hierarchy

Living organisms exhibit different degrees of diversity. The entire diversity cannot be classified into one group and hence, we recognise different taxonomic groups depending upon different levels of diversity. Biological classification is built up from populations and species at the very base. A *species* is a group of individuals which resemble each other very closely and are interbreeding. Those *species* which share many common characters are placed together in a larger group called *genus*; these in turn are assembled in a still larger group called *family* and so on. These different groups show a box within box arrangement although the size of the boxes may vary (Fig 2.1).

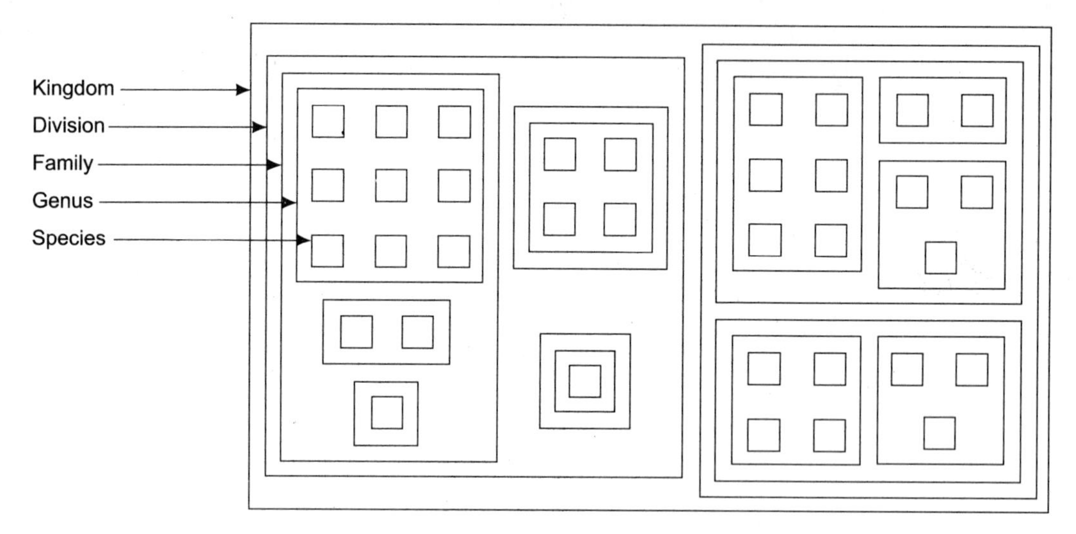

Fig. 2.1 Box-in-box arrangement of taxa in hierarchial order. Each rectangle represents a taxon.

Parallel with these different levels of groups is a series of categories forming a hierarchy. The hierarchy of categories is like a set of empty shelves arranged at different heights. Once formed, the groups have to be assigned to a particular category. It is the category into which we place a group, that determines the group name.

Taxonomic categories are artificial, highly subjective, and can be defined only by their position relative to other categories. The taxonomic groups are natural, can be defined and are objective. A taxonomic group is made up of lower taxonomic groups but a category is not made up of lower categories, as shown here (Table 2.1).

This arrangement of taxonomic categories in an ascending series ranging from the lowest to the highest category is *Taxonomic hierarchy*. The system in which different taxonomic groups are assigned to different taxonomic categories, depending upon the different levels of diversity of the organism, is known as *Taxonomic structure*. 1CBN has recognised 21 different taxonomic categories for classification of plants.

Table 2.1 Taxonomic Categories and Taxonomic Groups

Taxonomic Categories	*Taxonomic Groups*
Kingdom	Plant
Divisio	Spermatophyta
Subdivisio	Angiosperms, Gymnosperms
Classis	Dicotyledons, Monocotyledons
Subclassis	Polypetalae, Gamopetalae
Ordo (Order)	Ranales, Malvales, Parietales
Subordo	
Familia (Family)	Malvaceae, Tiliaceae, Bombacaceae
Subfamilia (Subfamily)	—
Tribus (Tribe)	—
Subtribus (Subtribe)	—
Genus	*Malva, Sida, Abutilon*
Subgenus	—
Sectio (Section)	—
Subsectio (Subsection)	—
Series	—
Subseries	
Species	*M. sylvestris, M. rotundifolia*
Subspecies	—
Varietas (Variety)	—
Subvarietas (Subvariety)	—
Forma	—

Properties of Taxonomic Hierarchy

(1) A plant may be a member of several taxonomic groups each of which is assigned to a taxonomic category, but is not itself a member of any taxonomic category (Table 2.2).

Table 2.2 Taxonomic Structure of *Potentilla glandulosa* var. *navadensis*

Division	Spermatophyta
Subdivision	Angiospermae
Class	Dicotyledonae
Subclass	Polypetalae
Order	Rosales
Suborder	—
Family	Rosaceae
Subfamily	Rosoideae
Tribe	Potentillinae
Subtribe	—
Genus	*Potentilla*
Subgenus	—

(Table 2.2 Contd.)

(Table 2.2 Contd.)

Section	Drynocallus
Subsection	Closterostylae
Series	—
Subseries	—
Species	*glandulosa*
Subspecies	—
Variety	*navadensis*
Subvariety	—
Forma	

The plant *Potentilla glandulosa* subsp *navadensis* belongs to all the taxonomic groups mentioned above but does not belong to any particular category or categories.

(2) The characters that are shared by the members of a taxon of a lower category, constitutute the characters of a taxon of the next higher category. For example,

Medicago lupulina
M. polymorpha
M. sativa
Twisted pods - characteristic of genus *Medicago*

Trigonella corniculata
T. foenum-graecum
T. polycerata
Straight recurved pods - characteristic of genus *Trigonella*

Genus *Medicago*
Genus *Trigonella*

[Trifoliate leaves
Toothed leaflets] > [Characteristics of Genus *Trigonella*]

Tribes Trifolieae, Vicieae etc.

[Papilionaceous corolla
Diadelphous stamens] > Characterstics of Family Papilionaceae

Families Papilionaceae, Caesalpiniaceae, Mimosaceae

[Monocarpellary ovary
Marginal placentation
Fruit a legume] > Characterstic of Order Leguminales

(3) The lower the rank of the taxon in the taxonomic hierarchy the fewer are its members and higher is the number of common characters. For example, in the above structure, Order Leguminales has 3 common characters, Families Papilionaceae, Caesalpiniaceae, and Mimosaceae have 5(3 + 2) common characters, Tribes have 7(3 + 2 + 2) and genera 8(3 + 2 + 2 + 1) common characters.

Order Leguminales	– 600 genera
Family Papilionaceae	– 427 genera
Family Caesalpiniaceae	– 133 genera
Family Mimosaceae	– 40 genera

Advantages of studying Taxonomic hierarchy are:

(1) Grouping of living organisms into different categories on the basis of degrees of diversity exhibited by them is possible.

(2) Such classification helps one to refer a particular group of plants into appropriate classes.

Taxonomic groups like Angiospermae, Monocotyledoneae or Leguminales are composed of larger number and diverse types of plants. These groups are known as Subdivision, Class and Order and are together called Major categories. The categories Genus, Species and Variety are called Minor categories and are of smaller magnitude.

Definition of each of these categories vary, up to a certain point according to indivdual opinion and the state of science. But their relative order must not be altered.

Major Categories. The whole plant kingdom is divided into Divisions although the number of divisions varies in different systems of classification, e.g. in Engler and Prantl's System there are 14 Divisions and the 14th Division is *Embryophyta siphonogama*. These taxa have a few common characters, e.g. the members of this Division have, (a) a more conspicuous sporophytic generation, (b) ovules, (c) produce seeds. Larger categories share less number of common features. According to the Rules of Nomenclature, any category may be divided into intermediate subordinate categories between itself and the next lower rank. And this is done by adding the prefix 'sub' to the name of the higher category.

The *2 subdivisions* of the Division Embryophyta Siphonogama are Gymnospermae and Angiospermae. But it is also possible to have a Division composed of *classes,* i.e. the next lower rank, e.g. Division Pteridophyta does not have any subdivision—and is directly divided into *classes*. The names applied to classes are 'Latin' and have the ending '-eae', e.g. Monocotyledoneae and Dicotyledoneae—the 2 classes of the subdivision Angiopermae, or Class Lycopodiineae, Filicineae and others under division Pteridophyta.

Class Dicotyledoneae comprises such a vast number of members that most workers prefer to divide it into several *Subclasses*. No hard and fast rule is there either for naming or separating taxonomic groups as *Subclasses.* For example, in Takhtajan's classification system, Class Liliopsida is divided into 4 Subclasses; Alismatidae, Liliidae, Commelinidae and Arecidae. But in Engler and Prantl's system there is no category like subclass; classes are directly divided into *Orders.*

Each Class (or Subclass) is next divided into *Orders* which is a rank higher than Family. *Order* name is also 'Latin' and the conventional ending is '-ales' although the ending '-ae' is permissible in some, e.g., Rubiflorae, Glumiflorae, Farinosae etc. Example of '-ales' ending are Ranales, Malvales, Parietales, Sapindales etc. There are more definite features which bind the components of an order together.

Very large orders are divided into *Suborders,* e.g. Potamogetonieae and Alismatinieae are the 2 Suborders of the order Helobieae. Conventional ending of suborders is '-nieae'. Order or suborder is further divided into *families.* Family is the last rank amongst 'major categories' and the customary ending is '-aceae' except for nine families where the earlier given names can also be used, because of their familiarity. These nine families are Capparidaceae, Cruciferae, Guttiferae, Papilionaceae, Umbelliferae, Labiatae, Compositae, Palmae and Gramineae.

Minor Categories. A minor category usually becomes a part of the name of the plant, e.g. *Genus* is a category as well as a part of the name of the plant—*Lathyrus aphaca*. This does not happen with major categories.

Amongst minor categories highest is '*Genus*'. It is next in rank to family. It is first of the two names making a binomial. Genus names are either 'Latin' or latinized from other languages. Genus (if large) may be divided into *subgenera* and these in turn into *Sections, Subsections* and *Series, Subseries* or a genus may include a homogeneous group of plants—of the category *Species*. A *species* is the basic unit of classification, Latin or latinized, and the second part of a binomial; usually have similar endings as the Genus name.

Infraspecific Categories. Any category below the rank of species is Infraspecific category. Most commonly used categories of this group are *Subspecies, Variety* and *Forma*. There are differences of opinion about the definitions of these categories.

Species Concept. The species is for most purposes widely accepted as the basic unit of taxonomy. Every biologist knows what is meant by the term 'species' yet it has been one of the most controversial terms used in taxonomical studies. That species is the lowest unit of biological classification is derived from Aristotalean logic and philosophy (Crombie 1950). Linnaeus believed in fixity of species and that species were divinely created. The idea that species are made of groups of individuals i.e. populations is also from Linnaeus. At the same time the belief of many evolutionary scientist is that species are formed as a result of the evolutionary processes. Varied types of definition of species have been put forward by many authors: (a) as a name in a book, (b) as a judgement, (c) as what a competent authority thinks it is, (d) as a group of individuals which in totality of their attributes resemble each other to a degree usually regarded as specific, (e) as the smallest group to which distinctive characters can be assigned, and f) as a discrete and immutable entity of divine origin. Although some of these features can be applied to define a species, all these features combined together do not give a true concept of species.

Why do we give so much importance to species? Neither their abominable number nor their absolute distinctness from each other make them so important. Species are important because, as Mayr (1942) commented, "they represent an important level of integration in living nature. This recognition is fundamental to pure biology. An inventory of the species of animals and plants of the world can form the basis for further research in biology." Whether one realises or not, every researcher of biological fields (including molecular biology) works with a species or part of species. He or she can communicate the result to the academic world only if the species he/she is working with is correctly identified.

Linnaeus believed in constancy of characters in a species and therefore absolute distinction whereas Darwin (1859) believed in variation and thus overlapping of characters in different species. This would result in an ever-changing pattern. Later, however, Darwin regarded species as an 'abstraction', purely arbitrary and subjective.

Dobzhansky (1950) was of the opinion that "species is the largest and most-inclusive reproductive community of individuals sharing a common genepool." According to Porter (1959), "A species is a recognizable and self perpetuating population that is isolated genetically as well as by its geographic distribution and its environment. Stebbins (1966) believed that a species is "many groups of related populations which live together in the same area, but which remain distinct from each other without exchanging genes." Mayr (1969) defined species as "groups of interbreeding natural populations."

Cronquist's (1968) treatment of a species is excellent: "Aside from infraspecific taxa, the only taxonomic category with an inherent rank is the species. An exact definition of the species is impossible, and the more precise one attempts to be, the larger the number of species which do not fit the definition. Still, the basic concept is simple enough. A species is the smallest population which is permanently (in terms of human time) distinct and distinguishable from all others. It is the smallest unit which simply cannot be ignored in the scheme of classification. It is the primary taxonomic unit, and it may also be thought of as the basic evolutionary unit. In sexual populations, gene exchange by hybridization within a species is ordinarily rampant, whereas such gene exchange between different species is restricted or even impossible. Interspecific hybrids are not always wholly sterile, but they are not so fertile and so competitively adapted as to swamp out the parents. Although there are some differences in interpretation, a reasonable degree of reproductive isolation from other species, under natural conditions, is an essential specific quality. Without such isolation, the population would lose its identity through interbreeding.

Many species have varieties which are partially segregated from each other, nearly always with a geographic or ecologic correlation, but connected by numerous intermediates. When parts of what had been a single species have diverged enough so that the distinction between them is reasonably sharp and prospectively permanent, the parts have become distinct species. It is only to be expected that at any particular time some species will be in the process of breaking up into separate species. The fact that polyploidy may introduce a barrier to interbreeding without any other significant difference, complicates the problem further. The line between strong varieties and weak species is necessarily an arbitrary one, involving subjective taxonomic judgement. The weak species of one taxonomist may be the strong varities (or subspecies) of another.

The difficulties in interpretation do not alter the fact that the species is an inherent taxonomic level, the smallest unit which is permanently distinct and distinguishable."

Now the question arises that why so many different definitions from different quarters? Much of this disagreement is artificial and arises mainly because everyone attempted to define what the species "really is". The main issue is to recognize what we should treat as species for any particular purpose. Another important fact is the failure to distinguish between the different meanings of the word, "species". According to Blackwelder (1962):

(a) "Species" without any article means groups of individuals (populations) recognised by any means,

(b) "A species" refers to one of these groups of individuals (or populations),

(c) "The species" denotes the general concept covering all the groups known as species, and

(d) "The species", "Species level", "Species category" etc. refer to the level or rank in the hierarchy.

It is understandable, therefore, that for different purposes, definition of species would vary and we have to determine or judge for ourselves which particular definition is suitable for our purpose. Rigorous definitions of species are not possible because the criteria may also change with the charcteristics of each group. On the basis of such different definitions for "species" proposed by different taxonomists, there are three types of species concept :

(1) Nominalistic,

(2) Taxonomical (including typological, morphological, and morphogeographical), and

(3) Biological

(1) Nominalistic Species Concept. For purpose of nomenclature all living organisms must be referred to species. A species is a category in taxonomic hierarchy recognised by ICBN. It can be recognised by the language of formal relation, e.g. species is a rank lower than category Genus in the taxonomic hierarchy.

(2) Taxonomical Species Conecpt can be either:

(i) Morphological,

(ii) Typological, or

(iii) Morphogeographical

(i) According to Morphological Species Concept—a species represents a group or assemblage of individuals with common morphological characters and separable from other such groups by a number of discontinuous characters or variations.

Table 2.3 Discontinuous Variations between *Rhynchosia minima* and *R. capitata*

Rhynchosia minima	*R. capitata*
1. Climber	Prostrate herb
2. Lax racemes of numerous flowers	Compact few-flowered racemes
3. Fruit–elongated pods	Fruit–rounded pods
4. Seeds without caruncle	Seeds with caruncle

The two groups (populations) of plants (Table 2.3) have discontinuous variations in four corelated characters. Since there is no member with intermediate features, these two assemblages can be treated as two distinct species.

(ii) Typological species concept. The term species is associated with the Latin word 'specere' meaning 'appearance'. This appearance of living things or individuals is highly variable. Aristotle rejected this concept of appearance. According to him Biological species are highly variable and every natural group of organisms has an "invariant generalised or idealised pattern shared by all members of the group". This concept is known as Typological concept. Most taxonomists recognised species on the basis of their observable morphological differences. It may also be said that, "Species are random aggregates of individuals and show the property or characteristic of the type.

(iii) Morphogeographical Species Concept. During post-Darwinian period, morphological species concept was modified. According to this concept—"all organisms exhibit both continuous and discontinuous variations as well as variations in geographical distribution." This was proposed by Du Reitz (1930). A species represents the smallest population which is permanently separable from other such populations by distinct discontinuous characters. According to this concept, therefore, ABCD and $A_1B_1C_1D_1$ as two distinct species must be genetically homogeneous groups as well as geographically isolated. (All the eight individuals of the two species together must not be genetically homogeneous!)

This concept is the Taxonomical Species Concept in true sense and is accepted by all leading taxonomists. This concept takes into account all available data/evidences from various fields of biology. The species so recognised are morphologically dissimilar as well as exhibit geographical distinction, e.g., *Protea* spp. from South Africa and Australia. *Brassica juncea* with petiolate cauline leaves and yellow seeds is morphologically distinct from *B. campestris* with sessile cauline leaves and brown seeds. Geographically they are distinct because the former grows on saline soil and the latter on acidic soil.

The Morphogeographical (or Taxonomical) Species Concept has been used in all the World Floras as the vast majority of plant species are recognised by employing this concept. It can also be used by experimental taxonomists as evidences from different fields of biology have been included. But at the same time this concept is highly subjective, lacks a proper scientific definition and requires a lot of experience to practice it.

It is intended to be a generally applicable concept and takes into account all available evidence—morphological, geographical, cytogenetics, and others, but insists that the species so recognised must be delimitable by morphological characters. Such morphologically defined species, as Heywood (1958) pointed out, "will, however, illustrate different kinds of evolutionary situation." Accordingly, species are only equivalent by designation, and not by virtue of the nature or extent of their evolutionary differentiation. Species, thus, are phenetic. Camp (1951) suggested that a different term *'binom'* be applied to such species and the term *'species'* be restricted to those units whose genetic relationship are known.

The purpose of a taxonomic species i.e. a species recognised by taxonomic species concept, is to be a part of a general purpose classification based largely on phenetic evidence to bring out a broad spectrum of the diversity of living things. Such a species concept is essential if the taxonomic system in general is to continue functioning. The taxonomic species is, therefore, regarded as the first step which may be subjected to experimental investigations and further information obtained.

The taxonomic species is often criticised by biosystematists and considered less important as compared to biological species.

Biological Species Concept. Löve (1964) commented that groups of non-interbreeding populations are "Biological Species" and groups of morphologically distinct populations are "Taxonomic Species" regardless of their breeding relationship. The Biological Species Concept as the theoretical basis for species developed during the first half of the 20th century, along with (1) genetical basis of variation, (2) reproductive mechanism, (3) hybridization, and (4) isolating mechanism. A Biological Species exhibits two important features: a) There should be interbreeding among individuals of a population and b) there should be reproductive isolation between members of two separate populations. On the basis of this Grant (1957) proposed the definition: "A species is a community of cross-fertilizing individuals linked together by bonds of mating and reproductively isolated from other species by barriers of mating."

Followers of biological species concept presume that there are objective discontinuities in nature which delimit the units that should be recognised as species. These discontinuities are caused by restriction of gene flow between actually or potentially interbreeding populations. A biosystematist, therefore, would define a species (or ecospecies) in terms of gene exchange—"if two populations are capable of exchanging genes freely under either natural or artificial conditions, they are conspecific." But if this is not possible due to internal barriers like incompatiblility or hybrid infertility, the populations are distinct ecospecies.

Such a definition is apparently simple, and provides an objective criterion to a species i.e. something which can be experimentally determined. It has a scientific or biological meaning as it demarcates the stage in the process of evolutionary divergence (Valentine and Löve 1958). And yet, there are some major theoretical and practical difficulties when we apply this concept to recognise a species.

(1) Reproductive isolation in a strict sense means internal genetics or genetic-physiological mechanisms (Davis and Heywood 1967). According to the biological species concept, species should be intersterile and the test for fertility or sterility must be done experimentally. To prove that two populations are reproductively isolated we have to carry out tests for fertility and sterility. But the genes responsible for morphological differences and those for fertility/sterility are independent of each other. As a result the correct position is often not clear and specific rank or status may be denied to markedly distinct, discontinuous groups because they are not reproductively isolated.

(2) The specific status may also be denied to morphologically distinct groups due to potential interbreeding in nature. For example, according to taxonomical species concept there are two species of *Salvia–S. apiana* and *S. mellifera.* But they hybridize easily and form interspecific hybrids. According to biological species concept these two are treated as one and only species i.e. Compilo species. The two species *Chlorophytum glaucum* and *C. glaucoides* are morphologically distinct but are not completely intersterile. According to taxonomic species concept they are two distinct species and according to biological species concept there is only one species. The reverse is also possible. Followers of biological species concept recognise two species of

Gilia–G. inconspicua and *G. tranemontana* as they do not interbreed. But according to taxonomic species concept, there is only one species *G. inconspicua* as there are no morphological distinctions. These are known as Sibling species.

(3) Another problem is what degree of sterility should be taken into account. The fertility/sterility test is not an all-or-nothing criterion. Every degree of inferility may occur between any two populations: (i) they may not hybridise at all, (ii) they may hybridise but F_1 is sterile, (iii) may produce seeds but seeds do not germinate, (iv) may produce very few seeds, (v) viability of the seeds restricted.

(4) In allopatric populations the fertility/sterility test is only of theoretical value. The possibility of interfertility does not arise because these populations have no occasion to cross. Even if they are shown to be interfertile under cultivation, this cannot be the only criterion to treat them as conspecific.

(5) This test is meaningless if the two populations hybridise easily or if they are apomictic groups.

(6) Sometimes the two populations may be effectively isolated, yet exchange of genes is possible through the intermediary of a third population.

(7) Lack of cytogenetical and experimental data is yet another problem. Such data is available for a minute fraction of the world's flora. Although Simpson (1961) terms it as "pseudo-problem", it is a major practical difficulty.

Although taxonomic species concept and biological species concept are based on different lines they coincide in practice in sexually outbreeding groups due to which the biological species concept is generalised. That means the biological distinctness come to be regarded as primary and morphological distinctions as secondary. The role of morphology in such a viewpoint has been clearly stated by Simpson (1961). He distinguishes between the *definition* of species in genetical terms and its *recognition* in morphological terms or in other words, morphology provides the evidence for putting the genetical definition into practice.

To be precise, we may define 'species' as "morphologically definable units, made up of groups of individuals (populations) which (it is assumed) are usually out-breeding and are expression of one or more gene pools."

It will be rather unrealistic if we abandon the practical, almost universal use of the term 'species' in a primarily morpho–geographical sense and favour the use of this term for impractical, largely theoretical units in terms of gene–pools and reproductive barriers. Acceptance of the concept is entirely different from acceptance in practice.

Definition of 'species' as proposed by Valentine and Löve (1958) is "a group of organisms capable of exchanging genes."

Numerical taxonomists (Sokal and Crovello 1970) opined that the biological species concept should be abandoned as it is neither operational nor realistic and has no practical value. Raven (1976) observed that—as species are not evolutionary units and gene–flow is not clearly explained for all species—the biological species concept is not of much value. Phenetic distinctness is more important in this regard and this should be considered as *Phenetic Species Concept.*

Genus Concept. Just like species—a genus is also a taxonomic category higher than species in the taxonomic hierarchy and is recognised by ICBN. According to the oldest concept—Genus is a group of living organisms bearing a name, e.g. *Sorghum, Pennisetum* and others.

Functions of a genus—A genus brings together closely related species into taxonomic groups which represent higher degree of diversity. Rollins (1953) defined genus as "a group of closely related species."

But this definition is not clear enough. Same definition can be put forward for subgenus, section or subsection also.

Mayr (1969) observed—"genus is a taxonomic category which contains either one species or a monophyletic group of species, separated from other genera by a decided gap." If a genus contains 1,000 species, the number of correlated characters which separate this genus from others is fewer i.e. larger the number of species lesser is the number of common characters binding them together.

If a genus contains an optimum number of species, the number of correlated characters are just sufficient to make it a natural taxon. If a genus contains a large number of species, there is possibility of many overlapping characters between them and hence, the difficulty to identify them easily, e.g. *Allium* (450 species). But at the same time it is also not possible to treat each species as a separate genus.

A genus will be a natural taxon if it contains an optimum number of species (ca 80–100). It is easy to delimit the species in such a genus, e.g. *Lycium* (80–90 spp.), *Atropa* (4 spp), *Litchi* (2 spp), *Dracaena* (80 spp).

Except for this one there is no other operational definition of a genus. Taxonomic groups above species rank, do not have any practical definition. However, both the definitions of Rollins and Mayr are vague. To recognise a genus, therefore, three criteria are followed:

(1) A genus should be a phylogenetic unit i.e., the species included must be monophyletic in nature. Monophyly (in modern taxonomy) is based upon cytogenetic and geographical evidence in relation to morphology and is very useful in deciding where the line of demarcation should be made between two genera.

(2) A genus must be an ecological unit, i.e. all the species should exhibit similar mode of life. For example, the genus *Utricularia* (250 spp) includes—(a) all insectivorous plants, and (b) all of them grow in N_2-deficient soil.

(3) A genus must be sufficiently different from other genera and should be separated by distinct discontinuous variations. For example, in *Ranunculus* the flowers are actinomorphic and fruits–a group of achenes. Whereas, in *Delphinium* the flowers are zygomorphic and fruits–a follicle.

If the classification is to be a natural one, genera must be delimited "not by a single arbitrary or artificial character but by the sum total of characters" (Sherff 1940) and also after a worldwide study of their species i.e., from the complete geographical range. When the deciduous and evergreen Rhododendrons of North America are compared, they appear to be two distinct genera. Some taxonomists did place the deciduous species in the genus *Azalea*. With the study of the genus *Rhododendron* from the complete range, the apparent discontinuities were broken down and it was clear that the genus *Rhododendron* includes both deciduous and evergreen species.

It is, however, to be seen whether it is practicable to recognise the group as a separate genus. If the degree of difference between the groups is small or insignificant, infra-generic rank may be more appropriate. In certain families such as Ranunculaceae or Berberidaceae, the genera are easily distinguishable on the basis of morphological distinctions. The same is not true for the genera in families like Compositae, Umbelliferae or Gramineae.

If the species of any two genera are not easily separable, it is preferable to have larger units of classification. In such a situation, ranks like subgenus or section may be introduced so that there is no change in the classification and at the same time relationship is expressed.

If in doubt, a genus may be split into two or more genera but this would mean nomenclatural change for some species. Such changes can be avoided by introducing infrageneric ranks because they do not require changes in binomials.

An ideal genus is the one in which the species included show more similarities than differences. Sometimes a genus may include only one species, e.g. *Leitneria floridana*. Such a genus is *monotypic*.

Family Concept. Unlike the concepts of genus and species, very little is known about family. It is a taxonomic category higher than genus and is recognised by ICBN. Much of the considerations applied for the concept of genus holds good for the concept of family also.

Family is a taxonomic category which contains either a single genus or a group of monophyletic genera and is separable from other families by distinct discontinuous features. When only one genus is included it is a monotypic family, for example, Illiciaceae, Cannaceae or Leitneriaceae.

As there is no proper definition for family — certain criteria must be followed to recognise a group of genera as a family:

(1) A family should be a phylogenetic unit, i.e. the included genera must be of monophyletic origin.
(2) It should be an ecological unit, e.g. members of the family Orchidaceae are mostly (i) epiphytic, (ii) show mycorrhizal association with fungi. Similarly all Cactaceae members are xerophytic plants.
(3) Must be sufficiently different and separable from other families by discontinuous variations. The family Capparaceae with its stamens raised on androphore and absence of replum is distinct from family Brassicaceae in which androphore is absent and replum is present.

The numerous parts of flowers and fruits of the angiosperms are not much affected by environmental changes and hence, families are constructed by using correlated characters of mainly flowers and fruits. Usually vegetative characters are liable to change with environmental modifications and therefore their use is limited.

Families are of two types—*definable* and *indefinable*. Definable families are distinct and markedly different from each other; are quite natural taxa. Such families contain a large number of genera which are difficult to identify because of many common and overlapping characters i.e., more similarities than differences, e.g. Compositae, Cruciferae, Umbelliferae, Gramineae etc. Indefinable families themselves show overlapping characters and are not markedly distinct from each other. Genera included are lesser in number and are easily identifiable. Predominantly woody family Magnoliaceae has many features common with the herbaceous family Ranunculaceae—free stamens, apocarpous ovaries and spirocyclic arrangement of floral parts. It is convenient to segregate such families into monotypic definable families from the point of view of evolution, e.g. Ranunculaceae s.l. can be split into Ranunculaceae s.str. and monotypic Paeoniaeae. But this process would lead to a great increase in the number of families.

Families like genera do not have a fixed number of members. The family Compositae (a dicot) includes about 20,000–25,000 species, Gramineae (a monocot) about 11,000 species, whereas monotypic Eucryphiaceae and Leitneriaceae (both dicots) contains 4 species and a single species respectively.

Families are often characterized by position of ovary (inferior in Cucurbitaceae), carpel number (pentacarpellary in Malvaceae), symmetry of the flower (zygomorphic in Labiatae), fusion of stamens (syngenesious in Compositae), leaf arrangement (opposite decussate in Asclepiadaceae) and even habit (parasitic in Cuscutaceae). Certain families are easily recognised even by amateurs, for example Compositae with head inflorescence, Umbelliferae with umbel inflorescence and schizocarpic fruits or Papilionaceae with papilionaceous corolla. Family Leguminosae has been treated variously by different taxonomists. Some divide Leguminosae into three families and treat it as an order. Others maintain one family with three subfamilies. This is a case where taxonomists agree where the line of distinction should be drawn but do not agree on the rank to be assigned to these groups.

3

Nomenclature

The purpose of giving a name to a plant is to provide an easy means of reference. Giving names to the newly acquired plants, or determining the correct name of already known plants follows a set of rules of nomenclature.

The elemental rules of nomenclature were first suggested by Linnaeus (1737, 1751; see Lawrence 1951). Augustine de Candolle's *Théorie Élémentaire de la Botanique* includes a detailed account of the rules for plant nomenclature (1813; see Lawrence 1951), and was the first significant work since the publications of Linnaeus. Later, these rules (also called de Candolle's rules) were adopted by the *International Code of Botanical Nomenclature (ICBN).*

The first International Botanical Congress was held in Paris in August 1867, when many botanists from various countries met and adopted a set of rules for the naming of plants. Subsequent Congresses made significant contributions in modifying and amending some of these rules. It was only at the Cambridge Congress in 1930, that "for the first time in botanical history, a code of nomenclature came into being that was international in function as well as in name" (Lawrence 1951). The rules adopted at the Cambridge Congrees (1930) were modified and amended, and the *International Code of Botanical Nomenclature (ICBN),* presently in use, appeared in 1978. It was adopted at the 12th International Botanical Congress held in Leningrad (Russia) in August 1975.

The aim of the ICBN is to provide a suitable procedure of naming various taxonomic groups and also to avoid or reject such names as are contrary to rules and not valid.

Binomial Nomenclature

Man, from time immemorial, has been a "nomenclaturist". For his convenience, he has given names to everything that has come his way—animals, plants, birds, or anything else. To begin with, the names given to plants were long descriptive sentences, e.g. *Gravellia robusta grandiflora australiana.* Although, these names were meaningful, it became impossible to remember such long plant names when the number of plants increased. Hence, this polynomial system did not continue for long. Gaspard (or Casper) Bauhin (1560-1624) came up with the novel idea of having only two names for every plant. He made a distinction between the generic name and specific epithet of the plants. This binary or binomial nomenclature, with which Carl Linnaeus's name is always associated, was, in fact, suggested by Casper Bauhin at least 100 years earlier.

However, the Swedish Botanist Linnaeus (1707–1778) was responsible for naming all living things from buffalo to buttercup, methodically applying two names to each, i.e. the binomial system. The vegetable kingdom named in this fashion was introduced in the book entitled *Species Plantarum* in 1753. The scientific names are in Latin, not the classical Latin, but a more or less popular Latin spoken by common people during the Middle Ages. One may raise an eyebrow and ask: "Why only Latin?" Well, Latin because: (1) It is specific, i.e. gives the precise meaning. (2) It is precise and concise and, therefore, (3) it is pertinent to the needs of descriptive phases of natural sciences. (4) Latin is written in the Roman alphabet and the confusion that will be created by the use of any other language of different scripts such as Chinese, Greek or Sanskrit, can be avoided. (5) Being a "dead" language now, it cannot arouse political controversy. Objections, however, have been voiced from many quarters against the use of Latin for plant names. Kelsey and Dayton (1942) in their *Standardised Plant Names* tried to introduce an English nomenclature using mostly common English names for plants, or anglicised Latin names. Using any spoken language and that, moreover, the national language of any country, is quite difficult. It is still more difficult, as no equivalent world flora has been published where English names or any other "vernacular" names for plants have been used.

Why not use the common names? There is no dearth of common names in any language. Benson (1962) pointed out the reasons why vernacular or common names cannot replace the Latin or Latinised botanical names:

1. Names in a common language are ordinarily applicable in only a single language; they are not universal.
2. In most parts of the world, relatively few species have common or vernacular names in any language.
3. Common names are applied indiscriminately to genera, species or varieties.
4. Often two or more unrelated plants are known by the same name, and frequently even in one language a single species may have two to several common names applied either in the same or different localities.

The same plant, *Piper nigrum,* is variously known as black pepper, white pepper, kali mirch, gole mirch etc. On the other hand, the common name lily is used for many genera of the Liliaceae *(Lilium. Erythronium. Hemerocallis),* Iridaceae *(Belamcanda, Nemastylis),* Amaryllidaceae *(Zephyranthes, Crinum)* and Zingiberaceae *(Hedychium).* Moss is a group of plants belonging to the Bryophytes, but reindeer moss is a lichen, Spanish moss is an angiosperm, and bogmoss is *Sphagnum. Spathodea campanulata* is known by four different names in the English language alone: squirt tree, scarlet bell, fountain tree or African tulip.

Every binomial consists of two parts The first part is the generic name, the second the specific epithet. The two parts should be in italics when in print, and underlined separately when typed or hand-written. The generic name always starts with a capital letter, whereas a specific epithet usually starts with a small letter, except in a few cases where the use of capital letter is permissible, e.g. *Pinus Roxburghii.* It is always a noun, also singular, and denotes the nominative case. There can be various sources, such as:

1. Names from many vernacular languages, e.g. *Salmalia* from shalmali (Sanskrit), *Madhuca* from madhukaha (Sanskrit), *Populus* from poplar (English), and others. Some names are based on the names in local languages of the areas where they occur, e.g. *Ginkgo* from the Chinese, *Tsuga* from the Japanese, *Nelumbo* from the Ceylonese and *Ravenala* from the Madagascarian.
2. Some names reflect the botanical character, e.g, *Trifolium* (with three leaves), *Cephalanthus* (flowers in heads), *Callicarpa* (with beautiful fruits), and *Liriodendron* (tree with lily-like flowers).
3. Many genera are named in honour of some famous botanist, e.g. *Bauhinia* (for Bauhin), *Hookera* (for Hooker), *Linnaea* (for Linnaeus); well-known scientists, e.g. *Einsteinia* (for Einstein); famous heads of state, e.g. *Victoria* (for Queen Victoria) and *Washingtonia* (for George Washington).
4. Some generic names are mythological in origin, e.g. *Narcissus* is after the famous Greek god Narcissus, *Circaea* or enchanter's nightshade refers to Circe, the famous enchantress, and *Nymphaea* refers to the water nymphs.
5. Some others are named after planets, e.g. *Mercurialis* after Mercury and *Neptunia* after Neptune.
6. Some are named after the name of country: *Salvadora* after El Salvador.

Specific epithets may likewise be derived from any source; it may be in honour of a scientist: *Pinus Roxburghii;* depicting some character of the plant: *Casuarina equisetifolia (Equisetum*-like leaves), *Jacaranda mimosaefolia (Mimosa*-like leaves); or geographical distribution: *Ocimum americanum, Camellia sinensis*; or simply after a vernacular name: *Psidium guajava* after guava. Often, the specific epithet is made up of two hyphenated words: *Hibiscus rosa-sinensis, Alisma plantago-aquatica.* .

The two parts of the plant name usually belong to the same gender and often have similar endings. When the specific epithet is an adjective, it must agree in gender with the generic name. Usually the generic ending *-us* is masculine, the ending *-a* is feminine and the ending *-um* is neuter. By convention, all trees are considered feminine for nomenclatural purposes but exceptions are permissible: *Quercus rubra. Pinus nigra.* Here the generic names are both masculine but the specific epithets are feminine. In Latin there are four main sets of adjectives used as name endings (see Table 5.1).

Table 3.1 Four Different Sets of Name-Endings used in Latin

	Masculine	*Feminine*	*Neuter*
1	-us	-a	-um
	sativus	*sativa*	*sativum*
2	-er	-ra	-rum
	niger	*nigra*	*nigrum*
3	-er	-ris	-re
	sylvester	*sylvestris*	*sylvestre*
	campester	*campestris*	*campestre*
4	-is	-is	-e
	humilis	*humilis*	*humile*
	occidentalis	*occidentalis*	*occidentale*

Given below are some common specific epithets and their meanings:

aphylla	–	leafless	nigra	–	black
alba	–	white	ochroleucus	–	yellowish-white
aquatica	–	in water	purpureus	–	purple
aureus	–	golden	palustris	–	of marshes or swamps
borealis	–	northern	repens	–	creeping
cerifera	–	wax-bearing	rara	–	rare
coccineus	–	scarlet	roseus	–	rose-coloured
communis	–	gregarious	serrata	–	with serrate margin
decumbens	–	reclining	scaposus	–	having a scape
dulcis	–	sweet	sulcatus	–	furrowed
edulis	–	edible	tridentata	–	with three spines
foetida	–	ill-scented	tenellus	–	slender, tender, soft
fluitans	–	floating	terrestris	–	growing on dry ground
grandiflora	–	large-flowered	tuberosus	–	tuberous
humilis	–	dwarf	uncinatus	–	hooked
linearis	–	narrow, linear	virens	–	green
magnus	–	large	viridis	–	green
minutus	–	very small	velutinous	–	velvety
mexicana	–	of Mexico	zeylanicus	–	of Ceylon

Another interesting feature of the specific epithets in honour of persons is that the proper ending is *i* or *ii* if the person honoured is a man: *agharkarii, baileyi*; it is *ae* if the person honoured is a woman: *rnargaratae*, *piersonae.*

Citation of Author's Name

To have a complete botanical/scientific name for a particular plant, it must be followed by the name of the person who identified and described the plant and suggested the name on the basis of this description, e.g. *Sesamum indicum* was identified, described and named by Linnaeus and hence, should be written as *Sesamum indicum* L. In all systematic/taxonomic work, it is essential to cite the authority of the scientific names. Citation of the author's name is helpful if ever confusion results from two plants having the same name. Author's names may be cited as full names or in abbreviated from, e.g. Roxb. for Roxburgh, Ait. for Aiton, Buch. -Ham. for Buchanan Hamilton, Cav. for Cavanilles, All. for Allioni, Wall. for Wallich, Bl. for Blume, Willd. for Willdenow. For two or more than two persons of the same family, different methods are adopted, e.g. William Hooker's name is abbreviated as Hook. and his son Joseph Dalton Hooker's as Hook. f., where f. stands for *filius* meaning son. In the de Candolle family, the father Augustin de Candolle is cited as DC., the son Alphonse as A. DC. and the grandson Casimir as C. DC.

If two persons have named a plant together, their names are joined by et or &, e.g. *Antigonon leptopus* Hook. & Arn.

When a name is proposed by one author and not validly published, and a second author has it published validly at a later date and ascribes it to the former author, the name of the former author, followed by the word ex, should be inserted before the name of the second author. For example, in *Cassia montana* Heyne ex Roth, Heyne proposed the name but did not publish it validly and the valid publication was done by Roth. The meaning of ex is "validly published by".

When a name proposed, described and diagnosed by one author is published in the work of another, the two names of the two author's are linked together by the word in. For example, *Hygrophila salicifolia* (Vahl) Nees in Wall.; *Euonymus indicus* Heyne ex Wall. in Roxb.

International Code of Botanical Nomenclature

Modern botanists all over the world use the *International Code of Botanical Nomenclature* (ICBN) which in a simple and precise manner deals with (1) terms which denote the ranks of taxonomic groups or units, and also (2) the scientific names which are applied to the individual taxonomic groups of plants (Greuter 1988). This code aims to provide a stable method of naming taxonomic groups, avoiding and rejecting the names that are ambiguous or create confusion.

Six principles form the basis of botanical nomenclature. The detailed provisions are divided into *Rules* and *Recommendations.* The objective of the rules is to put the nomenclature of the past into order and to reject the names that do not comply with the rules. The objective of the recommendations is to try to bring about greater uniformity and clarity, particularly in future nomenclature. Names that are recommended should await the formation of rules for their application. The *Rules* and *Recommendations* apply to all living organisms treated as plants (including fungi, but excluding bacteria), and also to fossils. The nomenclature of bacteria is governed by the *International Code of Nomenclature of Bacteria.*

The *principles* are the guidelines for the legitimate naming of any taxon:

I. *"Botanical nomenclature is independent of zoological nomenclature. The Code (Greuter* 1988) *applies equally to names of taxonomic group treated as plant whether or not these groups were originally so treated."* For the purpose of this code, "plants" do not include bacteria. As the code provides only for the nomenclature of plants, the same name may sometimes be assigned both to a plant and to an animal, e.g. *Cecropia* is the name of a moth, according to zoological nomenclature, and at the same time it refers to a tree belonging to the Moraceae.

II. *"The application of names of taxonomic groups is determined by means of nomenclatural types."* According to this principle, the name of each species is permanently associated with a particular specimen, the nomenclatural type. The type for a genus is a species, for a family it is a genus, and for an order it is a family. The following types are recognised:

The *holotype* is the one specimen or other element used or designated by the author in the original publication as the main nomenclatural type. Any type selected after the original publication is not to be regarded as a holotype. At present, it is essential that a holotype designated for a newly described species be deposited in a national herbarium.

An *isotype* is a duplicate specimen of a hololype. These are plants forming part of the same gathering as the holotype or growing with it and gathered at the same time. A *syntype* is one or two or more specimens studied and cited by the author, when the holotype is not designated by him. A *paratype* is a specimen cited wilh the original description in addition to the holotype. When the author fails to designate a holotype or the holotype is missing, a *lectotype* or a *neotype* is selected to serve as a nomenclatural type. A *lecotoype* is a specimen selected from those cited by the author with the original description. A *neotype* is a specimen selected from the material that was not cited by the author with the original description. A *neotype* is selected onIy when all the original specimens collected and cited by the author are missing.

Impatiens thomsonii Hook. f. is a member of the Balsaminaceae and its description is given in the *Flora of British India.* The author has cited tree specimens on which the description was based:

1. Collected by Thomson from Piti and Kunawar.
2. Collected by Strach. and Wint. from the Kumaon and Garhwal Hills.
3. Collected by J.D. Hooker from Sikkim.

For each, specimen–number, place and date of collection, and the name of the collector are given. Hooker stated that specimen no. 3 is the nomenclatural type and therefore holotype. Specimens 1 and 2 are the paratypes. If Hooker had not designated the 3rd specimen as holotype, then all three would have been syntypes. One of these syntypes can serve as a lectotype, if the holotype is missing. If all three specimens are destroyed for some reason, then a fourth specimen (collected by Wallich from Sikkim), which does not find mention in Hooker's description, will be treated as a neotype. Duplicate specimens collected by Hooker, along wilh the holotype, are treated as isotypes.

III. *"The nomenclature of a taxonomic group is based upon Priority of Publication."* According to this principle, each taxon should bear only one correct name and that should be the earliest published name. The rule of priority states: "For any taxon from family to genus inclusive, the correct name is the earliest legitimate one with the same rank, except in cases of limitation of priority by conservation." It also states: "The principle of priority does not apply to names of taxa above the rank of family."

To avoid confusion caused by the strict application of this rule of priority, certain specific, generic and family names are conserved in preference to the earlier published names, by resolution of the International Botanical Congresses. Conserved names are known as *nomina conservanda.* For example, Sterculiaceae Lindl. 1830 is a conserved name and not published earlier. The earlier published name for the same family is Byttneriaceae R. Br. 1814 (see Lawrence 1951). Conservation of specific names is restricted to names of species of major economic importance (Greuter 1981) and was adopted at the International Botanical Congress held at Sydney, Australia, in 1981.

Any association of a specific epithet with a generic name is known as *combination.* All those names which are presented in accordance with the rules of nomenclature are termed legitimate and the names contrary to the rules of nomenclature are termed illegitimate.

Priority of nomenclature for vascular plants is applicable from May 1st, 1753 which is the date of publication of Linnaeus' *Species Plantarum.*

When taxonomic study indicates that a species described in one genus is to be transferred to another genus, the specific epithet, if legitimate, should be retained. In the following example:

Sida cordata (Burm.f.) Borssum 1966
Melochia cordata Burm.f. 1768
Sida veronicifolia Lamk. 1787
Sida multicaulis Cav. 1785

According to the rule of priority, *Melochia cordata* Burm.f. is the valid name. A later study by Borssum revealed that the genus was *Sida* and not *Melochia.* As the specific epithet was legitimate, *cordata* was retained and a new combination *Sida cordata* was made by Borssum. *M. cordata* Burm.f. is now known as a basionym To write the present name correctly, the author's name of the basionym is placed in parentheses, followed by the name of the author who made the new combination.

When two or more taxa of the same rank are united, the earliest legitimate name or epithet is selected. For example, if the genera *Sloanea.* 1753, *Echinocarpus* Blume 1825, and *Phoenicosperma* Miq. 1865 are combined, *Sloanea* L. should be the correct name to be used, as it is the earliest name and the other two would be the *taxonomic synonyms.*

If the taxonomic revision of a genus reveals that it should be divided into two or more genera, the original generic name should be retained for the genus that includes the species designated as the type. For example, the genus *Aesculus* is divided into four sections; *Aesculus* sect. *Aesculus,* sect. *Pavia,* sect. *Macrothyrsus* and sect. *Calothyrsus.* If the last three are regarded as distinct genera, i.e. the four sections are treated as separate genera, then the name *Aesculus* should be retained for the first one—*Aesculus* sect. *Aesculus*, which includes the type species *A. hippocastanum* L. Similarly, when a species is broken into two or more than two parts." the specific epithet should be retained for the one which includes the type specimen/figure/description. For example, *Acer saccharum* Marshall was first described by Marshall. Later, Michaux considered it to comprise two species and, according to the rule, he retained the name *Acer saccharum* for the specimen that was described by Marshall and named the other specimen *Acer nigrum* Michx.f.

IV. "*Each taxonomic group with a particular circumscription, position, and rank can bear only one correct name, that is a validly and effectively published name.*"

All those names that are published in printed form in scientific journals and are available in botanical institutions with libraries, which are accessible to botanists in general, are the effectively published names. On the contrary, names that are published in nursery catalogues, newsprint or seed-exchange lists are not effectively published names. A plant name is not effectively published if printed on a label attached to herbarium specimens even if the specimens are widely distributed (Jones and Luchsinger 1987). For valid publication, a name must be effectively published; it must be accompanied with a description or a reference to a previously published description of that taxon. From January 1st, 1935, names of new taxa of recent plants (with the exception of algae and fossil) must be accompanied by a Latin diagnosis for valid publication. The description itself need not be in Latin, although it is recommended. The description and diagnosis (the distinguishing features of the taxon as mentioned by the author) of new taxa, published before January 1st, 1935, are treated as valid, even if they were in any modern language including Japanese, Russian or any other where Roman alphabets are not used.

The name of a taxon is not validly published if it is cited merely as a synonym.

Phalaris arundinacea Linn. Sp. Pl. 55, 1753. This name, given by Linnaeus, published in his *Species Plantarum* on p. 5 in 1753, has a Latin diagnosis and therefore a valid name.

Digitaria sanguinalis (L.) Scop. Fl. Carn. ed. 2, 1:52, 1772

Panicum sanguinale L. Sp. Pl. 57, 1753.

Scopoli discovered that the type specimen had the characters of *Digitaria* and, hence, the new combination was made by him and published in *Flora Carniolica.* He did not give a Latin diagnosis, as it had already been given by Linnaeus in *Species Plantarum* in 1753. The second name now becomes the *nomenclatural synonym.*

V. "*Scientific names of taxonomic groups are treated as Latin regardless of their derivation.*" According to this rule, generic names, specific epithets, as well as other names should be Latin or Latinised with the addition of prefixes and suffixes, whatever source they might have been taken from (cf. p. 16, 17).

VI. "*The rules of nomenclature are retroactive unless expressly limited.*" In connection with this principle, the principle of *Later Homonym* emerged. A name is a *later homonym* if it is spelled like a name previously and validly published for a taxon of the same rank, based on a different type specimen. Different genera (of the same family or different families) and different species of the same genus cannot have the same name. In such instances, the later-formed name or the later homonym is illegitimate and has to be rejected. For example, *Tapienanthus Boiss.* ex Benth. 1848 of Labiatae is a later homonym of *Tapienanthus* Herb. 1837 of Amaryllidaceae, and must be rejected; *Viburnum fragrans* Bunge 1831 and *Viburnum fragrans* Lois. 1824 both belong to the same family Caprifoliaceae, but the type specimens for them are different. Therefore, the later homonym, *V. fragrans* Bunge 1831, should be rejected. To indicate that a plant has a later homonym, the word non is used before the author's name of the later homonym and placed after the early homonym, e.g. *Viburnum fragrans* Lois 1824 non Bunge 1831 will indicate that this plant has a later homonym. If the plant name itself is a later homonym, the word nec is used before the author's name of the early homonym and placed after the later homonym, e.g.

Viburnum farreri Stearn 1966

V. fragrans Bunge 1831 nec Lois. 1824

The fifteenth International Botanical Congress was held at Yokohama, Japan in 1993. The]nternational Code of Botanical Nomenclature (called the Tokyo Code) adopted at this Congress is significantly different from the earlier Code (called the Berlin Code) which was adopted at the fourteenth Botanical Congress. Some of the important changes are as follows:

(i) the rules on Typification and Effective publication have been clarified by creating a logical arrangement of the Articles 7–10, and 29–31, respectively;
(ii) the proposals for (a) Conservation of species names, and (b) rejection of any name which would cause a disadvantageous nomenclatural change, were accepted by overwhelming majority;
(iii) an entirely new concept has been incorporated in the Tokyo code. This concerns the recognition of "Interpretative Type" to serve the requirement of typification when an established name cannot be reliably identified for the purpose of precise application of a name;
(iv) this Code permits the use of the term "phylum" as an alternate to "divisio";
(v) an extensive revision of Article 46 has clarified the use of the prepositions *'ex'* and *'in'* in author citations;

(vi) for valid publication of a new taxon of fossil plants, on or after January 1,1996, there must be an accompanying description or diagnosis in Latin or English, (or a reference to such earlier publication) and not in any language as before;

(vii) The 15th International Botanical Congress proposed that after January 1, 2000 and after approval by the 16th International Botanical Congress (in 1999), new names must be registered.

Nomenclature of Hybrids

A hybrid is an offspring of two different genera, or species of plants. Hybrids between two species of the same genus or interspecific hybrids are usually shown by a formula: *Verbascum lychnite* □ *V. nigrum* or *Verbascum lychnite* □ *nigrum,* or may be given a formal name: *Verbascum* □ *nigralychnites.* Intergeneric hybrids also can be named in the same fashion: either by a formula, e.g. *Cochloidea* □ *Odontoglossum; Cooperia* □ *Zephyranthes*, or by a formal name □ *Cooperanthes;* □ *Triticale.*

Some Basic Definitions Related to Nomenclature

Synonyms are different names for the same plants. It is a rejected name due to wrong application or difference in taxonomic judgement. When two or more than two names are given to the same taxon, based on the same type specimen, they are *nomenclatural synonyms.* For example, *Chilocarpus malabaricus* Bedd., and *Hunteria atrovirens* DC. are the nomenclatural synonyms of *Chilocarpus atrovirens* (G. Don) Blume, as the type specimen is the same. In another example, *Vernonia leiocarpa* DC. and *Vernonia melanocarpa* (Gleason) DC. are taxonomic synonyms, as the two are based on different type specimens.

Basionym. When a species is described in one genus but transferred to another later, the specific epithet, if legitimate, should be retained. *Myrobalanus bellirica* Gaertner 1791 is now known as *Terminalia bellirica* (Gaertner) Roxb. 1805. *Myrobalanus bellirica* is the *basionym* of *Terminalia bellirica.*

Homonym. When two or more identical names are given based on different type specimens, they are homonyms and the earliest published one amongst these is legitimate and should be retained. *Centranthera indica* (L.) Gamble 1924, if shifted to *Limnophila,* cannot be named *Limnophila indica* because there is already a species with the same name, i.e. *Limnophila indica* (L.) Druce 1914. In the event of doing so, these will be homonyms.

Tautonym is an illegitimate binomial where the generic name and specific epithet are exactly the same, e.g. *Sassafras sassafras* (L.) Karst 1882, is a tautonym and therefore a rejected name.

Autonym is a legitimate, automatically created tautonym for infrageneric or infraspecific taxa. For example, *Hypericum* subgenus *Hypericum* section *Hypericum; Sesbania sesban* var. *sesban.*

Nomen nudum is a name without description and so should be rejected. In some cases, the names are nomenclaturally *superfluous,* i.e. published as a substitute for an already published legitimate name. For example, *Sassafras triloba* Raf. 1840 was a superfluous name for *Laurus sassafras* L. 1753, and therefore rejected.

4
Identification

Taxonomic Literature

Determination of an unknown as being identical with a known specimen is called **identification.** Identification and determination of correct name go together. The methodology of identification can be achieved by using **taxonomic literature**.

What is **taxonomic literature**? Taxonomic literature is any botanical work associated with identification, classification and determination, and use of correct botanical name of a taxon. A good knowledge of existing literature is essential to a practicing taxonomist. Very often it is the lack of extensive libraries that hampers taxonomic work in developing countries in particular. It is possible for only very few institutions to possess a vast collection of the complete range of taxonomic literature. Moreover, language barrier between various countries also pose a problem. Research papers have been published in many languages such as Chinese, Russian, Hungarian and others (where even the script is different) in innumerable books and periodicals. No scientific subject depends upon accumulation of previously and presently published literature, as taxonomy does. Taxonomic literature can be classified as:

i. General Taxonomic Indices
ii. World floras and Manuals
iii. Monographs and Revisions
iv. Bibliographies, Catalogues and Reviews
v. Periodicals
vi. Glossaries and Dictionaries

i. General Taxonomic Indices—These are indices of plant names and help in locating the original publication of a name, the valid name, and the synonyms and to which family, subfamily, or tribe the plant belongs.

1. Index Kewensis plantarum phanerogamarum

Index kewensis is an index of all angiospermous plants in the world. An alphabetical index to the generic names and binomials, it is used to determine the new and changed names of seed-bearing plants of worldwide distribution and bibliographic references to the place of first publication. The original work was complied by B.D. Jackson during 1893–1895 under the supervision of Sir J.D. Hooker in two volumes and a supplement is added every five years. Last supplement (No. 21) was published in 2001. The first two basic volumes include plant names published between 1753 to 1885. All known specific epithets published for a particular genus are in alphabetical order under each generic name. Each name/binomial is followed by author's name, geographical origin of the plants and a brief literature citation. Synonyms are mentioned in italics in the first volumes. Names of infraspecific taxa are not included. Recent supplements do not include synonyms; instead references to illustrations have been added. In total, almost 968,000 records are included. In the beginning of 1980s, the data were transferred to a computer database. Approximately 6,000 entries are added every year. It was decided in 1993, to publish the complete Index Kewensis as a CD-ROM so that this data is easily available.

2. **Index Londinensis to Illustrations of Flowering Plants, Ferns and Fern Allies**—This work was complied at Royal Botanic Gardens, Kew under the direction of Royal Horticultural Society of London in six volumes. It is an alphabetical index of illustrations published between 1753 and 1920. A supplement of two volumes was published in 1941 and covered 1921 to 1935. These illustrations are arranged alphabetically by genus and species. Although now greatly out of date, it is still a very valuable source of reference. It has been partially updated by **Flowering Plant Index of Illustrations and Information,** complied by R.T. Issacson in 1979 in two volumes. This book enlists post-1935 colored illustrations only.
3. **Index Holmiensis**—This is an alphabetic list of distribution maps of vascular plants recorded in various taxonomic literature. Publication commenced in 1969 and not complete as yet. It is compiled and published in Sweden.
4. **Gray Herbarium Card Index**—This index is published from Harvard University, Cambridge, Massachusetts, USA. Complied in Gray Herbarium, it covers flowering plants and ferns of the western hemisphere (New World) and has index cards for all the new names and new combinations, i.e. genera, species and infraspecific taxa, published since 1873. It has now been set up on a database. Number of cards issued has already exceeded 287,225. The index has been published in ten volumes between 1893 to 1967. A two-volume supplement was published by G K Hall in 1978 and it covers the years 1967 to 1977.
5. **Index Nominum Genericorum**—This is a listing of all the generic names of plants of all groups, both fossil and extant. It was published in 1979 in three volumes of ***Regnum Vegetabile*—**vols 100-102. All the generic names from 1753 to 1975 have been included and it took 25 years to complete. Type specimens have been indicated and bibliographic and nomenclatural details of about 63,500 entries have been entered. The first supplement—***Regnum Vegetabile***, vol. 11, 1986 lists another 2500 generic names and corrected names of the previous publications.

   In addition to these indices of flowering plants, there are many for the pteridophytes, bryophytes, algae and fossils.
6. **Index Filicum**—Deals only with the members of Filicinae or True Ferns. Published in 1906 by the Danish botanist CFA Christensen from Copenhagen, Denmark, the original publication covered the names up to 1905. Thereafter five supplements have been published which included names up to 1975. Supplement 5 covers all pteridophytes, although the earlier ones treated only ferns. Similarly, there is Index Lycopodiorum for earlier Lycopodiales, Index Isoetales for Isoetales, Index Psilotales for Psilotales, Index Selaginellarum for Selaginellales and Index to Equisetophyta for Equisetales.
7. **Bryophytes—Index Muscorum** consists of five volumes of Regnum Vegetabile (1959-1969). All species and infraspecific taxa of mosses published up to the end of 1962 have been included in this alphabetical list. Biennial supplements are taken out in taxon (First one in vol. 26, 1977). **Index Hepaticarum** including Anthocerotales was commenced in 1962 by CEB Bonner and covers up to 1973. For this original work also, biennial supplements are prepared–the first appeared in Taxon, vol 27, 1978.
8. **Algae**—Complete indices of algal species are not yet available. However, diatoms have been covered in **Catalogue of the Fossil and Recent genera and species of Diatoms and their synonyms** (in seven parts, 1967 to 1978). An index of all the known algae, i.e. their names with classes and families and synonyms appeared in vol 103 of Regnum Vegetabile in 1980.

(ii) World Floras and Manuals—Floras are detailed systematic enumeration of a major group of plants of a particular region. Usually a flora gives an account of all the vascular plants only. One of the well-known systems of classification is followed for arrangement of families. Complete botanical name, author's name, original name with date and page number of the book where published, synonyms—are mentioned. Identification keys and descriptions are also usually given. Floras normally concentrate on providing a means of identification of the species occurring in a particular region.

Typically, there is not a single flora, which can be named as "World Flora". None of the floras accounts for every species even of angiosperms alone. The two books **Genera Plantarum** and **Die natürlichen Pflanzenfamilien** written by G Bentham and Sir JD Hooker and A Engler and K Prantl, respectively, are the only books considered as World Floras. **Genera Plantarum** (1862-1883) dealt only with seed plants. It described 200 families and 7569 genera and species thereof. Every genus and species were studied from actual herbarium specimens and accurate description was written from these observations. The dicotyledons were divided into three groups: Polypetalae (with free petals), Gamopetalae (with fused petals) and Monochlamydeae (with no petals). This work was in seven volumes, written and published over a span of 20 years from 1862 to 1883.

The other publication considered as World Flora is **Die natürlichen Pflanzenfamilien** written and published by Adolf Engler and Karl Prantl during the years 1887 to 1915. This work included the entire plant kingdom that was divided into 14 divisions and the 14th division "Embryophyta siphonogama" included the flowering plants. Identification of all the known genera of plants from algae to seed plants, their description and illustrations have been included.

Apart from these two extensive works there is none other, which can be considered as "World Floras."

Continental floras: In today's context, it may be difficult to write not only a world flora but also a Continental flora, i.e. a flora for a whole continent. However, G Bentham wrote Flora Australiensis as early as 1863 to 1878. TG Tutin et al. (1964-1980) wrote Flora Europaea.

Regional Floras. These floras are meant for distinct regions. Given below are a few examples:

G Bentham (1930) *Handbook of British flora*

J Hutchinson (1948) *British Flowering Plants*

CGGJ Van Steenis (1948) *Flora Malesiana*

KH Rechinger (1963) *Flora Iranica*

VL Komarov & BK Shishkin (1934-1964) *Flora SSSR*

E Blatter (1926) *Palms of British India and Ceylon*

NL Bor (1953) Manual *of Indian Forest botany*

JD Hooker (1872–1897) *Flora of British India* (7 Volumes)

A large number of floras have been prepared for different states of India:

JS Gamble (1915–1936) *Flora of Madras Presidensy*, 3 Volumes (Reprinted in 1967 by Botanical Survey of India).

PF Fyson (1915) *Flora of Nilgiri and Pulney Hill tops* (Reprinted in 1974–1975 by BSI)

JF Duthie (1903–1922) Flora of Upper Gangetic Plains and the adjacent Siwalik Ranges, 3 Volumes (Reprinted in 1960 by BSI).

UN Kanjilal et al. (1934–1940) *Flora of Assam,* 5 Volumes

HH Haines (1921–1925) The *Botany of Bihar and Orissa* (Reprinted in 1961 by BSI)

H Collett (1902) Flora Simlensis (Reprinted in 1971 by BSI)

CJ Saldanha and DH Nicholson (1976*) Floras of Hasan District (Karnataka*).

H Santapau (1953) Flora of Khandala on the Western Ghats of India.

MA Rau (1975) *High Altitude Flowering Plants of Western Himalayas.*

JK Maheshwari (1963, 1965) *Flora of Delhi*, 2 Volumes Vol 1 Text; Vol 2 Illustrations.

Overall, floras can be considered to be of three types depending upon their contents—

1. **Research Floras:** These floras are as good as regional revisions and therefore self-contained and provide excellent basis for anybody to persue taxonomic research. Their bulky nature prevents one to carry them in the field but the information therein is highly valuable for further revision work and identification work in the herbarium. Examples of such floras are: Flora Malesiana by CGGJ Van Steenis and Flora of Jamaica by W Fawcett and AB Rendle.
2. **Concise or Field Floras:** These floras are much shorter compared to the Research Floras. Description of the taxa is either very brief or lacking. Type specimens are not mentioned nor the herbarium specimens consulted (for writing) and a minimum number of synonyms included. Although, information on distribution and ecology and other unnecessary details are included. Van Steenis (1954) commented, "some modern floras are so full of non-phytographical text, on economics, ethnobotany, anatomy, history etc. that the descriptive text is suffocated by a mass of other information and is treated as of secondary rank". An example of such a flora, containing a great deal of everything except descriptive details is Flora of Egypt (1941-1954) by VT Tackholm and M Drar. Descriptive botany–essential for writing a good workable flora, depends mainly on the original critical observation by the author.
3. **Excursion Flora:** This is the third type of flora, and a very much shorter version. In these floras, space is saved by combining keys and descriptions as in Excursion flora von Deutschland (1958) by W Rothmaler, or by deleting or abbreviating certain information as in Excursion Flora of British Islets (1960) by AR Clapham, TG Tutin and EF Warburg. Such floras are useful for identification of plants in the field itself.

Numerous floras have been written for different countries, states and even smaller areas, all over the world. They differ in exhaustiveness, authoritativeness and completion. A guide to the available standard floras of all parts of the world, has been prepared by Frodin (1984).

A **manual** is as good as a regional flora. It will normally contain information on the area of coverage and keys and descriptions to the families, genera and species. Sometimes, much additional information is provided for each species, e.g. species name followed by the author's name/s, synonyms, common or local name/s, ecological and distributional data, and even illustrations and distribution maps.

The terms floras and manuals are, therefore interchangeable. Some of the well-known manuals are:

Manual of Botany (1950) by A Gray

Manual of Cultivated Plants (1949) by LH Bailey

Manual of Cultivated Trees and Shrubs Hardy in North America (1940) by A Rehder

Manual of Aquatic Plants (1957) by NC Fassett.

(iii) Monographs and Revisions—A **monograph** is a complete comprehensive account of a taxon of any rank—family, genus or species at a given time. It is a taxonomic treatise and synthesis of all known information about the taxon. This includes the existing taxonomic knowledge as well as the results of any original research work taken up by the author. Study of phylogeny is also included and hence, phylogenetic relationship of the taxon with other related taxa is known.

Usually there are some introductory chapters in which presentation and discussion of the original research done by the monographer are included. This is followed by a descriptive systematic treatment in which data from the field of morphology, anatomy, embryology, palynology, cytology, genetics and ecology are included. The geographical scope of a monograph is usually worldwide because it is impossible to discuss a taxon in detail if all the lesser taxa (that it contains) are not considered. Hence, all species and infraspecific taxa under a genus and all the genera in a family must be included in a monograph. In addition, one must be benefited from such scholarly work, if many other information such as: (i) Extensive literature reviews, and (ii) all nomenclatural information including designated type specimens, identification keys to all taxa, full synonyms, citations of specimens examined, distribution maps, a classification by the author for all the included taxa and a discussion on phylogenetic relationship amongst these taxa, are also included. Some of the good examples of monographs are:

The genus *Nicotiana* (1955) by TH Goodspeed

The genus *Datura* (1959) by AF Blakeslee et al.

Oats: Wild and Cultivated. A monograph of the genus *Avena* L. (Poaceae) 1977 by BR Baum

Monographie der Gattung *Oplismenus* (1981) by U Scholz

Revisions differ from monographs in the degree of scope and completeness. A taxonomic revision usually incorporates much lesser details of introductory material and a synoptic literature review. The systematic treatment is also less comprehensive. It may include a part, i.e. a section of a genus or it may be restricted to a taxon occupying a smaller geographical area. It may include an entire family also. Revisions normally incorporate a complete synonymy, keys to identify the included taxa, short descriptions, often only diagnostic features, distribution maps, a classification, and a brief discussion of supporting data. Illustrations, mostly in the form of line drawings are included both in monographs and revisions. Revisions may sometimes be based only on herbarium specimens.

It is important to come to a decision about the plants or plant groups that need a revision. If there is inconsistency of the characters within the taxon's geographic range and if there are difficulties in identifying various members within a taxon then such taxa need a revision. Young scientists interested in taxonomic studies should work towards writing up of revisions of important taxa. Many plant groups of the neotropics need revisionary studies. Taxonomic revisions based primarily on original research work form the central core of this important subject –the systematic botany.

Some of the important examples of revisions are:

Revision der Gattung *Ceropegia* (1957) by H Huber

The indigenous Old World Passifloras (1972) by WJJO de Wilde

A revision of the genus *Ptychosperma* Labill. (Arecaceae) (1978) by FB Essig

Below the level of revision is the **conspectus.** This is an outline of a revision in true sense. A conspectus includes listing of the all taxa with all or only the major synonyms, sometimes with short diagnosis or none and often with a brief mention of the geographical range of each taxon.

After conspectus, is the **synopsis**; this is a list of taxa with abridged diagnostic features to distinguish them from each other. Synopses mostly occur as the front few pages of a revision in the form of a summary of the contents.

(iv) Bibliographies, Catalogues and **Reviews.** The literature of systematic botany is so vast , that it is rather difficult for any one person to be conversant with it. Even for any institution, it is not always possible to maintain a card catalogue of the complete literature. With a large number of research papers being published from every corner of the world, there is need for consolidated information in the form of supporting literature. The three principal sources for such consolidated information are:

(a) **Bibliography**—these are published works that account for all the books and scientific articles published during a particular time. The International Association for Plant Taxonomy produces many of these. An exhaustive series of *Regnum Vegetabile* includes complete bibliographical details of literature. It helps in searching type material, priority of names, and dates of publication and biographic data of authors. Originally published in 1967, it is constantly under revision, and GL Stafleu and EA Mennega have published three supplements of the 2nd edition between 1992 and 1997.

(b) **Catalogues** account for books of libraries rich in botanical titles and are of especial value in taxonomic studies. For this, full name of the author/s should be known and also the exact title of the book, exact date of publication or when a particular edition is issued. For example, *Catalogue of the library of British Museum (of Natural History); Catalogue of the library of the Massachusetts Horticultural society.*

(c) **Reviews**—A review is a critical evaluation of a book or research paper often published in periodicals or journals. A person other than the author writes u(v)

(v) Periodicals or Journals—This is a type of publication issued at regular interval and carries the original research papers. Each issue is a number or fascicle or Heft (in German). The numbers published in one calendar year, together comprise a Volume or Band (in German). The journals may be published from a scientific organization or society, e.g. *Botanical Journal of Linnaean Society* or from an educational or non-profit research institution, e.g. *Phytomorphology* from Department of Botany, Delhi University.

Periodicals are usually entitled: Journals, Annals, Bulletins or Proceedings. Whenever a monumental work is published by a single author, it is treated as a Memoir or Transaction, e.g. *Memoirs of Torrey Botanical Society.* Titles of some periodicals are very long and it is customary to abbreviate them, e.g. *Pl Syst Evol* stands for *Plant Systematics and Evolution* (Denmark) Some of the periodicals or journals devoted mainly to taxonomic research are:

Botanical Journal of Linnaean Society (London)
Kew Bulletin (Royal Botanic Gardens, Kew, London)
Memoirs of Gray Herbarium (Harvard University, Cambridge, Massachusetts, USA)
Journal of the Arnold Arboretum (Harvard University, Jamaica Plain, Massachusetts, USA)
Botanical Magazine (Tokyo)
Systematic Botany (New York)
Taxon (International Association for Plant Taxonomy, Berlin)
Blumea (Leiden)
Journal of Indian Botanical Society (JIBS) (Bangalore)
Bulletin of Botanical Survey of India (Calcutta)
Journal of Bombay Natural History Society (Bombay)

(vi) Glossaries and Dictionaries—A glossary is an alphabetical list of terms together with the explanation of their meaning. Almost all modern manuals and books include a chapter on Glossary of the botanical terms used.

Dictionaries also include terms and their meaning but usually pertaining to a particular subject e.g. Dictionary of Biology, Dictionary of Horticulture, Dictionary of Ecology and Environment.

Most botanical dictionaries are of plant names and are sources for the history of origin and development of the Latin and vernacular names. Such knowledge could not be obtained from an ordinary or normal dictionary. A Dictionary of the Flowering Plants and Ferns, 8th edition , originally by JC Willis (1973) and later revised by HK Airy Shaw is one such example. This is an alphabetical list of all generic names published since 1753 and family names published since 1789. A similar book – The Plant Book 2nd edn by DJ Mabberley was published in 1997. This book in a dictionary form, attempts to present all currently accepted generic and family names and commonly used English names of flowering plants, gymnosperms and ferns (as well as other pteridophytes) excluding the fossil groups.

Botanic Gardens

Botanic gardens are living collection of plants from various families, genera and species. Study of botany developed along with the study of medicine. Although gardens were present during Egyptian and Mesopotamian civilization, they were meant only for growing food plants and ornamentals; or for pleasure and as status symbol. They were not the true botanical gardens where plants were collected, grown and maintained for scientific purposes. The first such botanical garden might be considered to have been that of Theophrastus. This garden was attached to his school—the "lyceum" near Athens in Greece. It was gifted to him by his mentor Aristotle.

Study of Botany developed with the study of medicine as it started with the identification of plants that are of medicinal value. The Romans maintained small gardens where they grew such plants and these gardens were meant for giving instruction to the medical students. Later on similar gardens were established in Europe also. During the medieval period gardens attached to the monasteries were also meant for giving medical training to the staff there. In a typical monastic garden vegetable, fruits and various herbs were grown. During the 16th and 17th centuries, physic gardens affiliated to the medical faculties of many universities were established where medicinal herbs were grown.

Luca Ghini established the first "modern" botanical garden in Pisa, Italy in 1543 as recorded in a letter dated July 4, 1545 by Ghini himself. The letter speaks of this garden laid out in Pisa "for the use of students " (Garbari 1987). Ghini was a Professor of medicine at Bologna when the Grand-Duke Cosimo 1 de'Medici of Tuscany called him to Pisa to teach botany. Immediately after he joined, he realized the necessity of a garden so that the students could be introduced to live plants, a herbarium where dried plants could be preserved and of accurate representation taken from life to ensure they were scientifically correct (Garbari 1980). Two other university botanical gardens were started at Padua and Florence in 1545. Many other important gardens came up all over Europe during these early years: Bologna, Italy (1567); Leyden, Netherlands (1587); Montpellier, France (1593); Heidelberg, Germany (1593); Copenhagen, Netherlands (1600); Strasbourg, France (1619); Oxford, England (1621); Paris, France (1653); Berlin, Germany (1646); Uppsala, Sweden (1655); Edinburgh, Scotland (1670); Chelsea, England (1673); Amsterdam, Netherlands (1682); Vienna, Austria (1754); Kew, England (1673); Cambridge, England (1762); Coimbra, Portugal; and Krakow, Poland (1783).

To start with the botanical gardens of Europe had collections mainly of indigenous European species, southern and southeastern Europe, and countries adjacent to Mediterranean Sea including Egypt. Some plants of these warmer regions had to be grown in large pots that could be placed in glass houses during winter. These glasshouses were simple rooms made of glass panels with windows facing south –another innovation of Luca Ghini.

During the next few decades, plants were introduced from southeastern Europe and the adjacent Asia—plants that were outstanding because of brilliance of colour or fragrance, e.g. hyacinths, tulips and various types of lilies. By the end of 1620's or so the gardens of Austria, Germany, and Netherlands excelled because of the shift of plant activity centre from Italy and Spain to these countries. With the decline of the Spanish sea ventures, French exploration became more promising and France turned out to be the new centre for flourishing activity in botanical sciences. During this period, while the Dutch started introducing plants from the tropics, the English from Virginia and nearby localities (areas) the French introduced plants from the Canadian wilderness. Botanical gardens in Paris became enriched in plants like sumac (*Rhus coriaria),* poison ivy (*Rhus radicans),* black locust (*Robinia pseudoacacia)* and many perennials such as black-eyed susan (*Thunbergia alata),* dutchman's breeches (*Dicentra spectabilis)* and goldenrod (*Solidago* spp*).* Collection and introduction of plants to different botanical gardens all over Europe continued unabated.

Sir Joseph Banks of the Royal Botanic gardens, Kew, accompanied Captain Cook on his first voyage (1768–1771) and thus began the introduction of plants from the southern hemisphere. Botanical gardens in tropical countries also started developing by this time and many of them became instrumental in introduction of various economically important plants. The first of these economic gardens was started on the island of St Vincent, British West Indies for growing spices and other commercially important plants. The India Botanic Garden, at Howrah, West Bengal, India, originally known as East India Company's garden was founded in 1787 for the cultivation of spices. Botanical gardens – Cultuurtium at Buitenzorg (West Java), Singapore Royal Botanical Garden at Singapore and Royal Botanical Garden at Penang (Malaya) were instrumental in the introduction of para rubber (*Hevea brasiliensis)* to the Far East.

The major botanical gardens in America include Missouri Botanical Garden (Shaw's Garden) at St Louis (1859), Arnold Arboretum at Harvard University (1872) and the New York Botanical Garden in New York City (1891).

Indian Botanic Garden, Sibpore, Howrah

For almost two centuries the Indian Botanic garden at Sibpore, Howrah, in West Bengal has been a pioneering institution for researches in systematic botany and horticulture. Established in1787 by Lieutenant Colonel Robert Kyd this is perhaps the largest and oldest botanical garden in southeast Asia. The garden formerly known as the East India Company's Garden is situated on the west bank of river Hoogly flowing through the twin cities of Calcutta and Howrah. The garden covers an area of 273 acres or 110 hectares. Robert Kyd, though a military officer in the East India Company was very much interested in plants. In his private garden at Shalimar in Howrah, he grew plants of economic importance. He proposed the establishment of a garden under East India Company's patronage for growing horticultural and economically important plants. During his tenure as an honorary superintendent of the garden, Robert Kyd, a great enthusiast, introduced many economically important plants –Cardamom (*Elettaria cardamomum*), pepper (*Piper nigrum*), nutmeg (*Myristica fragrans*), cotton (*Gossypium herbaceum*), tobacco (*Nicotiana tabacum*), indigo *(Indigofera tinctoria),* coffee (*Coffea arabica),* sago (*Cycas circinalis*) and teak (*Tectona grandis).* Many other exotic and economically important plants like tea, cinchona and rubber were introduced in later years.

Apart from Roxburgh, many other famous botanists (Thomas Henry Colebrooke (1813–1814), Buchanan Frances Hamilton (1814–1815), Nathaniel Wallich (1816–1817) and (1817–1842), William Griffith (1842–1845) have looked after this garden during its formative stage. During Dr Wallich's time the garden was one of the most beautiful gardens of the East.

From 1963 onwards the garden is under Botanical Survey of India, Ministry of Environment and Forests, Government of India. The garden is well-known for its collection of palms (unique of its kind in India), water lily (4 species and 30 varieties of *Nymphaea) and* the giant water lily –*Victoria amazonica* and *V. cruziana*; Bougainvilleas -141 cultivars of *Bougainvillea glabra* and *B. spectabilis*; orchids and cacti and of course the Great Banyan tree. The tree is about to be 250 yrs old (246 yrs in 2003), occupies an area of 1.40 hectares with a canopy circumference of 420 meters and more than 1825 prop roots. In 1925 the main trunk of the tree was removed due to fungal attack. Even then the tree is growing vigorously, looks like a mini forest and is a great attraction for the visitors.

Research programs on plant introduction, propagation or multiplication methodology, horticultural aspects and conservation techniques are carried out in the garden. Facilities for laboratory and library and also fieldwork are available for such research studies. The library within the campus has a valuable collection of about one-lakh books, periodicals and other botanical literature. The Central National Herbarium located in the garden has more than 1.5 million of herbarium specimens.

The Botanical Survey of India started preparing a list of rare, endangered and threatened species of plants in the Red data book of Indian Plants and also a Directory of Botanic gardens, parks and other gardens in India (Chakraverty and Mukhopadhyay 1990). The organization also initiated the introduction and propagation of rare, endangered, endemic and curious plants as well as wild plants related to any economic plant for germplasm collection. Experimental Botanic Gardens have, therefore, been established in the Eastern (Shillong), Western (Pune), Northern (Dehra Dun), Southern (Yercaud), and Central (Allahabad) Regional circles for this purpose. The Indian Botanic gardens and these Experimental gardens are the main conservation centers for endangered and threatened species of plants.

The National Orchidarium and Botanical garden at Shillong was established in 1959. In its premises there is a collection of large number of plants that includes ferns, conifers, palm-like gymnosperms and some medicinal, economic and ornamental plants. Some of the interesting plants of this garden are *Cupressus funebris, Cyathea* sp., *Cycas pectinata, Evodia roxburghiana, Illicium griffithii, Magnolia insignis* and *Taxus baccata.* The orchid house is in the center of the garden containing about 200 rare and endangered taxa.

Barapani Experimental Garden at Shillong, Meghalaya was established in 1996.A beautiful hill setting in the Eastern Khasi Hills, vast lakes, wonderful scenario and a pleasant climate, make this garden very attractive. Main aim of this garden is to collect rare and endangered species of northeastern India for cultivation and conservation. A number of rare orchids and the endemic insectivorous plant *Nepenthes khasiana* are reared here.

The Experimental Garden of Botanical Survey of India at Pune was established in 1960. Around 3000 trees and shrubs belonging to 400 species and 40 variety of economic and medicinal plants; and plants of taxonomic interest; and rare and threatened plants such as two species of *Ceropegia (C.attenuata, C.vincaefolia), Iphigenia stellata, Psilotum nudum, Rauvolfia serpentina, Santalum album* and *Vanilla wightiana* are being grown here.

Germplasm collection of Himalayan wild plants and medicinal plants and conservation of endangered plants have been taken up at the experimental Garden of Northern Circle at Dehra Dun. The Garden was established in 2 areas: in an area of 10.4 hectares at Nagdev at Pauri and another in 10 hectares of land at Khirsu- 22Km away from Pauri. The garden at Pauri maintains a plantation of a large number of conifers. In the upper ridges, ca 1.0 hectare of land is covered with natural vegetation of plants like *Myrica esculenta, Quercus leucotrichophora, Rhododendron arboreum,* numerous ferns and terrestrial orchids. The garden at Khirsu has been left as a nature reserve.

National Orchidarium and an Experimental Garden at Yercaud were established in 1964 in the same premises. The garden has an area of 18.4 hectare and is situated in a patch of semi-evergreen, subtropical hill forest. A large number of trees, shrubs, cacti and succulents, ferns and many orchids grow in this garden. Researches on conservation methods and improved cultivation techniques are carried out here.

Experimental Garden at Allahabad was started in 1960. A good collection of medicinal, economic and flowering ornamentals enrich this garden. Roses, croton and *Dioscorea* species have also been added.

Experimental Garden at Port Blair, Andaman and Nicobar islands, was established much later in 1980. Its chief objective is to introduce economically viable wild plants.

BSI has another experimental garden at Itanagar, Arunachal Pradesh. Epiphytes, ferns, bryophytes, lichens, beautiful orchids and rhododendrons makes this place truly a 'Botanist's Paradise'. On a land of 124 hectares at Sankie view, by the side of the river Sankie amongst the natural surroundings of tropical evergreen and mixed evergreen forests, this garden is situated. Some of the interesting plants maintained here are *Angiopteris erecta, Abroma augusta, Amoora* sp. *Boehmeria glomerulifera, Coptis teeta, Duabanga grandiflora, Dendrocalamus hamiltonii, Elaeocarpus* sp., *Livistona jenkinsiana, Mesua ferrea, Nepenthes khasiana, Paphiopedilum* spp., and *Rhododendron* spp.

Lloyd Botanic Garden, Darjeeling

Established in 1878, Lloyd Botanic Garden is situated at an elevation of about 6,000 ft in the Eastern Himalayas. The area of the garden is 40 acres of land donated by William Lloyd – a resident of Darjeeling in 1878.

The garden is roughly divided into three sections:

1. Upper indigenous section
2. Lower exotic section with many species from the temperate parts of the world.
3. Miscellaneous section with predominant species from Eastern Himalayas and various others from North-western India, Eastern India, Miyamarh and the Nilgiris of Southern India.

The beautiful alpine plants, geraniums, composites, rhododendrons, conifers, as well as rock gardens, conservatories, the herbaceous borders and annual bed of flowering plants are some of the colourful features of the garden. The spacious orchidarium houses more than 12,000 specimens of beautiful orchids from different habitats. Some of the exotic plants are *Callitris* sp. from Australia, two specimens of *Metasequoia glyptostroboides* from China—once known in fossil form and later in 1944 discovered in wild state in China, *Liriodendron tulipifera* the tulip tree—another living fossil from China and a collection of oaks (*Quercus* spp). A small herbarium, very useful for identification of local plants has also been added to this garden.

National Botanic Gardens, National Botanical Research Institute, Lucknow

Once a flourishing royal garden of Nawabs of Oudh, National Botanic Gardens is spread over 75 acres of land on the south bank of the river Gomti. It is a nicely decorated garden with a beautiful Rosarium having a large collection of rose plants, cactus house, palm avenue, an aquatic garden with lotus and different species of *Nymphaea,* Vanasthali or Woodland, an area thickly planted with trees, a large number of fruit orchards, a fern house and a palm house with various species from different parts of the world.

A small research laboratory for conducting researches on xerophytic plants and a 'hydroponicum' where soil-less cultures of several varieties of vegetables, fruits and ornamentals can be done, are also present in the garden premises. The largest collection of climbers in India is seen in the climber Pergola. There is a large herbarium with more than one lac specimens representing flora of India and adjoining countries.

Well-known Botanical Gardens of the World

1. **Royal Botanic Gardens, Kew, London.**

This famous garden is more than just an outstandingly beautiful place. It is one of the world's most important botanic gardens with an illustrious history of exploration and research in the world of plants. As pollution and habitat destruction continue to threaten natural environments on an unprecedented scale, the unique knowledge and expertise of the garden staff can help find answers to at least some of the world's most pressing conservation problems.

Wherever possible, scientists of Kew try to conserve plants in their environment, working both within Britain and with botanists overseas to identify areas most in need of protection. At Jordell laboratory, Kew scientists are involved in a range of studies for the medicinal uses of plants.

Kew's Herbarium is the largest in the world with more than 7 million preserved plant specimens. With 40,000 different kinds of plants, Kew Gardens and Wakehurst place, a beautiful woodland garden with a variety of natural habitats –woodland, meadow and wetland in West Sussex, continue to enchant visitors throughout the year. Here, in just one afternoon, one can walk through woodlands of North America, Europe and Asia. From the Elizabethan mansion a series of walks lead to the ravine of Westwood valley and from there to the Himalayan Glade, the Conservation Area with different habitats and even to the giant redwoods of Bloomers Valley or Horsebridge woods. Wealth of enjoyment at Kew includes – crocuses and bluebells in spring, rhododendrons and roses in summer, a blaze of colour in autumn and evergreen in winter. In the conservatories at Kew, plants from the tropics, deserts, mountains and oceans of almost every part of the world are grown.

Kew's botanists and horticulturists are able to solve various global problems, be it environmental management, conservation of endangered species or search for alternative crops as sources of food, fodder, fuel and medicine.

2. **New York Botanical Garden, Bronx, New York, USA**.

The New York botanical garden is one of the two largest and most active botanical garden in United States of America and maintains national and international program of research, education, horticulture, ecology and plant conservation.

This garden was the result of the determination of a young couple Nathaniel Lord Britton and Elizabeth Britton after they visited the Royal Botanic Gardens at Kew. With the backing of the Torrey Botanical Club and the Department of Botany, Columbia University, the land was acquired in 1891. The northern part of Bronx Park was converted into a Botanical Garden. Today, this garden is the only place where original forested areas remain uncut in New York City.

Once the land was acquired and funds were raised, Dr Britton lost no time in planning and landscaping the garden, in collection of living plants to grow here and also in educational and research programs. He remained the Director of the garden until 1929 and during these 38 years of untiring service, Lord Britton fulfilled his dreams of leaving one of the best Botanical Garden in the world. New York Botanical Garden has one of the largest botanical museums in the world, the largest herbarium and the best botanical library in the United States of America and the largest greenhouse. Enid A Haupt conservatory and Thompson Rock Garden are two very important areas in the garden. The Palm house is in the Central Dome of the conservatory and one third of the genera of palms of the world can be seen here. There are ten other houses on either side of this dome where different climatic zones of the world are maintained. Four of these houses are used for seasonal flower shows and other exhibitions. The Enid A Haupt conservatory is worth visiting for the wonderful combination of plants belonging to the tropical rain forests to the desert plants, seasonal displays and educational exhibits such as 'The Greenworld Grocery Store' where plants and their products are shown. The natural rock garden is another outstanding feature of this garden as it is rather difficult to grow such plants outdoor in the extremes of climate in this part of the world.

There are many other important horticultural exhibits and collections in this garden. Imparting education is a major activity and ranges from classes for young children to PhD programs in botanical science. The garden offers facilities for research in systematic botany, floristics, phytogeography, ecology and other similar fields.

The herbarium collection covers the entire plant kingdom; in addition to phanerogams, collections of fungi, bryophytes and ferns are also well represented. It is particularly rich in plants of the New World tropics—Caribbean, Venezuela, the Guiana and Brazil. Many Type specimens of historical importance have remained deposited here as these were collected even before the garden was initiated. The herbarium is very active in 'exchange and loan of specimen' programs with other institutions. Among the activities of New York Botanical Garden, is also the publication work. Many important floras (Flora Neotropica, North American Flora) and journals (Botanical Reviews, Brittonia, Mycologia, Economic Botany, Memoirs of the New York Botanical Garden) have been published from this place.

3. **Munich Botanical Garden, Munich, Germany.**

The Munich Botanical Garden is a multipurpose garden from the time of its initiation. It is not only a scientific institution for teaching and research at per with a University but also a place for information, knowledge and relaxation for general public. Its planning was based on both scientific views as well as aesthetic. The usual systematic arrangement of plants was changed in favor of plant geographical and ecological arrangements.

The garden includes an "Ornamental garden"—the most showy part where the colour of the blooms changes according to seasons. This is so because some plants will flower in summer, some in spring and so on. The greenhouses cover an area of 8,500 m^2 and well protected from the cold winter wind. While some of them contain large xerophytic plants especially the tree- like monocots and leafy succulents (*Dracaena, Yucca, Agave, Aloe, Kalanchoe, Crassula)* some others contain tropical economic plants or trees and shrubs of the Mediterranean, Australia, South Africa and Argentina and Chile (South America)

The collections of cacti and succulents, aroids, orchids and bromeliads as well as Rhododendron grove, Fern gorge, Alpine Garden, Palm house are all highly impressive. The walks planted with herbaceous and bulbous plants on the two sides make the garden of Munich very colorful and attractive. Insectivorous plant, *Sarracenia,* Spanish moss, *Tillandsia,* tree ferns, *Dicksonia*, palms and giant Bromeliads are also grown here.

The scientific work is mainly research in systematic botany although work in the field of morphology, physiology and genetics are also carried out. Researches have been done on Crassulaceae, some aroids and Primulaceae and also some groups of the genus *Oenothera.*

The garden is also used for teaching. Lectures of systematic botany and practical courses in the institutes of the University are run with the supply of plant materials from the garden. School children of primary and secondary classes also use this garden for practical lessons.

Munich Botanical Garden is used for studies, information, relaxation and recreation.

4. **Botanic Garden Berlin, Dahlem, Germany**

This garden originated in 1679 when the "Great Elector " Friedrich Wilhelm created a herb and kitchen garden for his court. Gradually it got converted to a botanical garden in modern sense.

The botanical garden of Berlin had been situated in two other places before it was transferred to Dahlem in 1880. The garden is unique for its fine living collections and beautiful layout. During its formative years between 1900 and 1914, the total of 42 hectares of land was divided into—plant geographical displays, where the plants are arranged geographically, a woody plants collection—an arboretum covering an area of about 13 hectares, a plant systematic display area where different families are arranged according to the classification system of Engler and Prantl, and some special collections such as medicinal plants, water plants, economically important plants and a collection of plants showing morphological adaptations.

The garden and the museum within the garden area were destroyed during the Second World War. These were reconstructed and the garden got improved with a rich museum and herbarium by 1987. There are large conservatories, good rock gardens and many decorative pools and ponds with aquatic plants. Professor Adolf Engler was one of the Directors of this garden who retired in 1921 and died in 1930. L Diels, a plant geographer and specialist in east–Asian and Australian plants succeeded him as Director. It was during Diel's tenure, that the garden got destroyed. Although many improvements have taken, one cannot forget the loss of more the 2 million herbarium specimens.

5. **Main Botanical Garden, Moscow**

This is the largest botanic garden in Europe with an area of 361 hectares. The living plant collection numbers more than 21 thousand. The garden provides facilities for research in taxonomy, protection of rare and endangered species, remote hybridization and study of physiology of resistance to abiotic and biotic environmental factors. There are a number of conservatories and tropical houses, a good laboratory and a library. Introduction and acclimatization of exotic trees and shrubs is one of the main features of this garden.

6. Kebun Raya or Botanic garden, Bogor, Java

A very old and famous garden established in 1817. It covers an area of 200 acres. This garden is very rich in vegetation. Native trees from southern hemisphere, Central America and eastern Mediterranean are represented here. Among other things this garden has a separate section—Kebun Raya of Tjibodas that represents a virgin rain forest. In this section nothing has been planted –a primeval forest. The garden has a good herbarium and library.

7. Arnold Arboretum, Boston, USA

Starting as a small garden attached to Harvard University, it now occupies an area of 265 acres of land. The arboretum has a collection of trees and shrubs of various species of *Rhododendron, Lonicera* and *Paeonia*. Native roses, *Cornus, Davidia involucrata, Philadelphus, Betula pendula* (silver birch), *Salix babylonica* (weeping willow), *Tsuga* spp (hemlock) are other attractions of this garden.

It contains a large number of green houses and a rich collection of "bonsai" trees, the oldest, 175 years of age.

8. Missouri Botanical Garden, St. Louis, USA

Popular as Shaw's garden after Henry Shaw, who set up this garden, Missouri Botanical Garden is in the suburbs of the city of St Louis and covers an area of 75 acres of land. However, outside the city the Grey Summit Arboretum occupies an area of 1600 acres.

The garden houses North America flora, climatron area, houses for xerophytes, floral display houses, Linnaean house, a rose garden, experimental green house and large lily pools with *Victoria* and other water lilies. It is famous also for Chrysanthemums; collection of orchids and succulents is excellent. A large library and a herbarium containing more than 2 million specimens are asset to the garden.

Missouri Botanical Garden is an important centre for taxonomic studies. Flora of Panama and collection of tropical American flora are some of the encyclopedic work taken up here. Paleobotany, experimental botany and development of economic plants are other areas of interest.

9. **Royal Botanic Garden, Paradeniya, Sri Lanka**

The botanic gardens at Paradeniya are laid on 61 hectares of undulating land surrounded by the longest river of this island country, the Mahaweli. The chief function of this garden, 1912 onwards, is the development of ornamental horticulture. Breeding and production of *Anthurium* and various orchids was taken up in 1971 for an export-oriented cut-flower industry. Nurseries are maintained in the garden for selling plants and seeds, and also for exchange of such items with other gardens and institutions.

The garden provides–

a. Technical know-how for layout of parks and gardens and on flower production;
b. Training facilities for growers of ornamentals; and
c. Facilities to foreign and local botanists to study and collect plant specimens.

The garden is actively engaged in plant conservation, plant tissue culture and research in floriculture. A large collection (4000 species) of useful and ornamentals of the tropical world is available here, which includes spices, orchids, cacti, flowering vines, cycads, medicinal plants and many more.

10. Singapore Botanic Gardens, Singapore

The Singapore Botanic gardens was funded by the Agro-horticultural society of Singapore, which received a grant, some convicts as laborers and 60 acres of land in 1822. Eleven acres of this land is preserved as primeval forest zone. Later, an associated economic garden of 102 acres was set up which helped trial, acclimatization, and multiplication of valuable plantation crops such as cinchona, coffee, tea, maize, sugar and rubber. It is from the 22 seedlings of rubber *(Hevea brasiliensis)* sent to this garden in 1889, after germinating the seeds at Kew, that the vast rubber plantations and rubber industry of the South East Asian region developed.

A very rich collection of orchids in the National Orchid Garden is located in the highest hill in the Singapore Botanic Gardens. Three hectares of its gentle eastern slopes were landscaped to grow the orchids. About 60,000 orchid plants belonging to 400 species and more than 2000 hybrids are grown here. There are terraced slopes, pathways, fountains and waterfalls to provide an effect of freshness in a lush-green tropical garden. Special features of this orchid garden include the VIP i.e. very important plants display; Tan Hoon Siang misthouse, Yuen-peng Mcneice bromeliad collection with a display of over 20,000 plants and the Burkill Hall at the top of the hill – once a home for the garden's directors. The house is encircled by a spectacular view that included the palm valley and the rainforest beyond. Hybrid orchids from different countries like Holland, Mexico, Australia, Japan, Papua-new Guinea, Thailand, and Philippines are housed in the misthouses. *Vanda* Miss Joaquim is the oldest natural hybrid of Singapore and Malaysia. From April 15,1981, it is the National Flower of Singapore.

Aims and objectives of Botanical Gardens

A botanical garden can be considered a living repository of plants with some scientific arrangement. Here the plants are usually labeled for identification. Early botanic gardens established in the sixteenth century were mostly renowned for the number of novelties they grew procured through exchange from distant lands.

The next stage in the evolution of botanical gardens was the study of systematic botany. Plants in botanical gardens were arranged on the basis of their diagnostic features and plants of a particular group were grown together. Gradually, centers were developed in these gardens for the study of diversities and curiosities of plant kingdom. The botanical gardens, henceforth, became the places of natural aesthetic beauty as well as taxonomic studies and education. Further evolution of botanic gardens had three major directives:

1. Comparative study of plants growing in a garden and the herbarium for researches in modern taxonomy and experimental botany.
2. As centers for the study of economically important plants. Botanical gardens served as acclimatization centers for economic plants introduced from one part of the world where it is native to, to another part of the world.
3. As centers of horticultural research. This included trials, selection, hybridization and release for cultivation of many new varieties of useful and ornamental garden plants.

These objectives were pursued by various botanical gardens for a long time till there was serious effect of environmental pollution, ecological imbalance and conservation of threatened plants. Botanical gardens of the present era are considered as main centers for conservation of rare and threatened plants from extinction. The aims and objectives, therefore, are recommended as:

1. Botanical gardens should be a living repository for the plants of that country where it is situated and selected exotic plants.
2. Should act as a safe abode for growing rare and endangered plants.
3. Should be a place for germplasm collection of various economic, medicinally important and ornamental plants and their progenitors.
4. Should act as a centre for educational programmes and research in experimental botany and horticulture.
5. Should be a proper place to grow species of different climatic conditions in conservatories, phytotrons, green houses or glass-houses and also conducting researches on propagation of rare and endemic plants and various different species for afforestation, energy, and alternative food and fodder plants.
6. As centers for introduction of economic and commercially exploitable species, their acclimatization and release for cultivation.
7. As a data bank for information and documentation of the plants growing in different botanical gardens of the country.
8. As institutes to generate awareness about value of plants, to interest people into growing beautiful and curious plants, to organize flower and foliage shows and exchange seeds, seedlings and saplings with other such institutes.

Basic requirements of Botanical Gardens

1. Planning a botanical garden should be in an area of 100 to 175 hectares of land of good soil.
2. Soil must be tested beforehand so that there is not any type of deficiency in minerals, water, sub-soil water level or any other.
3. At least 10 per cent of the area should be covered by well-maintained water-bodies for growing aquatic plants. Humidity within the garden can also be maintained by the presence of such water-bodies.
4. A natural source of water nearby is preferable, e.g. India Botanic Gardens at Howrah (West Bengal) is situated on the west bank of river Ganga.
5. A botanical garden should be well planned to accommodate the following:

(a) Taxonomic gardens—Plants should be grown familywise according to any well-known classification system and each family must be represented by a few important species.

(b) Medicinal plant gardens—Important medicinal and aromatic plants can be grown in a demarcated area. Within the campus of Indian Botanic Garden, at Howrah, there is a medicinal plants garden named as 'Charaka Udyan'. It has a collection of 1000 plants belonging to 450 species. Medicinal and aromatic plants have been grown in the same enclosure to represent ancient Indian culture of using these natural resources in human ailments.

(c) Germplasm collection—Every botanical garden should have germplasm collection of plants that are suited to the climate.

(d) Arboretum—Adequate space should be demarcated for arboretum right at the initiation of the garden.

6. Glasshouses, greenhouses, phytotrons and conservatories are must for any botanical garden. These should be architecturally attractive and well- maintained so that rare and endangered plants as well as plants of ornamental and horticultural importance, orchids, certain cacti and succulents can be grown under climatically suitable conditions.
7. Botanical gardens should be maintained in such a way, so that, not only the scientists but also the plant-lovers and general public visit them. To achieve this, aesthetic and recreational spots, monuments, fountains, plant-houses, ornamental gardens and many others can be added to a botanical garden.
8. A Botanical garden should be planned within an easy reach. Without the easy means of transport and conveyance, full utilization of such a valuable resource will not be achieved.
9. A well-equipped laboratory, a library and a herbarium within the campus of the garden make it more attractive.
10. The garden management staff should preferably have their residential accommodation adjacent to the garden although never within the garden complex.
11. If possible special guided tours within the garden area should be arranged for the visitors.

Botanical Gardens of today and tomorrow

Although the early botanical garden's aim was to provide information on food plants, medicinal plants, ornamental and exotic plants, and propagation of the plants introduced from various parts of the world, today's botanical gardens play an important role as centers for research, education, conservation (of threatened plants) and public service.

1. **Researches** in various fields such as systematic botany, reproductive biology, breeding and hybridization can be conducted in botanical garden. A well-maintained botanical garden is a conservatory of living plants that provide material for such studies.
2. **Education**—Many botanical gardens in association with academic institutions help in teaching various courses like systematic botany, horticultural methods, plant propagation, dendrology and others. Institutions with natural vegetation in the vicinity, can conduct studies on plant communities, natural hybridization, variations amongst species; and others with experimental plots and extensive collection of plants from various families growing there can start courses in plant phylogeny and classification. The Glasgow botanic garden was a physic garden meant for teaching medicine, associated with the University throughout the eighteenth century. A wide

range of educational and information provision for the general public and students have been developed here.

3. **Conservation**—The increasing role of botanical gardens in conservation has been emphasized by the Botanic Gardens Conservation Congress of November 1985. This resulted in the formulation of 'Botanic Gardens Conservation Strategy' implemented by many gardens. Today many botanical gardens are deeply involved in the conservation of threatened plants by –1. growing such plants under protection; 2. making public aware about the disappearance of many plants; 3. putting a stop to the destruction of natural habitats of the rare and threatened plants.

   Many botanical gardens carry out 1. improved cultivation techniques for endangered and threatened plants; 2. discovery and documentation of wild populations of such plants and; 3. development of cloned and cultured material for reproductive biological studies. Rare plants are removed from threatened habitats and transplanted in protected areas controlled by botanical gardens.

4. **Public service**—For the identification of native as well as exotic plants species, botanical gardens are most useful. In addition, for information about methods of propagation, pest and disease control of new cultivars, when and how to plant some species and to solve some other similar problems, staff of botanical gardens are experienced and helpful. Annual sale of plant materials like seedlings, seeds, bulbs, corms and rhizomes are often arranged by many botanical gardens as part of public service.

Botanical Gardens of Future

During the last few decades gene erosion has been taking place at an alarming rate. Botanic gardens of future should provide basic prerequisites to counter this through their biological resources, technical equipment and scientific and gardening experiences. Conservation of complete range of genetic diversity is necessary, and hence, new methods such as establishment of conservation gardens and protected areas like biosphere reserves, nature reserves, germplasm centers and management of genetic resources in natural communities are needful.

Botanical gardens should be involved in propagation and cultivation of rare species, experimental improvement of threatened communities, monitories of disappearing populations, analysis of unique habitats and study of reproductive biology of such taxa.

Botanical gardens of this nature have the potential to be excellent centers for systematic, genetic, ecological, physiological, horticultural and evolutionary research. These institutes should be able to impart training to the future generations in understanding and appreciating the roles of plants and animals in our environment, everyday lives and biological heritage.

Botanical gardens can play a useful role for conservation of endangered species. In these days of habitat loss, these gardens are acting as nurseries or refugia of plants that would otherwise have been extinct. As ex situ centers for conservation, modern tools are available for the preservation of seeds and propagules in seed banks. A network of botanical gardens cultivating rare, vulnerable and endangered plants and exchanging information and plant material would go a long way in preserving the floristic and genetic germplasm which we have inherited as products of millions of years of evolution (Nayar 1987).

Botanical Survey of India

India is a vast country with much variation in her geographical features-the lofty Himalayas in the north with its swift- flowing rivers and valleys and snow-capped mountains; table lands and plateaus in the

Vindhya hills and the central part of India; deserts in the western part; the Western and Eastern ghats in the peninsular region; deep valleys of Assam; and the deltas of various rivers. In no other country, in all probability, so much variation in geographic region and accordingly variation in forest type and flora can be experienced.

Exploration of the plant resources and preparation of floristic accounts of various parts of this country, therefore, started as early as the middle of the nineteenth century. Several botanists and naturalists got involved in this type study. After a few years of such individualistic work, it was felt that co-ordination of these studies was necessary and for which a central organization would be most beneficial. Accordingly a scheme for the organization of Botanical survey was drawn up and with the approval of the then secretary of state for India (under British regime), it was established on the 13th of February 1890.

Functions of the survey were: Exploration of the vegetable resources of the country and coordinating the botanical researches of the workers in different parts of the country. Sir George King, who held the position of superintendent of the Royal Botanic Gardens (Indian Botanic Garden), Sibpur was given the charge of the Botanical Survey of India as its first Director.

The Botanical Survey of India was reorganized in 1954 and also expanded. Four regional circles were set up under the survey at Dehra Dun, Poona, Coimbatore and Shillong with the headquarters remaining in Calcutta. A botanical laboratory under the supervision of a Director was established at Lucknow which was later shifted to Allahabad and finally to Sibpur in the Botanical Garden premises. At present a joint Director heads this Laboratory—the Central Botanical Laboratory. It has under it the following sections—Plant Physiology, Cytology, Economic Botany, Pharmacognosy, Ecology, Palynology and Plant Chemistry.

The Central National Herbarium also functions under the Botanical Survey of India. At present it is housed in the building constructed during 1965 to 1969. It contains more than 2,500,000 specimens of Indian as well as foreign origin. A Deputy Director heads the Herbarium, which is a repository of all authentic, and type collections of Indian plants and of the Flora of the World.

The Industrial section of the Indian Museum in Calcutta is another unit of the Botanical Survey of India. The office of the Director, who is in charge of this section, is in one of the wings of the Museum. Function of this unit is collection and preservation of rare, endemic and economically valuable species and their proper exhibition to public.

The publication section publishes two important journals devoted to plant taxonomy:

1. Records of the Botanical Survey of India, and
2. Bulletin of Botanical Survey of India.

In addition to all these units within the country the survey has one botanist posted in the Kew Herbarium (UK) for reference work of the Botanical Survey

Objectives of the Botanical Survey

1. Survey of plant resources of the country for preparing the inventory of plant wealth in the form of National and Regional floras.
2. Development of the Central National Herbarium and the various Regional Herbaria as repositories of the types and other authentic specimens.
3. Development of the Indian Botanic Garden, the Regional Experimental Gardens and the National Orchidaria, for the study, introduction and conservation of flora.
4. Development of the Central Botanical Laboratory for investigations on the plants through various disciplines of botany.

5. Development of the Industrial Botanical section of the Indian Museum on modern lines for studies on economically important plants.

The Regional Centers of Botanical Survey

The regional centers were started to carry on botanical explorations in the respective regions of the country, preparing regional, state and district floras.

1. Botanical Survey of India, Northern circle at Dehradun.
2. Botanical Survey of India, Eastern circle 'Woodlands' at Shillong.
3. Botanical Survey of India, Southern circle at Coimbatore
4. Botanical Survey of India, Western circle at Pune
5. Botanical Survey of India, Central circle at Allahabad
6. Botanical Survey of India, Arid zone circle at Jodhpur
7. Botanical Survey of India, Andaman and Nicobar circle at Port Blair
8. Botanical Survey of India, Arunachal Field station at Itanagar (Arunachal Pradesh)
9. Botanical survey of India, Sikkim Field station at Gangtok (Sikkim)

The experimental Gardens at Barapani under Eastern circle, Pauri under Northern circle, Yercaud under Southern circle, Mundhwa under Western circle and Allahabad under Central circle look after the introduction of economic and horticultural plants of the respective regions.

Apart from usual activities of the Botanical Survey of India the organization also works in collaboration with various schemes of Government and other such organizations for carrying out various research projects and programmes, for example, schemes sponsored by UNESCO, Monographs and Research schemes of CSIR, and ICAR and Tea Research under the Tea Board.

In conclusion, it may be said that the Botanical Survey of India after independence and reorganization has done useful service to the Indian Flora from the point of view of Systematic and Economic studies. Although understaffed to begin with, the various sections are now strengthened well so that the preparation of flora of India and the Regional Floras are expedited.

Herbarium

Herbarium—is also a type of botanical literature; helpful in identification and classification of various taxa. The word herbarium, in its original sense, refers to a book about medicinal plants. Tournefort used the term for a collection of dried plants. This usage was taken up by Linnaeus under whose influence it superseded the earlier terms such as 'hortus siccus' (Sharma 2003)

A herbarium is a collection of dry pressed and preserved specimens of plants affixed to paper, which are arranged systematically for the purpose of reference and identification. The credit of invention of herbarium goes to an Italian –Luca Ghini (1490-1556). His students disseminated the art of herbarium-making all over Europe. Collection and preservation of specimens was started as early as 1532, by one of Ghini's students, Gherards Cibo. Herbarium technique was a well- known practice at Linnaeus's time and he practiced mounting the dried specimens on single sheets of standard size and stored them horizontally almost similar to the practice today (Stearn 1957). However, this method, although general during the second half of the eighteenth century, was by no means universal. During the early days of herbarium – making, the mounted specimens were presented in bound volumes. Even as late as 1833 Asa Gray offered bound volumes of grasses and sedges for sale (De Wolf 1968). The practice of depositing the collected specimens in some established institutions as well as exchange of specimens developed quite early. As a result, when some of the early herbaria were destroyed by fire, insects, war and because of ignorance, whatever remains of them today, are the duplicates sent to other institutions on exchange.

Today, herbaria serve various purposes:

1. Classification of world's flora is primarily based on herbarium and the literature associated with it.
2. A herbarium has certain advantages over living collection because, only in a herbarium one can compare all related species of a genus in the same place, same state and same time.
3. A modern herbarium is a storehouse of data that must be retrieved and used. A researcher finds the collection to be adequate for the comprehensive characterization and effective delimitation of taxa, intra- and intertaxon relationship and understanding the evolutionary trends and also distribution pattern of the taxa. A physician treating a patient for allergic reactions with some particular pollen might get the information about the geographic distribution of the taxon and also its flowering time. The patient might then be asked to move to a region where the plant is absent or its flowering time is different. A taxonomist interested in the nature of habitat of a particular species might get to know this by seeing the specimens already studied and collected by a student from a similar region. An environmentalist can use a herbarium as a database for information about an unusual or rare community whose voucher specimens are available here. A chemist, who has discovered an insecticide, or fungicide from a species would also like to use a herbarium for detailed information about the species—its correct name, its distribution and if it can be used commercially.

Well, all these and many other type of informations can be obtained from a herbarium although expectations of many workers are occasionally not realized .Two chief functions of a herbarium are: a). Accurate identification; and b) ⅓ –taxonomic (descriptive) research.

Based on the purposes they serve, Davis and Heywood (1967) divided all herbaria into three groups:

1. Major or National herbarium
2. Regional or local herbarium
3. Working herbarium

Major herbaria—These herbaria contain collections from different parts of the world. They serve for identification and research. Many National herbaria are national trusts managed by the Government of that country. Herbarium of the British Museum (National History) is one such herbarium. Some important National herbaria are-

1. Museum National d'Histoire Naturelle (Museum of Natural History), Paris   10,000,000 specimens
2. Royal Botanic Gardens, Kew, London   6,500,000 specimens
3. Komarov Botanic Institute, St. Petersburg (Leningrad)   5,700,000 specimens
4. Conservatoire et Jardin Botaniques (Conservatory and Botanical Garden), Geneva   5,000,000 specimens
5. Combined Herbaria, Harvard University, Cambridge, Massachusetts includes—Arnold Arboretum, Farlow Herbarium, Gray's Herbarium, Oaks Ames Orchid Herbarium   5,000,000 specimens
6. New York Botanical Garden, New York   5,000,000 specimens
7. US National Herbarium (Smithsonian), Washington, D.C.   4,100,000 specimens
8. British Museum of Natural History, London   4,000,000 specimens

9. Natural History Museum, Vienna — 3,500,000 specimens
10. Missouri Botanical Garden, St. Louis — 2,900,000 specimens
11. Central National Herbarium, Calcutta — 2,500,000 specimens
12. Royal Botanical Garden, Edinburgh — 1,700,000 specimens

Functions of national herbaria could be grouped under four headings:

(i) Their own research programmes; (ii). service as repositories of type specimens and other important specimens; (iii). loan of specimens for study at other institutions and (iv) training of graduate students.

(i) Researches taken up at the national herbaria are usually difficult or sometimes impossible to perform with limited resources. It requires the examination of large number of specimens, of many species and from a wide geographical range. It may not be difficult to take specimens on loan from other herbaria or to travel to different herbaria but what is required is one's own necessary tools at hand for day to day work. Hence, work for major floras, e.g. Continental or regional floras and broad scale studies of any family can be effectively carried out at a national herbarium. It is obvious that an author writing a flora cannot do monographic work for each family or genus but if he is working at a national herbarium, the emanating result will be more authentic reflect more original work in the definition of taxa and establishment of names. Instead of mere repetition of earlier work, this new data would improve our knowledge about various groups of plants.

Broad scale studies of families and researches in biosystematics are also carried out at the national herbaria, especially those which are also university at the same time, for example, the National Herbarium of Harvard University. Floristic work on tropical and subtropical America is largely concentrated in the national herbaria of United States of America.

During the last few decades, the use made by systematists of characters and techniques from the fields of anatomy, embryology, palynology, cytology, ultrastructure, biochemistry, phytogeography and others, in solving taxonomic problems has added diversity to the collections. Although pressed, dried and mounted specimens are still most common, many herbaria house specimens in liquid medium, frozen condition as well as collections of photographs and illustrations.

(ii) As a rule all the Type specimens should be deposited in any one of the national herbarium of the world. The importance of large herbaria as repositories of types and historical materials need no documentation. The scattered smaller herbaria could not perform this function nearly so well even if their safe–keeping could always be guaranteed.

(iii) The national herbaria also give specimens on loan to other institutions for taxonomic research.

(iv) Training graduate students in taxonomic work is yet another function of these herbaria. A major botanical library is a necessary adjunct to a major herbarium. The specimens of a herbarium should not only be thoroughly studied but the publications of earlier years must also be consulted.

Regional Herbaria—These represent collections from one country or a part of a country. These are generally looked after by private institutions or by a university. Herbaria of this category are engaged in preparation of country or state floras. The university herbaria are also used for teaching and postgraduate research. Some of these herbaria contain special ecological collections, economically important collections and collection of voucher specimens of cytological and various experimental studies. The university of Michigan herbarium initiated the experiment of incorporating photographs of fossil plants as "plant collection" (Heywood 1967).

Working Herbaria—These herbaria are designed primarily for identification work, and for classroom studies. They consist of only a few replicas of a species from a particular area arranged within that genus. The genera are arranged either alphabetically or according to the flora of that region.

Families are generally arranged according to a particular system of classification. Often such working herbaria form part of a major herbarium e.g. The European reference collection at the British museum (Natural History), London.

The activities of a herbarium are listed here:

1. Providing a standard reference collection for verifying the identification of newly collected plants . This is a major function of many small herbaria.
2. Serving as a reference collection for plant taxonomy and other courses of study.
3. Training graduate and undergraduate students in herbarium practices.
4. Documenting the presence of a species at a particular location and providing data on its geographic range.
5. Providing samples for the flora of an area. A taxonomist writing the flora of a particular region can consult the herbarium collection to ascertain which species are represented in that region.
6. Pointing out the existence of classification problems. An examination of herbarium specimens may, indicate that additional research is required for certain taxa whose position or nomenclature is doubtful.
7. Offering data on various fields–vegetative and floral morphology, pollen samples for palynological studies, leaf and stem samples for anatomical studies and chemical analysis; data for preparation of distribution maps; ecological data; and economic values.
8. Preserving Type Specimen and specimens of historical importance and serving as a repository of chromosomal, chemosystematic and experimental voucher specimens.

Herbarium Methods

Collection of specimens—"What to collect" is of prime importance for a collector. Advice might be sought in this regard from the persons working in a local herbarium. It may be—either as much as one can collect (particularly in little-known areas) keeping in view the tiring work later on; or to collect only certain groups at a time. This may be according to the collector's interest or better representation in the herbarium. Common local plants are often not well represented in a herbarium and particularly those, which are difficult to press because of bulky nature of some parts (fruits of *Datura innoxia*) or heavily armored plants (thistles- *Cirsium vulgare).*

Herbarium specimens should:

(i) Be true representative of their populations; range of variation should be shown and exceptional variants should be recorded;

(ii) Be collected in flower and fruit; if dioecious, both male and female plants with flowers/and fruits should be collected;

(iii) In herbaceous plants, the rootstock should be collected whenever possible—and also bulbs, corms, tubers. To quicken the drying process, the fleshy underground parts can be cut into halves; dipping them in boiling water or immersing them in alcohol can kill bulbs;

(iv) Tall and bulky herbs can be cut into pieces and represented on two or three herbarium sheets; or they may be folded in the form of V, N, or W to represent on one sheet. In biennial plants sterile rosettes as well as flowering shoots should be pressed. When within the press, padding of cotton, paper or cardboard can be used to even out the pressure on leaves and flowers.

Collection

For collection of plant specimens from the field on an excursion trip, the following equipments are necessary; a plant press, a secatior or pruning knife, a small shovel, a vasculum or polythene bags, drying papers—coarse blotting papers or old newspapers, a field-note book and a hand-lens.

Plant press is the most important equipment for pressing plants. A good and efficient press is the one which can hold the specimens under constant and firm pressure, can dry the specimens to a degree less than crispness, and can retain the natural colour of all the parts as far as possible. It normally consists of a pair of wooden or metal frame held together with the help of leather straps or belts or strong rope. Its standard size is 12" ▯ 18". The specimen to be pressed is placed in between the folds of newspaper or pressing paper of any other quality with leaves spread well, apical portion with the leaves and inflorescence erect and avoiding any overlapping of various parts. Then the press is arranged in such a way that every pressing paper or newspaper with specimen alternates with a blotter. The leather straps are now tightened to the maximum possible so that parts of the specimen is flattened and it comes as close to the moisture - removing blotter as possible. Changing of papers is necessary after 24, 48 or 72 hours depending upon the season or the prevalent weather. Greatest efficiency from the pressing paper and the press is obtained when the specimen covers maximum surface of the paper. Normally only one specimen should be placed in the newspaper fold or pressing paper but if the specimens are small, a number of them can be pressed together. If artificial heat is to be used for drying the specimens then a sheet of corrugated material should be placed between each pressing paper with its specimen. Quick drying is necessary to avoid moulds on the specimens.

The field notebook is another very important object. It is a small notebook that can be carried to the field with ease. Each page of the notebook is meant for one plant. On this page the collector should write down the following; (a) the place of collection; (b) altitude of the area, (c) date of collection, (d) habit of the plant, (e) flower color, (f) fragrant or not, (g) any other feature that may not be seen once the specimen is dry.

A brief ecological note regarding the soil type, amount of annual rainfall in the area, growing on slope or not and other plants of the community may also be useful. The pages of the field-note-book are numbered and each page has some extra numbered strips, which are detachable. Whenever a plant has been pressed and the details about it written on the field-note book, the numbered strip from this page should be attached to the pressed plant. Same procedure is followed for the duplicates of this specimen.

A secatior or pruning knife is necessary to remove some overlapping parts from the specimen. This should be done judiciously and a basal portion of the part removed should be left behind to indicate the location of the pruned part.

A shovel is used to remove small plants with their roots and other underground parts.

Sometimes the plants are too small in size as some species of *Utricularia.* They cannot be seen without a hand lens. As far as possible collected plant should be pressed immediately in the field itself. However, sometimes it becomes necessary to postpone pressing of the plants at the later time. A vasculum is a useful instrument in this regard. It is a metallic box with a tightly fitting metal lid, usually with shoulder sling to carry it with ease. Specimens can be stored in this box for a few hours only before pressing them.. Bulky parts like rootstocks, rhizomes, bulbs can also be stored in this box. Generally the box is painted white to deflect heat and to be easily detectable if left in the field. Its bulky nature prevents many scientists to use it and now a days it has been replaced by polythene bags. The polythene bags are easy to handle; being almost weightless a large number of these bags can be carried to the field and made airtight with the help of rubber bands.

Mounting

For mounting the specimens special mounting papers or rag papers are used. These are not so thick and bulky but are tough and hard. Each sheet is usually 29 cm □ 42 cm (11.5" □ 16.5"). The specimen (after poisoning) is mounted in three different ways:

(1) It may be glued to the paper and some portions like the stem or branches, leaf petioles, flower-pedicels are stitched with needle and thread. This was the earlier practice.

(2) Presently use of adhesive linen, paper or cellophane strip are found to be much easier and faster method for mounting the specimens.

(3) Use of liquid paste or glue is also prescribed. The glue is applied to the backside (if it is discernible) of the specimen to be mounted and pressed on the surface of the mounting sheet; or first an outline of the specimen is sketched on the mounting sheet and glue applied to this area and then the backside of the specimen is pressed on this glued part. In both these methods, the specimen is allowed to dry for a few hours in pressed condition. The method is slow but economical.

(4) In another method, the glue or paste is smeared on a plastic or glass sheet measuring 35 cm □ 40 cm (14" □ 20"). The specimen to be mounted is placed with face upwards on this sheet so that all parts of the lower surface come in contact with the glue or paste. Then the specimen along with the field label is lifted up with the help of a pair of forceps and placed on the mounting sheet. The position of the specimen on the mounting sheet should be visualized in advance or marked with pencil beforehand, because after once laying the specimen on the sheet, it should not be moved. If done so, marks of the glue left by the specimen will make the sheet untidy. After its placement on the mounting sheet, the specimen should be pressed firmly by placing a blotting paper to remove excess glue. After using once-coated glass plate for mounting 2 or 3 specimens, it may be recoated with glue. This is a more efficient and quicker method but not so economical. Some insect-repellant chemical like a small quantity of mercuric chloride, thymol crystals or copper sulphide should be added to the glue or paste before using it. And in that case, one has to be extra careful while using this glue.

Herbarium Labels

Once the mounting has been completed, each sheet should have the herbarium label pasted on to it, size of the sheet is 8 cm □ 12 cm (2.75" □ 4.25") but it can even be up to 4" □ 6".

In general herbarium label should include the following data-

1. Name of the institute to which the herbarium is attached—a research institute, university or a college
2. Collection number
3. Date of collection
4. Name of the genus and species complete with the author's name
5. Name of the family to which the specimen belongs
6. Locality of collection

7. Brief description /remarks/notes
8. Collector's name
9. Vernacular name and uses if any.

The herbarium label is fixed at the bottom right corner, normally 1cm away from the edges of the mounting sheet. It may also be printed directly on the mounting sheet. However, writing the above details is difficult as the mounting sheets are quite rough and typing is impossible. Label should be written with such ink that should stay for a long time. It may be typed also. A specimen without a label, even if it is well pressed and neatly mounted is of no value.

Identification is the next step, once the specimen has been mounted. For identification the scientific method is to first study the characters of the plant –both from the specimen as well as the label attached to the mounting sheet and check them with the flora of that locality or region if any, and compare with the description and illustration. For a correct identification one must go through the process of working with family, genus and species keys and then come to a conclusion. Use of correct name is equally important and therefore, effort should be made to look for the latest accepted name for the taxon.

Accession: Mounted, labeled and identified specimens are then stamped with a distinctive mark of the herbarium, usually on the top right hand corner of the mounting sheet. Stamping may be done with rubber stamps or printed labels. The stamp carries the name of the institution, a serial number called the herbarium accession number and sometimes the date of accession.

Filing: The specimens are now ready for filing and they are sorted out family, genus and species wise. All the sheets of one species are placed in covers of lighter colour and are called species- cover or folders. All the species (in species-cover or folders) of one genus are placed in one or more than one folders of heavy paper of a different colour called the genus- cover. Use of such covers is helpful in protection of specimens, in arranging them according to some system of classification and convenience in handling. Name of the species and genus are written on the covers.

These genus-covers are now placed in the pigeonholes of the steel almirahs. These pigeonholes are marked with the names of the families in bold letters, and preferably in black ink.

Poisoning and preservation

If the plants are poisoned before pressing them, they are killed immediately and the abscission layer formation is prevented and therefore, the plant parts remain intact. Poisoning is generally done by dipping the whole plant in the poisoning solution for about 15-20 seconds depending upon the thickness of the plant. The specimen is then placed on a blotter so that the excess solution is removed and then the specimen is pressed in a drying plant-press. Chemicals used for this purpose are:

(a) $HgCl_2$ or Mercuric chloride—It is a corrosive solution and therefore enamel trays (no metal) should be used. Use of hand gloves is advised and fingers should not get dipped in the liquid. Forceps made of bamboo or wood is recommended. A person having cuts or open wound in the hands must not handle this chemical, as it is a deadly poison.

(b) LPCP—Lauryl pentachlorophenate is sometimes used in place of mercuric chloride. Handling this chemical is comparatively safe.

(c) Formalin—Use of formalin as a poisoning substance is suitable in tropical countries. Whenever the day's collection is numerous, it is not possible to keep changing the blotters particularly if the area is very humid. In such a situation, the plants are spread on old newspapers and a large number of them are bundled together and placed in a large polythene bag. Formalin of 10% concentration is now poured over the bundle—an amount sufficient to soak the entire bundle but not in excess. And then the bag is made airtight. No further change is necessary for the next few days and sometimes even over 3 to 4 months when it is a large expedition. This method has certain advantages:

(a) Labor and time for daily changing, pressing and drying the blotters is saved. (b) Large number of blotters and heavy plant-presses need not be carried to the field. (c) These collections need not be poisoned further, before mounting on herbarium sheets. (d) Formation of abscission layer is prevented and therefore, leaves and other parts of the plants do not get detached.

Pest Control

Most specimens submitted to the herbarium are quite dry. But even then some treatment is necessary to protect them against insect pests such as beetles and silverfish. Pest control measures can be placed under three heads:

i. Heating: Some herbaria use electric heat treatment for the specimens. Using specially insulated cases with a central heating element at the bottom does this. The system is effective but the specimens become brittle.

ii. Deep-freezing: In many herbaria, the heating practice (to control insect pests) has been replaced by deep-freezing .A temperature of -20º to -60º C is maintained in these herbaria.

iii. Microwave ovens: Use of microwave ovens has been in practice in some herbaria. But it is not a very effective method because of the following reasons:

(a) Sudden vaporization of water in the thicker parts of the plants can make it burst;

(a) Metallic staples on the sheets may burn the sheets because of overheating;

(b) Often the seeds from herbarium materials are used for growing plants for research purposes. Because of heat the embryos may be killed and become useless for this purpose.

iv. Fumigation: This is an effective method for killing pests (silverfish, beetles etc) in mounted as well as unmounted duplicate specimens. Any one of the volatile poisonous liquids such as methyl bromide (CH_3Br), carbon disulphide (CS_2) or carbon tetrachloride (CCl_4) is used. These are placed in small Petri-plates or saucers in the herbarium cases that are kept closed for about a week or so. A mixture of ethylene dichloride and carbon tetrachloride in the ratio of 3:1 was once used as a common fumigant .The former is explosive without the latter and carbon tetrachloride is highly toxic for human being. Hence, there is ban on the use of this fumigant.

Downfume -75 can be used as a safer fumigant in herbaria as it has a clearance from Environmental Protection Agency. Another method is the use of vapona resin strips or raid strips. One third of a strip is placed in each herbarium case for 7 to 10 days twice a year .The cases are not to be opened during this period (Singh GC 1999).

Paradichlorobenzene (PDB) or Naphthalene (powdered) in small cloth bags is sometimes placed in the herbarium cases for the same purpose. Paradichlorobenzene, however, is used as an insect-repellant and as it is hazardous to human health, workers should not be exposed to this fumigant for long hours. Fumigants do not kill egg or pupae of insect and therefore, fumigation must be done at regular intervals.

Herbarium Ethics

Herbaria are meant for use regardless of the type of collected material—for research, identification, teaching or any other. Maximum efficiency in locating materials without endangering the collection is expected. For this a few rules are to be followed.

(i) Instructions for the users—An introductory guide to the collections should be provided to those who wish to use a herbarium. Such a guideline should introduce the user to the systematic and geographic arrangement of the collections; any colour code adopted for the folders; a list of families with their numbers and the case numbers in which they are kept; an outline diagram of the arrangement of the cases; details (a list, location and arrangement) of any special collection such as ecological or economically important plants' collection and a list or catalogue of all the genera. Any other information about the herbarium such as proper handling of specimens, use of equipments such as microscopes while the specimen is in use, loan procedures, should also be included in the guideline.

All these and any other additional information can be published in printed form and given to the users to save time of both the herbarium staff and the visitors.

(ii) Herbarium specimens are preserved over long periods of time and anybody using these should do so with utmost care. Scientifically, these specimens are invaluable and irreplaceable. For proper handling of the specimens a few suggestions are:

(a) Herbarium sheet should be kept flat; these should not be shuffled through while in a folder, like the pages of a book. Specimens are brittle and easily damaged.

(b) Proper storage—either in herbarium–cases or in shelves is necessary. Too many specimens in a box can be damaging.

(c) Books and other heavy objects should not be placed on dry and mounted specimens.

(d) Loose, recognizable fragments of any specimen should be placed in packets or envelopes.

(e) Often it is necessary to dissect flowers and fruits from the specimens; also materials for anatomical, palynological and chemical studies are required. Materials for such studies should be used sparingly.

(f) Long-armed dissecting microscopes are useful as the entire specimen can be examined without bending the sheet.

(g) The support of any strong material such as a stiff board is necessary to carry the specimen even to a short distance.

(h) Nobody should write anything on the herbarium sheets

(i) An investigator wishing to use a herbarium on days other than weekdays and at unusual hours such as evening should seek permission to do so in advance.

(j) Most herbaria give materials on loan to other institutions. Reciprocal agreements between institutions are often made for this.

(k) With time, the herbarium specimens become brittle. Such old specimens should be handled with care.

Identification Keys

Floras normally include a **diagnostic key** with the help of which unidentified plants may be identified. A key has been defined as an orderly arrangement of a series of contrasting statements of which one is to be accepted and other rejected so that the possible names in the key are divided into smaller and smaller groups. Each time a choice is made, one or more taxa are eliminated. A more correctly used term for such a device is **determinator**. Keys use only the most prominent features which may not always be taxonomically the most important feature. Also, a key does not offer the description of the plant concerned. It is **artificial** in the sense that the pairs of contrasting characters are selected and therefore, the sequence of taxa is often quite unnatural.

In the past, particulary during the classical and medieval period, plants were the subject of description and illustration. For the purpose of accurate identification. During the seventeenth century, descriptions of plants were accompanied by bracketed diagrams, which functioned both as classificatory device or **conspectus** as well as an identification aid. The use of modern "keys"—derived from the Latin word *clavis*—for identification purpose is credited to Lamarck (Jones and Luchsinger (1987).

The artificial keys are of two main types: **single-access** or **sequential** keys, and **multi-access** keys

Single-access or Sequential Keys

Statements used in such keys are based on the characters of the plants to be identified. A key may be so constructed as to indicate natural relationship of affinity or it may have a purely artificial basis, i.e., regardless of any natural or phylogenctie relationship. Keys may sometimes be illustrated; simple or complex; may be based only on floral features or only vegetative features or both. Even different environmental requrements may be included in constructing a key.

Single-access or sequential keys are always written in the form of **dichotomons** keys. Such a key consists of a number of **couplets**. Each couplet has a pair of contrasting or contradictory statement, each of which is known as a **lead**. The two leads of a couplet are arranged in **yokes**. The two leads of couplet are identified by the same number or letter. Key can be as simple as only one pair of leads or very complex.

1. stamens 6; capsules long-beaked, usually less than 1.5 cm long. .... *Cleome brachycarpa*
1. Stamens 12 or more; capsules short-beaked, usually more than 2.5 cm long. .... *C. viscosa*

Depending upon the arrangement of the couplets, there can be a **bracket** key or an **indented** key. In bracket key (also known as **parallel** key), the couplets are arranged either with alternate indentation or all starting from the common margin. Numbering or lettering is consecutive or continuous. In this type of keys, the two leads of couplet are always together, and they are more or less of same length.

In indented (or **yoked**) keys, each successive couplet is indented under the one preceding it. In this the two leads of a couplet may not remain together all the time. This key is better understood because similar elements are grouped in such a way that they are visible at a glance. Examples of both these types of keys are shown below for which the same or identical data has been used :

Bracketed Key

1. Inflorescence resembles a bottle-brush; flowers in spikes or heads .... 2.
1. Inflorescence not as above; flowers solitary, clustered or in umbels .... 3.
    2. Stamens free .... *Callistemon*
    2. Stamens united into bundles opposite the petals .... *Melaleuca*
3. Calyx lobes and petals united to form an operculum or cap;
   fruit a dehiscent capsule .... *Eucalyptus*
3. Calyx lobes and petals distinct; fruit a berry .... 4.
    4. Flowers large, white, on 1 or few-flowered peduncles; ovary 4 or 5 celled .... *Psidium*
    4. Flowers small, greenish white; terminal, axillary or lateral cymes;
       ovary 2-celled .... *Syzygium*

A bracket key for the 5 genera of Myrtaceac that occur in Delhi.

Indented Key

1. Inflorescence resembles a bottle-brush; flowers in spikes or heads
    2. Stamens free .... *Callistemon*
    2. Stamens united into bundles opposite the petals .... *Melaleuca*
1. Inflorescence not as above; flowers solitary, clustered or in umbels or heads.
    3. Calyx lobes and petals united to form an operculum or cap; fruit a dehiscent capsule ....
    .... *Eucalyptus*
    3. Calyx lobes and petals distinct; fruit a berry
        4. Flowers large, white, on 1 or few-flowered peduncles; ovary 4 or 5 celled. .... *Psidium*
        4. Flowers small, greenish-white, in terminal, axillary or lateral cymes; ovary 2-celled ....
        .... *Syzygium*

An indented key for the same 5 genera of Myrtaceae using the same set of characters.

For preparing an identification key, certain rules are to be followed.

1. Only contradictory statements should be taken into account, so that one of these fits and the other is rejected.
2. The smaller group of the two should be dealt with first, followed by the larger group.
3. The same character should not be used repeatedly.
4. As far as possible, a negative statement should not be chosen, e.g. instead of a statement like "stamens not diadelphous", "stamens free" should be used.
5. Each lead of a couplet must start with the same word.
6. Use of discontinous characters in more beneficial than continuous or overlapping characters. Generalised characters must also be avoided, e.g. leaves 5–10 cm broad and leaves 3–6 cm broad or pedicel short, leaves with broad, white margin and pedicel long, leaves with narrower margin.
7. Only those morphological features should be considered that can be seen with the naked eye or at best with a hand lens. If microscopic features like chromosome number, or anatiomical or embryological characters are used to form a key, it will not be possible to identify unknown herbarium specimens.
8. While making a key for dioecious plants, characters of both the male and female plants must be incorporated as male and female plants may not be available at the same time.

Keys are the traditional method of identification in taxonomy. If keys are well-written keeping in view all the features mentioned above, if there is an adequate number of specimens available and if the person working with the key is careful, then the specimens can be identified with sufficient ease. However, there are disadvantages with the use of such keys:

(1) Use of certain characters become necessary even though the character is not evident in the unknown specimen.

(2) In **bracketed** key, the disadvantage is that the couplets are no longer in visual groups and particularly so if the key is a very long one.

(3) In **indented** key, the draw-back is that it is difficult to locate the alternate leads of the initial couplets as they may appear on subsequent pages.

(4) If an **indented** (or **yoked**) key is running through a number of pages, it becomes more and more sloping with increase in total number of couplets. This reduces the space available for writing the

leads and also wastage of page-space is there. Some pages may apperar to be almost blank with one of the two leads forming the bottom line.

Multi-access Keys

Attempts have been made to improve the standard dichotomous keys with one entry point (single-access or seqential keys) and as a result, today there are uses of **ploycIaves** and **compter techniques.** Polyclaves are **multi-access** or multientary, order-free keys. Such keys are user-oriented, and can be commenced at any position. It is different from dichotomous keys in the sense that here the user of the key is in a position to select the characters (to be used) and not the author. The rigid format of the traditional dichotomous keys need not be followed; instead the user has total freedom to decide any character in any sequence. Even if the information about a few characters is inadequate or insufficient or unavailable the user can proceed with identification. It has been observed that identification is sometimes achieved even without making use of all the characters available. Multi-access keys are usually produced in the form of punched cards and in this form they were first succesfully used by ATJ Bianchi (1931) and SH Clarke (1938). Clarke prepared such a multientry perforated key for the identification of hardwoods. There are two main types of such perforated or punched keys:

(1) Edge-punched key

In this type of keys, there is one card for each taxon showing the combination of characters of that particualar taxon. Each attribute is represented by a punched hole around the perimeter of the card (Fig. 4.1). When a taxon possesses a particular attribute the hole is clipped or punched to form an open notch. For actual identification, a list of attributes of the unknown specimen is prepared and for each attribute the specimen possesses, the hole is punched. All the appropriate cards are stacked together and a needle is inserted through one of the holes representing one of the attributes possessed by the taxon. The needle is then lifted horizontally and shaken gently so that all those taxa possessing the attribute fall away from the whole stack. The cards that remain on the needle are the taxa which do not possess this attribute and are kept aside. The cards that fall down are gathered again and the process is repeated several times so that ultimately only one card will fall from the stack. The taxon represented on the card is the correct identification of the unknown taxon. The attributes can be selected from various fields as shown in the illustration (Fig. 4.1).

(2) Body-punched key

These keys are named so, as the holes are punched in rows on the main body of the card. Each card represents one character-state (or attribute). Numbers are printed on the card to point out the standard position of each taxon; if the taxon possesses that particular attribute (mentioned on the card) its position is punched out. For actual identification, as in the previous method, a list of attributes possessed by the unidentified taxon is prepared and the appropriate cards are selected. When all the cards of the attributes possessed by the unknown taxon have been placed one above the other, after punching of the holes, the taxon which possesses all the attributes (that were listed) will show a hole running through all these cards. Position of the hole on the card will give the correct identification of the taxon.

The advantage of multi-access keys is that as the attributes are selected by the user, if there is any unusual attribute in a taxon, the rest of the taxa are easily eliminated. But there are disadvantages also: firstly, the use of cards are relatively cumbersome, time-consuming and expensive to construct and secondly, they require plenty of space for storage. To overcome these disadvantages there have been use of written version of multi-access keys (Hedge and Lamond 1972). Other methods based on the principle of multi-access keys are Tabular Keys (Singh 1999) and Lateral Keys (Stace 1989).

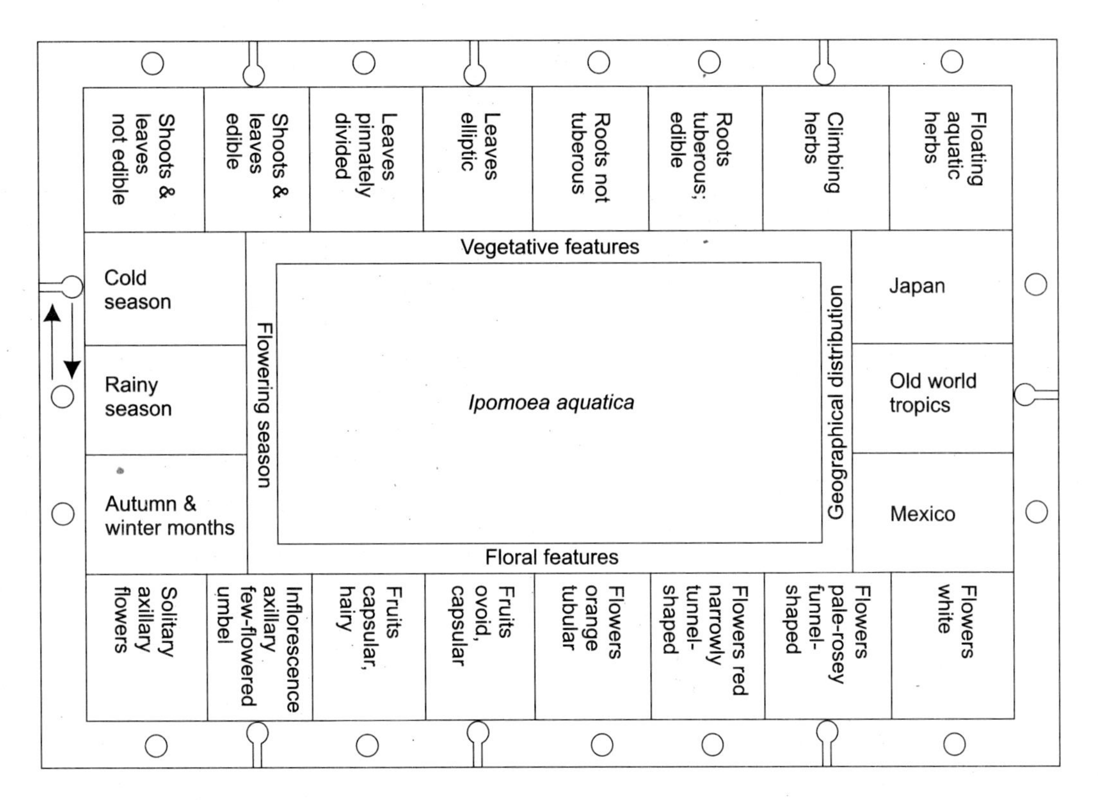

Fig. 4.1 An edge-punched card

It is not surprising that so many types of multi-access keys are in existence. Use of traditional dichotomous keys is certainly difficult because—they are inflexible as there is a sequence and if one character is missing, progress is not possible. Even a minor mistake can lead to wrong result and that too will be realised only at the end of the whole operation.

Keys have their limitations. Even at their best, they are only selected assortment of a few characters of the groups of plants or plant families involved. Variation is one of the significant laws of biology. A traditional dichotomous key usually includes the most outstanding or the most usual feature. With multi-access keys it may be possible to include all the exceptions encountered.

5
Classification

History of Classification

Whenever we are confronted with a mass of material, evidence or data in our everyday life, we try to classify them into convenient groups for better appreciation of their nature, an analysis of their characteristics and for future reference. It can thus be said that classification of infomtation is a basic human activity. Whatever we come across or observe, we try to classify. Man started classifying plants even when he was still only a food-gatherer. He classified which plants were edible and which were not, which could be used for medicinal purposes, and which were poisonous and so on.

Plant classification is the placement of plants or groups of plants in separate compartments at different levels according to phenetic similarities, phylogenetic relationships or mere artificial criteria. Thus, individuals resembling each other very closely may be grouped under a "species"; a few species sharing many common characteristics under a "genus"; similar "genera" under a "family" and so on.

In actual practice, classification deals more with the placing of a plant group in its proper position within the system than the placing of an individual plant in one of the several minor categories.

The history of plant classification is a fascinating subject. One learns how the different systems have evolved during the various stages of development, and also about the people responsible for them.

The **Pre-Linnaean Systems**. The systems of classification proposed by different systematists before Linnaeus can be broadly divided into two groups; (a) artificial classifications, and (b) mechanical classifications.

(a) The **artificial systems** of classification were based on the habit of the plants classified.

Theophrastus (370–285 B.C.) is considered the Father of Botany and was a student of the great Greek philosopher, Aristotle. He classified all plants on the basis of form and texture. He recognised trees, shrubs, undershrubs and herbs, and distinguished between annual, biennial and perennial plants. He also recognised determinate and indeterminate inflorescences; hypogynous, perigynous and epigynous flowers; apetalous, polypetalous and gamopetalous condition; the distinction between monocots and dicots; and the distinction between roots and rhizomes. He considered trees to be the most highly developed plants; his groups were strictly artificial and used no categories as we do today. His writings include *Enquiry into Plants* and *The Causes of Plants.* He had a botanic garden of his own, attached to his Lyceum near Athens, and all the classification work was the result of the studies conducted on the plants of this garden. In his *Historia Plantarum* he roughly classified ca. 480 types of different plants.

Dioscorides (60 A.D.) was a physician. He dealt with medicinal plants and identified ca. 600 different species. His book *De Materia Medica* included information on roots, stems, leaves and sometimes flowers of different plants of medicinal value.

Albertus Magnus (1193–1280 A.D.) was Bishop of Ratisbon. He recognised differences in stem structure of dicots and monocots anatomically, even with the help of the crude lenses available at that time. He also distinguished between leafy and non-leafy plants; leafy plants were further divided into monocots and dicots, and dicots into herbaceous and woody plants. In other major respects, he agreed with the classification of Theophrastus.

Otto Brunfels (1464–1534), a German, was one of the renowned herbalists of his time. He wrote an illustrated herbal based on material from the works of Theophrastus, Dioscorides and Pliny. However, real credit goes to him for recognising the Perfecti and Imperfecti groups of plants, i.e. plants with and without flowers, observable by keeping the plant or branch at arm's length. The three volumes of *Herbarium Vivae Eicones* by Otto Brunfels are well known for the illustrations of living plants. Contemporary with Brunfels came **Jerome Bock** (1498–1554), **Leonard Fuchs** (1501–1566), a Bavarian physician; **P. Mathiola** (1501–1577), an Italian herbalist; **Mathias de l'Obel** (1538–1616), a Dutch herbalist; **John Gerard** (1545–1612), an English surgeon and botanist, and a Flemish botanist, **Charles de l'Ecluse** (1526–1609). All these herbalists contributed to the descriptive phase of systematics and to the botany of medicinal plants.

(b) The **mechanical systems** are based on one or a few selected characters of mainly morphological nature. During this period, the hierarchy of categories was improved and more natural groups were recognised. These systems were published between 1580 and about 1760.

Andrea Caesalpino (1519–1603), an Italian botanist, has often been referred to as the first plant taxonomist. He classified plants first on the basis of habit,—woody or herbaceous—and further divided them on the basis of fruits, seeds and embryo characters. He recognised inferior and superior ovary, bulbs present or absent, sap milky or colourless, and the number of locules in an ovary. His famous work *De Plantis* (1583) included descriptions of some 1520 plant species arranged into woody and herbaceous groups. His most important conclusion was that the flowers and fruits were more reliable characters than the habit. However, he never ordered his ideas into outline or synopsis and they were therefore not adopted by his immediate successors; but many later workers, like Tournefort, John Ray and Linnaeus, were influenced by his work.

Jean Bauhin (1541–1631) and **Gaspard Bauhin** (1560–1624). Jean Bauhin is well-known for his comprehensive work carried out on about 5000 plants. The descriptions included were good diagnoses. *Historia Plantarum Universalis* was an excellently illustrated work, published in three volumes in 1650, after his death. His brother, Gaspard Bauhin, made the distinction between genera and species and described ca. 6000 species in his work *Pinax Theatri Botanici* (1623). He followed l'Obel and Gerard, the herbalists, for classifying the plants. To many of these plants, he gave a generic and specific epithet. Thus, binary nomenclature, with which Linnaeus' name is associated, was actually founded by Gaspard Bauhin about a century before its use in the *Species Plantarum* of Linnaeus.

John Ray (1628–1705) was an English philosopher, theologian and naturalist. His classificatory system was based on form relationship and about 18000 species were included. These were broadly divided into two groups—Herbae, i.e. herbaceous, and Arborae, i.e. woody. The Herbae were divided into Imperfectae, i.e. basically cryptogams, and Perfectae. The group Perfectae was further divided into dicots and monocots. The Arborae were also divided into dicots and monocots, which were further classified on the basis of fruit types as cone-bearing, nut-bearing, bacciferous, pomiferous, pruniferous and siliquous. He closely followed Caesalpino and pointed out that characters from all parts of the plant should be used for classification. He distinguished between inherited and imputed-differences, and noted the significance of the former for taxonomy. He also made use of anatomical characteristics that were collected by others like Grew and Malpighi. He recognised plant groups like the composites, umbels, mints, crucifers, legumes and a few others. In many respects, his system was superior to the Artificial System of Linnaeus.

Joseph Pitton de Tournefort (1656–1708) while working in Paris as a professor of botany, travelled widely in Europe and made collections of plant materials. He followed some of his predecessors in suggesting a system of classification based on form relationship. He broadly classified all the flowering plants into two categories—trees and herbs. Each of these categories was subdivided on the basis of apetalous/petalous/gamopetalous corolla, and then regular/irregular flowers. This system was used widely in different parts of Europe until 1760, when the system of Linnaeus appeared. In France it was used until 1780, when it was replaced by the system of de Jussieu. Tournefort is also given credit for being the father of the modern genus concept. He recognised 698 genera and 10146 species, and provided generic descriptions of these. Many of the generic names used by him are still in use, for example, *Salix, Populus, Fagus, Betula, Castanea, Quercus, Ulmus* and many others. However, this system was too artificial, as it was based more on vegetative characters, and was considered inferior to that of John Ray. No distinction was made between phanerogams and cryptogams or between monocots and dicots.

The Pre-Linnaean era ends here with the system of Tournefort.

Linnaeus' System of Classification and Thereafter

Carolus Linnaeus is regarded as the father of taxonomic botany, and also of zoology. He was Swedish, born on May 23rd, 1707. While working in the University of Uppsala as a student, he published *Hortus Uplandicus* in Latin, which was an enumeration of the plants in the Uppsala Botanical Garden. The plants were arranged according to Tournefort's system in the first edition. In the second edition, Linnaeus gave his own system of classification. He spent about 3 years, from 1737 to 1739, in Holland, and this was the most important period of his life. He published 14 treatises during this period, including *Genera Plantarum, Flora Lapponica* and *Hortus Cliffortianus.* He was also very popular as a teacher. He died on January 10, 1778 at the age of 70.

The Sexual System of Linnaeus, as it is usually called, appeared at a time when it had become absolutely necessary. For two centuries, plant materials of various forms had been collected, but they could not be identified or classified, as there was no logical system of classification available. Thus, when the scientific world was eagerly waiting, Linnaeus published his System by which these plants could be easily classified.

In short. the System has 24 classes for all the plants known at that time, based on number, union and length of stamens. Each class was then subdivided into orders on the basis of number of styles in the flowers. The system was first published in the second edition of *Hortus Uplandicus* (1732). Later, it was revised and formed the basis of *Genera Plantarum* (1737). Descriptions of 935 genera are included in this publication.

Species Plantarum (1753) is the most important work of Linnaeus in the systematics of vascular plants. In this book he included the identification and description of nearly 6000 species belonging to 1000 genera. Linnaeus is to be remembered as having given a precise system of plant nomenclature. *Species Plantarum,* published in May 1753, in two volumes, was the first book in which binomial nomenclature was used for describing all the plants. Each plant was given a generic name, followed by a trivial name and then a specific phrase-name. These were followed by references to previous publications or his herbarium specimens and then native habitat of the plant concerned. The trivial name was in the true sense the specific name and, together with the generic name, formed the binomial for most plants. The third part of name, that is the specific pharse-name was considered by Linnaeus to be the specific name and was in the form of a polynomial descriptive phrase. The binomial nomenclature, formed of a generic name and a specific epithet, was formulated by Gaspard Bauhin in 1596. Bauhin, however, did not use this system in his work very consistently, whereas Linnaeus followed it carefully throughout his *Species Plantarum.* For this very reason, the taxonomists of today have chosen the 1st of May, 1753, as the starting point of current botanical nomenclature. Any name published before this date is not valid.

The post-Linnean period can also be divided into two parts.

(a) Period of Natural Systems (from 1760 to 1880). Now that the easy method of identification was at hand (the Linnaean system of classification), many naturalists and explorers went on expenditions to collect plants from different parts of the world.

John Clayton (1685–1773) was probably the first taxonomist to go to America in 1705. A lawyer by profession his interest in botany was unusual. He collected plants, wrote descriptions, and tried to identify them by corresponding with **J Gronovius**, who was a professor at Leiden in the Netherlands. Later, in 1739, on the basis of these writings, *Flora Virginica* was published by **Gronovius**. Another Englishman, **Mark Catesby** (1680-1749), also explored parts of North America and published a voluminous *Natural History of Carolina, Florida, and the Bahama Islands* (1731–1743).

Almost at the same time another botanist, **George Eberhard Rumpf** (1628–1702), was studying the tropical plants of Indonesia and the island of Amboina. Unfortunately, both his collections and manuscript were destroyed, and it was only much later that the *Herbarium Amboineuse* could be published by **J. Burmann** (1741–1755) on the basis of a new manuscript compiled by him. About 1750 species of plants were described, of which 1060 were also illustrated in this book.

A large group of enthusiastic students of Linnaeus also explored different parts of the world. They were **Peter Kalm**, who went to North America, **Christopher Ternstrom**, who went to the East Indies, **Frederik Hasselquist**, who visited Palestine, **Carl Peter Thunberg**, who went to the Cape of Good Hope, and **Peter Forskal**, who went to Egypt and Arabia. Many other botanists explored North and South America, Africa and Australia. Quite a few of the European botanists also came to India and therefore most of the earlier literature on floristic explorations of different zones of this country was written by English botanists such as **Sir JD Hooker, D Prain, JF Duthie, HH Haines** and **JS Gamble**.

Michel Adanson (1727–1806) was a French botanist who explored places such as tropical Africa. He rejected all the artificial classifications that had been published before his time, as he believed in giving equal significance to all the characters. This is known as the Adansonian Principle and became important in the classificatory systems of the 20th Century. He is often referred to as the grandfather of numerical taxonomy.

By means of all these explorations throughout the globe, botanists came to know about many more plants than were included in Linnaeus' System. As a consequence, the search started for a natural system of classification. Adanson tried to form a new system, but was not successful.

Jean BAPM de Lamarck (1744–1829), also a French botanist, came up with a brilliant idea. This is known as Lamarckism, and means that changes in the environment bring about changes in the structure of the organisms.

All three brothers, **Antoine** (1686–1758), **Bernard** (1699–1776) and **Joseph** (1704–1779) **de Jussieu** were botanists. **Bernard de Jussieu** arranged the plants in the botanical garden at Le Trianon, Versailles, according to a new system of his own, which was neither based on the Aristotelian concept of habit nor as artificial as that given by Linnaeus. The work was published by another botanist, his nephew, **Antoine Laurent de Jussieu** in *Genera Plantarum* (1789). In this system he divided all the plants into 15 classes, and these classes were subdivided into 100 orders. There was a solitary class Acotyledons that included all the known cryptogams. Class 15, i.e. Diclines Irregularis, was also not exclusively angiospermic, as Coniferae were included; but all the other 13 classes were of flowering plants alone.

AP de Candolle (1778–1841) coined the term taxonomy for the first time, meaning the arrangement of plants. His important works include *Théorie Elémentaire de la Botanique* (1813), in which his approach to plant classification was explained. According to him, anatomy should have been the basis of taxonomic classification rather than physiology. Later, while staying at Geneva, he wrote one of the classics in botany: *Prodromus Systematis Naturalis Regni Vegetabilis* (1806–1893). This was a compilation of the descriptions of the then known families, genera and species. It ran in 17 volumes, of which 7 volumes were written and produced by himself. The other 10 volumes were written by specialists and published under the supervision of his son, **Alphonse de Candolle**.

The first half of the 19th Century was important in the history of taxonomy, as a number of systems of classification were put forward during this time, and also numerous botanical explorations were undertaken.

Robert Brown (1773–1858): The main contribution of this Scottish scientist was the recognition of gymnosperms as a group distinct from angiosperms, and that they were characterised by the presence of naked ovules. He also carried out floral morphological studies of many families like Asclepiadaceae, Euphorbiaceae, Polygalaceae and others.

Bentham and Hooker: George Bentham (1800–1884), an Englishman, was a very critical and well-trained taxonomist. **Sir Joseph Dalton Hooker** (1817–1911), son of a botanist father, Sir William J. Hooker, was the then Director of Royal Botanical Garden at Kew, London. He was also a very enthusiastic plant explorer and phytogeographer. Bentham and Hooker's System of Classification first appeared in a three-volume work in Latin named *Genera Plantarum.* The system is based on groups of plant characters which are correlated with each other. For example, if a specimen consistently shows syngenesious stamens, inferior, bicarpellary, unilocular ovary with basal placentation, then it can confidently be assigned to the family Compositae. His work includes names, description and the classification of all the seed plants then known. Basically, it is on the same basis as that of de Candolle but with some modifications such as greater emphasis on the free and fused condition of petals.

The Dicotyledonae were divided into three groups—Polypetalae, Gamopetalae and Monochlamydeae. Polypetalae includes three series— Thalamiflorae, Disciflorae and Calyciflorae, of which the first and third are mentioned in de Candolle's work. Similarly, Gamopetalae and Monochlamydeae were the other two subclasses, Corolliflorae and Monochlamydeae in de Candolle's system. Gamopetalae is subdivided into three series—Inferae, Bicarpellatae and Heteromerae. Each of these series is subdivided into a number of cohorts, which are equivalent to the families in the present-day classificatory systems. Each of the cohorts comprises a few related families. The Monochlamydeae are arranged in eight series of unequal value. For example, the Curvembryeae was a natural group, but the Unisexuales included families which later showed quite different affinities.

The monocots are divided into seven series, starting with epigynous orders, Orchidaceae and Scitaminae, then the petaloid hypogynous Liliaceae, then orders where the perianth is not petaloid, and then orders like Padanaceae and Aroideae with aborted perianth. Next is the series Apocarpae, where the carpels are free. The last order is the Glumaceae, where the floral structure is highly reduced.

Although this system was accepted throughout the British Empire, in the United States of America and also in some Continental countries, its shortcomings were realised. Placing the Gymnospermae between the Dicotyledoneae and Monocotyledoneae had also been criticised, the critics trying to keep all the gymnosperms together instead of mixing them with the dicot families (as was done by de Candolle). Another point of criticism is the inclusion of certain families like the Chenopodiaceae in the Monochlamydeae, when they have clear affinities with Caryophyllaceae of Polypetalae.

Whether the simple flower structure is a primitive feature or a reduced variation of a highly evolved form has always been a matter of controversy. Monochlamydeae, a group showing reduced flower structure, should therefore be a homogeneous group of all the families showing reduced floral structure.

Among the monocotyledonous families, the relative position of ovary and perianth has been overemphasised. For this reason, related families like Liliaceae and Amaryllidaceae are placed in two different orders and unrelated families like Juncaceae and Palmae are placed together.

However, in spite of all its disadvantages, the Bentham and Hooker system is still very popular in many countries, because identification of plants in the field is comparatively easy. It is followed in the Kew Herbarium for the arrangement of plant specimens, as also in many other herbaria, particularly those of the Commonwealth countries.

(b) **Systems Based on Phylogeny.** Towards the end of the 19th Century, Darwin's work on the theories of evolution and origin of species was also published coinciding, incidentally, with the publication of *Genera Plantarum* by **Bentham and Hooker** (1862–1883). Darwin's theories were so well-documented that the concept of natural classification was dropped. Even Hooker wanted to revise their system of classification, but Bentham deterred him from doing so, as he could not agree with Darwin's ideas.

Nevertheless, the classificatory systems that were written after the publication and general acceptance of Darwin's work were all based on theories of descent and evolution. Everyone waited to account for the true phylogenetic relationship of plants. Two schools of thought, therefore, developed and remained side by side for a considerable time, regarding the *nature of the primitive flower.* One of these was the Englerian school, proposed by a German taxonomist, **Alexander Braun**, in 1859 in his book *Flora der Provinz Bradenburg*. He classified monocotyledons and showed a progression from more simple to complex forms. He treated the naked, unisexual flowers of Lemnaceae as the most primitive, and gradually increasing in complexity, reached the most complex family, Orchidaceae. Amongst the dicots, he treated the Apetalae as the most primitive and, through gradually more complex forms like Sympetalae and Eleutheropetalae, ended with the Leguminosae.

His successor, **AW Eichler** (1875), published a modified form of this system in his *Blütendiagramme.* In 1883, he elaborated it further, and this system gradually replaced that of de Candolle.

Eichler divided the plant kingdom into two subgroups: Cryptogamae, containing flowerless plants, and Phanerogamae, including seed plants. The former was divided into three main parts: Thallophytes, including Algae and Fungi; Bryophytes, including Hepaticae and Musci; Pteridophytes, including three classes—Equisetineae, Lycopodineae and Filicineae. The Phanerogamae were divided into two groups: Gymnospermae and Angiospermae. The latter were subdivided into two classes: Monocotyledoneae and Dicotyledoneae. In these two groups, plants were arranged according to increasing complexity of floral structure.

Adolf Engler (1844–1930) published a classificatory system (based on Eichler's system) in 1892 in the form of a guide to the plants of Breslau Botanic Garden. The phanerogams were named Embryophyta Siphonogama, and were divided into Gymnospermae and Angiospermae. The Angiospermae were further divided into the classes Monocotyledonae and Dicotyledonae. In both these classes, flowers with simple structures were treated as primitive forms. The class Monocotyledonae and the two subclasses of the Dicotyledonae were next divided into orders, each order comprising a few related families. The Monocotyledonae were divided into 11 orders and 45 families. Subclass Archichlamydeae of the Dicotyledonae comprises 37 orders and 227 families; subclass Metachlamydeae includes 11 orders and 64 families (as revised by **Melchior** 1964).

Engler himself did not consider his system of classification to be a phylogenetic one; but he maintained and expanded the concept of the primitive flower as proposed by **Alexander Braun** (1859). In addition to the fact that the unisexual and naked flowers borne on catkin-like inflorescences were the primitive flowers, Engler also pointed out that these flowers were wind-pollinated, as in the gymnosperms. It was assumed that in all probability the primitive angiospermous flowers originated from a gymnospermous ancestor bearing a unisexual strobilus. This was recognised as the Englerian concept of primitive flower. The Englerian school further proposed a polyphyletic origin for angiosperms.

Engler and Prantl applied this system to plants all over the world, and published a 23-volume work, *Die natürlichen Pflanzenfamilien,* during the years 1887 and 1915. In this work, identification of all known plants from algae to angiosperms was taken up, and all the taxa were fully described. The original work was subsequently revised by many botanists. The latest revision was by Melchior (1964), who retained the basic structure of the system and accepted the concept of primitive flower. The descriptions of individual taxa now include the compilation of data from various fields other than morphology and embryology (considered in earlier revisions), so that, the position of some of the orders and families have been altered.

In many American and Continental European herbaria, Engler and Prantl's system is followed for the arrangement of plant specimens.

AB Rendle (1865–1938) is known for his two-volume work, *Classification of Flowering Plants* (1904, 1925). It is again based on Engler and Prantl's system of classification, with certain minor modifications. He, too, did not regard his system as a phylogenetic one, but rather a system of convenience. The monocots were treated as more primitive than the dicots. In this group the taxon Palmae had a controversial position. Engler placed it in the order Principes, but Rendle treated it as a family under Spathiflorae. The dicots, on the basis of flower structure, were divided into three groups: Monochlamydeae, Dialypetalae and Sympetalae. Salicales was the most primitive order, and Umbelliferae the most advanced amongst the dicots. Rendle is also known for his taxonomic studies of the Gramineae, Orchidaceae and Naiadaceae. His descriptions are very clear and complete, with beautiful diagrams, a record of exceptional features and discussion of the relationship of families with each other.

Richard von Wettstein (1862–1931), an Austrian botanist, proposed a classificatory system which, although based on Englerian concepts, was more a phylogenetic system. Like Engler, he also presumed that unisexual perianth-less flowers were more primitive than the bisexual flowers with elaborate perianth. He believed in the monophyletic origin of angiosperms from *Gnetum* or a *Gnetum*-like ancestral form of gymnosperms. He also treated the dicots as more primitive than the monocots and stated that the monocots were probably derived from Ranalian stock of the dicots. Further, according to Wettstein, herbaceous plants were more advanced than woody plants, and numerous flowers in an inflorescence was more primitive than a solitary or few-flowered inflorescence. Naturally, therefore, his arrangement of individual taxa was quite different from that of Engler or Rendle, inclining more towards phylogenetic classification. Moreover, his descriptions included data from a much larger range of fields, even from serology. In general, he considered (a) the presence of abortive reproductive parts in a flower as evidence of reduction; (b) adaptive modifications as indicating advancement, and (c) the spiral arrangement of flower parts as primitive as compared to cyclic arrangement.

Although this system was not accepted on a large scale, it is known that many of Wettstein's ideas were adopted in other phylogenetic systems of later years. A revision of his work was published posthumously in 1935: *Handbuch der Systematischen Botanik* from Leipzig.

August A Pulle (1878–1950), from the Utrecht Botanical Museum, The Netherlands, published in 1938, a phylogenetic system of classification based essentially on Englerian concept. In this system, however, the division Spermatophyta comprised four subdivisions (instead of two). These subdivisions were:

(a) Pteridospermae, a subdivision of two extinct families,

(b) Gymnospermae, comprising the classes (i) Cycadinae, (ii) Bennettitinae (extinct), (iii) Cordaitinae (extinct), (iv) Ginkgoinae and (v) Coniferae;

(c) Chlamydospermae (Gnetales and Welwitschiales), and

(d) Angiospermae (Monocotyledoneae and Dicotyledoneae).

The Monocotyledoneae was subdivided into ten orders, each comprising a number of families. The Dicotyledoneae was divided into eight series; each series comprised a few orders, and each order had a few families.

Pulle's classification of Gymnospermae was more acceptable, as it was in closer accord with phylogenetic trends. Although the arrangement of orders and families of the Angiospermae was more or less on the same lines as that of Wettstein, it showed some originality also. He regarded Sympetalae as a polyphyletic group, and rearranged its components. Also, Wettstein did not consider Sympetalae and Monochlamydeae to be natural taxa.

Carl Skottsberg (1880–1963), a Swedish professor, gave a modified Englerian classification, utilising some of the concepts of Wettstein as well (1940). He considered the monocots as having originated from some unknown dicot. Amongst the dicots he considered the Casuarinaceae to be the most primitive, but not more primitive than other dicots. In his view, apocarpy and polycarpy were primitive; syncarpy and monocarpy were advanced; apetalous forms polyphyletic. He adopted some of the concepts of the Benthamian-Bessey school, and placed the taxon Amentiferae after the Rosales. He differed from Pulle, as he retained the Primulales in the Sympetalae.

Charles E Bessey (1845–1915), a student of **Asa Gray**, was the first American to bring out a system of classification. He is well-known also as the originator of the Ranalian concept of primitive flower, which was later modified by Hutchinson (1926).

According to this concept, the primitive angiosperm flower bore a resemblance to a certain gymnosperm with a bisexual strobilus having microsporophylls below and megasporophylls above. In the course of evolution, the lower sporophylls developed into sepals and petals by progressive sterilization, and the upper sporophylls into stamens and carpels. The central axis became shortened into the thalamus. Eventually, the primitive flower has numerous free perianth lobes and free stamens, and carpels spirally arranged on the receptacle (or the thalamus). Amongst the living angiosperms, the genus *Magnolia* of the family Magnoliaceae is the nearest to this structure.

Bessey's idea was to formulate a classificatory system for the flowering plants that would reflect their evolutionary relationship. To achieve this, he proposed a set of dicta based on empirical evidence, to distinguish between primitive and advanced features. Primitive features were those which were expected in ancient plants, and advanced features were expected in the more recent taxa. However, it must be made clear that primitive and advanced are not the same as simple and complex. An advanced taxon may be structurally simple due to reduction, or it may have an elaborate structure and therefore be highly complex.

Bessey's classificatory system resembled that of Bentham and Hooker. It may be said to be a reorganised system of Bentham and Hooker, on the basis of phylogeny and evolution. He considered the seed plants to have had a polyphyletic origin and to comprise three distinct phyla, of which only the Anthophyta (Angiosperms) was dealt with. Anthophyta was divided into two classes: Alternifoliae (or monocots) and Oppositifoliae (or dicots). Both these classes were divided into two subclasses: Strobiloideae, including orders with superior ovary, and Cotyloideae, including orders with inferior ovary. In the dicots, each subclass was further divided into superorders, then orders, and thereafter families.

Bessey's system received increasing support from all over the world. He may be criticied for assuming perigynous and epigynous forms to constitute a single evolutionary line.

Hans Hallier (1868–1932), a German botanist from Hamburg, independently developed ideas similar to those of Bessey. His system of classification was based on data synthesised from various fields like palaeobotany, anatomy, serology and ontogeny, herbarium material and previous literature. He did not agree with the Englerian concept of a primitive flower. Instead, he also proposed the idea of a strobiloid ancestory for a primitive angiospermous flower, as did Bessey. The Ranalian concept is therefore also known as the Besseyan-Hallierian concept. He considered the angiosperms as having originated monophyletically from a Bennettitalean type of ancestor. Dicots were regarded as more primitive than monocots. Polycarpy and spiral arrangement were considered primitive, and syncarpy and cyclic arrangement advanced. His realignment of certain families displayed much thoughtfulness, e.g. Salicaceae and Flacourtiaceae are related to each other; similarly, Cactaceae and Aizoaceae are aligned, and so also Cucurbitaceae and Passifloraceae. He also recognised the family Amaryllidaceae as an unnatural assemblage of plants, and divided it into the Agavaceae, Alstroemeriaceae and Amaryllidaceae.

In his 1912 publication, 213 families were recognised. However, in spite of such an advanced outlook, his system did not receive much recognition.

John Hutchinson (1884–1972) was the Keeper of the Royal Botanic Gardens, Kew, England. He was concerned with the phylogeny of angiosperms alone, and a classificatory system for this plant group was published in his two-volume work: *The Families of Flowering Plants* (1926, 1934), and later revised in his *British Flowering Plants* (1948), *Evolution and Phylogeny of Flowering Plants* (1969) and *The Families of Flowering Plants arranged according to a New System Based on Their Probable Phylogeny* (1973).

This classificatory system has close affinities with those of Bentham and Hooker, and Bessey. According to Hutchinson, the gymnosperms formed a lineal monophyletic series starting from the primitive Cycadaceae and developing into an ascending order of Ginkgoaceae, Taxaceae, Pinaceae and finally Cupressaceae.

Hutchinson considered the angiosperms to have originated monophyletically from the hypothetical Pro-angiosperms. The angiosperms were divided into two major groups: Dicotyledons and Mocotyledons. A cleavage of the group dicotyledons into predominantly woody Lignosae and predominantly herbaceous Herbaceae was the next step. These two groups were considered to be the two different evolutionary lines, Lignosae stemming from Magnoliales, and Herbaceae stemming from Ranales, and they developed parallel to each other.

There are 83 orders and 349 families included in the dicotyledons, and 29 orders and 69 families in the monocotyledons in the latest revision of his system. No one order was directly derived from another; derivation was shown to be from the ancestral stock. The monocots separately formed another evolutionary line and were derived from Ranales at a very early stage, starting with the Butomales and ending in the Graminales. On the basis of the nature of the perianth, the monocots were divided into three distinct groups: Calyciferae, Corolliferae and Glumiflorae.

The main objection to this system is the early distinction of the dicotyledons into arborescent and herbaceous groups. The splitting up of many families is also inacceptable to some botanists.

Hutchinson's system was not followed on a large scale, but it encouraged a number of phylogenists to revise their system of classification.

Oswald Tippo, an American from Illinois University, published an outline of a system in 1942 which was claimed to be phylogenetic. It was based on two previously published works—one by GM Smith

(1938) for non-vascular plants, and the other by AJ Eames (1936) for vascular plants. In dealing with the angiosperms, Tippo considered Magnoliales to be a very primitive order, and Amentiferae a highly advanced taxon.

Karl Christian Mez (1866–1944), a professor of botany at the University of Koenigsberg, Germany, proposed a classificatory system based on serological studies of various families of angiosperms. According to him, the relationship between the different groups of angiosperms could be ascertained by analysis of their protein reactions. This experimental method, also called serum diagnosis or serodiagnosis, consisted of mixing an extracted plant protein with blood serum from some experimental animal. This would result in the formation of antibodies in the inoculated serum. To this was added the protein extract of the second plant whose relationship with the first was to be established. Depending upon the quantity of precipitate formed, the degree of relationship between different plants could be judged.

This method is very useful to determine the relationship of plants of unknown or doubtful affinities.

Lyman Benson of California published his work in 1957 in his book *Plant Classification.* It was basically a Besseyan-type system and the relationship between the different groups was indicated in a unique manner. Instead of producing a phylogenetic tree, he represented the whole system in the form of a two-dimensional horizontal chart. The Dicotyledoneae were divided into five groups which were informal polyphyletic grades representing stages in floral evolution. The woody members of the Ranales were regarded as the most primitive taxa, and it was presumed that the ancestors to the angiosperms also retained the features of this group, in addition to other primitive features. The position of any order in this chart shows its approximate degree of specialisation and also the degree of departure from the features characteristic of the order Ranales amongst the dicots, and of the order Alismatales amongst the monocots. This system was not accepted widely because of certain demerits such as: (a) the primitive families under Ranales do not appear to be a natural group; (b) division of Sympetalae into Corolliflorae and Ovariflorae is artificial, and (c) inclusion of Proteales, Cactales and some other taxa in the Calyciflorae is similarly artificial.

Armen Takhtajan (1910–1997) of the Soviet Academy of Sciences, Leningrad. USSR, proposed his system of classification in 1954 (translated into English in 1958). Some modifications were made in his later revisions in the years 1969, 1973 and 1980. Hallier's synthetic evolutionary classification of flowering plants had made a great impression on him. According to Takhtajan (1980), Hallier had a deep insight into the morphological evolution and phylogeny of flowering plants, and his system was a synthetic one.

In Takhtajan's system of classification, Magnoliales was suggested to be the most primitive taxon that has given rise to all the other groups of angiosperms—both dicotyledons and monocotyledons. It is based on all data available from various fields of study and therefore is a better phylogenetic system. Instead of the conventional terms, he used Magnoliophyta for the angiosperms, Magnoliopsida for the dicots and Liliopsida for the monocots. Magnoliopsida is divided into 7 subclasses, 20 superorders, 71 orders and 333 families. Liliopsida is divided into 3 subclasses, 8 superorders, 21 orders and 77 families (1980). He introduced a supplementary rank of superorder between the subclass and order. The superorder names have – anae endings. He derived Magnoliophyta monophyletically from Bennettitalean ancestors.

Arthur Cronquist (1919–1992) of the New York Botanical Garden, published his classificatory system in his book, *Evolution and Classification of Flowering Plants* (1968). It was later revised in 1981 in the book *An Integrated System of Classification of Flowering Plants.* In the later system, Cronquist replaced the usual terminology—Dicotyledoneae and Monocotyledoneae—with Magnoliatae and Liliatae, respectively.

The class magnoliatae is further divided into six subclasses, of which the subclass Magnoliidae is assumed to be the basal complex from which all other subclasses have been derived. This is the most primitive subclass. The members of subclass Hamamelididae are mostly wind-pollinated families with reduced, chiefly apetalous flowers, often borne on catkins. The Caryophyllidae members are characterized by free-central or basal placentation and the occurrence of betacyanins in many of them. The Rosidae and Dilleniidae are supposed to be parallel groups with very little morphological distinction. The difference between the two groups lies in the centripetal development of stamens, uniovulate locules and the presence of a nectariferous disc in the Rosidae, and centrifugal development of stamens, multiovulate locules and absence of nectariferous disc in the Dilleniidae. The Asteridae are the higher sympetalous families with stamens, usually as many as corolla lobes. These are mostly derived from Rosidae.

The class Liliatae consists of four subclasses, and is derived from primitive Magnoliatae of herbaceous habit with apocarpous flowers, ordinary perianth, and uniaperturate pollen.

Takhtajan's (1969) four subclasses of Liliopsida were adopted in Cronquist's system with some changes in the Liliidae and Commelinidae. The most primitive Alismatidae is an aquatic group. The Arecidae are characterized by the large petiolate leaves, arborescent habit, and flowers in spadix. Commelinidae members have reduced floral structure accompanied by wind pollination. The most advanced Liliidae have modified stem and insect pollination.

CR de Soo (1975) presented a system slightly modified from those of Takhtajan and Cronquist, chiefly based on phytochemical studies. In this system the subclass Dilleniidae was replaced by the Malvidae, and a few families were excluded from the former. Also, he changed the terminology—Dicotyledonopsida for dicots, Monocotyledonopsida for monocots and Eucomiidae for Hamamelididae.

Robert Thorne (1920) presented a synopsis of his classificatory system in 1968, and the detailed system was published in 1976. This system was claimed to be considerably different from those of Takhtajan and Cronquist, and included 21 superorders, 50 orders, 74 suborders, 321 families and 432 subfamilies. His aim was to stress relationships rather than differences. However, there have been criticisms also from various quarters.

RMT Dahlgren (1932–1987) published a system of classification (1975a, 1977a) which agreed with the major features of Takhtajan's and Cronquist's systems. He, however, tried to represent the system in the form of a three-dimensional phylogenetic shrub in transection instead of the traditional two-dimensional representation. In this system (as revised in 1980a), the angiosperms or Magnoliopsida have two subclasses—Magnoliidae (dicots) and Liliidae (monocots). The former is divided into 24 superorders, 80 orders and 346 families, and the latter into 7 superorders, 26 orders and 92 families. An attempt was made to utilise as much data as possible to bring out the distinctions between different groups. He used the ending -florae instead of -anae for the nomenclature of superorders.

In the recent past, many workers have published classificatory systems based on phytochemical data (**Dahlgren** 1983a, b, **Gornall et al.** 1979, **Soo** 1975, **Young** 1981). Cronquist (1983) warned against constructing phylogenetic classificatory systems based on "concealed evidence" such as serological data, amino acid sequence data, biosynthetical pathways of secondary metabolites, and minute details in embryology and other microstructural evidence. In doing so, we may deviate to such an extent from a phenetically useful classification that the practical taxonomist would be happy to completely ignore our classification in favour of an artificial one that will prove useful in the field and in a modestly equipped laboratory. There may some day be one type of classification for pheneticists and another for phylogenists. In fact, this is already happening—in floras and practical handbooks, mostly Englerian classification is followed, while in academic institutions the approach is more phylogenetic.

All classificatory systems, however phylogenetic they may be at the moment, are only temporary or transient in character. They are not permanent because new findings will always bring about changes. However, it is always useful to study the historical background to aquaint oneself with the basic principles of the modern classificatory systems, which may otherwise appear to be arbitrary.

Classificatory Systems: A Comparative Study

It is an impossible task for anyone (person or group) to study all the plants that occur on this earth. Even if we consider the angiosperms alone, the situation is no better, because this is the most dominant group on the global surface. We therefore have to classify them into groups on the basis of their similarities and dissimilarities, and then arrange these groups in different levels or categories. This arrangement—or classification, as it is termed—is exceedingly important.

From time to time, taxonomists have proposed different classificatory systems: Artificial Systems, Natural Systems, and Phylogenetic Systems.

An **Artificial System** classifies organisms usually by one or a few characters irrespective of any relationship amongst them. It is only for the sake of convenience and as an aid to identification.

A **Natural** or **Formal System** as it is sometimes called, is based mainly on morphological features which can be examined with the naked eye. The use of correlated characters is also important in these systems. Another important feature is the use of as many characters as possible so that the closely related taxa can be placed together or close to each other. This so-called Adansonian Principle (after Michel Adanson 1727–1806) is the basis of the Natural System of Classification. A natural classification aims to arrange all known plants into groups, which are graded according to the degree of resemblance so that each species, genus, tribe, family and order stands next to those it resembles in most respects. This system is based on data available during a particular period and the relationship amongst the taxa is a function of the overall similarities and dissimilarities. This is also known as Phenetic Relationship and is understood easily without referring to ancestral groups. The advantages of the Natural System are: *(a)* plants alike in hereditary constitution are grouped together *(b)* a great deal of information is obtained and *(c)* additional information can be easily incorporated.

Phylogenetic System. Darwin (1859) proposed the theory of evolution, in which he brought forward the fact that the present-day plants originated from some ancestral ones after undergoing periodical modifications due to environmental changes; therefore, all living plants today are related to each other in one way or the other. Thus, later classificatory systems are mostly phylogenetic, showing the presumed evolution of different plants or plant groups. and usually constructed on the basis of natural classification.

Natural Classification is often known as horizontal classification, i.e. based on data available during a particular time period. Phylogenetic classification, on the other hand. is known as vertical classification, as it mainly depends upon evolutionary relationship, or presumed ancestry. Sometimes this is designated Evolutionary Classification.

The major goal of taxonomic studies is to have a truly phylogenetic classificatory system. One of the best pre-evolutionary natural systems of classification was that of George Bentham (1800–1884) and Sir Joseph Dalton Hooker (1817–1911). The generic descriptions written from actual herbarium specimens were models of completeness and precision. Even now, this system is the most convenient for identification of plants in the field.

Two German botanists. Adolf Engler (1844–1930) and Karl Prantl (1842–1893), published their classical work *Die natürlichen Pflanzenfamilien* during the post-evolution era. This system was widely accepted over a long period of time. It is a well-illustrated work with phylogenetic arrangement and modern keys and provides data for identification of all the known genera of plants from the primitive algae to the advanced seed plants. However, they did not recognise the significance of reduction and, therefore, simplicity of structure was equated with primitiveness. The catkin-bearing families of "Amentiferae" were treated as primitive and placed before the petaliferous families like Ranunculaceae, Magnoliaceae and others. Another criticism is the placement of the monocotyledons before the dicotyledons.

John Hutchinson (1884–1972) considered the angiosperms as having originated monophyletically from the hypothetical Pro-angiosperms. Though similar to natural systems, an attempt was made to arrange the plant groups according to their routes of descent (Jones and Luchsinger 1987); but Hutchinson made the error of splitting the dicotyledons into two linear evolutionary lines, the Herbaceae and the Lignosae. This unnatural division resulted in separating closely allied families. For example, Saxifragaceae was separated from the related family Rosaceae, and the Araliaceae from Umbelliferae. Also, families like the Euphorbiaceae and Papilionaceae include a large number of herbs and shrubs as well as trees. Hutchinson was, however, well-versed in the families of angiosperms all over the world, and therefore his treatments of these families are most informative.

Armen Takhtajan's System was more natural than any other system published so far and the inclusion of evidence from all branches of botany (anatomy, embryology, palynology, chromosome number, vegetative and floral morphology, chemical features and geographical distribution) made it a phylogenetic system. With further progress in our knowledge, this system is liable to change. Derivation of the monocots from the Nymphaeales is another point of criticism. Similarities between the two taxa are probably due to convergent evolution, and the ancestors of both are completely obscured in geological history (Stebbins 1974).

Dahlgren's System is a modification of Takhtajan's system and has similar approach. This system of classification was revised in 1975, 1977, and 1980, and yet, according to Dahlgren, it is still provisional. Monocots (Liliidae), according to him, are possibly a monophyletic group due to *(a)* single cotyledon, and *(b)* characteristic triangular protein bodies in the sieve tube plastids. Use of the termination -florae (e.g. Rosiflorae) for superorders is a point of criticism. Like Takhtajan's System, with further increase in our knowledge, this system, too, is liable to change.

A comparative account (Table 5.1) of some important classificatory systems follows:

Table 5.1 Comparative account of five classificatory systems—Bentham and Hooker, Engler and Prantl, Hutchinson, Takhtajan and Dahlgren.

Bentham and Hooker's System
1. First published between 1862 and 1883 as *Genera Plantarum* in Latin, a three-volume work.
2. This system essentially deals with seed plants, and around 97000 species are described.
3. It was modelled directly on the system developed by de Candolle but is refined.
4. A system based on form relationship and corelated characters.
5. The seed plants are divided into three major classes—Class I—Dicotyledons, Class II—Gymnosperms, and Class III—Monocotyledons. Monocotyledons are treated as the most advanced.
6. Each class is further divided into subclasses, each subclass into series, and each series into cohorts; the cohorts include the families.
7. One of the unusual features of this system is the position of the gymnosperms between the Dicotyledons and the Monocotyledons.

(Table 5.1 Contd.)

(Table 5.1 Contd.)

8. Although the system was more natural than that of the earlier workers like de Candolle, a number of taxa could still not be classified satisfactorily. Such orders as could not be accommodated anywhere were placed under "Ordines anomali".

Engler and Prantl's System

1. First published as *Die Natürlichen Pflanzenfamilien,* a 23-volume work, between 1887 and 1915.
2. The entire plant kingdom, i.e. from algae to angiosperms has been described.
3. The classification of Eichler was adopted and modified.
4. This system emphasised that the incomplete or unisexual flowers were primitive. The subsequent addition of other floral whorls was an advancement.
5. In this system the angiosperms were treated as a division—Angiospermae—divided into classes—Monocotyledons and Dicotyledons. Monocots treated as primitive.
6. The Class Monocotyledonae includes 11 orders, some of which are further divided into suborders and families. The class Dicotyledonae is divided into two subclasses—Archichlamydeae (33 orders) and Sympetalae (11 orders) which in turn are divided into orders and families.
7. The Gymnospermae were placed before the monocots and were presumed to be the progenitor of the catkin-bearing Amentiferae. Angiosperms were considered to be polyphyletic; derived from seed ferns as well as the gymnosperms through Amentiferae.
8. Although intended otherwise, the system proved to be more natural and less phylogenetic. Interpretation of simple unisexual flowers (Amentiferae) as primitive is one of the main demerits.

Hutchinson's System

1. First published as *The Families of Flowering Plants,* in two volumes—Vol. I Dicotyledons in 1926 and Vol. II Monocotyledons in 1934.
2. This system mainly concerns the angiosperms.
3. Based on the principles of Charles Bessey's dicta on the relative primitivity and advancement of plant characters.
4. Based on the principles that evolution is both progressive and retrogressive and that all parts of a plant may not be involved in evolution at the same time.
5. The angiosperms were considered to originate from hypothetical Pro-angiosperms, and were classified into two major groups—Dicotyledons and Monocotyledons. A cleavage of the Dicotyledons into a herbaceous (Herbaceae) and a woody (Lignosae) group; two evolutionary lines were suggested. The Monocots have been derived (at an early stage) from the Ranales of Herbaceae.
6. Amongst Herbaceae, the most primitive order is the Ranales, and the most advanced order is the Lamiales; amongst Lignosae the most primitive order is the Magnoliales and the most advanced is Verbenales. In the Monocots, three groups—Calyciflorae, Corolliferae, and Glumiflorae—are based on the nature of the perianth; Butomales is the most primitive and Graminales the most advanced order.
7. Gymnosperms form a linear monophyletic series beginning with the primitive Cycadaceae and ending in the most advanced Cupressaceae.
8. This system is phylogenetic in nature, but has not been followed widely. The controversial taxon Amentiferae is regarded as an advanced group.

Takhtajan's System

1. First published as *Die Evolution der Angiospermen,* in one volume, in 1959.
2. Mainly concerned with angiosperms.
3. Influenced by Hallier's work, who attempted to create a synthetic evolutionary classification of flowering plants.

(Table 5.1 Contd.)

(Table 5.1 Contd.)

4. The important features of this system are—(*a*) Magnoliales s.l. is the most primitive group that gave rise to all the branches of angiosperms, and (*b*) Monocotyledons and Nymphaeales are derived from a hypothetical common dicotyledonous ancestor with vesselless wood and monocolpate pollen.
5. The angiosperms are considered monophyletic in origin, from primitive fossil orders like Bennettitales. Angiosperms are termed Magnoliophyta, which is classified into classes Magnoliopsida and Liliopsida; Liliopsida or the monocots are derived from the Nymphaeales.
6. Magnoliopsida includes 7 subclasses, 20 superorders, 71 orders and 342 families, and Liliopsida includes 3 subclasses, 8 superorders, 21 orders and 77 families (1980). Amentiferae is considered to an advanced group; the naked, unisexual inflorescences are derived from multiwhorled, bisexual flowers and inflorescences.
7. Gymnosperms are not considered.
8. Data on anatomical, and embryological features, vegetative and floral morphology, chromosome numbers, chemical features and geographical distribution have been considered. With further progress in our knowledge, this system is liable to change.

Dahlgren's System

1. First published in a text book of angiosperm taxonomy, *Angiospermenes Taxonomi* (in Danish) in 1974.
2. Mainly concerned with angiosperms.
3. A modification of Takhtajan's System of Classification.
4. An imaginary phylogenetic shrub in transaction has been prepared to show the placement of different orders in accordance to their "closeness" to each other. The size of these figures roughly corresponds to the number of species in the orders.
5. Angiosperms are considered monophyletic in origin because the combination of important features (8–nucleate embryo sac, formation of endosperm, companion cells in phloem, etc.) would hardly have evolved independently in different groups of gymnosperms.
6. Angiosperms (class Magnoliopsida) is divisible into two subclasses, Magnoliidae and Liliidae; Magnoliidae includes 24 superorders, 80 orders and 346 families, and Liliidae includes 7 superorders, 26 orders and 92 families.
7. Gymnosperms are not considered.
8. As in Takhtajan's System of Classification, data from different branches of botany have been considered to make this system both natural and phylogenetic With further progress in our knowledge, this system, too, is liable to change.

None of the present classificatory systems is ideal. An *ideal* system can be synthesized by including *all* the merits (demerits can be discarded) of different systems of classification proposed so far. In this ideal classificatory system, attention should be paid to (*a*) primitive and advanced features, and (*b*) interrelationships between different taxa (at species, genera. families and orders level) based on evidence from (*i*) vegetative features, (*ii*) floral features, (*iii*) anatomy, (*iv*) embryology, (*v*) cytology, (*vi*) breeding behaviour, (*vii*) chemotaxonomy, (*viii*) physiology, (*ix*) biochemistry, cell and molecular biology, (*x*) histochemistry, (*xi*) ultrastructure, (*xii*) numerical and computer taxonomy, (*xiii*) ecology, and (*xiv*) distribution patterns. In the case of conflicting evidence, it must be examined whether it is due to specialisation and/ or reduction. It must also be realised that parallel, divergent and convergent evolution has taken place and an ideal classificatory system should also reflect evolutionary relationships. Only when the taxa (in different taxonomic ranks) represent truly evolutionary groups, is it possible that new information, from characters as yet unstudied, fall into the pattern that has been established on a relatively limited amount of information. The dilemma of an ideal system of classification will continue as long as the taxonomists continue to integrate and disintegrate the taxa.

6

Characters and Sources of Taxonomic Evidence

In taxonomic studies, characters of an organism are defined as features, which refer to form, function and behaviour. These features or attributes can be expressed qualitatively and quantitatively. From this definition of 'features' or 'characters', it is clearly understood that the differences, similarities and discontinuities between plants are reflected in their characters (features or attributes). Such characters are determined by observing or analysing samples of individual plant materials and recording the observations methodically or by conducting some experiments on these materials. A herbarium contains a large collection of plant specimens which can be a ready source of such materials for descriptive or alpha-taxonomy. Each specimen can reflect numerous characters, which can be studied and compared with other such specimens.

For the purposes of classification, morphological and anatomical characters are observed with the aid of the "eye", dissecting microscope, Scanning and Transmission Electron Microscopy (SEM and TEM). Modern instrumentation (Spectrophotometer, Fluorescent spectrophotometer) helps in comparative studies of physiological processes—such as photosynthesis and also analyses of chemical compounds (Electrophoresis, Gel Electrophoresis) present in plants (or synthesized by plants). Use of evidences from cytology, biogeography, palynology, phytochemistry, population biology, molecular biology and ultra-structure, for the purpose of classification has become common during the second half of the twentieth century.

Cronquist (1977) observed that because of constraint of time and tedious effort required to obtain molecular data, morphology, in the broad sense will have an edge over other branches for taxonomic classification for some more years to come. He also suggested that may be a day will come when there will be two sets of a classification systems—one for practical purposes based largely on morphology (in a broader sense) and the other for the use of the intellectuals—based on molecular evidences. Cronquist was not in favour of molecular evidences.

Turner (1977) disagreed with Cronquist's viewpoint as he neglected the recent applications of chemical data for classification.

Characters have many uses such as providing information for the construction of taxonomic systems, supplying characters for construction of keys for identification, providing data useful in the description and delimitation of taxa and enabling scientists to formulate new classification systems.

Characters are abstractions and taxonomists deal with their expression. Characters are something like 'leaf width', 'stamen number', 'corolla shape', 'carpel number' or 'placentation'. A taxonomic character is made up of two or more **Character-states.** For example, while describing a leaf—'ovate' and 'dentate' are the character states of the 'character' margin; and 'acute, 'obtuse' etc. are the character-states of the 'character' apex. Similarly a leaf-width of 5 cm and basal placentation are also character-states. When used in description, or identification, characters are said to be **diagnostic** or **key** characters. Characters of constant nature, which help to define a group, are **synthetic** characters. Quantitative characters are the ones that can be easily counted or measured or assessed in any other way. Qualitative characters are the ones like flower colour, odour, leaf shape or pubescence.

Another often-used term is **good character**. This means that the character is genetically determined, largely unaffected by the environment and relatively constant throughout the populations of the particular taxon. The more constant a character, greater is its reliability and importance.

Each organism has a large number of potential characters and taxonomist may use only a few characters amongst them for the purpose of classification. Some of these characters are found at random and some others in groups of 2 or 3 or more. This association of characters is known as **co-relation**—and the characters as **co-related** characters. For example, syngenesious stamens, inferior ovary and capitulum inflorescence together will derive at the family Compositae or gamopetalous corolla, epipetalous stamens and obliquely placed ovary with persistent calyx will indicate that the family is Solanaceae.

Such co-related characters are quite useful in identification. In the family Papilionaceae, two species, i.e. *Indigofera astragalina* and *I. hirsuta* can be separated on the basis of two co-related characters. In *I. astragalina* the length of pedicel of flower is less than 25 mm and breadth of pod is 3 mm. In *I. hirsuta* length of pedicel is more than 25 mm and breadth of pod is 2 mm. Similarly, the two species of *Cassia*, i.e. *C. tora* and *C. obtusifolia* (Caesalpiniaceae) although much alike, can be distinguished on the basis of: a) leaf apex, and b) position, shape and colour of extrafloral (or foliar) nectaries. In *C. tora* the leaf apex is obtuse and the nectary is present between the two pairs of leaflets, greenish yellow and narrowed at both ends. In *C. obtusifolia*, the leaf apex is mucronate, extrafloral nectary is between the lower pair of leaflets only, deep orange and club-shaped.

Bentham and Hooker's System of Classification is constructed on the basis of co-related characters.

Taxonomic evidences. Plant taxonomy was founded on various characters of the plant groups studied and observed from time to time, and they provide the bulk of taxonomical evidences. Source of such evidences are from various fields— morphology, anatomy, palynology, embryology, cytology, genetics, chemistry, and ultrastructure. All parts of a plant provide taxonomical characters (or evidences) at various stages of its development and, therefore data should be assembled from as many diverse disciplines as possible. Earlier there have been biased opinions on the relative taxonomical values of the evidences obtained by specialists in various fields. However, there is no reason to believe that certain characters are more significant than some others. Symmetry of the flower is not more important than the type of stomata or the type of embryogeny is not less significant than chromosome morphology. Of course, this type of claims has been made sometimes but from the taxonomic point of view, all characters have equal weightage and must not be *a priori* or deduced.

Morphological Evidence. Morphological characters of plants have long been used for the purpose of classification and they are still indispensable. The features of floral morphology in particular, have provided many important characters for the classification system of Linnaeus, Bentham and Hooker, Bessey, Engler and Prantl, Hutchinson and others. As these features are easily observable they have been used in the classification systems, identification keys and general descriptions. The general terminology used in plant morphology has been given in the Appendix at the end of this book.

The general constancy of the reproductive features of flowers, fruits and seeds make them important source of evidence. Example of the use of these features can be given from all taxonomic levels or ranks.

Inflorescence, bracts and floret types of Gramineae and Compositae are so distinct that they can be easily identified; the stamens of Papilionaceae (diadelphous) and Malvaceae (monadelphous) are unique. Likewise, the corollas and stamens of Labiatae and Scrophulariaceae are also important. Characters of fruits are important in Cruciferae and Umbelliferae even to identify at the generic level. Similarly seed structure is of significance in Caryophyllaceae to distinguish different genera and species; in *Euphorbia* to identify different species. Duke (1961) pointed out that shape, colour and sculpturing of seed coat help in identifying the species of *Drymaria* (Caryophyllaceae).

Vegetative features in most taxa are not so useful, but in some others they can be of significance. For example, oblique leaf base is characteristic of Solanaceae members; leaf shape is useful in identifying different species of *Ulmus*; size, shape, and colour of the leaves can be helpful in distinguishing between the two species of *Morus – M. alba* and *M. nigra*. Similarly spcies of *Dalbergia—D. latifolia, D. sissoo*, and *D. sympathetica* can be easily recognised by size, shape and arrangement of leaflets on rachis (Naik 1984).

In trees especially of temperate regions, the small flowers for short duration may not provide significant taxonomic evidence and hence, the vegetative features such as buds, thorns, spines and others, and leaf type, arrangement, and venation pattern may be of great significance.

Anatomical Evidence. Anatomical features have played an increasingly important role in elucidation of phylogenetic relationships. Bailey (1923, 1944a,b) and his students Frost (1930a,b, 1931) and Kribs (1935, 1937), Cheadle (1943) and Metcalfe and Chalk (1950) brought into light the significance of wood anatomy in taxonomical studies. Others, whose names must be mentioned for highlighting the importance of anatomical studies in taxonomy, are Auguste Mathew for describing the anatomical features of wood of various forest plants in his book *Florae forestiere* (1858) and H Solereder for his classic book *Systematische Anatomie der Dicotyledonen* (1899), later translated to English in 1908.

Wood Anatomy. The data from wood anatomy are of considerable taxonomic significance. Trends of evolution in dicot woods are seen in various parts of the xylem: the vessels, rays, parenchyma and others. These trends are as follows:

1. **Shortening of fusiform cambial initials**—Vessel elements elongate little as compared to the length of fusiform cambial initial from which they were derived. Shorter vessel elements are usually seen in species from drier areas (Carlquist 1966).
2. **Increase in F/V ratio**—The length of primitive vessel elements (V) are only slightly less than that of the accompanying tracheids (or fibre-F). The ratio between lengths of these two elements, i.e. F/V is between 1:00 to 1:20 for primitive angiosperms and can be up to 4:00 in most specialized woods (Carlquist 1975).
3. **Perforation plate morphology specialization**—Primitive vessel elements appear like a scalariformly pitted tracheid with the end walls highly inclined, elongated and with numerous scalariform perforations. Transition from a vesselless (i.e. primitive) to vessel-bearing (i.e. advanced) condition involves loss of pit membranes in perforations. Vessels of many primitive dicots show such transitional stages (Carlquist 1992), which can be determined by counting the number of perforations (Carlquist 1997).
4. **Lateral wall specialization**—Frost (1931) observed that the pitting type of lateral walls of the vessels shows a transitional trend from scalariform to opposite to alternate circular pits. Such a trend represents an improvement in mechanical strength of woods (Carlquist 1975). In some succulent members of the family Crassulaceae, pits in vessel walls show highly increased pit size. This reduces the mechanical strength, which is not required in succulent plants. Similarly, scalariform pitting (giving maximum mechanical strength) has been observed in families like Rhizophoraceae and Vitaceae (much advanced) (Carlquist 1997).
5. **Transition from angular outline (in transection) to circular outline in vessels**—Primitive vessels in dicots is angular in outline and rounded or circular in outline as specialized or advanced feature (Frost 1930a).

6. **Vessel abundance**—This is yet another feature of taxonomic significance. Solitary vessels are observed in primitive dicot families. Distribution of vessels in transaction is also significant. Diffuse-porous wood is considered to be more primitive as compared to ring-porous wood.
7. **Vascular rays**—The evolutionary trends seen in vascular rays are: 1) presence of both multiseriate and uniseriate rays, as against the loss of one or the other, 2) heterogeneity versus homogeneity of a ray i.e. ray cells both procumbent and upright or only procumbent, and 3) presence of long, uniseriate wings on multiseriate rays as opposed to loss of the wings (Kribs 1935). Taxonomically important features of vascular rays are – abundance, cellular composition and width, degree of wall thickness of ray cells and pitting in ray parenchyma cells.
8. **Axial parenchyma**—These are the parenchyma cells derived from the cambium in secondary tissue. It is used as against ray parenchyma. These are of two types—(i) apotracheal, i.e. without any specific relation to vessels and (ii) paratracheal, i.e. in close association with vessels. Absence of axial parenchyma is a primitive feature, e.g. the family Winteraceae. When present, their diffuse arrangement is more primitive (Kribs 1937).
9. **Storied wood**—Wood in which the axial cells and rays are arranged in horizontal series on tangential surfaces is storied wood. Bailey (1923) showed that nonstoried woods have pseudotransverse divisions of fusiform cambial initials to increase cambial circumference, whereas the plane of division changes to vertical in woods with storied woods. From evolutionary point of view, storied woods are more specialized or evolved as compared to nonstoried wood.
10. **Other valuable criteria of wood that are usable as important taxonomic features**—(i) laticifers in secondary xylem. (ii) secretion of gums and resins, and (iii) crystals—their types and locations.

Wood anatomy has been used to solve taxonomic problems at all taxonomic levels or ranks.

1. Magnoliales and Amentiferae: Engler and Prantl treated Amentiferae as the most primitive taxon amongst the dicots, whereas Bentham and Hooker, Hutchinson, Cronquist and Takhtajan believe that the Mangnoliales is the most primitive group. Wood anatomical studies have shown that the Mangnoliales include some vesselless genera (pr), vessels are long with narrow diameter, oblique end walls, and with scalariform pitting on lateral walls (pr), scalariform perforation plates (pr) and diffuse-porous wood (pr). On the other hand, Amentiferae shows many specialized features: all members with vessels (adv); vessels short and circular (in transaction), with opposite or alternate pitting on lateral walls (adv), simple perforation plates (adv), and ring-porous wood (adv). Thus, the concept that the Amentiferae are a primitive group is rejected.
2. Gottwald (1977) constructed structural groups on the basis of wood anatomical studies on a large number of species of 32 families of the order Magnoliales s.l. which show marked gradation from primitive to advanced stages. It has been pointed out from this investigation that there is no compelling evidence to support those phylogenetic schemes in which Magnoliales s.l. is considered as the only ancestral group for all recent dicots.
3. The family Rhoipteleaceae has been variously placed in the orders Urticales and Juglandales. The anatomical studies of Withner (1941) concluded that Rhoipteleaceae belong to the Juglandales. Scalariform perforation plates occur in both these taxa but not in Urticales.
4. The genus *Illicium* was originally included in the family Winteraceae. Anatomical studies revealed that Winteraceae members are vesselless and *Illicium* is vessel-bearing (Carlquist 1992). *Illicium*, therefore, was separated and placed in a family of its own, Illiciaceae.

Leaf Anatomy. Although not so widely worked out as wood anatomy, leaf anatomical studies have been useful in some of the taxonomically difficult groups such as Euphorbiaceae, Cyperaceae and Gramineae. In all these families, the flower structure is minute and this requires the help from other branches of study. Brown (1958) pointed out that tissue arrangements are extremely diverse in the major taxonomical groups. These tissues are the endodermal sheath, parenchyma sheath and the mesophyll, particularly the arrangement of the cells outside the parenchyma sheath (Davis and Heywood 1967). Metcalfe (1968) pointed out a number of anatomical features of leaves to distinguish the family Cyperaceae from Gramineae.

Epidermal characters of leaves are also taxonomically significant features. Stace (1965, 1981) found that in family Combretaceae, trichome anatomy is an important feature in taxonomic delineation at all levels, starting from circumscription of the family to the separation of species as well as variety.

Important work on trichomes has also been done by Ramayya (1969, 1972) and Ramayya and Rajgopal (1968, 1971) on Compositae, Portulacaceae and Aizoaceae. The presence or absence of trichomes, their distribution pattern and the structure, i.e. the size, shape and arrangement of cells making up the trichome are diagnostic features.

Stomata on both leaf and stem surfaces are significant taxonomic features. Different stomatal types are recognised which are produced by characteristic arrangement of the guard cells and subsidiary cells, i.e. the epidermal cells immediately surrounding the stomates. Thirty five different types have been reported in vascular plants as a whole, some amongst which are reported to be only in the Pteridophytes (Stace 1989). In the angiosperms, four main types of stomata are known: (1) Ranunculaceous or anomocytic, without subsidiary cells, (2) Cruciferous or anisocytic, with unequal subsidiary cells, (3) Rubiaceous or paracytic, with two subsidiary cells parallel to the long axis of the guard cells, and (4) Caryophyllaceous or diacytic, with two subsidiary cells at right angle to the long axis of the guard cells. Stebbins and Khush (1961) report four other types of stomata amongst the Monocots. These are: (1) four or more subsidiary cells of equal size as in *Rhoeo* (Commelinaceae), (2) four subsidiary cells of which the two lateral ones are larger than the two polar ones as in *Commelina* (Commelinaceae), (3) only two subsidiary cells that are parallel to the long axis of the guard cells—most common amongst the Monocots, and (4) subsidiary cells absent as in most Liliales. Guard cells are mostly bean-shaped, except in Gramineae where they are dumbbell-shaped.

Stomatal characters are sometimes helpful at higher taxonomic levels. For example, stomata are all diacytic in the members of family Acanthaceae but in closely related Scrophulariacae they are only anomocytic. However, stomatal features are not always so reliable. Paliwal (1965, 1969, 1970) and Paliwal and Kakkar (1971) have observed that there is occasional difference in the mature stomatal types in different organs of the same plant or even on the surface of the same organ. In *Streptocarpus* and *Saintpaulia* (Gesneriaceae) the cotyledons bear anomocytic stomata and the mature parts have anisocytic types (Sahasrabudhe and Stace 1974). Twenty three different types of stomata have been reported in *Catharanthus roseus* (Baruah and Nath 1996). *Lippia nodiflora* is an unusual member where all the four types of (mature) stomata have been reported to occur on the same leaf (Pant and Kidwai 1964).

Phloem. Phloem is not as important as the xylem in taxonomic studies. But Behnke's (1969, 1975, 1976a,b, 1982, 1995) work on sieve tube plastids is important in phylogeny of angiosperm. Special types like inter-and intraxylary phloem are taxonomically important. For example, intraxylary phloem is characteristic of the order Myrtales.

According to many authors, anatomical features are conservative. But Carlquist (1969,1997) regards these features as always related to adaptation and physiological functions and therefore, a dynamic functioning system.

Floral Anatomy. The number and distribution of vascular bundles within the receptacle and floral parts have long been recognized as useful criteria in studying floral morphology. Flower is comparable to a determinate shoot tip that bears the fertile appendages— the sporophylls and sterile appendages. The sterile appendages are the sepals, the outer, green leafy structure together forming the calyx and petals, the inner ones varying in size, shape and colour together forming the corolla. In some members the sepals and petals are not distinguishable and are known as tepals

Typical vascular ground plan of flowers shows that a) sepals receive three traces each b) petals and stamens receive one trace each, and c) the carpels are typically three-traced each.

Puri (1952, 1962) pointed out that during specialization and evolution of the flower the vascular plan undergoes considerable modification through—(i) Reduction, (ii) Amplification, (iii) Cohesion, and (iv) Adnation of vascular bundles. Since floral morphological characters are slow to change, it is natural to suppose that their vascular supply is also conservative and this knowledge has been helpful in solving certain taxonomic problems. It has been particularly helpful in consideration of the nature of the 'simple' flower, i.e. whether this simplicity is due to reduction and specialization or due to actual primitive nature. Eames (1961) summarized the evolutionary trends found within the Angiosperm flower as:

(1) From many parts, indefinite in number, to few, definite in number.
2) From three or four sets of appendages—perianth, androecium and gynoecium—to one.
(3) From spiral to whorled arrangement of appendages.
(4) From free floral parts to fusion.
(5) From actinomorphy to zygomorphy.

Floral anatomy has been significant in solving taxonomic problems at different ranks. In Ranunculaceae s.l. sepals receive three and petals one trace each. The genus *Paeonia* has sepals and petals with a few to many traces. Development of stamens are centripetal in all Ranunculaceae s.l. but centrifugal in *Paeonia*. Number of traces supplying the carpels of *Paeonia* is numerous but in other members of Ranunculaceae only three traces. Thus, floral anatomical studies support the separation of the genus *Paeonia* from family Ranunculaceae s.l. Similarly the genus *Trapa* was earlier included in the family Onagraceae and was later removed to a family of its own—Trapaceae. Floral anatomical studies justify this separation.

The genus *Cyrtandromoea* was removed to family Scrophulariaceae from the family Gesneriaceae (Burtt 1965). Singh and Jain (1978) supported this separation on the basis of floral anatomy. In Gesneriaceae the sepal laterals arise conjointly with petal traces whereas in *Cyrtandromoea* they arise independently and this is also true in Scrophulariaceae members. Also, as in Scrophulariaceae, this genus shows the presence of several lateral traces in the carpels and a bilocular ovary and absence of disc at the base of ovary.

Likewise, *Hydrocotyle asiatica* was changed to the new combination—*Centella asiatica*. Mittal (1955) showed that in all other species of *Hydrocotyle* (except *H. asiatica*), the ovular traces are derived from placental strands whereas in *H. asiatica*, the ovules receive their vascular supply from the alternate bundles in each carpel as in *Centella*.

Floral anatomy is also helpful in determining the interrelationship between taxa of higher ranks like the families. Floral studies of the families Papaveraceae, Capparaceae/Cruciferae and Moringaceae show that their gynoecia have the same typical ground plan. In all these families, placentation is parietal; the placental strand which occur on the inner side of the secondary marginals are inversely oriented with reference the floral axis. On the basis of this feature the family Moringaceae was included in order Rhoeodales. This anatomical peculiarity has been found later in the families Cucurbitaceae and Passifloraceae also.

It is, however, certain that floral anatomical data can only be a supportive data and not a confirmative one. Correlation between floral anatomy and secondary wood anatomy can often produce profitable data (Davis and Heywood 1967).

Micromorphology and Ultrastructural Evidence. Use of electron microscopy in higher plants is comparatively new. In recent years advanced technology in microscopy, particularly scanning electron microscopy (SEM) and Transmission Electron Microscopy (TEM) have been useful in studying finer details of external features (micromorphology) such as trichomes, other surface features of various organs and pollen grains and similar details of internal cell structure (ultrastructure) of higher plants.

Micromorphology. Scanning Electron Microscopy (SEM) has been useful in comparative studies of a large number of micromorphological features so that these have now become standard taxonomic characters. Spores and pollen grains, details of leaf surfaces, especially the stomatal architecture and fruit and seed surfaces have so far been considered.

Epidermal studies are important as it forms a universal covering of the complete plant body. Even herbarium specimens can be used for these studies. Various species of *Cassia* can be identified on the basis of types of stomata and cell walls—straight, wavy or sinuate (Kotresha and Setharam 2000). Epidermal micromorphology of nine species of *Albizia* found in Nigeria were studied and compared by Ogundipe and Akinrinlade (1998). The trichomes in these plants are non-glandular, uni- or multicellular and with raised or sunken base. The epidermal cells are isodiametric or polygonal with straight, arcuate (curved like a crescent) or sinuate anticlinal walls. Stomata are anomocytic and/or paracytic. An identification key can be prepared for these nine species on the basis of SEM studies for epidermal micromorphology.

SEM studies on ovarian nectaries of *Cassia occidentalis* are also interesting. In this species, several cup-shaped nectaries are present on the entire surface of the ovary. These are metabolically highly active and rich in metabolites. SEM studies also reveal that there is a cuticular border of the nectaries and rudimentary stomata (without guard cells) on the ovarian surface. As the flowers of *C. occidentalis* are polypetalous, the floral parts are exposed to the environment. Hence, such features of the nectaries may be helpful in protecting floral parts from desiccation (Schmid 1988, Sarangpani and Shirke 1996). Similar studies have brought to light that *Myosotis* shows an unusual development of anther wall in the family Boraginaceae (Strittmatter and Galati 2001).

In the family Balsaminaceae, the seed surface by SEM studies reveals that two species of section Acualimpatiens—*Impatiens acualis* and *I. scapiflora*—are characterized by short and thick hairs on seed while other two species—*I. modesta* and *I. stocksii* (section Orchimpatiens) by long and slender hairs. This feature might be one of the important characters to differentiate between these two sections (Shimizu et al. 1996).

Gattuso (1995) studied the extrafloral nectaries of various species of the genus *Polygonum* that grow in Argentina. The observations reveal the peculiar structure of the nectaries of *Polygonum convolvulus*, which comprises a few glandular hairs enclosed in a pit with a subcircular single aperture. This feature has highly diagnostic value since out of 23 species in Argentina only *P. convolvulus* possesses this characteristic extrafloral nectary.

Glandular and nonglandular trichomes are located on the surface of anther lobes in *Tecoma stans* of family Bignoniaceae (Singh and Chauhan 1999). Such trichomes are otherwise not reported in rest of the members of this family.

Importance of epidermal characters in general and those of trichomes in particular has been widely recognized in systematics of angiosperms. Arrangement and shape of epidermal cells and type of cell wall are features of prime importance. In the genus *Crotalaria* of Fabaceae (= Papilionaceae) different species can be identified on the basis of type of cell wall—straight, sinuate, wavy, or curved, presence or absence of trichomes on abaxial/adaxial/or both surfaces of leaf and stomatal index (Nazneen Parveen et al. 2000).

Low temperature scanning electron microscopy has proved to be an adequate technique to study the wall and cell contents of pollen grains. Studies of pollen features with freeze sections with SEM have helped us to understand the stratification of the pollen wall and the morphology and structure of columellae. This technique is particularly useful in differentiating between granular and rod-shaped (cylindrical) columellae. Columellae forms have been misinterpreted as granular in several species. Granular exines (alvedate exine) occur both in gymnosperms and angiosperms hypothetically derived from atectate exine. This technique (Cry – SEM) has been used to study the tapetum and microspores of *Betula pendula*. The result shows a well-organized network of strands connecting the tapetal cells with the microspores. The strands are generally connected to the microchannels of the exine.

The exine structure of *Alnus, Betula, Fagus* and *Rhododendron* pollen grains and *Lycopodium* spores have been investigated by Atomic Force Microscopy (AFM) and Scanning Tunneling Microscopy (STM)—two more new techniques. With the technique of Electron Micrography, immunocytochemistry, rapid-freeze fixation and substitution and monoclonal antibody application have also been used to localize the subcellular sites of allergens in *Betula pendula* pollen and tapetal cells. The allergen is predominantly located in the starch grains and to a lesser extent in the exine.

Various seed-morphological studies of the leguminous taxa have been performed from time to time (Gunn 1981, Lersten et al. 1992). 'Lens' is a part of seed-coat modified in the form of pit, groove, scar or mound situated between hilum and the chalaza. Earlier researchers were of the opinion that 'lens' is present only in the seeds of Fabaceae members. SEM studies by Sahai (1999) conclude that 'lens' is a constant feature of the seeds of *Cassia*—a Caesalpiniaceae member.

Seed-coat structure is also helpful in distinguishing the three subfamilies of Cactaceae. According to Barthlott and Voit (1979), SEM studies of the seed-coat structures support its division into three subfamilies: Pereskoideae with simple unspecialized testa is in the ancestral position; Opuntioideae have seeds with hard aril and therefore isolated from the other two groups; Cactoideae that includes 90 percent of the species of the family Cactaceae with highly specialized seed-coat structure has an advanced position. This last subfamily is further divided into a number of tribes and subtribes on the basis of distinctive seed-coat structure.

Ultrastructure. Ultrastructure of various angiospermous tissues have been studied extensively in the last two decades through Transmission Electron Microscopy (TEM), and valuable taxonomic information have been obtained. Amongst these, studies of sieve tube elements of phloem tissues have been important. Various types of plastids that occur in the sieve tube elements of phloem tissues, defined on the basis of presence and form of protein–accumulations are highly diagnostic and can be utilized to distinguish between different taxa. Behnke (1974, 1975, 1981, 1982, 1985, 1995) and Behnke and Barthlott (1983) have worked on this aspect in various families. Sieve element plastids contain starch grains that vary in number, shape and size. Protein accumulates in some specific plastids and may be in the form of crystalloids or filaments. Hence, two types of plastids are recognized: P-type, which accumulates protein and S-type that does not accumulate protein. P-type plastids may be of very primitive nature.

Polymotosy. The trimission and scanning electron microscopes add important contribution to our understanding of the ultrastructure of the pollen wall and the development of the pollen grain.

Computer Software specialists distribute a wide range of scientific and technical software, which helps to the construction of 3-dimentional graphics that facilitate the ultrastructure and interpretation of our transmission and scanning electron micrographs. Scanned images can be converted into 3-dimension graphics that give more detailed information of the structure.

Palynological Evidence. Palynology is the science of pollen and spores and its application in various fields of study. Pollen grains have been found to occur in glacier ice, in the air, at the poles as well as over the oceans. Fossil spores and pollen have been in peat and various sediments, in lignite, coal and shales and even in strata laid down in precambian era hundreds of millions of years ago. So they are more widely distributed in time and space than any thing else in the plant kingdom. The pollen grains in the living angiosperms vary in size from 5 by 2 Nºin some members of Boraginaceae to spheroidal bodies of 200 Nº or more in some Cucurbitaceae and Nyctaginaceae.

Pollen grains are provided with a covering called—sporoderm. The outer layer of the sporoderm is exine, which is resistant to decay and is furnished with various types of undulations, spiny projections, pits or thickenings. Plants can usually be identified with the help of various features of the pollen grains.

Use of pollen morphology in taxonomical studies has been appreciated as early as 1935 when Wodehouse wrote the book—*Pollen Grains,* now a classic work. Erdtman G (1943) published a treatise on pollen analysis and later in 1952 his more important work—*Pollen Morphology and Plant Taxonomy* was published. Erdtman G (1966) gave an excellent review of the systematic applications of palynology. The significance of pollen morphological studies in taxonomy has been further realized with the work of some other workers like, Wagenitz 1975, Stix 1960, Chanda 1966,1969 and Nair PKK, 1966, 1969, 1970.

Records on pollen morphology are also found in world floras like *Die natürlichen Pflanzenfamilien* by Engler and Prantl and in the regional floras like the *Flora of British India and Ceylon* by Hooker (1872–1897). Hooker observed pollen differences in *Nymphaea alba* and *N. speciosa*. Description of pollen grains in *Euryale ferox* (Nymphaeaceae), *Parkia* (Mimosaceae), *Passiflora* (Passifloraceae) and several members of the families like Malvaceae, Apocynaceae, and Asclepiadaceae have also been included. In one of his earlier studies too, Hooker (1859) (see Turill 1953) mentioned about pollen grains in peat. This suggests that he already had an understanding of the morphostructure of pollen and its use in taxonomical and floristic studies.

Pollen characters such as size and general shape, number and position of furrows, number and position of apertures and the details of sculpturing of the exine are of taxonomic value.

Size and shape. There is a considerable variation in the size of pollen grains in different plants—and they may be very small, small, medium, large, very large and gigantic depending upon their length. The shape also varies in different views. In polar view, a pollen grain may be circular or triangular; in equatorial view, its shape may be peroblate, prooblate, spheroidal, subprolate, prolate or perprolate. The bilateral pollen are planoconvex, concavo-convex or biconvex in lateral view.

Furrows. Two basic types of pollen grains—monocolpate and tricolpate occur among the angiosperms. Monocolpate pollen grains are provided with a single furrow on one side away from the point of contact in the tetrad. These are common in some primitive dicot families like Piperaceae and Chloranthaceae and some woody members of the Ranales, and the monocots. Tricolpate pollen grains have three meridionally placed furrows. These are characterstic of most dicots. A large number of derived types are observed amongst the dicotyledons. There may be an increase in the number of furrows and pores and therefore pancolpate or a reduction in the number of furrows often to none and therefore acolpate pollen grains.

Apertures. Form, number, distribution and postion of apertures are also taxonomically important. Number of apertures varies from one to many.

Exine Sculpturing. The pollen grains of anemophilous families are smooth and thin (as in Gramineae), whereas the ones disseminated by insects or birds have various types of sculpturing on the exine.

Nuclear condition of angiosperm pollen at the time of anthesis is also of stystematic value, i.e. whether these are shed at 2 – celled or 3 – celled stage. Brewbaker (1967) published a detailed study of this aspect. Amongst the angiosperms, the 2 – celled condition is considered as more primitive than the 3 – celled one. The monocot families possess 2 – celled pollen grains, the apetalous and polypetalous dicot families are also 2 – celled (except the Centrospermae, where the pollen grains are uniformly 3 – celled), and gamopetalous dicots are all 3 – celled. Sometimes, both 2 – celled and 3 – celled pollen grains occur in the same genus, e.g. *Burmannia*, *Ipomoea* and *Lobelia.*

Pollen characters have helped in differentiation of various taxa at different levels.

Family level. The families may be 'stenopalynous' or 'eupalynous' or 'eurypalynous'. In the former, pollen type is constant, e.g. Cruciferae with tricolpate and reticulate pollen or Ericaceae with pollen tetrads. In the later group, pollen may vary in size, aperture number, exine ornamentation etc., e.g. there are two families—Araceae and Lemnaceae in the order Arales (of Hutchinson). The family Araceae is stenopalynous with uniporate and spinous pollen and the family Lemnaceae is eurypalynous with 1-2-4 colpate, triporate or inaperturate pollen with various types of exine ornamentation.

Family Nyctaginaceae is characterized by large-sized pollen grains (up to 160 *N*º in *Mirabilis jalapa*), thick exine and conspicuous strata.

Pollen structure and morphology provide a significant contribution to the systematics of the Papilionoideae both at the tribal (apertures and exine stratification), generic and specific level (size, shape and exine ornamentation). Pollen structures found to be associated with bird pollination either have verrucate ornamentation (rather than reticulate) or a complex exine stratification (rather than simple).

Tribe level. The tribe Bombaceae of the family Malvaceae has been recognized as a separate family Bombacaceae. Palynological study supports this view. The exine of Bombacaceae pollen is reticulate but in Malvaceae, it is spinous.

Genus level. The genera *Salix* and *Populus* of the family Salicaceae can be distinguished on the basis of pollen characters. *Salix* has long and narrow, 3–furrowed pollen grains as compared to the spherical grains without distinct apertures in *Populus.*

The thickening of exine around the pores makes a distinguishing character for different genera of the Betulaceae. In optical section it is knob-like in *Betula*, club-shaped in *Corylus*, unexpanded in *Carpinus* and the genus *Alnus* is characterised by internal arch-like interporal thickenings.

Members of the family Epacridaceae show a variety of pollen grains (Erdtman 1952). These are mostly single, smooth and solitary, e.g. *Brachyloma*. But in *Epacris* and *Leucopogon* the pollen is in tetrads. Pollen characters have been extensively employed in classifying the genera of the Acanthaceae and the Primulaceae.

Species level. Various species of *Anemone* can be distinguished on the basis of characters of germinal aperture of pollen grains. Variations in aperture and exine ornamentation help in the identification of various species of *Bauhinia*. Pollen grains are pilate in *B. acuminata,* striate in *B. krugii,* spinulate in *B. malabarica,* reticulate-tuberculate in *B. retusa.*

Size of pollen grains is helpful in distinguishing the two species of the genus *Malva –M. rotundifolia* has pollen grains measuring 74-84μ and in *M. sylvestris,* they measure 105–126μ.

Trisyngyne is an interesting genus; there are separate male and female plants. The male specimen, on the basis of only morphological features was placed in the family Euphorbiaceae. When the female specimen was discovered, it was transferred to the family Fagaceae. If, however, the pollen grains of the male specimen were examined earlier, this controversy would not have arose. The pollen grains of *Trisyngyne* are exactly the same type as those of *Nothofagus*, also a member of Fagaceae.

Varietal level. Pollen characters are sometimes helpful in distinguishing between taxa at the varietal level, e.g. *Caltha palustris* var. *alba* has pantoporate pollen grains and *C. palustris* var. *normalis* has tricolpate grains.

Palynology, although not universally applicable, provides important evidence in taxonomy. To begin with mostly the external characters of the pollen grains were used for taxonomic considerations. However, in recent years, internal structure, structure of pollen wall and chemical composition has turned out to be equally important. Many workers like Kuprainova (1965, Amentiferae), Nowicke (1970, 1975, Nyctaginaceae, Centrospermae), Nowicke and Skvarla (1981, 1983, Berberidaceae, Corynocarpaceae), Skvarla and Nowicke (1976, Centrospermae), Simpson and Skvarla (1981, Krameriaceae) and others are engaged in this type of study. The development and availability of SEM have brought revolution in the study of exine structure of pollen grains. Pollen samples are easily available from herbarium specimens and numerous relatively rapid techniques are available for the study of pollen grains. This has made it possible to make palynological survey of large number of taxa within a short time.

Embryological Evidence. Embryology has also played a significant role in systematic considerations. According to Davis (1966), every details of embryological study is of potential taxonomic value and to learn about its actual value in this regard, routine examination of the same process in all the apparently related plant species must be taken up. Many embryologists have taken up this type of study in various angiospermous plants. The two publications of Johri (1984) and Johri et al. (1992) are encyclopaedic work in this field.

Embryological features are important in classification of larger taxonomic units like Genera and Tribes and in determining the relationship amongst the families.

P Maheshwari (1950) listed 12 features, which can be used for taxonomic considerations. These are:

1. Number and arrangement of loculi in anther, structure and thickenings of endothesium, and nature of tapetum.
2. Mode of partition of microspore mother cell into the four microspores.
3. Development and organization of pollen grain.
4. Development and organization of ovule.
5. Form and extent of the nucellus.
6. Origin and extent of the sporogenous tissue in the ovule.
7. Time of wall formation in megasporogenesis, arrangement of megaspores and development of the embryo sac.
8. Form and organization of mature embryo sac.
9. Fertilization—pathway of pollen tube and interval between pollination and fertilization.
10. Type of endosperm and presence or absence of endosperm haustorium.
11. Form and organization of mature embryo.
12. Certain abnormalities—parthenocarpy, apogamy, adventive embryony, and polyembryony.

There are certain families which are specially marked out by their embryological features, e.g. Podostemaceae members are characterized by the presence of pseudoembryo sac, bisporic type of embryo sac, occurrence of pollen grains in pairs, tenuinucellar bitegmic ovule where only the outer integument forms the micropyle.

In Onagraceae, Oenothera type of embryo sac is of universal occurrence. In all taxa of Cyperaceae only one functional microspore is formed in each pollen mother cell instead of four.

The family Loranthaceae has been usually divided into two subfamilies—Loranthoideae and Viscoideae. Embryologically these two subfamilies are quite distinct from each other as regards the mode of development of embryo sac, endosperm and embryo and in the location of the viscid layer. P Maheshwari et al. (1957), Johri and Bhatnagar (1960,1972), Barlow (1964), Kuijt (1968,1969) and Bhandari and Vohra (1983) all agree to treat these two subfamilies as two distinct families.

The genus *Paeonia* was originally a member of the family Ranunculaceae. Embryological features support its removal and realocation in a distinct family Paeoniaceae. Presence of large and elongated generative cell, a coenocytic phase of embryogenesis, massive seed coat and hypogeal germination of seeds are not observed in any other member of the Ranunculaceae. Similarly, removal of the genus *Trapa* from Onagraceae to Trapaceae is also juistified because it has Polygonum type of embryo sac (not Oenothera type), no endosperm (Nuclear endosperm in Onagraceae), Solanad type of embryo development (not Onagrad type) well-developed embryonal haustorial suspensor and extremely unequal cotyledons (Ram 1956).

The studies of Narang (1953) reveal that embryologically, the family Stackhousiaceae is closest to Celastraceae and Hippocrateaceae, rather than to the Scrophulariaceae, Lobeliaceae or Selaginaceae. Hence, there is no serious objection in placing the Stackhousiaceae in the order Celastrales.

Usually, within a family there is not much variation in the embryological features. To this, the family Euphorbiaceae is an exception where a large number of different types of embryo sacs are observed in different members.

P Maheshwari (1958) emphasized, "while the embryologist lays no claim to erect a phylogenetic system of his own, embryological data need to be considered alongwith information from other sources in order to approach a natural system of classification".

Cytological Evidence. Cytology as such, refers to the study of structure and function of the cell and cytotaxonomy is the utilization of cytological characters for solving taxonomic problems. However, mostly the information about the chromosomes is taken into account, i.e., number, shape and size of chromosomes. Cytogenetics is the study of genetics by visualization of chromosome and chromosomal observation usually through light microscopy. Davis and Heywood (1967), Solbrig (1968) and Stebbins (1971) have discussed in detail about the two topics of cytotaxonomy and the use of cytogenetical evidences in solving taxonomic problem. Ehrendorfer (1964) has reviewed the literature pertaining to taxonomic role of chromosome studies.

Chromosome Number. Chromosome number is an important taxonomic character as it is probably one of the most constant single feature employed. Usually, within a genus all the individuals have the same chromosome number, e.g. all known species of *Platanus* have $x = 7$ (basic chromosome number); all species of *Ascarina* have $n = 14$ (haploid chromosome number); all species of *Quercus* have $n = 12$. But there are exceptions also e.g. in *Moringa* the haploid chromosome number may be $n = 11, 14$. Sometimes, even within a species, the members may show different chromosome numbers. Snoad (1955) reported the occurrence of 23 to 83 chromosomes in the root tip cells of four plants of *Hymenocallis calathinum* of family Amaryllidaceae. K Jones (1955) is of the opinion that there can be variation in the chromosome number between the plants raised artificially and the ones collected from the field under natural conditions. To obtain correct data regarding chromosome number, correct identification of the material is necessary. Counting made from a single specimen is not enough and therefore a large sampling should be done. Some species have very wide geographical distribution and these must be adequately sampled. In some instances, the older counts might have been from faulty techniques and therefore the result might have been erroneous. All these points must be kept in mind to obtain error-free result, which can be used taxonomically.

Chromosome numbers vary from as low as n = 2 in *Haplopappus gracilis* (Compositae) to as high as n = 154 in *Morus nigra* (Moraceae) and n = 263-265 in *Poa litorosa* (Gramineae) (Hair and Beuzenberg 1961). If the chromosome number is constant, it is not of much value for taxonomic discrimination within the group. However, the constant number can be a useful character to distinguish such a group from other nearby groups.

Often it is observed that within a group, the basic chromosome number is in multiples in various members. The lowest diploid number (2n) by doubling gives a tetraploid (4n) and then an octoploid (8n). Intermediate triploids (3n) and hexaploids (6n) are formed through hybridization between the different levels. Individuals of this type, where the somatic chromosome number is in exact multiples of the basic number, are termed as **euploids.** The genus *Taraxacum* is an example where there are different species with 2n = 16, 24, 32, 40, 48—a polyploid series.

Sometimes, however, the chromosome number within a group does not show any simple numerical relationship and these are termed as **aneuploids.** An example is the genus *Carex* in which there are species with n = 6 to n = 112 with multiples of 5, 6, 7 and 8 (Heilborn 1939).

Chromosome size and structure. In addition to variation in number, chromosomes vary in form, size, and volume and in the amount of distribution of heterochromatin. The size of the chromosomes of a karyotype is more or less a constant feature in species. It is usual to consider the length of chromosome to characterise the size of the karyotype. In most plants, chromosome length varies from 0.5 to 30 μ (Warmke 1941). In general the monocotyledonous plants have larger chromosomes as compared to the dicotyledonous plants but exceptions are also there. For example, the genus *Paeonia* (formerly of the family Ranunculaceae) has large chromosomes. *Paeonia* has already been placed in a separate family, Paeoniaceae and this is an additional feature to support this.

Many different forms of chromosomes are recognized; they may be V – shaped (the two arms equal and centromere placed medianly), J-shaped (asymmetrical with one arm much longer than the other) or I-shaped or rod-shaped (with a subterminal centromere) with many intergradations also.

Different forms of chromosomes sometimes help to work out the evolutionary trends within a group of taxa. Karyotypic study of *Aconitum* (family Ranunculaceae) reveals that there is variation in chromosome length and the position of centromere from species to species. Most specialised karyotype is seen in the most highly and recently evolved annual species. In the tribe Helleborae of family Ranunculaceae, most asymmetric karyotypes are reported from *Delphinium* and *Aconitum*, the two most advanced genera (Levitszky 1931).

Some groups of plants possess very distinctive types of chromosomes, which provide important taxonomic data. For example, small chromosomes without discernible centromeres, often said to have diffuse or non-localized centromeres, occur in the Juncaceae and Cyperaceae, two monocotyledonous families that are now considered closely related, although previously placed far apart because of their quite different flower structure. Since such chromosomes do not depend upon the presence of discreet centromere for regularity of mitotic and meiotic behavior, their fragmentation is not necessarily deleterious and irregular chromosome numbers are characteristic of many species (Stace 1989). In the *Luzula spicata* group, for example, plants can have 2n = 12, 14 or 24, but the total chromosomal volume is about the same in all and indicates that the higher numbers are probably derived by fragmentation (agmatoploidy). In other cases different chromosome numbers occur in different cells of one root tip (mixoploidy) (Stace 1989).

In general, therefore, asymmetry in karyotype and evolutionary specialization (in habit and morphology) in plant are correlated.

Family level. From classificatory point of view, chromosome number does not have much value at higher levels like family level or above. The two monocotyledonous genera *Agave* and *Yucca* were originally placed in Amaryllidaceae (with inferior ovary) and Liliaceae (with superior ovary) respectively. Hutchinson placed the two genera in a distinct family Agavaceae because of their overall similarity. Cytologically, both the genera posses 5 long and 25 short chromosomes. Their karyotypic similarity justifies their inclusion in the same family, Agavaceae.

The genus *Biebersteinia* was earlier included in the family Geraniaceae. Takhtajan (1987) recognized a separate family Biebersteiniaceae (for this genus) on the basis of presence of endosperm and basic chromosome number x = 5 (in Geraniaceae it is x = 7–14).

Generic level. In the family Ranunculaceae, chromosome number and chromosome morphology have provided basis for more natural arrangement of genera and tribes. Two major tribes of the Ranunculaceae are Helleboreae and Anemoneae as recognized by Engler have genera with base chromosome numbers 7, 8 and 9 and both have genera with large and small chromosome types. The genera *Aquilegia, Isopyrum* and *Thallictrum* have base chromosome number 7 and all small type of chromosomes. These genera have, therefore, been placed in the tribe Thallictreae. Two other genera *Adonis* and *Nigella*, with base chromosome number 6 are also distinct from other genera of this family.

Cytological data has proved to be useful in the taxonomy of the genus *Chlorophytum* of family Liliaceae (Naik 1977). Three species: *C. bharuchae*, *C. glaucum* and *C. glaucoides* are very similar to each other morphologically and therefore it is difficult to identify them easily. Cytological investigations revealed that *C. bharuchae* has 2n = 16 and the other two species have 2n = 42. These two species with 2n = 42 differ in their karyomorphology.

There are several other examples where the chromosome numbers support generic status. The genera *Cistus* and *Halimium* with base chromosome number x = 9 are more closely related to each other than either with *Helianthemum* having base chromosome number x = 8. This supports the recent classification of these genera in which *Halimium* is treated as a separate genus near *Cistus* and not included in *Helianthemum* (Davis and Heywood 1963).

The two species *Cicendia filiformis* and *Microcala pusilla* of the family Gentianaceae were previously included in the same genus *Cicendia.* Their base chromosome numbers are x = 13 and x = 10 respectively and this support their separation into two distinct genera.

However, chromosome number is of little value if both aneuploidy and polyploidy occur in a genus, as in *Crepis.*

Species and infra-specific level. At the species level, karyotype data has been very useful as is exemplified by the studies on *Clarkia* (Lewis 1951, 1953; Lewis and Raven 1958), *Nicotiana* (Goodspeed 1945, 1947), *Gilia* (Grant 1950, 1952a,b, 1954a,b), *Viola* (Pettet 1964) and many others

Monotropa hypopitys (Monotropaceae) was treated as one species with two varieties—*M. hypopitys* var. *hirsuta* and *M. hypopitys* var. *glabra.* When examined cytologically, the former was discovered to be a hexaploid with 2n = 48 and the later one is a diploid with n = 16. The hexaploid was retrained as species—M. *hypopitys* and the diploid one was raised to species rank and renamed – M. *hypophegea.*

Chromosomal data has limited use in taxonomic studies. Very often reliable data is not available. However, with increasing knowledge in karyotypic studies and with improved techniques for these studies, chromosomal data will become useful in the field of taxonomy.

Chemical evidence. The term 'chemotaxonomy' is old and rather loosely used term. Some authors use 'biochemical' or even 'chemical' systematics. But chemists do not like the second terminology. Some even prefer to use 'molecular' taxonomy because it is definitely molecules in relation to taxonomy.

Chemosystematics or chemotaxonomy or biochemical systematics has been practiced by human cultures as they recognized and classified plants by tastes, smell, colour, whether they were poisonous or of any medicinal value.

By 100 AD, Dioscorides had recognized the aromatic mints and grouped them together.

European adventurers and merchants circumnavigated the earth in search of the exotic spices, which were valued for their food preservation quality, and also for masking the odour and taste of spoiled meat and other foodstuff.

Nineteenth century taxonomists often recognized the chemical characters and used them in their work even before the compounds were identified completely.

Although morphological characteristics are of practical importance for the determination of species, chemical characters can be used for establishing relationship between species, genera and families. Application of chemical data to systematic problem is known as chemosystematics and it is one of the rapidly expanding interdisciplinary fields. The development of several types of chromatography such as 'Paper chromatography', 'Thin layer chromatography (TLC)', 'Gas-liquid chromatography (GLC)', 'Double diffusion technique', various types of 'spectroscopic' and other instrumentation techniques since 1940s have all simplified the survey procedures. As a matter of fact chemical knowledge can provide significant systematic information not available from any other approach and can solve sometimes a systematic problem that has not been solved by cytological, anatomical or morphological techniques.

Plants produce many types of natural products and quite often the biosynthetic pathways producing these compounds differ fom one taxon to another. These data sometimes have supported the existing classificastion or in some instances they have contradicted the existing classification based on morphology and in still others, chemical data have provided decisive information in absence of adequate data from other fields.

The large number of natural plant products is divided into two major groups according to their molecule size:

(1) Compounds of low molecular weight, i.e. 1000 or less are termed "micromolecules". These are of two types—Primary metabolites and Secondary metabolites. **Primary metabolites**—These chemical compounds are usually of universal occurrence and are part of vital metabolic pathways. For example, Aconitic acid (isolated from *Aconitum)* or citric acid (from *Citrus* spp) participates in the Kreb's cycle. These compounds are, therefore, present in all aerobic organism and hence their presence or absence is not of any taxonomic value. Amino acids constituting the plant proteins and sugars participating in photosynthetic carbon cycle are also primary metabolites.

Sometimes, however, difference in quantity of such metabolites may be useful taxonomically.

Secondary metabolites are the bye-products of metabolism. They usually perform non-vital functions or functions that are not universal, and therefore less widespread amongst plants. However, this feature itself makes the secondary metabolites of taxonomic importance. Compounds that are included in this group are terpenoids, flavonoid pigments and other phenolic compounds, amino acids, alkaloids, oils and waxes, fatty acids, cyanogenic compounds and glucosinolates.

(2) Compounds of molecular weight above 1000, e.g. proteins, DNA, RNA, cytochrome C, complex polysaccharides are termed "macromolecules". The information carrying molecules are called **semantide.** DNA in a primary semantide, RNA a secondary semantide, and proteins are tertiary semantides, following from the sequential transfer of the genetic code from the primary genetic information (DNA).

Apart from these there are some compounds that are directly visible such as starch grains, crystals, raphides etc. These secondary products are often stored, sometimes in large quantity, within the living cells or may be deposited in glands or resin ducts or in tissues like bark or heartwood. The anatomists knew about these for a long time. These directly visible compounds are rather restricted in distribution and have been used in taxonomical studies by many earlier workers.

Starch grains. Fourteen types of starch grains have been recognized in the angiosperms and these have been used to delimit taxa at all levels. Starch develops as grains in plastids and may be simple or compound. Starch grains occur most commonly in the family Gramineae, and have been used as an aid to classification. Tateoka (1962) reported that the typical members of the Tribe Hordeae have compound grains. Hence, the members like *Lolium, Nardus* and *Parapholis* are out of place in this tribe as they contain simple grains. Amongst the three commonly used cereals, wheat (*Triticum*), maize (*Zea mays*), and rice *(Oryza), Oryza* is distinct from the other two as it contains compound starch grains.

Raphides. These are acicular (needle-like) crystals of calcium oxalate in form of bundles and are commonly seen in aroids and aquatic plants like *Pistia* (water lettuce) and *Eichhornia* (water hyacinth).

The order Scitamineae can be divided into four groups on the basis of presence and absence of raphides and character of stomata:

(1) Heliconiaceae Musaceae Strelitziaceae	Raphide sacs present; Guard cells symmetrical
(2) Costaceae Marantaceae Zingiberaceae	Raphide sacs absent; Guard cells asymmetrical
(3) Cannaceae	Raphide sacs absent; Guard cells symmetrical
(4) Lowiaceae	Raphide sacs present; Guard cells asymmetrical

This classification is suggested by Tomlinson (1962).

Presence or absence of raphides has been employed for a natural classification of family Rubiaceae, along with testa structure, degree of development of albumen and floral type (Davis and Heywood 1967). Members, which contain raphides, are all included in tribe Rubioideae.

Similarly, as raphides are present in all members of Onagraceae except *Trapa,* it is justified to remove it to a separate family Trapaceae. Another genus *Hydrangea,* which was formerly included in the family Saxifragaceae, is now placed in Hydrangeaceae because of absence of raphides in it.

Druses. Often large number of crystals may remain conglomerated together to produce a more or less rounded crystal aggregate. These are called sphaeroraphides or druses and are commonly seen in families like Caricaceae *(Carica),* Apocynaceae *(Nerium)* and many others. These are calcium oxalate crystals. Calcium carbonate crystals are less common; usually occur in the epidermis impregnating a portion of the cell wall. These crystal aggregates are called cystoliths and are seen in various species of *Ficus.* Crystals of $CaSO_4$, $2H_2O$ or gypsum occurs in some plants. Families Tamaricaceae and Fouqueriaceae are placed in the order Tamaricales by Hutchinson (1973). Removal of Fouqueriaceae from Tamaricales (Dahlgren et al. 1976, Takhtajan 1987) is justified because it contains the crystals of calcium oxalate whereas crystals of $CaSO_4$, $2H_2O$ occur in Tamaricaceae.

Silica. Silica occurs in many dicotyledonous families as well as in members of Gramineae and Palmae of the monocotyledons. These are important in taxonomy of grasses (Naik 1984).

Chemical Test Characters. These are produced as metabolic bye-products. Amongst all the secondary metabolites, phenolic compounds have been used most widely in chemosystematic studies. This is a group of rather loosely connected class of chemical compounds, which have the basal structure of phenol, C_6H_5OH. Most of them are highly complicated, have several aromatic rings and side chains. Many of them are important flower pigments. Bate-Smith (1962), Harborne (1963, 1967) and Ribereau-Gayon (1972) have written reviews and monographs on this subject.

Taxonomically most important phenolics are the flavonoids, because of:

(a) Their almost universal occurrence in nearly all plants,

(b) Their great structural variations—more than 2,000 flavonoids have already been isolated from various plants,

(c) The proved fact that these variations have genetic basis,

(d) It is easy to isolate and identify these chemicals even from very small quantity of plant materials,

(e) Chemical stability of flavonoids makes it possible to analyse the plant materials collected number of years ago, i.e. even from old dry herbarium specimens, and

(f) Flavonoid data is valuable at all taxonomic ranks.

Effective use of flavonoid data in systematics was shown by the work of Alston and Turner (1963) on *Baptisia.* Natural hybridization occurs wherever two species of *Baptisia* occur together. The hybrids resulting from this accidental crossing are often so different from either parent that their identification becomes confusing. These researchers studied the flavonoid profiles of the two parents and also the hybrids and demonstrated that the interspecific hybrids could be identified on the basis of 'additive profiles', i.e. certain species-specific compounds from each parent were found together in the hybrids. This work is so well documented that it is often considered as the standard for flavonoid work.

However, some later experiments with other materials have shown that hybrids do not always exhibit 'additive profiles' of the parents. Crawford and Giannasi (1982) have shown that compounds not found in the parents may occur in the hybrids. These experiments were done with natural and induced or artificial hybrids of *Rhododendron* species of Eastern North America.

The genus *Psilotum* is understood to be the living representative of the most primitive group of Pteridophytes. But flavonoid data speaks otherwise. The occurrence of 'biflavonyls' in *Psilotum* brings it closer to some lycopods and gymnosperms and not the leptosporangiate ferns as had been proposed.

Taxonomic affinity of the two families (of Violales) Bixaceae and Cistaceae is in dispute. Cronquist (1981) and Takhtajan (1980) place them in the Violales. They accumulate sulfated flavonoids—a character shared with Tamaricaceae and Frankeniaceae (Harborne 1975), which are also treated in or very close to the Violales. Dahlgren (1980) and Thorne (1981), however, include these two families (Bixaceae and Cistaceae) in the Malvales because of the presence of myricetin and 8-OH-flavonoids (Gornall et al. 1979).

Flavonoid data in connection with distribution of betalain pigments have been useful in solving another taxonomic problem. More than 20 years before the term 'betalain' itself was coined, it was recognized that this group of pigments is widespread in the order Centrospermae (=Caryophyllales) of the angiosperms (Gibbs 1945).

The flavonoids are of two series: a) blue to purple to crimson series is called anthocyanins, and b) yellow to orange to scarlet are anthoxanthins. The visually comparable betalain groups are called betacyanins and betaxanthins. These are N-containing pigments, interconvertible and structurally related. They occur in only ten families of angiosperms, which are treated under the Centrospermae because of similar morphological and anatomical features.

It has been observed that anthocyanins and betalains never occur together. Mabry (1976) recognized two groups amongst these families—one containing anthocyanins and the other betalains—derived from a common "Centrospermous" ancestor. The two groups are best treated as members of one order Caryophyllales, consisting of one betalain suborder, Chenopodiineae (Achatocarpaceae, Aizoaceae, Amaranthceae, Basellaceae, Chenopodiaceae, Didiereaceae, Dysphaniaceae, Nyctaginaceae, Phytolaccaceae and Portulacaceae) and one anthocyanin suborder, Caryophyllineae (Caryophyllaceae, Gyrostemonaceae and Molluginaceae) (Mabry et al. 1963, Behnke and Turner 1971). Cronquist (1981) includes Cactaceae also in this order, as they contain two types of betalain pigments—betanin and phyllocactin (Iwashina et al. 1986). It is interesting from taxonomic point of view that betalains are common to Centrospermae and Cactales, which have long been considered to be phylogenetically related.

Another important flavonoid study is that of the family Lemnaceae by McClure and Alston (1966). A number of flavonoids occur in the four genera—*Spirodela*, *Lemna*, *Wolffia* and *Wolffiella*. Lemnaceae is an aquatic family of the monocots. The included species are highly reduced both in size and structural complexity. Some of these plants measure less than 2 mm across, and without roots and apparent stems or leaves. They simply resemble minute, green, vegetative balls with undifferentiated thalloid plant body, with unisexual flowers—one or two male flowers with a single stamen each and a unicarpellate female flower with 1 to 6 basal, erect ovules.

Most phylogenists presume that the Lemnaceae is derived from terrestrial ancestors and that within the family there has been a phyletic trend towards increasing simplicity through reduction. There is a reduction series from *Spirodela* through *Lemna* to *Wolffia* and *Wolffiella*. This has been corroborated by flavonoid data. *Spirodela* contains anthocyanins, flavones, glycoflavones and flavonols; *Lemna* contains anthocyanins, flavones and glycoflavones; *Wolffia*, flavones and glycoflavones; and *Wolffiella*, only flavonols. The flavonoid data also suggests that *Wolffia* is biphyletic; some of the species are derived through *Lemna* – like elements and some others from *Wolffiella* - like elements.

Terpenoid. This group of chemicals is widespread in plants. The range of secondary metabolites covered by this term is unusually large. All these substances have a common biosynthetic origin and are based on the isoprene molecule – CH_2=C (CH_3)–CH=CH_2. Union of two or more of these C_5 units forms the diverse types of terpenoids. The main classes of plant terpenoids are shown in a tabular form (Table 6.1) (after Harborne 1998).

Monoterpenoids—The volatile essential oil in various plant species which is sometimes characteristic, comprises monoterpenoids. Such chemicals are obtained by steam-distillation method and are commercially used as the basis of natural perfumes, spices and flavours in food preparations and medicines. Geraniol from *Geranium*, limonene from *Citrus* fruits, ⅓- and ⅔-pinene from pine oil are some of the examples.

Table 6.1 Different classes of plant terpenoids

Number of isoprene units	*Number of C molecules*	*Class of terpenoid*	*Occurrence*
1	C_5	Isoprene	In leaf of *Hamamelis japonica*
2	C_{10}	Monoterpenes	(a) Essential oils from plant families: Labiatae, Rosaceae, Rutaceae, Umbelliferae, etc. (b) Iridoids or monoterpenoid lactones, e.g., nepetalactone from *Nepeta cataria.* (c) Tropolones in gymnosperm woods, e.g., ®-thujaplicin, in certain fungi and in heartwood of Cupressaceae members (Erdtman and Norin 1966)
3	C_{15}	Sesquiterpenoids	(a) Sesquiterpene lactones in Compositae (b) Abscisins e.g., Abscicic acid
4	C_{20}	Diterpenoids	(a) Diterpene acids in plant resins (b) Gibberellins (e.g., gibberellic acid)
6	C_{30}	Triterpenoids	(a) Sterols (b) Triterpenes (c) Saponins (d) Cardiac glycosides
8	C_{40}	Tetraterpenoids	Carotenoids (e.g., ⅔-carotene)
n	C_n	Polyisoprene	Rubber (e.g. *Hevea brasiliensis* latex)

From taxonomic point of view, essential oils of terpenoid type are not very useful. Plant families rich in essential oils are Compositae (*Matricaria*), Labiatae (*Mentha* spp.), Myrtaceae (*Eucalyptus*), Pinaceae (*Pinus*), Rosaceae (*Rosa*) Rutaceae (*Citrus*, *Ruta*), and Umbelliferae (*Carum*, *Foeniculum*, *Coriandrum*, *Anethum*).

In some plants the oil patterns can be utilized for identification of species. Surveys in *Monarda* (Scora 1967) and *Salvia* (Emboden and Lewis 1967) have brought to light that volatile oils of the leaves are species-specific.

Iridoids—These are monoterpnoid lactones with a single isoprene unit. These bitter-tasting chemicals are reported from about 70 families of 13 orders amongst the angiosperms.

A typical iridoid is 'loganin' reported from the seeds of *Strychnos nux-vomica* of Loganiaceae. Second group of iridoids—seco-iridoids are common in the family Rubiaceae. Jensen et al. (1975) concluded from their researches that all the 13 orders (containing iridoids) are monophyletic. Iridoids usually characterize a group of morphologically related orders and families, and hence, this feature can be useful in proper placement of taxa of uncertain affinities. For example, family Fouquieriaceae is usually placed near Tamaricaceae in Violales (Melchior 1964) but some authors have placed it in five other orders also. Dahlgren et al. (1976) conclude that Fouquieriaceae should be placed in the iridoid-rich order Ericales as it contains the iridoid—loganin.

In the family Labiatae it has been noted that the subfamilies and tribes with binucleate, tricolpate pollen contained iridoids in the leaves and those with trinucleate, hexacolpate pollen lacked this chemical (Kooiman 1972).

Sesquiterpene lactones—About 90% of the recognized sesquiterpene lactones occur in the members of family Compositae (Herz 1977, Fischer et at. 1979, Seaman 1982, Harborne and Turner 1984, Harborne 1998).

These lactones are C_{15} terpenoids commonly known as 'bitter principles'. The presence of this chemical might, therefore act as anti-feedants in plants at least for the mammalian group of herbivores. 'Lactupicrin' from the leaves of wild species of *Lactuca* and 'absinthin' from *Artemisia absinthum* are the two well-known bitter-tasting lactones. Sesquiterpene lactones also have anti-tumour properties, e.g. 'bakkenolide-A' from *Petasites* and allergic effects causing contact dermatitis in man, e.g. 'parthenin' from *Parthenium hysterophorus* (Rodriguez et al. 1976). 'Vermeerin' from *Geigeria* sp. can cause cattle poisoning (Harborne and Turner 1984).

Sesquiterpene lactones have been reported from some other families too, e.g., Magnoliaceae (*Liriodendron, Michelia*), Lauraceae (*Neolitsea, Lindera, Laurus*) and Umbelliferae (*Ferula, Laser, Laserpitium, Smyrnium, Melanoselinium).*

Sesquiterpene lactones are usually located in glands on the leaves and inflorescences. In *Parthenium hysterophorous* the glandular trichomes contain this chemical. They are reported to occur in roots, wood, flowers and root barks also.

Taxonomic Utility—Sesquiterpene lactones are taxonomic markers of many species of the Compositae. Heywood et al. (1977) noted that synthesis of these chemicals is a well-established character in more primitive tribes of Compositae, which disappeared gradually during evolution and more or less absent in more advanced tribes. Many Compositae members viz *Ambrosia chamisonasis* and *Artemisia tridentata* have chemical races. But as Payne et al. (1973) reported, no correlation could be established between these chemical races and leaf morphology. Similarly in *Artemisia tridentata*, no apparent relationship is present between chemistry and changes in chromosome number (Kelsey et al. 1975, Bierner 1973 a, b).

The genus *Vernonia* of the tribe Vernonieae (family Compositae) has two centres of origin. The American species have germacranolides such as glaucolide A; the African species have a different lactone pattern—eudesmanolide vernolepin. This suggests that there are two separate lines of origin. Even flavonoid studies (Harborne and Williams 1977) and chromosomal evidence (SB Jones 1977) support this view. However, a few North American species of *Vernonia* have similar sesquiterpene chemistry as the African species (Gershenzon et al. 1984) and hence, the two groups are related (Turner 1981).

Diterpenoids—A chemically heterogeneous group with C_{20} i.e., four isoprene units, they have very limited distribution in plant resins and latex. The resins and latex have a protective function and animal and microbial predators do not attack plants that yield such substances. Gymnosperm resins are rich in tricyclic diterpenoids, e.g. abeitic and agathic acid. The 'copal' resins of leguminous trees are also rich in diterpenes.

The diterpenes from the leaves of certain Ericaceae members like *Kalmia, Leucothoe* and *Rhododendron* are highly toxic. Those from *Croton* spp. (Euphorbiaceae) and *Daphne mezereum* (Thymeleaceae) are not only toxic but highly irritant and carcinogenic as well. 'Columbin' from *Jateorhiza palmata* (Menispermaceae) and 'marubiin' from *Marrubium vulgare* (Labiatae) are extremely bitter.

Gibberellic acid—This is yet another diterpenoid commonly present in all green plants although in very low concentration.

Diterpenes occur in Gymnosperms such as Pinaceae, Cupressaceae and Podocarpaceae. Among the angiosperms, the families Ericaceae, Euphorbiaceae, Labiatae, and Leguminosae—Caesalpiniaceae in particular and Thymelaeaceae are rich in diterpenes.

Taxonomic use of the diterpenes has importance at the lower levels of classification in gymnosperms. *Tetraclinis articulata* of Cupressaceae is usually considered a Southern Hemisphere floristic element (Gough 1966), although it has been reported to occur in North Africa. Chemical nature of its resins is markedly different from its nearest neighbour *Widdringtonia* (from South Africa) and *Callitris* (a native of Australia) with which it was classified earlier. Diterpenoid data support its treatment as a Northern Hemispheric element (Harborne and Turner 1984).

Triterpenoids and Steroids—Triterpenoids are compounds with a carbon skeleton of six isoprene units (C_6). C_{30} hydrocarbon—squalene biosynthetically gives rise to the triterpenoids. These can be divided into four groups: true triterpenes, steroids, saponins and cardiac glycosides.

A large number of triterpenes commonly occur in plants as reported by Connolly and Hill (1991). Pentacyclic triterpenes ⅓- and ⅔- amyrin and the derived acids ursolic and oleanolic are reported from the waxy coatings of leaves and on fruits like apples and pears. They probably have a protective function against insect and microbial attack. In resins and barks of trees and in latex of plants like *Euphorbia* and *Hevea*, too, triterpenes are reported to occur.

'Limonin', the lipid-soluble bitter principle of *Citrus* fruits is another pentacyclic triterpene. Yet another group of triterpenes—cucurbitacin is characteristic of the seeds of Cucurbitaceae members. These compounds have been reported from four other families: Cruciferae *(Iberis)* (Curtis and Meade 1971), Primulaceae (*Anagallis arvensis*) (Yamada et al. 1978), Rosaceae (*Purshia tridentata*) (Dreyer and Trousdale 1978), and Scrophulariaceae (*Gratiola officinalis*) (Lavie and Glotter 1971).

Sterols—Sterols have recently been detected in plant tissues. The three phytosterols are 'sitosterol', 'stigmasterol' and 'campesterol'. A not-so-common one is '⅓-spinasterol' from spinach *(Spinacia oleracea)*, alfalfa *(Medicago sativa)* and senega (*Polygala senega)* root. Ergosterol is confined to many fungi and fucosterol in many brown algae (Phaeophyceae) and also in coconut (*Cocos nucifera*) (Harborne 1998).

Recent discoveries of the occurrence of animal sterols in plant tissues are rather interesting: animal estrogen, 'estrone' in date palm *(Phoenix dactylifera)* seed and pollen (Bennett et al. 1966) and pomegranate (*Punica granatum)* seeds. Even cholesterol is a trace compound in several higher plants (date-palm) and many red algae (Rhodophyceae) (Harborne 1998).

Saponins—These have been detected in more than seventy angiosperm families (Hostettmann and Marston 1995). They are glycosides of triterpenes and sterols and have soap-like properties such as foaming with water and haemolysing blood cells. Triterpenoid saponins are more common amongst the dicot families; whereas steroidal saponins occur mainly in four monocot families—Amaryllidaceae, Bromeliaceae, Dioscoreaceae and Liliaceae and the only dicot family—Scrophulariaceae.

'Hecogenin' from *Agave* and 'Yamogenin' from *Dioscorea* spp. have therapeutic importance. Saponins of alfalfa (*Medicago sativa*) are toxic to cattle and that of liquorice *(Glycyrrhiza glabra)*—'glycyrrhizin' is sweet to taste and used medicinally against cough and cold.

Cardiac glycosides or Cardenolides—Commonly occur in various plant families—Apocynaceae, Asclepiadaceae, Moraceae and Scrophulariaceae. Cardiac glycoside 'oleandrin' is obtained from *Nerium oleander*. All of them are toxic in nature and have pharmaceutical activity.

Taxonomically the distribution pattern of the triterpenes is often helpful at family, genus and species level. Bisset et al. (1966) and Bandaranayake et al. (1977) studied the distribution pattern of triterpenoids in Dipterocarpaceae. The presence or absence of twelve triterpenoids in bark, timber or resin of 4 genera and 92 species of this family was considered. The genus *Doona* was included in *Shorea* on morphological basis. Triterpenoid data of the two genera are strikingly different to treat them as a single genus.

Distribution of steroidal saponins in *Dioscorea* roots is also taxonomically interesting. Akahori (1965) observed that *Dioscorea* spp. with opposite leaves, stems twining to the right and edible roots, lacked saponins. On the other hand, species with alternate leaves, stems twining to the left and no bulbils, contained 'diosgenin'. The only species *D. bulbifera,* which belonged to the second group but lacked 'diosgenin', was morphologically distinct in possessing globose tubers.

Family Rutaceae is rich in 'limonin' and related chemicals. Dreyer (1966) detected this in 26 species of *Citrus* and three other genera *Poncirus, Microcitrus* and *Fortunella* (all of subfamily Aurantioideae) and six genera of subfamilies Toddalioideae and Rutoideae. 'Limonin' is reported to be absent from four other subfamilies of Rutaceae. Members of these two groups of subfamilies exhibit morphological distinctions also.

Tetraterpenoids—Carotenoids are C_{40} tetraterpenoids, i.e. with eight isoprene units and are widely distributed lipid-soluble pigments that occur in all plant groups from algae to angiosperms.

Over 600 different carotenoids are recognized (Britton et al. 1995a,b, Harborne 1998) but only a few are common in higher plants. Some of the common carotenoids are ⅓, ⅔, and ® carotene, lycopene, rubixanthin, zeaxanthin, violaxanthin, crocin, etc. Many yellow flower colours, as in many Compositae, are due to carotenoids. Rubixanthin occur in fruits of *Rosa* spp and crocin in *Crocus sativus.*

However, carotenoid data has not been found to be a useful taxonomic marker (Goodwin 1967, Harborne and Turner 1984). Fruit carotenoids may be chloroplast carotenoids as in *Cucumis* spp, Lycopene-based as in *Diospyros kaki* and some others. Even these data are not taxonomically important.

Amino acids. Plant amino acids are divided into two groups: the proteinic amino acids and non-proteinic or unusual amino acids (more than 300 or so). Twenty amino acids are the building blocks of proteins and are found in acid hydrolysates of plant proteins. Some of the protein amino acids are cysteine, aspartic acid, glutamine, arginine, lysine and others. In general, glutamic acid and aspartic acid and their acid amides, glutamine and asparagine are present in large quantities as they represent a storage form of nitrogen. On the other hand histidine, tryptophan, cysteine and methionine occur in very low amounts in plant tissues.

Amongst non-protein amino acids, ®-aminobutyric acid is regularly present in plants. The rest of the amino acids of this group have a more or less restricted occurrence and have taxonomic significance. Lathyrine is known only from *Lathyrus.* Canavanine occurs only in Fabaceae.

Most of the 'non-protein' amino acids are structural analogues of one or the other of the 20 protein amino acids, e.g., pipecolic acid with an additional methylene group is an analogue of protein amino acid 'proline' and azetidine- 2-carboxylic acid with one or less methylene group is another analogue of 'proline' (see Table 6.2). Pipecolic acid is mainly present in certain leguminous seeds and azetidine-2-carboxylic acid occurs in many members of Liliaceae.

Amino acids are colourless ionic compounds; have high melting point and are all water-soluble. Degree of solubility may, however vary, e.g. aromatic amino acids such as phenylalanine and tyrosine are sparingly soluble. These are a group of plant toxins particularly associated with seeds and quite common in members of the family Fabaceae (=Leguminosae). Their presence in the seeds is important as a protection from insects and herbivorous animals. They also serve as a storage form of ammonia nitrogen in seeds, the nitrogen being liberated during the metabolism of the germinated seedlings (Rosenthal 1982).

Toxicity of some of these unusual amino acids can cause diseases in human and animals. In many Asian countries, people in famine-stricken areas consume seeds of *Lathyrus sativus*—a legume—and suffer from neurological disorder "lathyrism". Other species of *Lathyrus* causing lathyrism are *L. tingitanus* (western Mediterranean)., *L. cicera* and *L. clymenum* (Mediterranean). The disease-causing amino acid is ⅓-amino ⅔-oxatyl-aminopropionic acid. Another toxic amino acid is hypoglycin A—present in unripe fruits of *Blighia sapida.* Consumption of these fruits causes a condition called "hypoglycaemia" in which blood sugar levels are seriously lowered and may even cause death.

Table 6.2 Some non-protein amino acids of taxonomic significance in plants (after Harborne and Turner 1984)

	Amino acid and its source	*Taxonomic significance*
1	$NH_2CONH–CH_2–CH\ (NH_2)–CO_2H$ Albizzine; seeds of *Albizia julibrisin* (glutamine*)	Characteristic of *Acacia* spp (but absent in series Gummiferae); also in *Albizia, Mimosa.*
2	CO_2H NH Azetidine 2-carboxylic acid; in rhizomes of *Polygonatum officinalis* (proline*)	Taxonomic marker for Liliaceae and Amaryllidaceae; also found in sugarbeet (*Beta vulgaris*) of Chenopodiaceae.
3	H_2N $C–NH–O(CH_2)_2–CH(NH_2)–CO_2H$ HN *Canavanine* in seeds of *Canavalia ensiformis* (arginine*)	Only found in Fabaceae (= Papilionaceae) of the Leguminales
4	HO HO $CH_2–CH–CO_2H$ NH_2 *Dopa* in seeds of *Mucuna pruriens* (tyrosine*)	Taxonomic marker as major seed constituent (6–9% dry wt) of genus *Mucuna*; also in lesser quantity in other legumes such as *Vicia faba.*
5	$H_2C = C$ $CH–CH_2–CH–(NH_2)–CO_2H$ H_2C *Hypoglycin A*, in unripe fruits of *Blighia sapida* (isoleucine*)	Characteristic in both Sapindaceae and Hippocastanaceae.
6	H_2N N $CH_2–CH–(NH_2)–CO_2H$ N *Lathyrine*; in seeds of *Lathyrus tingitanus* (arginine*)	Characteristic of *Lathyrus*; absent from closely related *Vicia.*
7	NH CO_2H *Pipecolic acid* in seeds of *Phaseolus vulgaris* (proline*)	Characteristic of *Phaseolus,* absent from closely related *Vigna*; fairly widespread in other plants.
8	N N $CH_2–CH–(NH_2)–CO_2H$ β-*(Pyrazol-1-yl) alanine* in watermelon (*Citrulus lanatus*) seeds (histidine*).	Characteristic of Cucurbitaceae.

* The most closely related protein amino acid.

Presence of non-protein amino acids has been reported from various members of plant kingdom. Apart from Leguminosae they are regularly found in families such as Cucurbitaceae, Euphorbiaceae, Iridaceae, Liliaceae, Rosaceae and Sapindaceae; amongst gymnosperms in Cycadaceae and in algae, fungi and bacteria. Ibotenic acid is one such compound from the fungus *Amanita muscaria* (fly-agaric) with insecticidal properties.

Taxonomic uses

It has been indicated that non-protein amino acids can be useful taxonomic markers at subgeneric, generic or tribe levels in families where they are consistently present (Table 6.2). For example, the widespread occurrence of canavanine in the family Fabaceae (= Papilionaceae or Leguminosae) is of systematic importance. Amino acid surveys have been done for the two genera *Lathyrus* and *Vicia.* In seeds of various species of *Lathyrus* seven distinctive non-protein amino acids including 'lathyrine' are found, whereas the seeds of *Vicia* spp. contain six additional amino acids including 'canavanine'. The two genera are closely related and it is difficult to separate them morphologically. This is an example where chemical data has been helpful in their identification.

Similar studies have been conducted with the seeds of *Vigna* and *Phaseolus* of Fabaceae (=Leguminosae/Papilionaceae). It has been noted that 'pipecolic' acid is absent in *Vigna* spp. but present in eight out of ten species of *Phaseolus.* On the basis of morphological studies the two species of *Phaseolus*, *P aureus* and *P. mungo* are similar to *Vigna* spp and have been transferred to *Vigna* subsequently (Verdcourt 1970). In this example also, the chemical data support the transfer. This re-assignment of the two species of *Phaseolus* to *Vigna* has been supported by serological evidence too (Chrispeels and Baumgartner 1978).

Cyanogenic glycosides—Some other chemicals of taxonomic importance are cyanogenic compounds—particularly cyanogenic glycosides that yield HCN on either enzymic or non-enzymic hydrolysis. The pathway is shown in Fig. 6.1. About 14 cyanogenic glycosides are known and are of restricted occurrence. The most common ones are 'linamarin' and 'lotaustralin'—usually found together in plants such as *Linum usitatissimum* (flax), *Trifolium repens* (clover) and *Lotus corniculatus* (birdsfoot trefoil). It is 'amygdalin' and 'prunasin' in *Prunus* seeds (bitter almonds) and 'dhurrin' in *Sorghum.*

The first step is enzymic hydrolysis and the specific enzyme—a hydrolase occurs in the plant itself. The second step i.e., liberation of HCN occurs spontaneously.

Cyanogenic glycosides may be present in some parts of a plant and absent in others, e.g. in *Anthemis* (Compositae) these are absent from leaves but are present in seeds. It is well known that bitter almonds from *Prunus amygdalus* trees produce the toxic gas prussic acid or hydrogen cyanide (HCN). Its presence can be detected easily by its smell of bitter almonds and more safely with the use of sodium picrate paper. This test may be performed with herbarium specimen but use of fresh material is better. A few pieces of

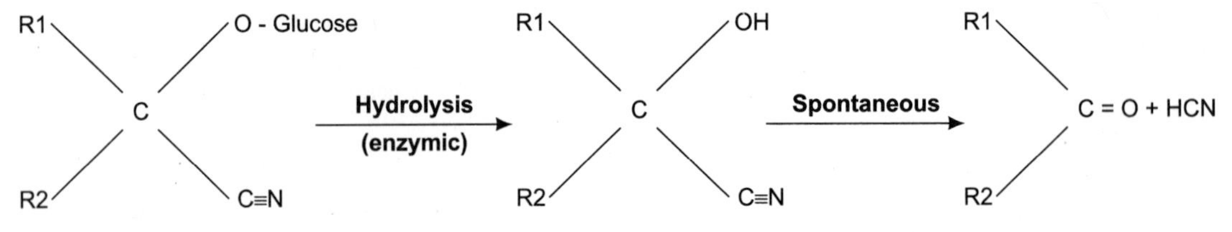

Fig. 6.1 Release of HCN from cyanogenic glycosides (after Harborne 1998)

fresh leaf are crushed and taken in a test tube. Over this a filter paper soaked in picric acid is suspended and held in place by a cork stopper. With the release of HCN, the yellow colour of picrate paper changes to red or brown. HCN is released as a result of enzymic hydrolysis of the cyanogenic glycosides (both present in the leaf tissue). Sometimes, when the enzyme is absent the result may be negative. But if the test set-up is left undisturbed for 24 hours, delayed positive result may be observed due to slow non-enzymic release of HCN from the leaf tissue (Harborne and Turner 1984).

Taxonomic Considerations. Cyanogens are characteristic of the Rosaceae. Lindley (1830) and Endlicher (1836) had used the presence or absence of this compound to distinguish between the tribes Amygdaleae (with cyanogens) and Chrysobalaneae (without cyanogens). The tribe Chrysobalaneae has now been raised to the family rank as Chrysobalanaceae and the former included in subfamily Prunoideae. This shows the early use of cyanogens for taxonomic purposes.

The distribution of cyanogens in the family Rosaceae is complex and surprisingly there is a correlation between cyanogenesis as a leaf character and the basic chromosome numbers of different subfamilies. Cyanogenesis is reported in Spiraeoideae ($x = 9$), Prunoideae ($x = 8$), Kerrieae ($x = 9$), and Maloideae ($x = 17$) but not in Rosoideae ($x = 7$). This supports the view that Rosoideae evolved by aneuploidy from Spiraeoideae losing the original mechanism of cyanogenesis in the process.

Analysis of cyanogenic compounds reveals that evolution of cyanogenic types followed exactly the same pathway in the superorder Magnoliidae as well as in the Liliatae (of the Monocotyledoneae). This similarity favours the assumption that the Liliatae evolved from ancestors resembling the present day Magnoliidae (Hegnauer 1973b).

However, it is not clear what is the function of cyanogenic glycosides in plants. Many food and fodder plants such as cassava (*Manihot esculenta*), clover (*Trifolium repens)* and great millet (*Sorghum bicolor*) contain cyanogens and cyanide poisoning in man or livestock may follow consumption of such plants. DA Jones (1972) reported that cyanogenic glycosides in clover (*Trifolium repens)* protect young seedlings from being eaten away by slugs and snails.

Glucosinolates—Another group of important chemicals are the **glucosinolates** or **mustard oil glucosides.** These are sulphur-containing organic compound naturally occurring in plants. Glucosinolates are the *in vivo* precursors of the mustard oils and yield acrid volatile flavours after enzymic hydrolysis.

About 70 glucosinolates are known. They occur universally in the family Brassicaceae and other related families of the same order i.e., Capparaceae, Resedaceae, Tovariaceae and Moringaceae. Another family Papaveraceae is often closely associated with Brassicaceae because of similar floral structure, parietal placentation and similar type of fruits. Absense of glucosinolates from Papaveraceae is good chemosystematic evidence for separation of Papaveraceae along with Fumariaceae into a distinct order Papaverales (Kjaer 1963, 1966).

The function of glucosinolates in plants still remains a mystery. They are known for their antibacterial properties and some are feeding attractants to caterpillars and aphids feeding on crucifers (Harborne 1998).

Macromolecules—The macromolecules of plants are distinct from other chemicals present in plant body because of their high molecular weight. This may vary from 10,000 to over 1,00,000. Molecular weight of other plant metabolites is rarely above 1000. Chemically macromolecules are made up of long chains of small structural units (= building blocks) linked together. For example, proteins are long chains of amino acids joined together by peptide (-CO – NH -) links. Polysaccarides are derived when small units of sugar such as glucose are joined by glycosidic (- O -) links. The nucleic acids are more complex type of macromolecules with three types of structural units i.e., purine and pyrimidine bases, pentose sugars and phosphate groups. Apart from these, some mixed polymers are also known such as the glycoproteins, which contains both sugars and amino acids in covalent linkage.

Proteins—Proteins along with DNA and RNA are known as **semantides** and are involved in information transfer within the plant body. Amongst these, DNA is **primary semantide**, RNA is **secondary semantide** and proteins are **tertiary semantides.** These are the macromolecules well distributed in the plant body. Proteins are important for chemosystematic studies because:

(a) They are large, complex molecules, which might show little qualitative variation in changing environmental factors.
(b) They are universally distributed.
(c) They are present in large quantities and many of them are relatively simple to extract.
(d) There are many simple and rapid methods of protein analysis available.

During the last few decades, much data on protein sequences have been accumulated which might be used for genealogical studies. Since proteins connect two important evolutionary levels, i.e. genetic expression and natural selection (and hence, a part of the phenotype) they contain very valuable information which may be used to understand phylogeny and taxonomy (Jensen et al. 1983).

For phylogenetic research, the methods of analysis of protein characters (to be compared) must be easy. Different serological techniques and application of amino acid sequence data are some of them (Lester et al. 1983, Boulter 1983). [see Table 6.3].

For taxonomic research, proteins from adequately larger number of taxa have to be compared and the techniques must not be time-consuming. Amongst the most useful techniques are—immunological, electrophoretic and immunoelectrophoretic methods.

In the process of electrophoresis, proteins from different plants when applied on an agar gel surface, will show movement under the effect of an electric current. This is usually carried out in a slab or column of acrylamide gel. It is normally composed of two discontinuous gels—one of larger pore size placed above another of smaller pore size. Hence, this is also known as disc electrophoresis. Protein separation depends upon the sieving effect of the pores in the gels. A broad separation is done in the upper region with larger pores and complete separation into different bands in the lower zone with smaller pores. After the movement stops, the separated proteins are stained with a marker dye. Those proteins, which move to the same place in the gel and form similar coloured bands upon staining, are the homologous proteins.

Taxonomically, electrophoretic data have been useful mainly at and below the genus level. For example, while working with storage proteins in cereals, Johnson (1972) concluded that *Triticum*

Table 6.3 Protein Characters and their utilization in Plant taxonomy (After Harborne and Turner, 1984)

Properties	*Protein Studies*	*Taxonomic application*
1. Amino acid sequencing	Cytochrome C, Plastocyanin and ferredoxin of fresh leaves	Not yet of established utility; but of potential phyletic importance
2. Subunit variation	Fraction I protein of leaves	Valuable for identifying species hybrids
3. Enzyme types (isozymes)	Enzyme extracts of young seedlings	Chiefly useful at subspecific and species level
4. Serological interactions	Total protein fraction or major storage protein of seed	Most useful at family or order level
5. Storage protein complexity	Seed or pollen proteins	Useful at species and genus levels

aestivum contained all those proteins that were present in the two diploid species, supposedly ancestral to it, on the basis of morphological and cytological evidences.

However, there are some drawbacks too in using such data of protein analysis. It has been discovered that selected proteins are not always representative taxonomic markers of the genome and therefore, cannot be used in the construction of phylogenetic trees that are comparable with similar trees based on morphological evidence (Sengbusch 1983). Also, there is difficulty in obtaining pure material (of protein) in sufficiently large quantity to start an experiment. The process of purification of proteins itself is lengthy and time-consuming.

The methods applied for isolation, study and comparison of semantides—include amino-acid sequencing, serology, and DNA hybridisation.

Amino acid sequencing—Only about 2% of the dry weight of plants is nitrogen (N_2) whereas 40% is carbon (C). But, in spite of this fact a large number of organic substances from plants contain nitrogen. Nitrogen is available to plants in the form of ammonia produced either from N_2 fixation in (leguminous or other) roots with the help of symbiotic bacteria or by enzymic reduction of absorbed nitrate in shoot and leaf. First organic form of N_2 in plants is glutamine – an amino acid. Amino acids are involved in the biosynthesis of almost all other nitrogen-containing organic compounds known in plants: alkaloids, amines, cyanogenic glycosides and many others.

Plant amino acids may be the 'protein' amino acids or 'non-protein' ones. The former are twenty (20) in number and are common in plants although their concentration may vary in different tissues depending upon the metabolic status of the particular plant. These are colourless ionic compounds, water-soluble and with high melting point.

Amino acid sequencing aims at identification of pure proteins down to the atomic level. It is possible to separate the amino acids one by one from a polypeptide chain, followed by their identification by various techniques. This way the complete sequence of amino acids can be built up, step by step. To begin with, a large polypeptide chain is broken up into smaller fragments and then each of these fragments are sequenced separately.

Paper chromatography and Thin layer chromatography (TLC) are mostly used for separation and quantitative estimation of the protein amino acids. In paper chromatography, a concentrated aqueous alcoholic plant extract is applied directly to the chromatographic paper. The best solvent pair used is n-butanol-acetic acid-water (BAW) and phenol-water. The same solvent (pair) can be used for TLC on silica gel G (Harborne 1998). For TLC on microcrystalline cellulose, replacement of BAW by chloroform-methanol-2m NH_4 OH (2:2:1) was suggested by Brenner et al. (1969).

Electrophoresis is yet another method for screening plant tissues for amino acids. Combining both electrophoresis and TLC is, however, the best method for clear separation of the amino acids (Harborne 1998).

Amino acid sequencing of various plant proteins for taxonomic purposes and the phylogenies obtained from this, require computation work. Such computer derived data and their implication in systematics is often controversial than simple chemical information.

After the first amino acid sequencing of insulin was carried out, sequencing of many other proteins have been reported and subsequently with comparative study of these sequences, study of biological evolution at a molecular level has been possible (Bryson and Vogel 1965, Fitch and Margoliash 1967). Amino acid sequencing investigates the variation in the precise sequence of amino acids in a single homologous protein, i.e. presumably of monophyletic origin, in a large group of organisms. Matsubara and Hase (1983) studied the amino acid sequencing of chloroplast-type ferredoxins—a group of acidic proteins—isolated from angiosperms, a fern, horsetails (*Equisetum spp.),* Chlorophyta, Chrysophyta, Rhodophyta and Cyanophyta. Amino acid differences indicated that the algal groups were close to each other and distant from the higher plants.

Among the higher plants, taxa that belonged to the same family or order did not necessarily have closer relationship amongst them as compared to the taxa of different families or orders. Ferredoxin structures, therefore, did not reflect any proper taxonomic relationship among the higher plants.

Rubisco, another protein with abundant occurrence in photosynthetic organisms, has been found useful in plant phylogenetic and evolutionary research (Wildman 1983). Polypeptic composition of rubiscos can even act as indicator of comparative age of genera and species.

In Brassicaceae, particularly in the genus *Brassica,* identification of species is rather difficult on the basis of only morphology and/or anatomy because of overlapping characters. Cytological investigations, and later phytochemical and serological studies have been useful in separating them.

Sequencing studies on rubisco conducted in various members of Brassicaceae (Robbins and Vaughan 1983) show that this protein may be a taxonomic marker particularly when the species might have a hybrid origin. Rubisco is the most abundant protein in the biosphere, particularly in the chloroplast. It consists of two subunits, each encoded by a separate gene; the smaller subunit rbc S (encoded by nuclear gene) and the larger subunit rbc L (encoded by chloroplast gene).

Assigning the correct taxonomic position to the aquatic monotypic family Hydrostachyaceae has been difficult because of highly reduced reproductive and vegetative morphology and absence of any diagnostic chemical feature. It has been placed variously in the Lamiales (Dahlgren 1980a), Scrophulariales (Takhtajan 1987), Callitrichales (Cronquist 1981), and Bruniales (Thorne 1992b). Hempel et al. (1995) worked on rbcL sequence data and presently the family Hydrostachyaceae is placed close to the Cornales.

During the last two decades, ample work has been done towards the determination of rbcL sequencing data for various families (Downie and Palmer 1992, Hempel et al. 1995).

There are certain disadvantages too, in using amino-acid-sequencing data for taxonomic purposes. There is always difficulty in obtaining sufficient quantity of protein in pure form to start an experiment. Another problem faced is the laborious and highly expensive procedures for amino-acid-sequencing—be it the earlier methods or the use of the machine, i.e. amino-acid-sequencer.

Serology and Taxonomy—Serology is defined as a part of biology concerned with nature and interactions between antigens and antibodies. Antigens occur in Red Blood Cells (RBC) and antibodies originate in the blood serum of the mammals. Both these substances cause clotting of blood. Extracts of some plant proteins can destroy (or haemolyse) the red blood cells. When such protein extracts from plants are injected into mammalian bodies like those of rabbits or mice, they bring about haemolysis in the beginning but after sometime the mammalian blood produces antibodies (B_1 in Fig 6.2) in the serum that inhibits haemolysis. Such blood serum (B_1) or antiserum can be used as reagent to identify similar or identical antigens. Serological reactions between this antiserum and antigenic material from different sources result in the formation of a precipitate. This is called *precipitin reaction.* Quantity of precipitate (p in Fig 6.2) with different antigens from different plant proteins indicate how closely related they are. For e.g., in Fig 6.2. A and C are more closely related than either A and B or B and C. The degree of precipitin reaction would depend on the similarity between the proteins of the two taxa.

Precipitin reaction was first reported by Kraus in 1897 and was first employed by Bordet in 1899 in his work on birds. Plant taxonomists soon utilized it.

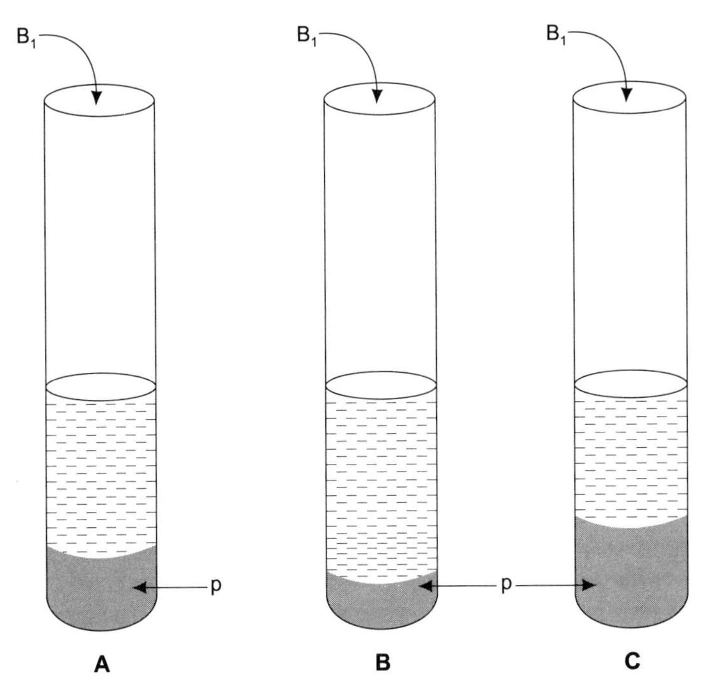

Fig. 6.2 Precipitin reaction. Tubes **A, B** and **C** contain different plant proteins. Reaction with antiserum (B_1) results in the formation of precipitate (p). Quantity of precipitate indicates **A** and **C** are more closely related than either **A** and **B** or **B** and **C**.

Serology does not usually involve identification of particular proteins. It may be disadvantageous for significant interpretation of result when such unknown proteins are compared. Sometimes, therefore, attempts are made at precise identification by running parallel electrophoretic separation on similar samples (Stace 1989). Serological investigations have been made using various tissues of the plants such as leaves, fruits, tubers, seeds, pollen and spores, because different parts of any plant would contain different proteins. Most work has been done using such organs or tissues where protein occurs as a food reserve, e.g. pollen grains, seeds and tubers.

The precipitin reaction was a very crude method for comparing the proteins of various taxa. Presently refined sampling and recording techniques have made this method more fruitful. Various methods are being used by which individual antigen-antibody reactions can be studied. Some of the methods are discussed below:

1. *Double–diffusion serology*—In this method the antigen mixture and antiserum are allowed to diffuse towards one another in a gel (Fig. 6.3). Different proteins travel at different rates and therefore, different reactions occur at different places on the gel. Hence, comparison of precipitin reactions of several antigen mixtures from different taxa simultaneously on the same gel is possible (Fig. 6.3A)
2. *Immunoelectrophoresis*—In this method the antigens are first separated unidirectionally in a gel by electrophoresis and then allowed to travel towards antiserum (Fig 6.3B) A better separation of the constituent reactions is obtained by this method but the disadvantage is that only one antigen mixture can be dealt with on a single gel.

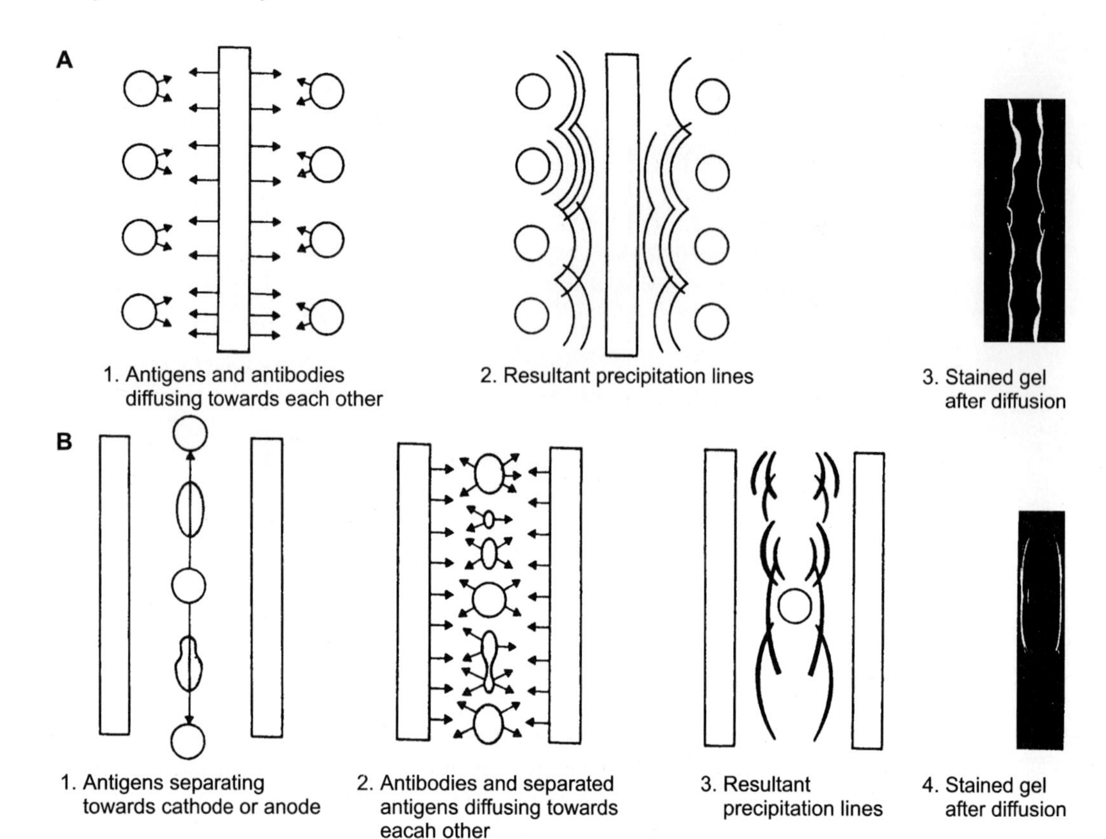

Fig. 6.3 **A, B** Protein separation in various species of *Bromus*. **A** double diffusion serology; **B** Immuno-electrophoresis (after Smith 1972)

3. *Absorption*—Large number of common proteins—especially those involved in common metabolic processes—are present in protein mixture. Antibodies for these antigens are first removed from the antiserum so that comparison of the remaining ones with those of other taxa is easier.
4. *Radio–immunoassay* (*RIA*)—In this method antibodies or antigens are labelled with radioactive molecules so that detection is easy even when these are in minute quantities.
5. *Enzyme–linked immunosorbent assay* (*ELISA*). Here either the antibodies or the antigens are labelled or linked with enzymes so that detection is possible even in very small quantities.

Serotaxonomy or serological taxonomy developed and became popular in Germany. Carl Mez and Ziegenspeck (1926) produced a "Family tree" for the plant kingdom based largely on serological results.

During the last four decades or so, serological studies have contributed enormous useful information for use in plant taxonomy at different ranks (Dahlgren 1983b). However, certain important points should be taken care of for using such data.

1. How authentic is the serological method and the technique used? Hence, different data resulting from source, composition and condition of injected material should be considered while evaluating the data.

2. Is the serological data useful if the antigenic material originates from different tissues of the plants compared? For example, in *Nymphaea* there is copious perisperm, very little endosperm and small embryo, whereas in *Nelumbo*, the large embryo is the only source of storage seed proteins.
3. Choice of taxa and number of taxa chosen too is important. It is possible not to have adequate seed or pollen material from the most suitable taxa. Therefore, such taxa may be omitted although their study is more relevant and the results might not be satisfactory without them.
4. Lastly, the serological data alone may not be sufficient to draw any phylogenetic or taxonomic conclusion. It would be better to look for other characters also such as morphological, embryological, and cytological and others that accord with serological findings.

Following are a few examples where serological data has been useful for establishing relationship at intergeneric and interspecific levels.

Serological evidence has been helpful in establishing the relationship of Nymphaeales and Nelumbonales (Simon 1970). *Nelumbo* antiserum reacted very little (or none) with antigenic material from different taxa of Nymphaeaceae and reacted better with other taxa of Magnoliiflorae. Antiserum of various species of Nymphaeaceae resembled those of *Nelumbo* antiserum but the reaction with *Magnolia* seed protein was weaker. It has been suggested from this study that *Nelumbo* is more closely related to Magnoliales and not Nymphaeales. Other evidences from the fields of anatomy, embryology, morphology and chemistry also support this viewpoint. Plently of experiments have been conducted with Amentiferous taxa. Cronquist (1981) and Takhtajan (1980,1987) retain majority of them as subclass Hamamelidae. It is to be seen whether these families have converged because of adaptation to wind-pollination or they form a homogeneous group of families. Hence, whether Juglandales and Myricales are close to Fagales is doubtful. According to Thorne (1981) there is link between Juglandales, Myricales and the wind-pollinated Anacardiaceae because of similar growth habit, foliar morphology, pollen grains, fruits, seed anatomy and basic chromosome number. Serological studies using *Carya* and *Juglans* of Juglandaceae, *Myrica* and *Comptonia* of Myricaceae, *Quercus* and *Fagus* of Fagacee and *Rhus* and *Toxicodendron* of Anacardiaceae revealed serological similarity between the taxa of Juglandaceae and Myricaceae, and better reaction of the members of these two families with *Quercus* than with *Rhus* (Peterson and Fairbrothers 1983). However, there was no reaction with *Fagus* (Fagaceae) and *Toxicodendron* (Anacardiaceae). Peterson (1983) concluded from further studies that the orders Juglandales and Myricales are close to fagalean families, i.e. Fagaceae, Corylaceae, and Betulaceae and remotely related to Anacardiaceae. These data support the classifications of Cronquist (1981) and Takhtajan (1980, 1987).

The serological reaction was tested between a number of glucosinolate-containing families such as Brassicaceae, Capparaceae, Moringaceae, Resedaceae, Tropaeolaceae and Tovariaceae. Mutual reaction was observed between the taxa of Brassicaceae, Capparaceae, Resedaceae and Tovariaceae but not between any of these and members of Moringaceae or Tropaeolaceae (also glucosinolate-containing). It appears that these two families should not be included in the order Capparales. Their floral structure and pollen morphology are also different. But the similarity in seed characters, presence of myrosin cells and glucosinolates support their placement in the Capparales (Kolbe 1978).

Serology is a useful taxonomic tool even at lower ranks like species and subspecies. Smith (1972,1983) studied the relationships of various species of *Bromus* (brome-grass, Gramineae) relating immunoelectrophoretic data to those from morphology, cytology and cytogenetics. He could throw some light on species problems, diploid-tetraploid relationships and hybrids. A new species *Bromus pseudosecalinus* was recognized by him (on the basis of serological and cytological distinctness) which was previously known as a variety of *B. secalinus.*

Nucleic Acids—Nucleic acid sequence analysis is yet in its infancy with higher plants so that it is difficult as yet to assess its impact on plant taxonomy. The approach is obviously one of considerable potential. The evolutionary significance of genomic differences is at present difficult to evaluate, although a number of models have already been proposed. The question remains whether the enormous investment in equipment and man-hours required for DNA sequence determination will eventually pay off in new taxonomic insight.

There are three methods for measuring differences in DNA between organisms:

The simplest method is to compare base composition, where a simple measurement of the different nitrogenous bases formed on hydrolysis is done. Cytosine (C) and guanine (G) are complementary to each other and therefore they are present in equal quantities. Similarly adenine (A) and thymine (T) being complementary to each other also occur in equal quantities. However, the quantities (of the bases) of the two pairs (G + C) and (A + T) vary in different DNAs. These variations are usually expressed as a percentage of the total base content, e.g. as (G + C)%.

Values of (G + C)% of DNA shows wide variation amongst bacteria: may be 30% in *Clostridium*, 50% in *Escherichia coli* and 80% in *Streptomyces*. Similar range of variation of (G + C)% value has been seen amongst fungi also. Amongst angiosperms, however, the variation is much less— between 36 and 40% (G + C) except in the family Gramineae where the value is much higher—48–49%. (G + C) compositions are between 37-41% in DNA of ferns and its allied forms except in *Selaginella* where this value is 45–50% (Harborne 1998).

Use of nucleic acids in taxonomy started with the second method, i.e. DNA/DNA hybridization. In this method the natural double-stranded DNA is separated into single-stranded polynucleotide chains. Single strands of DNA from one taxon are now allowed to reanneliate with similarly treated DNA from another taxon. This process is known as DNA/DNA hybridization and is done in vitro. Comparison is made with controls consisting of reanneliated strands from a single species (i.e. showing maximum closeness). The percentage of reanneliation of DNA between two taxa will indicate how closely related they are. The amount of recombination is measured by changes in UV absorbance. Only 50% reanneliation of DNA has been observed between *Vicia villosa* and *Pisum sativum* whereas only 1/5th reanneliation occurs between *Phaseolus* and *Pisum* (Bolton 1966).

Mabry (1976) concluded on the basis of this technique that although the family Caryophyllaceae do not contain betalain, it is quite close to the betalain-containing families of the same order i.e., Centrospermae. However, the betalain-contining families are more close to each other.

Presently, with various technical advancements, it is possible to break or cleave DNA at specific points. For this restriction endonucleases are used which can produce highly characteristic restriction fragments of DNA and better results are expected by this method. The fragments of DNA can be separated by gel electrophoresis and hybridized with DNA or RNA 'probes'. DNA-DNA reanneliation experiments using single-copy DNA have been conducted by Belford et al. (1981) and using repetitive DNA by Flavell et al. (1977). Encouraging results have been obtained by using single-copy DNA in *Atriplex, Secale* and many legumes (Stace 1989).

The third method of comparison is by sequence analysis. In angiosperms, most attention has been given to chloroplast DNA or cpDNA. Chloroplast genome is small and compact. It is relatively abundant in leaves so that extraction and analysis is easier. It is at the same time relatively stable in gene content and order (Palmer 1991). As a result different plant species do not show too many changes. Also, evolutionary processes do not alter it. Complete chloroplast sequences are available for *Marchantia polymorpha*—a non-vascular plant, *Nicotiana tabacum*—a dicotyledonous angiosperm, and *Oryza sativa*—a monocotyledonous plant (Doyle JJ & Doyle JL 1999).

Many scientists have worked with the larger subunit of the photosynthetic enzyme Rubisco (the rbcL gene). Sequences for more than 200 plant species are already available for comparative studies (Soltis et al. 1992).

According to Stoebe et al. (1999) "Plants possess a notoriously incomplete fossil record, yet they also possess chloroplast DNA, a molecule well-suited to studying plant evolution." For any group of organisms the larger the molecular data-matrix analysis is done, the more reliable would be their evolutionary history. Several chloroplast genomes have been sequenced but there has not been much utilization of this data for the purpose of reconstructing plant evolution.

Lipids—The lipids, together with proteins and carbohydrates, form the bulk of organic matter of plant tissue. Their occurrence and distribution, therefore, have been a potential source of taxonomic evidence.

Oils, fats and waxes together form the lipids—a more or less heterogeneous group, completely or partially soluble in various organic solvent, but insoluble or only feebly soluble in water. Fats are solid and oils liquid at normal room temperature.

Simple lipids are made up of carbon, hydrogen and oxygen and are esters of fatty acids with glycerol ($CH_2OH–CHOH–CH_2OH$)– a trihydric alcohol. Such esters are known as 'triglycerides' or the neutral fats.

Fatty acids in plants occur mostly as fats or lipids and comprise up to 7% of the dry weight in leaves of higher plants. They are important as membrane constituents in the chloroplasts and mitochondria. Lipids occur in all parts of the plants and in considerable quantities in storage organs like fruits and seeds of different plants and can be utilized as stored energy during seed germination. Of commercial importance are the oils extracted from the seeds of olive (*Olea europaea*), coconut palm (*Cocos nucifera*), oil palm (*Elaeis guianeensis*) and groundnut (*Arachis hypogaea*). They form droplets suspended in the cytoplasm. These oils are rich in unsaturated fatty acids and are used as cooking medium, in soap manufacture, and paint and varnishes industry.

Since one part of the fat or oil molecule is always glycerol, differences in the various fats and oils are due to different fatty acids with which glycerol is combined. These fatty acids have varying carbon chain lengths. Those containing double bonds are unsaturated, e.g. oleic, linoleic, linolenic etc. Fats containing unsaturated fatty acids have lower melting point. palmitic, lauric and stearic acids are some of the well-known fatty acids that are saturated (Table 6.4).

Esters of fatty acids with more than 18 carbon atoms are characteristic constituent of waxes. In addition there are a large number of rare or unusual fatty acids known as lipid components, e.g. erucic, petroselinic, sterculic, vernolic and others, which are characteristic of some families.

Taxonomic Uses. Plant lipid analysis is mostly done by chromatographic techniques: thin layer chromatography (TLC) for separation and purification of the lipids and gas-liquid chromatography (GLC) for identifying the fatty acids produced on saponification (Harborne 1998).

Surveys at the family level bring to light that unrelated families can have similar fatty acid contents and therefore this feature cannot be used as a taxonomic marker. Lauraceae (a dicotyledonous family) and Palmae (a monocotyledonous family) are distinct and distant from each other in any classificatory system, but both of them are rich in myristic and lauric acids. On the other hand, presence of petroselinic acid in *Petroselinium* of Umbelliferae (Hegnaner 1973) and *Aralia spinosa* of Araliaceae (Shorland 1963) supports the close affinity between them. It has also been noted that all species belonging to a family do

Table 6.4 Common fatty acids, their formula and natural source

Fatty acids	*Formula*	*Natural Source*
Lauric	$CH_3\ (CH_2)_{10}\ COOH$	Coconut (*Cocos nucifera*)
Myristic	$CH_3\ (CH_2)_{12}\ COOH$	Nutmeg (*Myristica fragrans*)
Palmitic	$CH_3\ (CH_2)_{14}\ COOH$	Oil palm (*Elaeis guianeensis*)
Stearic	$CH_3\ (CH_2)_{16}\ COOH$	Seed fat of members of Dipterocarpaceae, Sapotaceae & Sterculiaceae
Oleic	$CH_3\ (CH_2)_7CH\ CH(CH_2)_7\ COOH$	Olive (*Olea europaea)*
Linoleic	$CH_3(CH_2)_3\ (CH_2CH{=}CH)_2(CH_2)_2\ COOH$	Linseed (*Linum usitatissimum)*
Linolenic	$CH_3(CH_2CH{=}CH)_3(CH_2)_7$	Many Euphorbiaceae seeds.
Cerotic	$CH_3\ (CH_2)_{24}\ COOH$	Beeswax

not necessarily have the same fatty acid profile. For example, in the family Cruciferae, Appelqvist (1976) reported that erucic acid has been found to occur in 75% of the total number of species surveyed although erucic acid was earlier considered to be a taxonomic marker of this family. In another family Compositae, a number of rare fatty acids occur variously in different tribes (Heywood et al. 1977). Another rare fatty acid—sterculic acid—occurs in *Sterculia* spp. of Sterculiaceae and the closely related Malvaceae. Saturated fatty acids—oleic, linoleic and linolenic—are commonly found in seeds of many Euphorbiaceae members such as *Antidesma diandrum, Bischofia javanica, Euphorbia heterophylla, E. marginata, Mercurialis annua* and others (Shorland 1963).

Many attempts have not been made to utilize fatty acid data to solve taxonomic problems particularly for taxa of lower ranks (below family level). Vickery (1971), while working with seed fats in 26 species of the family Proteaceae, reported that more fatty acids were present in the subfamily Grevilleoideae than in the Proteoideae.

Fatty acid profile has been useful in solving the problem of the "handkerchief" tree or *Davidia involucrata.* This taxon has been placed variously: a) In the genus *Nyssa,* b) as a separate genus in family Nyssaceae, c) as a separate family Davidiaceae, and d) as a separate genus in Actinidiaceae. The fatty acid data reveal a close relationship between *Nyssa* and *Davidia,* in addition to some other similarities (Hohn and Meinschein 1976). Presently, *Davidia* belongs to a distinct family Davidiaceae and placed near Nyssaceae.

Plant Waxes. The plant waxes are esters of long-chain alcohols (24 to 36 C atoms) usually with long- chain fatty acids. These alcohols are relatively insoluble at room temperature. C_{36} alcohol is only sparingly soluble in hot chloroform, an organic solvent. This is the reason why waxes normally are solids at room temperature. Plant waxes occasionally contain free fatty acids, free alcohols as well as high molecular weight aldehydes, ketones and hydrocarbons and hence have a heterogeneous chemical composition. The alkane fraction of the waxy coatings on leaves and other parts of the plant body is a mixture of hydrocarbons of similar properties. Alkanes also occur in fungi and other lower plant groups (Weete 1972). Biosynthetically, these hydrocarbons are related to the fatty acids and are usually formed from them by chain elongation and decarboxylation. Alkanes in the cuticular wax have a protective function as well as water- repellant properties and probably also disease resistance (Harborne 1998).

About utility of alkanes in solving taxonomic problems, there is varied opinion. Eglinton el al. (1962) worked with the members of subfamily Sempervivoideae of the family Crassulaceae and observed that although there was apparent correlation (*Sedum* with C_{33} alkane as dominant), there was no proper distinction between most genera.

However, Scora et al. (1975) compared the alkane distribution in *Persea* (22 taxa) and *Beilschmiedia meirsii* of the family Lauraceae. In the former taxon $C_{33} H_{68}$ is the dominant alkane and in the later $C_{27} H_{56}$ is the dominant one. Here is an example, where alkane pattern can be utilized to distinguish between two genera of a family.

At species level alkane pattern is distinctive and hence, useful in delineating them. It is not only the leaf-alkanes, but often alkanes from other parts of plants are also useful. For example, Mecklenburg (1966) observed that in the tuber-bearing species of *Solanum,* alkanes from inflorescence were more useful to distinguish different species than leaf-alkanes. Nordby and Nagy (1977) studied the alkane patterns in various species of *Citrus* and could distinguish between mandarins (*Citrus reticulata*), oranges (*C. paradisii)* and limes (*C. aurantifolia*), on the basis of peel-wax pattern.

Alkaloids. This group of chemicals has ethnobotanical, medicinal, poisonous, chemical and systematic value. Both structurally and biosynthetically alkaloids are a difficult and highly heterogeneous group of compounds and comprise the largest single class of secondary metabolites in plants. There are about 10,000 alkaloids known till date (Harborne 1998). Alkaloids are organic bases that contain one or more N_2 atoms, often in combination with a heterocyclic ring. Alkaloids usually have toxic effect when administered in human system in higher concentration but they may be used as medicine when administered at lower concentration.

Isoquinolines form one of the largest groups of plant alkaloids and they include a number of valuable clinical agents such as codeine, morphine, emetine and tubocurarine. Alkaloids are colourless, mostly crystalline, (rarely liquid e.g. nicotine) at room temperature and usually bitter to taste—quinine is one example. A large number of them are specific to a particular family or a group of closely related families, e.g. the tropane alkaloids are characteristic of the members of the family Solanaceae and the related Convolvulaceae. Isoquinoline alkaloids occur in the closely linked families Fumariaceae and Papaveraceae.

Chemically heterogeneous nature of the alkaloids makes their identification by simple chromatographic method difficult. To identify an alkaloid from a new plant source, its chemical nature (or type) must be known. To begin with, the alkaloids are extracted from plants (leaves, roots, bark or any other part) in a weakly acid-alcohol solvent and then concentrated ammonia is used to precipitate them. This crude extract is further used in paper chromatography or TLC to recognize the type of alkaloid.

Presence of a particular alkaloid can be detected by using different 'alkaloid reagents' (Harborne 1998). Both dried (Hultin and Torssell 1965) and fresh (Lüning 1967) plant tissues can be used for extraction of alkaloids. Lüning (1967) observed that in orchids, alkaloids often get destroyed while drying the tissues and hence, recommended the use of fresh tissue.

Recognition (or identification) of alkaloids is very important particularly in forensic medicine and a useful procedure for this has been given by Clarke (1970).

Taxonomic utility. Although alkaloids occur most commonly in the angiosperms (only about 30 families), a few of the other plant groups also contain them, for example, families Taxaceae and Cephalotaxaceae of the gymnosperms, *Equisetum* and *Lycopodium* of the pteridophytes, and *Amanita* and

Claviceps of the fungi. The toxin present in the insect "ladybird" (*Coccinella septempunctata*) for defensive mechanism is also an alkaloid: conccinelline.

Amongst the angiosperms, some families are very rich in alkaloids—Loganiaceae, Magnoliaceae, Papaveraceae, Ranunculaceae, Rutaceae and some others (see table 6.5)

Most of the true alkaloids can be related to different groups of parent bases like pyridine, piperidine, isoquinoline, tropane, and quinolizidine and indole alkaloids.

According to Hegnauer (1963,1966), alkaloids can appear in rather remotely related taxa possibly due to convergence and the metabolic pathways are independent of each other. For example, the tropane

Table 6.5 Families rich in alkaloids

Family	*Genus*	*Alkaloids*	
1. Annonaceae	*Annona squamosa* *Xylopia brasiliensis*	anonaine xylopin (Bhaumik et al. 1979) liriodendrine lenuginosine	
2. Apocynaceae	*Vinca rosea*	vincristine vinblastin	
3. Berberidaceae	*Berberis* spp.	berberine	
4. Compositae	*Haplopappus* sp.	pyridine	
5. Erythroxylaceae	*Erythroxylon coca*	cocaine	
6. Himantandraceae	*Himantandra baccata* *H. belgraveana*	himabacine himabaline himandravine himangravine	Hegnauer 1963
7. Leguminosae	*Lupinus*	lupinine cytosine sparteine	
8. Loganiaceae	*Strychnos nux-vomica*	strychnine	
9. Menispermaceae	*Chondodendron tomentosum* *Coscinium fenestratum*	tubocurarine berberine	
10. Magnoliaceae	*Liriodendron tulipifera*	liriodendrine	Hegnauer 1963
11. Nymphaeaceae	*Nuphar japonicum*	desoxynupharide (a pseudoalkaloid)	Hegnauer 1963
12. Papaveraceae	*Papaver somniferum*	morphine, codeine thebaine	
13. Ranunculaceae	*Aconitum napellus*	aconitine, ranunculin	Ruijgrok 1968
14. Rubiaceae	*Cinchona officinalis* *Coffea arabica*	quinine caffeine	
15. Solanaceae	*Atropa belladona* *Hyoscyamus niger* *Solanum tuberosum*	atropine (tropane alkaloids) hyoscyamine, solanine, chaconine	
16. Umbelliferae	*Conium maculatum*	coniine	

alkaloids (Romeike 1978) though occur mainly in the families Solanaceae and Convolvulaceae, are found in a wide range of angiosperm families, which in no way are related to each other. They occur in Cruciferae, Dioscoreaceae, Elaeocarpaceae, Erythroxylaceae. Euphorbiaceae, Orchidaceae and Rhizophoraceae as well.

Another example is the pyrrolizidine alkaloids, which are found chiefly in the family Boraginaceae but are reported to be highly characteristic of the tribe Senecioneae of the Compositae and the genus *Crotalaria* of the Leguminosae (Culvenor 1978). Their occurrence is reported also in the families—Apocynaceae, Celastraceae, Orchidaceae, Ranunculaceae, Rhizophoraceae, Santalaceae, Sapotaceae and Scrophulariaceae (Harborne & Turner 1984)

Alkaloids have been used comparatively less frequently at lower ranks of taxonomic hierarchy, for example, in taxonomic revision of a species. Sometimes a single taxon may contain a large number of alkaloids, e.g. *Vinca rosea* (Apocynaceae) contains 72 different structures (Taylor and Farnsworth 1975). Mears and Mabry (1971) mentioned about the possible taxonomic use of alkaloids in the Lguminosae. Three genera *Genista, Ammodendron* and *Adenocarpus* contain 'ammodendrine hystrine' alkaloids. *Genista* and *Adenocarpus* are included in tribe Genisteae and *Ammodendron* in Sophoreae. Alkaloid pattern suggests inclusion of *Ammodendron* in the tribe Genisteae.

Presence of isoquinoline alkaloids in the families Fumariaceae and Papaveraceae indicates close relationship between them.

The latex of Opium Poppy *Papaver somniferum* provides the medically used alkaloids papaverine, codeine, morphine, noscapine and thebaine. Codeine is monomethyle ether of morphine and thebaine is dehydrodimethyle ether of morphine. Biosynthetically, thebaine is synthesized first—then codeine and eventually morphine. Before this pathway was known, all the three alkaloids were reported to be present in varying quantities in *Papaver somniferum.* This discovery of the biosynthetic pathway of the said alkaloids during 1950's, suggested that some mutant strains of *Papave*r should exist which produce only thebaine, due to a genetic block in the final demethylation to morphine. Researches were taken up during 1960's to locate any related species or variety that contained thebaine alone—as this alkaloid is of medical interest. *Papaver bracteatum* is one such thebaine-rich species, one strain of which produces cent percent thebaine. This strain is now cultivated on a commercial basis for the 'safe' alkaloid thebaine for medicinal use. In this instance, alkaloid chemistry has been helpful in distinguishing the two species—*Papaver somniferum* and *P. bracteatum.*

The benzyltetrahydroisoquinoline alkaloids occur in 18 dicot orders—most of which are primitive.

Sometimes identical or chemically closely related alkaloids, occur in biologically quite unrelated organisms, e.g. 'nicotine' occur in *Nicotiana* and *Equisetum* and 'anabasine' in *Nicotiana* and *Anabasis.* The genus *Erythrina* consists of a range of some 108 species of magnificent orange or red-flowered trees, shrubs and herbaceous plants. The close botanical relationship between all the members of genus *Erythrina* is paralleled by the presence of a series of spirocyclic isoquinoline alkaloids which are found in all parts of the plants.

7

Numerical Taxonomy

A classification method based on the numerical analysis of the variation in a large number of characters of a group of organisms is known as Numerical Taxonomy. It is assumed that classification will be more predictive if the number of characters on which it is based is more. It is also assumed that, to begin with, each character is of equal weightage although some characters may later be treated to have more weightage. Initially, a matrix of data is compiled of operational taxonomic units (OTU's) against characters so that for every OTU the state of each of perhaps 50 or more characters is recorded. The matrix can be subjected to a variety of mathematical analyses, which provide a measure of the similarity or dissimilarity between all the OTUs. The end product is usually one or more dendrograms. Numerical methods in taxonomy are not new. Simple statistical methods like standard deviations, t-tests and chi-squares have been used for several years. The use of computers by taxonomists has established an interesting modern trend called **Numerical Taxonomy** or **Taximetrics.** Mathematical and statistical evaluation of taxonomic information and computation of this data has provided taxonomists with new approaches to understand classification. Numerical taxonomy provides methods that are objective, explicit and repeatable, and is based on idea put forward by M Adanson (1763). He proposed that a classification should use a vast range of characters covering all aspects of the plants, and in construction of a classification system all characters must be given equal importance. The idea forms the basis of modern numerical taxonomy, also called **Neo-Adansonian** classification.

Numerical taxonomy is an organized method of evaluating data in a repeatable manner enabling comparison of many characters from many populations of plants. The use of computers has made it possible to compare a large number of characters from many organisms with relative ease. After comparison of the organisms they are grouped according to overall similarity or dissimilarity and wherever necessary presented graphically.

Robert R Sokal and Peter Sneath (1963) described Numerical taxonomy as the numerical evaluation of the affinity or similarity between taxonomic units and the ordering of these units into taxa on the basis of their affinities.

Heywood (1967) defines taximetrics as a numerical evaluation of the similarity between groups of organisms and the ordering of these groups into higher ranking taxa on the basis of these similarities.

Principles of Numerical Taxonomy

Numerical taxonomy is based on seven principles:

1. The greater the content of information in the taxa of a classification is and the more characters on which it is based, the better a given classification will be.
2. Every character is of equal weight (=importance) in creating natural taxa.
3. Overall similarity between any two entities is a function of their individual similarity in each of the many characters for which they are being compared.
4. Distinct taxa can be recognized because correlations of characters differ in the groups of organisms under study.

5. Phylogenetic inferences can be made from the taxonomic structure of a group and from character correlations, given certain assumptions about evolutionary pathways and mechanisms.
6. Taxonomy is viewed and practiced as an empirical science.
7. Classifications are based on phenetic similarity.

Procedure Adopted by Numerical Taxonomy

As organisms are classified on the evidences obtained from their characters, it becomes imperative/ essential to employ **all** the characters for the ideal or a natural classification. But since, each individual may possess thousands of characters it becomes impracticable to use all characters. The number of characters studied in numerical taxonomy are usually about 50-100 from approximately the same or greater number of organisms. It is presumed that the greater the number of characters the more valid the classification is.

The term organism in this field refers to individuals, populations, species, genera or any other taxonomic category and for this reason they are generally called as Operational Taxonomic Units or OTUs.

The different conditions in which the identification characters occur are known as "character- state". A particular organ may be absent or present, functional or non-functional. In such simple cases they are said to be two state characters.

Many traits exhibit a number of possible states and so are termed as multistate characters. Both two-state characters and multistate characters may be qualitative or quantitative.

Since numerical taxonomy is an operational science, the procedure is divided into a number of repeatable steps allowing the results to be checked at every step.

Choice of units to be studied: First step is to decide what kind of units to study. In numerical taxonomy, the basic unit of study is called the "Operational Taxonomic Unit". Thus, the OTU can be an individual plant if the taxonomist is studying a single population of plants to find out the range of variation in its characters. Similarly, one may treat an entire population of plants as an OTU if the study is on a single species represented by different populations existing in nature; or the OTU may be different species when a genus is evaluated. Therefore, in numerical taxonomy, the OTU varies with the material being studied, and this helps the taxonomists in making an objective study.

Selection of Characters (Attributes)

After selecting the OTUs it is necessary to select characters by which they are to be classified. The characters, which vary greatly amongst the OTUs, are clearly more useful in numerical taxonomy. A sufficiently large number of suitable characters are selected. It is usually recommended that not less than 50 characters should be used. Preferably a minimum of 60 and generally 80 to 100 or more characters are needed to produce a fairly stable and reliable classification. According to Sneath and Sokal (1977) these characters should be from all parts and all the stages of life cycle. The selected characters have to be coded or given some symbol or mark. There are two methods of coding taxonomic information. Taxonomic characters are binary, or qualitative or quantitative multi-state type. What is regarded as a good character is a matter of scientific judgment and experience of a taxonomist. These must be –

1. unit characters,
2. heritable,
3. must not be susceptible to wide environmental modifications,

4. should not be affected by experimental or observational uncertainties, and
5. denote clear discontinuities from other such characters and character-states.

Binary coding or two-state coding: This is the simplest form of coding adapted in numerical taxonomy where the characters are divided into + and – or as 1 and 0. The positive characters are recorded as + or 1 and the negative characters as – or 0. It is possible to use this method of coding for all characters studied. In case a particular character is not present in an OTU being examined, the symbol or code NC is used, indicating that there is no comparison for that character. However, we find that by using this method of coding, we tend to increase our work because there are large variations in the plant, and very often a single character such as colour of flower can be represented in a wide range i.e. we can have white, pink, red, yellow and other colors in roses. If we are to use this data in a binary coding, then we will have to use each color as a character and it would be coded as + or – as the case may be (see Table 7.1).

Sometimes it is not clear which feature should be treated at +ve and which as –ve, e.g. plants lead-tolerent and lead-resistant and an arbitrary decision is necessary.

Table 7.1 Data matrix for binary or two-state coding with hypothetical *t* OTUs and *n* characters. For presence + and for absence – marks are used; NC code stands for characters not comparable for the particular OTU. It is known as *n* ´ *t* table

Character states	*A*	*B*	*C*	*D*
1	+	+	+	NC
2	+	+	+	+
3	+	+	+	–
4	–	+	NC	NC
5	+	+	NC	+
6	+	+	+	+
7	+	+	+	NC
8	NC	–	+	+
9	+	+	+	+
10	+	+	+	+

Multi-state Coding Method

A character having three or more different states:

Qualitative multi-state characters—(disordered multi-state characters) possess three or more contrasting forms each ranking equal. They can be analysed in two ways:

1. To convert them to a series of binaries or
2. Each entity is offered a choice of one of n series (n = number of states assumed by the attribute). The degree of similarity is measured in terms of dissimilarity, i.e. less dissimilar entities are more similar.

Quantitative multi-state characters represent measures of size of a continuous scale such as weight or length. They can be recorded into several two-state characters or arbitrarily dividing the scale into two parts that need not necessarily be equally long. However, all the characters selected are given equal weight when creating taxonomic groups. Data processing is to estimate the similarity between pairs of OTUs and the basic data are recorded in a data matrix.

Data Matrix is tabulation of the data, such as taxonomic characters, to show difference between categories (taxa) often in a machine-readable form. Data matrix is also known as comparison chart, data chart, data table etc.

Once the OTUs have been selected and character-states and their subsequent coding has been determined, the data is presented in the form of primary data matrix or t ´ n matrix where 't' represents the OTUs and 'n' the characters (Table 7.1)

If we have used 50 OTUs and scored 100 characters from each, then we will obtain 50 ´ 100 = 5,000 units of information. Thus, the large amount of information obtained makes the use of computers almost absolutely necessary in numerical taxonomy.

Measurement of Similarity—There are many methods for estimating the phenetic resemblance (similarity) between the taxonomic entities analysed .Overall similarity is calculated by comparing each OTU with every other and is usually expressed as a percentage, 100 percent S for identical and 0 per cent for no resemblance. A similarity table or matrix (Table 7.2) is then constructed by tabulating the S coefficients for each one of the OTU. The simplest form of Coefficient of Associations used by Sneath (1957) is a numerical index–S for similarity between each pair of organisms examined. It is derived as $S = \frac{Ns}{Ns + N_d}$ where Ns stands for the number of +ve features shared by any 2 OTUs and N_d stands for number of features +ve in one OTU and –ve in the other.

Table 7.2 Similarity matrix constructed by tabulating similarity coefficient for each pair of OTUs. It is known as *t* ´ *t* table

OTUs ↓	*A*	*B*	*C*	*D*	*E* →
A	100				
B	90	100			
C	60	60	100		
D	57	50	50	100	
E	90	90	60	58	100

Similarity index is a measure of the similarity in species composition of the communities such as Jaccard's coefficient, Kulezinski's coefficient, and Sorensen's coefficient. Similarity coefficient is a measure of the association of character states of two specimens or taxa.

Jaccard's coefficient is a measure of the similarity in species composition between two communities (A and B) calculated as $Si = \frac{C}{a + b - C}$ where C the number of species common to both , and a and b are the number of species occurring only in communities A and B respectively.

Kulezinski's coefficient is an index of similarity in species composition between two communities (A and B) which is relatively little influenced by dissimilar sample sizes; calculated as Sk = C (a + b) where C is the number of species common to both; a and b are the number occurring in communities A and B respectively.

Sørensen coefficient (Sφrensen index). Here 'S' is a measure of the similarity between species composition of two communities, calculated as 2C/a + b where a and b are the number of species in communities A and B respectively and C the number common to both; usually expressed as a percentage.

Gower (1971) has proposed a general coefficient of similarity applicable to mixed qualitative and quantitative characters

Gower coefficient $$(GS) = \frac{\sum^{n} Sijk}{n}$$

where '*Sijk*' is a score on character i for comparison of OTUs J and K such that *Sijk* = 1 – (1x – xI/R) Here R is the range of character i in overall *t* OTUs. If all characters are qualitative (1,0) then SG becomes 'S'.

Taxonomic Distance (d) is an expression of the relationship between individuals or taxa in terms of multidimensional space, where each dimension represents a character, based on quantitative estimates of dissimilarity.

Correlation Coefficient is a measure of the linear relationship between quantitative variables indicating the degree to which they vary; denoted by "r"values range from –1 (perfect negative correlation) through 0 (no linear relationship) to +1(perfect positive correlation).

The following are the most commonly used coefficient for quantitative character:

Taxonomic Distance (d)—

$$d_{jk} = \sqrt{\sum^{n}(X_{ij} - Y_{ik})\frac{2}{n}}$$

Where the character state of OTUj for character i is X_{ij} and that of OTUk is X_{jk}. The symbol XX indicated the sum over *n* characters. The value of d is the distance in a phenetic space divided by XXn. The distance with 1,0 characters is related to Ssm as indicated below:

$$d = \sqrt{(1 - Sim)}$$

The correlation coefficient (r)—

$$rjk = \frac{\Sigma^{n}[(X_{ij} - \overline{X}_{i})(X_{jk} - \overline{X}_{k})]}{\sqrt{\left\{\left[\Sigma^{n}(X_{ij} - \overline{X}_{j})^{2}\right]\left[\Sigma^{n}(X_{ik} - \overline{X}_{k})^{2}\right]\right\}}}$$

The symbol $\overline{X}_j$ and $\overline{X}_k$ are the average values of x overall *n* characters for OTUs J and K respectively.

Large Scale Scientific computation

The increasing speed of computers and advances in numerical methods have made it possible to solve most small problems rapidly by means of readily available software, and the attention of many numerical analysts has turned to the solution of problems so large that they require inordinate amount of computer time.

A variety of clustering methods have been employed in numerical taxonomy including sequential, agglomerative, and hierarchic and nonoverlapping clustering.

Cluster Analysis—After making a similarity table (Table 7.2), it is then rearranged so that OTUs whose numbers have the highest mutual similarity are brought together. This can be done by several methods and related taxa or groups are recognized (Fig. 7.1). Computer sorts out (cluster) the OTUs according to their overall similarity, that is according to the number of attributes or characters in common. These clusters are called phenons and can be arranged hierarchically in a tree diagram or dendrogram.

Unlike phylogenetic classification, in which shared derived features group taxa, phenetic classification is the grouping of taxa by overall similarity, regardless of whether these similarities are symplesiomorphous or synapomorphous in a phylogenetic sense.

Dendrogram: Dendrograms are most commonly employed to represent taxonomic structure resulting from cluster analysis. These have the advantage of being readily interpretable as conventional taxonomic hierarchies.

Once similarity has been calculated various techniques are adopted to group the taxa. First we group those taxa, which are at least distance to form groups and then proceed to form groups of groups, and a structure formed by this method is known as a dendrogram or phenogram.

Advantages

1. Numerical taxonomy has the power to integrate data from the variety of sources, such as morphology, physiology, chemistry, affinities between DNA strands, amino acid sequences or proteins and more. This is very difficult to do by conventional taxonomy.
2. Greater efficiency is promoted through the automation of large portion of the taxonomic process. Thus, less highly skilled workers or automation can do much highly specialized work.
3. The data coded in numerical form can be integrated with existing electronic data processing system in taxonomic institutions and used for the creation of description, keys, catalogues, maps and other documents.
4. The methods, being quantitative, provide greater description along the spectrum of taxonomic difference and are more sensitive in delimiting taxa. They should give better classifications and keys than can be obtained by conventional methods.
5. The creation of explicit data tables for numerical taxonomy has forced workers in this field to use more and better-described characters. This necessarily will improve the quality of conventional taxonomy as well.
6. A fundamental advantage of numerical taxonomy has been the re-examination of the principles of taxonomy and of the purposes of classification. This has benefited taxonomy in general, and has led to the posing of some basic questions.
7. Numerical taxonomy has led to the re-interpretation of a number of biological concepts and to the raising of new biological and evolutionary questions.

Disadvantages

1. For the believers of "Biological Species Concept" numerical species recognized by this method is unacceptable unless some "genetic " or "crossability" evidences are incorporated.

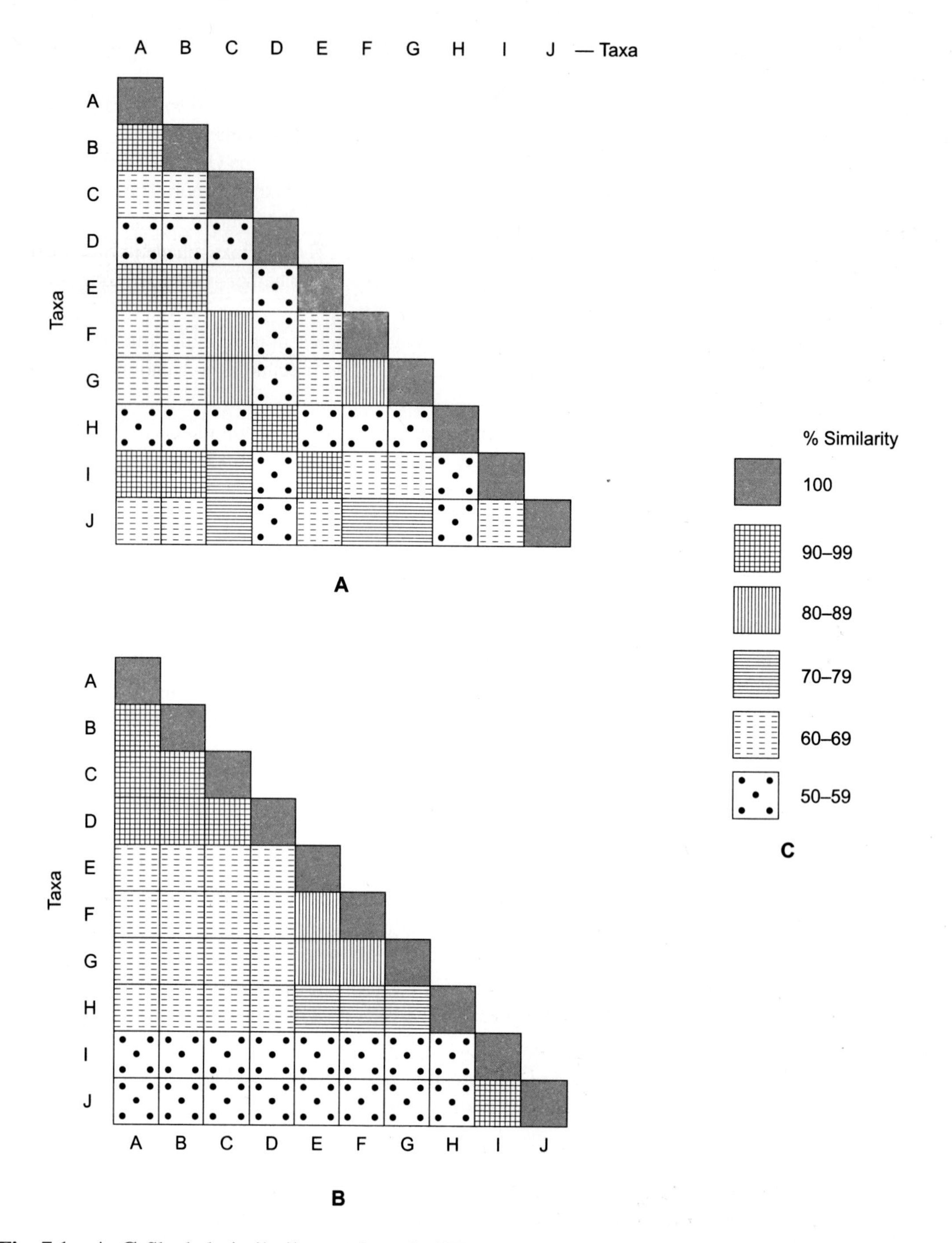

Fig. 7.1 **A–C** Shaded similarily matrices **A** OTUs arranged haphazardly in $t \times t$ matrix. **B** OTUs rearranged so that the similar phenons can be recognized visually. **C** Percentage similarity.

2. Orthodox taxonomists feel that they are more successful than a mechanical computer fed with non-relevant selection of characters.
3. Selection of characters also poses problems.

Examples of studies made by numerical methods are *Apocynum* (Apocynaceae), *Cucurbita* and hybrids (Cucurbitaceae), *Crotalaria* (Fabaceae), *Salix* (Salicaceae), Lamiaceae, Verbenaceae and allied families. *Zinnia* (Asteraceae), *Silene* (Caryophyllaceae), *Quercus* (Fagaceae), *Oenothera* (Onagraceae), *Solanum* (Solanaceae), wheat cultivars, Poaceae, Bromeliaceae, Maize cultivars, barley cultivars and hybrids (Sneath and Sokal 1973).

8

Origin of Angiosperms

Angiosperms are well-known for their incredible diversity in species number, range of habitat, and morphology. Currently there are at least 250, 000–300,000 living species of flowering plants (Crane et al. 1995). This number exceeds the total number of species of all other photosynthetic land plants and algae combined (Sporne 1974, Prance 1977). With the exception of coniferous forests and moss-lichen tundra, angiosperms dominate not only in number of species they also occupy a much larger range of habitats. There is an unparallel array of life-forms from parasites, epiphytes, saprophytes to normal mesophytes as well as xerophytes and hydrophytes. The range of size can be exemplified from tiny *Wolffia*, floating duckweed (about 0.5 mm in diameter) of the Lemnaceae to the lofty *Eucalyptus amygdalina* (450–480 ft tall) of the Myrtaceae. Corresponding diversification in vegetative and floral characters are also exhibited,

The angiosperms are geologically a very young group. They have dominated the earth's vegetation since the mid-Cretaceous. Yet their origin and early evolution have remained enigmatic for over a century. One part of the enigma lies in the difficulty of identifying the earliest angiosperms and other involves the uncertainty among extant and fossil gymnosperms. Despite this late appearance, angiosperms have developed a bewildering diversity exceeding by far that of any other plant group. Timing of angiosperms origin and early diversification is still controversial, but new finding of exquisitely preserved Cretaceous flowers (Fig. 8.1) have greatly extended our knowledge of early angiosperms. It is clear from the floral record that extensive extinctions have taken place in several angiosperm lineages and that the total angiosperm diversity is much higher than what is seen today.

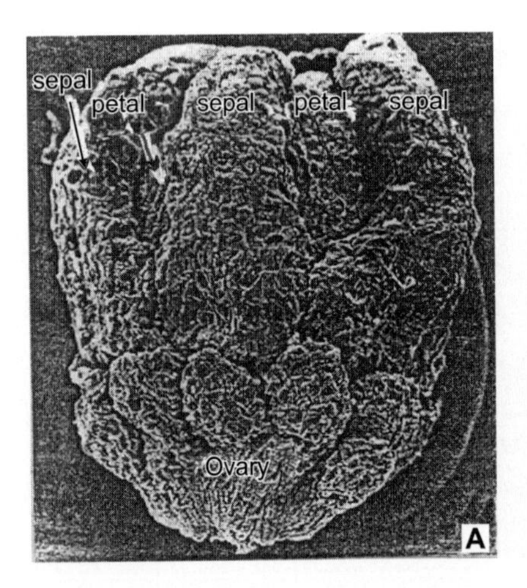

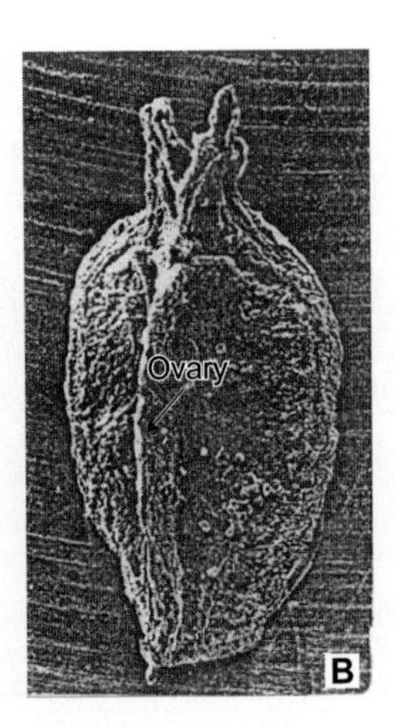

Fig. 8.1 Fossil floral structures from the Cretaceous period. **A.** *Scandianthus costatus* from southern Sweden (Late Cretaceous). Perianth of distinct sepals and petals borne on top of the ribbed ovary. **B.** Magnoliid flower (Chloranthaceae) from Portugal (Early Cretaceous). Simple perianth-like structures at the top of the ovary. **C** Stamen of magnoliid flower from Portugal (Early Cretaceous).

Phylogeny–Problems and Approaches

One aspect of phylogeny that has been important for understanding the angiosperm's origin is their relationship to other seed plants. In tracing back the origin and evolution of flowering plants, it is obligatory to establish the phylogenetic relationship between extant and primitive extinct species, and to characterize the ancestral form of the group.

Earlier attempts to learn about phylogenetic relationship (of angiosperms) were largely based on the knowledge of living groups. However, more recently, availability of some valuable fossil evidences (like pollen, leaves, and fructifications), along with a constant input of data from the fields of floral form and reproductive biology of some of the presumed basal (ancestral) group of extant angiosperms, has led to renewed analysis of this issue. Phylogenetists today are utilizing the data available not only in the fields of morphology but also in anatomy, embryology, palynology, phytochemistry and many others for comparative studies and also for interpretation of the fossil materials. More recently, the data on protein and nucleotide sequences (Sytsma and Baum 1997) and from mitochondrial, plastid and nuclear genomes (Yin-Long Qiu et al 1999) have also been made available to the phylogenetists. A plethora of these data ranging from fossils evidence, biogeographical distribution, comparative morphology and anatomy, and molecular evidences has led to reinvestigation of the phylogenetic relationship (of angiosperms) proposed in the past. The queries that arise now are:

1. Are the angiosperms monophyletic or polyphyletic?
2. Which are the closest gymnosperm/s?
3. What are the basal angiosperms?
4. What is the appearance of the earliest angiosperm?
5. What is the time of their origin?
6. What is the place of origin? And
7. Why there is so much diversification?

Monophyletic or Polyphyletic: The use of these two terms is relative depending upon how far the ancestry of a particular taxon can be traced back. Monophyletic taxon is the one that is derived from a single ancestor, and polyphyletic taxon is derived from more than one ancestral population. It is rather difficult to conclude whether the angiosperms are mono- or polyphyletic. The earlier ideas of the origin of angiosperms were mainly based on living (extant) plants in the absence of adequate fossil record. But recent discovery of well-preserved fossils gives a different picture and much food for thought. In any case, all organisms, higher or lower, are ultimately monophyletic in origin. It is only that the possible or probable ancestral population has to be located either from extinct or extant members. Hutchinson (1959) believed in monophyletic origin of angiosperms from ancestral group of Gymnosperms parallel to Cycadaeoideae.

Probable ancestors of angiosperms: Recent phylogenetic studies show that the Gnetopsids are the closest living relatives of the angiosperms whereas the closest extinct group is the Bennettitaleans (Fig. 8.2). Various theories have been proposed from time to time, regarding the probable ancestors of the angiosperms and these could be placed in the two groups –Euanthial and Pseudanthial theory

(A) **Euanthial theory**, also known as *Anthostrobilus* theory: According to Arber and Parkin (1907), the angiosperms originated from various gymnospermous stocks termed as Hemiangiosperms. This group of plants bore unbranched bisexual strobilus with spirally arranged pollen– and ovule-bearing organs. Carpel is regarded here as a modified megasporophyll. Keeping this general principle in view, various authors have regarded different gymnospermous groups as ancestral to angiosperms and suggested different hypotheses.

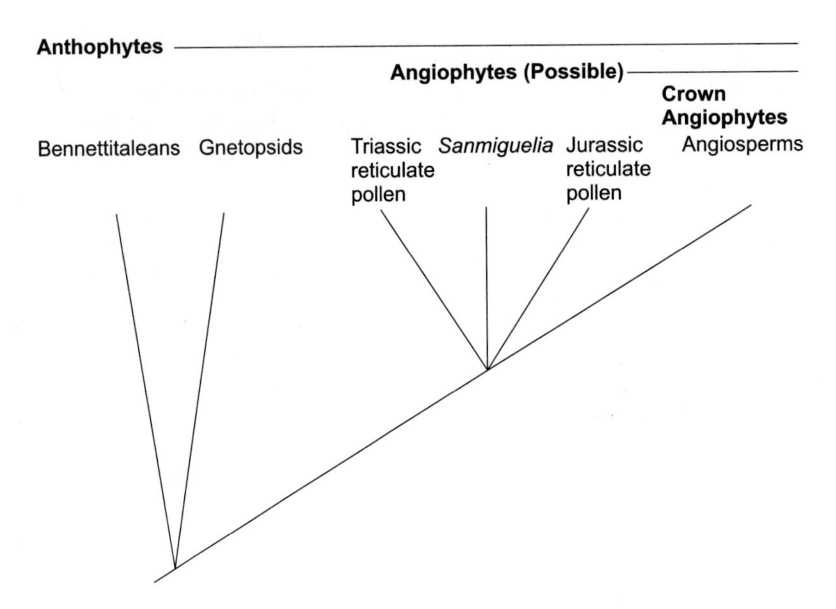

Fig. 8.2 A phylogenetic diagram to show the sister groups of angiosperms. Gnetopsids are the closest living groups and Bennettitales the closest extinct group (as proposed by Taylor and Hickey 1997).

(I) Bennettitalean hypothesis – Lemesle (1946) regarded Bennettitales as the probable ancestor, which appeared in the Triassic and disappeared in the Cretaceous (225 to 110 myr[1]). Considering Ranales as the primitive group of the angiosperms, Bessey (1915), Hallier (1912) and Hutchinson (1926, 1934) also presumed Bennettitales to be the ancestral group. The strobili of the Mesozoic (178myr) genus *Cycadaeoidea dacotensis* (of the Bennettitales) and the flowers of *Magnolia* are structurally similar as both are bisexual and appear as an elongated axis surrounded by bracts (outermost or lowermost), microsporophylls (middle), and megasporophylls (innermost or uppermost) (Fig. 8.3). However, this is only superficial resemblance. The microsporophyll (or stamens) of *Magnolia* are free (polyandrous) and spirally arranged, and those of Cycadaeoidea (Bennettitales) are in whorls and often basally connate. The megasporophylls (or carpels) of *Magnolia*, are free (apocarpous) and spirally arranged, and in *Cycadaeoidea* they are highly reduced, simple, stalk-like structures which bear a solitary, erect, terminal ovule arranged in whorls. It is not likely that these could be the ancestral form of the carpels in primitive angiosperms. Moreover, interspersed with the megasporophylls in the Bennettitales, there are sterile, interseminal scales, that are protective in function. No such organ has been observed in *Magnolia*. Other differences include– (i) Micropylar tube formed in the ovules of *Cycadaeoidea* and pollen grains are shed on the stigma; (ii) seeds non-endospermic with a large embryo in *Cycadaeoidea* and seeds with copious endosperm and small embryo in *Magnolia*; (iii) stem with a large pith, a thin vascular cylinder and a broad cortex in *Cycadaeoidea* and stem with a small pith, broad vascular cylinder and a narrow cortex in *Magnolia*. The superficial similarity between the two taxa could be due to a common ancestry. Both originated from the Pteridosperms or the seed-ferns (Arber and Parkin 1907), and there was parallel evolution.

[1]Myr – Million years ago before present

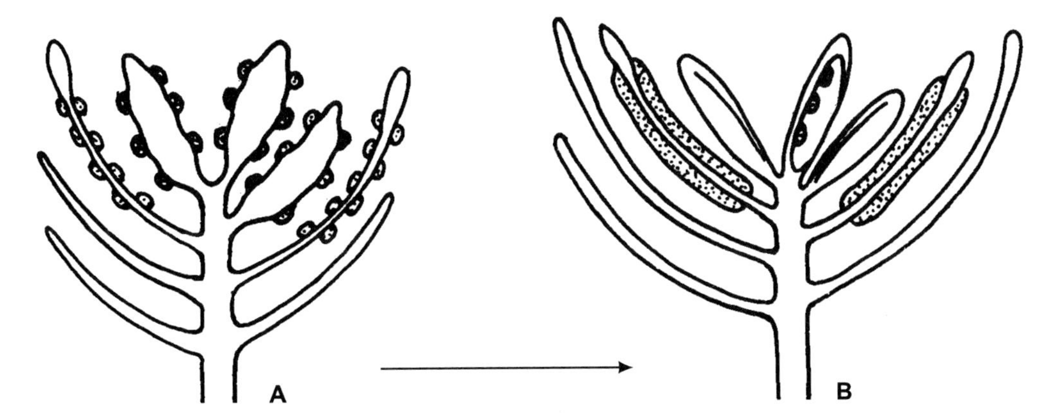

Fig. 8.3 Diagrammatic representation of uniaxial strobilus of *Cycadeoidea dacotensis* A. Giving rise to flowers of *Magnolia* B. (After Hickey and Taylor 1997).

(II) Ancestral groups of Gymnosperms: Armen Takhtajan (1969) is of the opinion that some very ancient group of Gymnosperms was the probable ancestral form (although he does not specify any particular group). Because of insufficient fossil records one would normally look for relationship between angiosperms, and the modern as well as fossil gymnosperms. Studies show that at one time or the other, almost every group of fossil gymnosperms has been considered a probable ancestor of the angiosperms.

Among the members of Cycadophytina only the Cycadales belonging to class Cycadopsida are represented by extant members. The rest are known in fossil forms of the Triassic to Cretaceous period (ca 246 to 65myr). Some fossils like the order Medullosales is reported from the Lower Carboniferous to Permian period (350 to 270 myr). Fossil Cycadales are also known from the Triassic period (225 myr). None of these groups can be considered to be the possible ancestor/s because each one of them shows some specialization, not observed in the Protoangiosperms.

1. Order **Caytoniales** (class Lyginopteridopsida). This Mesozoic (225 to 135 myr) group of fossil seed plants has often been presumed as an ancestral population of the angiosperms. The anthers of these plants were produced in groups or singly, on pinnately-branched sporophylls. These are comparable with the branched stamens of *Ricinus*.The ovules were half-enclosed in small pouches, which are pinnately arranged along the axis of a megasporophyll. Although these pouches have been compared with angiosperm ovaries, they are not homologous structures. Moreover, in primitive angiosperms, one megasporophyll forms only one ovary.
2. Order **Glossopteridales** (class Lyginopteridopsida) has been considered an angiosperm progenitor. This fossil group from Permian to Triassic period (270 to 225 myr) possessed well-preserved seeds borne on the surface of megasporophyll, its margin often rolled inwards (Fig. 8.4). The 'cupule'- like structure of the glossopterids appear to be similar to the outer integument of an angiosperm ovule. However, the pollen grains of the glossopterids bear no resemblance with those of angiosperms.
3. Order **Lyginopteridales** (class Lyginopteridopsida) This group, commonly known as the seed ferns is the nearest to the primitive angiosperms because of the following characters:

   (a) Large, usually pinnately-compound leaves and some species with reticulate venation .

   (b) Stem simple or branched; when branched, branching monopodial.

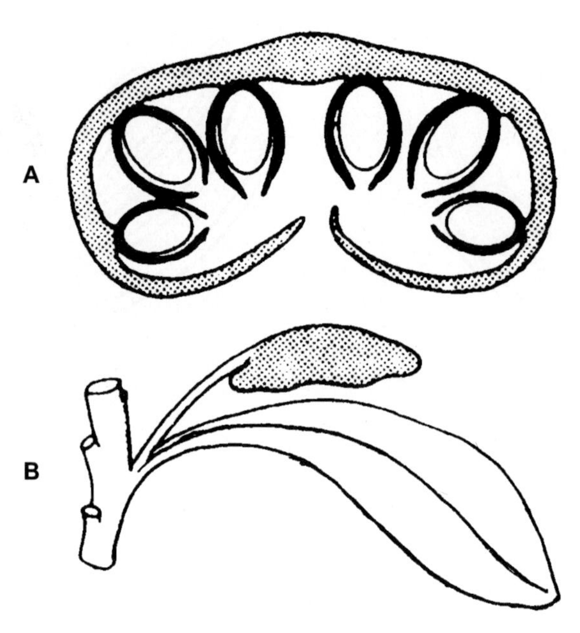

Fig. 8.4 Ovuliferous structure of a Glossopterid—*Dictyopteridium* showing the incurved margin of the megasporophyll (After Jones and Luchsinger 1987)

(c) Many species had a definite layer of cambium.

(d) Mega- and microsporophyll were often borne on the same plant. These were not aggregated into strobili but as strobili are known to occur in other groups like Bennettitales and Cycadales, it is presumed that there was a potentiality to form strobili.

(e) Sporophylls were compound, but again, there was capacity for reduction.

(f) The ovules of Lyginopteridales, or Pteridosperms, were mostly borne on the margins of the megasporophylls, and in some members on the abaxial or the adaxial surface. Either position is acceptable to consider the origin of angiosperms from this group.

(g) Absence of vessels in Pteridosperms (seed ferns) does not pose any problem because some of the primitive angiosperms are vesselless.

The fossils of seed ferns (Pteridosperms) are abundant in the Carboniferous (350 to 325 myr) deposits, but some occur as late as the Jurassic (180myr) age, when perhaps, there was the first appearance of the angiosperm fossils

4. Order **Cycadales** (class Cycadopsida) The family Dirhopalostachyaceae of this order comprises fossils from late Jurassic and early Cretaceous (180 to 135 myr). This family has also been suggested as a probable ancestor, on the basis of ovule-bearing structure. However, in this group (Cycadales), the vegetative parts exhibit sympodial branching and these are dioecious plants with well-developed male and female cones on separate plants.

(B) **Pseudanthial theory**. This theory was first proposed by Wettstein (1907) and is commonly associated with Englerian school of thought about the most primitive living angiosperm. According to this view, the taxon Amentiferae including families with highly reduced unisexual flowers borne in catkins (aments) are the most primitive angiosperms.

(I) Subdivision **Pinophytina.** Morphological features of various members of this group have been studied and there is no reason to believe any one of them to be the possible ancestor of the angiosperms:

(1) The xylem structure in members of Pinophytina is unique as it is made only of tracheids and is not comparable to that of the angiosperms.

(2) The angiosperms are phyllosporous, i.e. the carpels are modified leaves. Seeds-bearing organs of Pinophytina are variously modified but never phyllosporous.

(3) Primitive angiosperm flowers are bisexual but cones of all Pinophytina groups— Ginkgoöpsida, Cordaitopsida and Coniferopsida are unisexual.

(4) None of the members of this group has leaves with reticulate venation, which occurs in most angiosperms and also in some ferns and seed ferns.

(II) Subdivision **Gnetophytina:** This group resembles the angiosperms in many respects and at one time was seriously considered to be the ancestral form of angiosperms because of the following reasons;

(1) Both Gnetophytina members and primitive angiosperms are dicotyledonous.

(2) The Gnetophytina members have vessels in secondary xylem.

(3) The female gametophyte of this group is highly reduced.

(4) In *Gnetum*, particularly, there is no archegonium, and fertilization takes place when the female gametophyte is still in partly free-nuclear condition.

(5) The ovules of Gnetophytina resemble those of the angiosperms as they have two integuments.

(6) *Gnetum* itself, with the leaves showing reticulate venation, gives a general appearance of an angiosperm. In fact *Gnetum gnemon* can easily be mistaken for *Coffea arabica* (when not in flowers).

(7) The inflorescence of many members of the Gnetophytina resembles a catkin – an inflorescence of families like Salicaceae, Betulaceae, Fagaceae and others.

But detailed studies taken up later are not in favour of accepting Gnetophytina as an ancestral form of angiosperm, because:

(1) Vessels of Gnetophytina develop in a different way (from tracheids with circular pitting) although the ultimate structure is similar. This similarity is due to convergence and one has not developed from the other. Moreover, there are several vesselless members amongst extant angiosperms.

(2) Similarity between the inflorescence of Gnetophytina and the Amentiferae (Catkin-bearing families) is also surprising because Amentiferae is an advanced taxon with highly reduced flowers (according to Besseyan school of thought). The primitive angiospermous flower is bisexual with well-developed perianth, numerous stamens, and numerous, free (apocarpous) carpels. A Gnetophytina strobilus cannot give rise to a bisexual angiospermous flower. Tricolpate pollen grains also represent an advanced condition. This resemblance is therefore superficial.

(3) Although most angiosperms are dicotyledonous, it is also possible that they might have developed from polycotyledonous ancestors. One primitive angiosperm—genus *Degeneria* – has three or four cotyledons in most embryos.

(4) The reduction and loss of archegonium too is an interesting feature as it can be traced even in the extant genera. It has been emphasized that *Ephedra* has normal archegonia; *Welwitschia* has a specialized cell homologous with the archegonial initial which functions directly as an egg; and in *Gnetum* fertilization takes place even when the gametophyte is partly free-nuclear. The reduction

took place within the subdivision itself. The similarity with angiosperms, in this respect, is due to parallelism and shows the general tendency towards reduction of the gametophyte among the vascular plants as a whole.

Hence, Gnetophytina cannot be the probable ancestor/s of the angiosperms. But this theory has been supported by Young (1981). He emphasized that angiosperms were originally vesselless and these were lost in several lines during early stages of development. Mohammad and Sattler (1982) also presumed that angiosperms were derived from the Gnetales as they reported the presence of scalariform perforations in vessel elements of this group. However, this observation was nullified by that of Carlquist (1997), who examined large samples as compared to those of the earlier workers.

The importance of Gnetales in origin and phylogeny of angiosperms has increased further with the discovery of *Welwitschia* – like fossil from Late Triassic period in Texas, USA. Cornet (1997) described and named it as *Archaestrobilus cupulanthus*. It is regarded as a gnetophyte primitive to the extant genus *Gnetum*.

Fossil Evidences for the Origin of Angiosperms

The earliest history of flowering plant is poorly documented. Some of the sparse data from fossils have been accommodated into current phylogenetic models. Early ideas on the origin of land plants (including liverworts, hornworts, mosses and vascular plants) were based on living groups, as it was easier to study and compare. But after the discovery of exceptionally well-preseved fossil plants in the early Devonian (405myr) Rhynie Chert, researches have been taken up almost exclusively on the fossil record of vascular plants (Kenrich and Crane 1997). The discovery of fossils similar to angiosperms, or their sister groups (Bennettitaleans and Gnetaleans) can further the understanding of homologies among these groups. Two basic hypotheses have developed (from such studies) regarding the form of the ancestral angiosperm: (a) The Magnolialean hypothesis, and (b) the Herbaceous Origin hypothesis.

(a) **The Magnolialean Hypothesis:** This hypothesis suggests that the ancestral angiosperm was a woody arborescent plant with large, many–parted flower (Cronquist 1988, Takhtajan 1991, Thorne 1997) (Fig. 8.5). Based on fossil record, Doyle and Hickey (1976) and Hickey and Doyle (1977), expanded on Stebbin's (1974) concept and suggested that the first angiosperms were "shrubs of semixerophytic origin which entered mesic areas as colonizers of unstable habitats—the "weeds" of the early Cretaceous (135 myr)". The floral portion of the hypothesis finds its basis in the work of Arber and Parkin (1907). They viewed the transition of the angiosperm flower from the bisexual strobiloid reproductive organs of the Bennettitaleans. Several recent phylogenetic analyses support this hypothesis in some form or the other (Donoghue and Doyle 1989, Loconte and Stevenson 1991, Crane et al. 1995, Loconte 1997).

Recent molecular analyses of the angiosperms have often been characterized as placing *Amborella* at the base or as the first branch of the angiosperms (Fig. 8.6) *Amborella* is not itself the first branch anymore than the rest of the angiosperms minus *Amborella* (monotypic) on one side and a few hundred thousand species (the remaining angiosperms) on the other side.

In term of primitive or derived features, *Amborella* might be more advanced (with derived features) than numerous persisting species of the other larger branch of angiosperms and should not be viewed simply as 'the most primitive' angiosperm. *Amborella* might rather be considered as the 'most unique' angiosperm since it is the sole representative of one of the two basal clades (based on extant species only) of angiosperms and in that sense is equivalent of the remainder of the angiosperms. Generalization developed during the onset of the 20th century, emphasized large multi-parted flowers (like those of

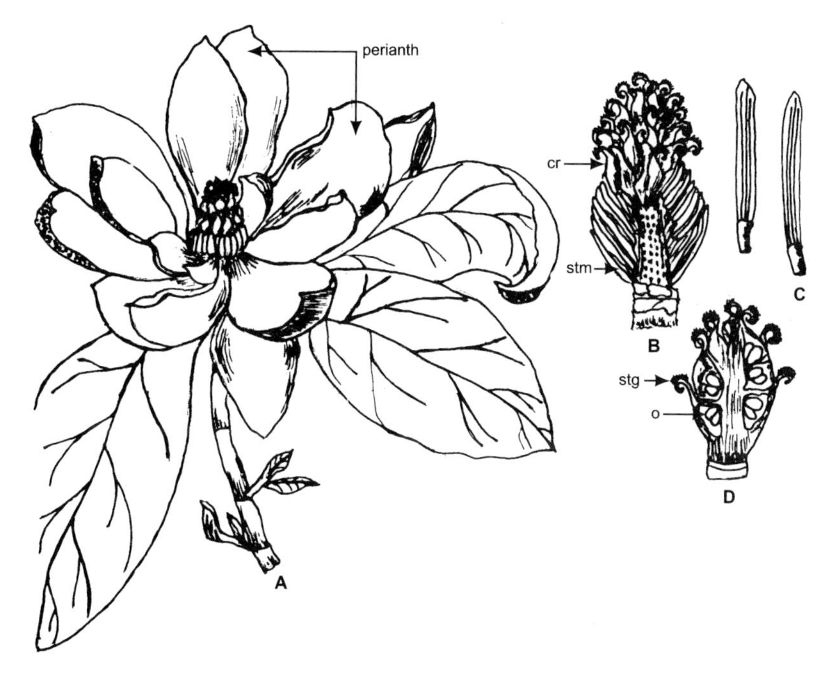

Fig. 8.5 Magnoliaceae flower structure. **A** Flowering twig **B** Androecium (half removed) and gynoecia on an elongated receptacle. **C** Stamens **D** Gynoecium in vertical section (*cr* carpel, *o* ovule, *stg* stigma, *stm* stamen).

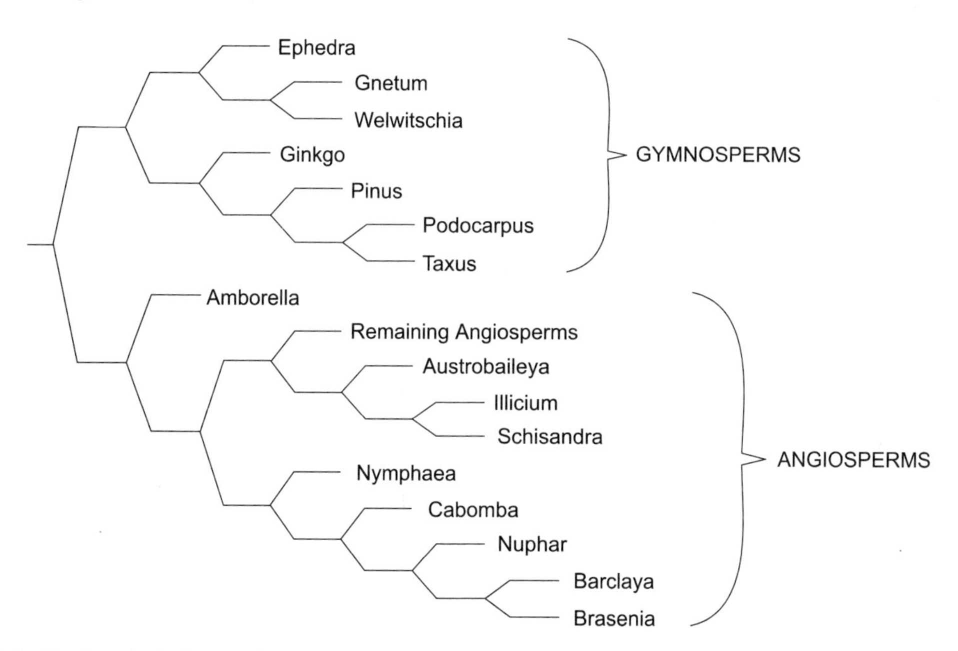

Fig. 8.6 Reduced phylogenetic tree based on a recent large- scale molecular parsimony analysis of seed plants relationship. *Amborella* is shown here as the extant sister taxon of the remaining of the angiosperms. The tree does not indicate that *Amborella* is 'the most primitive' angiosperm.

Magnolia) to be the starting point for angiosperm floral evolution (Arber and Parkin 1907, Bessey 1915). But in the subclass Magnoliidae, there are a large number of unspecialized angiosperm features (Crane et al. 1995), for example, parts generally free, and lack of differentiation within the perianth. And, at the same time they have several previously unrecognized floral features such as valvate anther dehiscence (Endress and Hufford 1989), and ascidiate carpel development. The probable basic condition of angiosperm flower is linked with the recognition of phylogenetic patterns among Magnoliidae members. However, there is an emerging agreement that too much and undue importance has been focused on extant Magnoliaceae and related families.

Herbaceous Origin Hypothesis: The alternate Herbaceous Origin Hypothesis suggests that the ancestral angiosperm was small in size and had many few-parted flowers (Burger 1981, Taylor and Hickey 1992) (Fig. 8.7). This hypothesis is similar to the Pseudanthial theory proposed by Wettstein (1907, 1935). According to this theory early angiosperm flower developed from a compound gymnosperm strobilus where a central axis is surrounded by numerous secondary axes subtended by bracts as seen in the Gnetopsids (Fig. 8.8)

Ge Sun et al. (2002) reported a new basal angiosperm family of herbaceous aquatic plants—Archaefructaceae. This family comprises two fossils, *Archaefructus liaoningensis* and *A. sinensis* sp. nov.

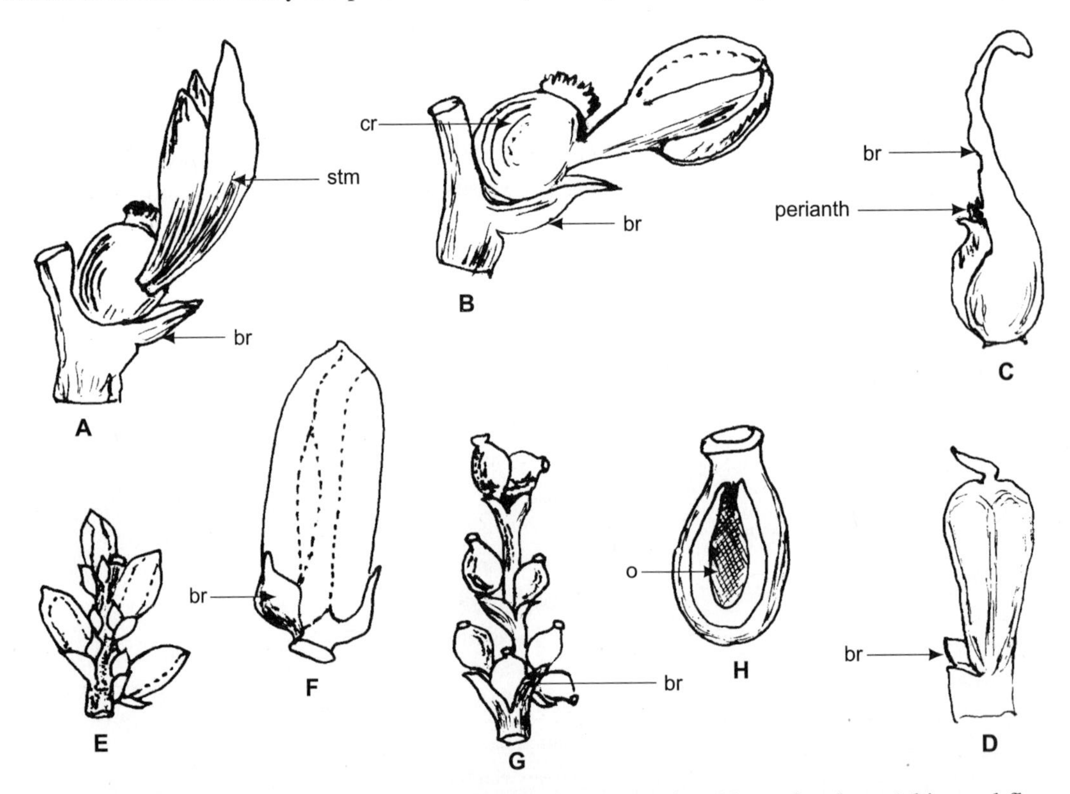

Fig. 8.7 Reduced floral structure of Chloranthaceae members. **A.** *Chloranthus henryi*, bisexual flower with a subtending bract, a trimerous stamen and a single pistil, **B** *Sarcandra glabra*, bisexual flower subtended by bract. **C, D** *Hedyosmum orientale*, pistillate **(C)** and staminate **(D)** flowers. **E–H** *Ascarina lucida*, staminate inflorescence **(E)**, staminate flower **(F)**, pistillate inflorescence **(G)** pistillate flower, vertical section **(H)** (*br* bract, *cr* carpel, *o* ovule *stm* stamen).

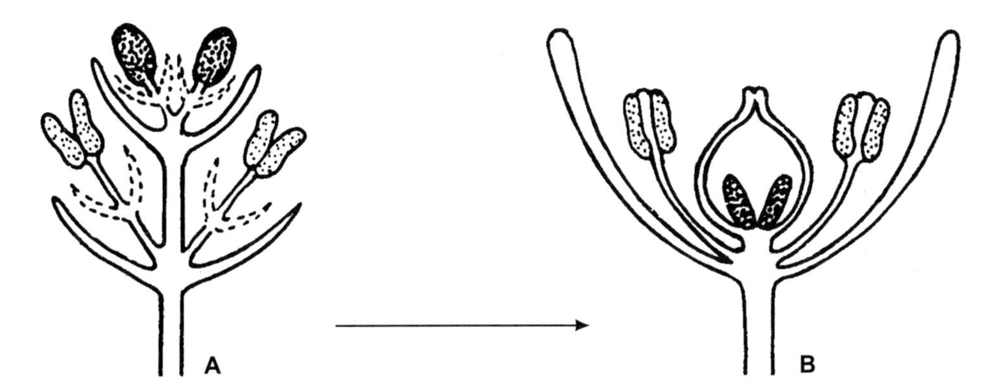

Fig. 8.8 **A, B** Compound gymnospermous strobilus **(A)** giving rise to herbaceous magnoliid flowers **(B)** (After Hickey and Taylor 1997).

Complete plants from roots to fertile shoots are known.The fossils were recovered from the lower part of the upper Jurassic/Lower Cretaceous Yixian Formation (124.6myr) in Beipiao and Lingyuan of Western Liaoning, China. *Archaefructus* spp. are a sister clade to all extant angiosperms when their characters are included in a combined three-gene molecular and morphological analysis (Fig. 8.9). *Archaefructus* spp. appear to support the pseudanthial theory, as their reproductive branches lack sepals and petals and bear stamens in pairs below the conduplicate carpels; stalks bearing paired stamens, perhaps are the remnants of an earlier branching system while the petals and sepals have not yet evolved from associated subtending leaves. Doyle and Donoghue (1993) suggest that herbaceous angiosperms are a successful but derived group. In general, herbaceous plants show higher speciation rates than woody plants (Eriksson and Bremer 1992). Higher diversity is also reported in disturbed areas and these are mostly perennial herbs (Taylor and Hickey 1997). Combined morphological (perennial, rhizomatous, herbaceous plants of small size with rapid vegetative growth) and molecular (DNA sequence data; analysis of 18s rRNA data) data indicate that taxa with trimerous, or still simpler flowers are basal in the angiosperms (Doyle et al. 1994, Les et al. 1991, Taylor and Hickey 1992).

Taylor and Hickey (1997) suggest that the ancestral angiosperms were small herbaceous plants with a rhizomatous to scrambling perennial habit. These plants had simple leaves with reticulate venation, and leaf base laterally extended to form a sheath. The vegetative anatomy included sieve-tube elements and elongated tracheary elements with both circular-bordered and scalariform pitting, and oblique end walls. Secondary growth (wherever present) was limited.The flowers were arranged in cymose or racemose inflorescences; the stamens basified with an apical extension of the connective, four microsporangia and two loculi. Pollen grains would have been small, monosulcate with perforate to reticulate sculpturing; and carpels were free with one or two proximally attached, orthotropous, bitegmic, crassinucellate ovules, and dicotyledonous embryos.

The major difference between the Herbaceous Origin Hypothesis and Pseudanthial Theory is that in the former the ancestral plant is considered to be a rhizomatous perennial herb instead of a tree.

A large number of phylogenetic analysis of morphology (Taylor and Hickey 1992) and molecular sequences (Zimmer et al. 1989, Hamby and Zimmer 1992, Chase et al. 1993, Qiu et al. 1993, Doyle et al. 1994, Sytsma and Baum 1997) support the Herbaceous Origin Hypothesis.

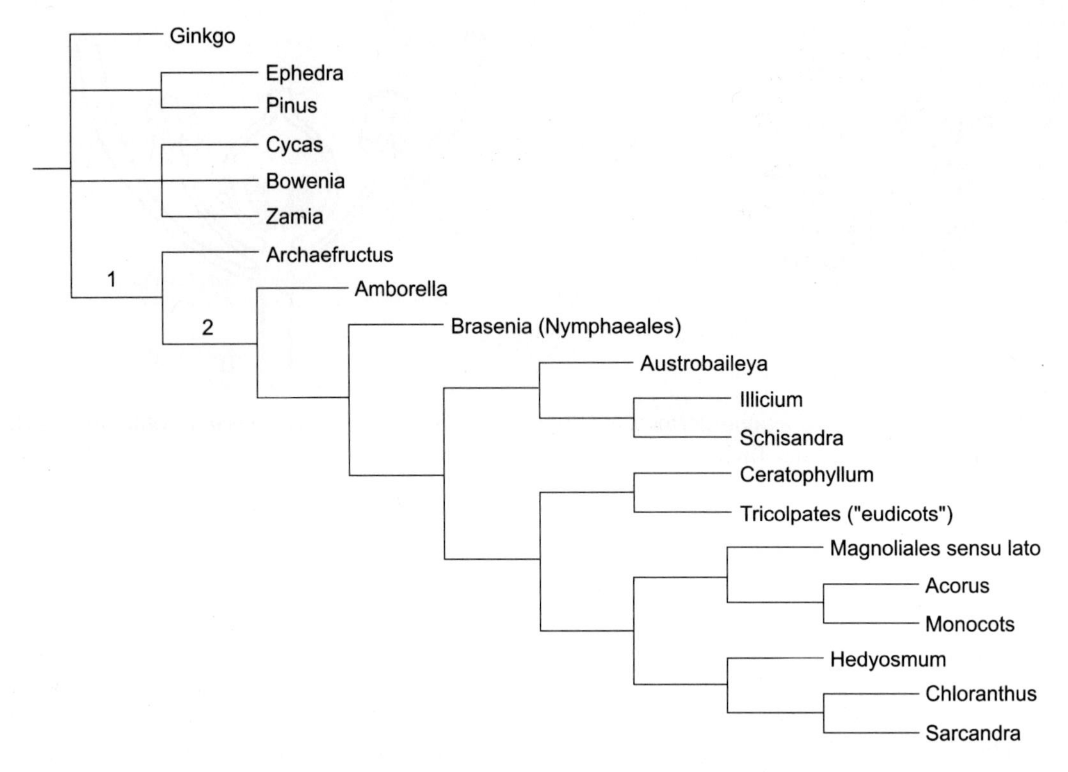

Fig. 8.9 Cladogram based on phylogenetic analysis of extant (living) seed plants and fossil *Archaefructus* where both molecular and morphological features have been used. *Archaefructus* is best considered a sister taxon to extant angiosperms.

Time of Origin. The time of origin of the angiosperms and their divergence from other seed plants are debatable. Some botanists suggest the origin of flowering plants during pre-Cretaceous period such as Jurassic (180 myr) or Triassic (225 myr) or even as early as Paleozoic (270 myr) era. According Wolfe et al. (1975), however, there is "no unequivocal evidence that indicates pre-Cretaceous origin for the angiosperms". Evidence from fossil record indicates the origin of flowering plants during the Early Cretaceous (ca 130–135 myr) period. Although, some angiosperm fossils have been reported from latter stages of Jurassic (180 myr), these were probably represented by a very small group of angiospermous plants. Pre-Cretaceous records have invariably revealed problems regarding their dating (or determination). The earliest angiosperm fossils are dispersed pollen occurring sporadically in pollen and spore assemblages of Valanginian – early Hautervian age (ca 140–135 myr). In these assemblages, angiosperms constitute less than 1% of the total spore and pollen diversity, but the rise of angiosperms to ecological dominance is dramatic and rapid. However, this sudden dominance of the angiosperms during the Upper Cretaceous (ca 90 to 100 myr) is rather surprising. Within a time period of only 10 million years, they occupied all the available spaces from the equator to the poles. Competition with the gymnosperms and ferns or oceanic and mountainous barriers did not pose any problem for their wide distribution during this period although in Lower Cretaceous they are, to some extent overshadowed by the fossils of gymnosperms and ferns.

The earliest angiosperm fossil record has been reported from the rocks of Valanginian age (141myr) of Lower Cretaceous in Israel (Brenner 1997). The monocolpate pollen is the fundamental type for the angiosperms, and is a feature inherited from the gymnospermous ancestry. These fossilized pollen show negligible morphological diversity from the typical inaperturate, reticulate, circular grains, with a tectate-columellate sexine and very small in size (Brenner 1997).

Place of Origin and Early Diversification

The origin of any taxonomic group can be determined only by a thorough search of their earliest fossil record. There are very few records of wood and leaves from the earliest phases of angiosperm history. Dispersed pollen as well as the new reproductive fossils indicate that angiosperms were diverse and at least locally abundant early in the Cretaceous; that the lack of leaves and wood may by due to lack of large trees; and that early angiosperms were perhaps smaller plants producing little wood and delicate leaves with low fossilization potential.

Earlier belief was that the angiosperms arose in the Arctic region and subsequently migrated southwards. AC Smith (1970) located the general area of South-east Asia, adjacent to Malaysia as the site where angiosperms evolved when Gondwanaland and Laurasia were undergoing initial separation. The first Cretaceous flora with abundant angiosperm reproductive organs was discovered in 1979 in southern Sweden. The fossils occurred in late Cretaceous (Santonian/Campanian, about 80 myr old) sediments and could be extracted from the sediments by sieving in water. They are small, charcoalified flowers, fruits and seeds, often with their three-dimensional form retained. Following this discovery, intensive search for similar sediments and fossils resulted in finding of numerous floras form different parts of the world. Most significant of these are a number of floras from the Early Cretaceous of Portugal and North America. The North American floras are slightly younger and embrace a longer time interval. Doyle and Hickey (1976) reported that the earliest angiosperm fossils from U.S. Barremian/Aptian (132/125 myr old) localities appeared near streams. According to Retallack and Dilcher (1981a), Western U.S. Cenomanian (97 myr) deposit support a fluvial[2] or coastal paleoecology. Available sedimentological data indicate that early angiosperms were small and rapidly growing to survive on unstable but nutrient-rich area. These sites were characterized by frequent loss of plant cover due to periodic disturbances.

Tilmann (1988) notes that plants which grow on "nutrient-rich " and "high plant-loss" sites have the following features: low total biomass, light penetration through plant body high, small size at maturity, reproduction at young age, low allocation to stems, and high allocation to leaves.

According to Stebbins (1974), the expected features of the plant groups that grow on "unstable environment" are certain physiological specializations like dormancy mechanism, rapid vegetative growth during favorable period, specialized resting organs and dispersal methods, and reduction in size. According to Taylor and Hickey (1997), this "unstable environment" of Stebbins (1974) can be compared with "climatically stable but periodically disturbed " environment, common in floodplains.

It has also been observed that perennial herbs are well-adapted to grow in unstable environments. Studies of Harper (1977) show that such plants with their typical characters (flexible vegetative and reproductive systems) can exploit a variety of habitats. Occurrence of both rhizomes and small seeds allows them to occupy and colonize distant and favorable sites. Rhizomes also help them in addition, to survive against herbivorous animals and frequent inundation. As the same plant species occupy different niches, its life history can vary from simple colonizing herbs to tall shrubby herbs competing with true shrubs or the perennial herbaceous habit.

---

[2]Fluvial – Areas that are frequently inundated.

Hence, it is suggested that the early angiosperms originated in the unstable environment of the coastal plains and could survive against physical disturbances as well as herbivorous animals including dinosaurs (Bakkar 1978, 1986). Although the angiosperms became dominant in some niches by mid-Cretaceous (110 myr) overall dominance as trees was reached almost towards the end of Cretaceous period (65 myr).

The presence of inaperturate pollen grains from Upper Valanginian (130 myr) to Lower Hauterivian (127 myr) sediments from coastal plains of Israel provides additional data for possible geographic locus of early angiosperm diversification. Brenner (1976) suggested the Northern Gondwana Province could have been the site of the origin of earliest angiosperms. New center of diversification arose when the basal angiospermous taxa with typical monosulcate pollen grains spread to other areas from this original site. Brenner (1997) further suggests that the ancestral angiosperms occurred first in the early Cretaceous (146-97myr) of the Eastern Gondwana Province (India–Antarctica–Australia).

The inaperturate pollen (as described for the ancient angiosperms) have some resemblance to those of certain extant (or living) members of the Piperales. These resemblances therefore support the viewpoint of Taylor and Hickey (1992, 1997) that the angiosperms have an herbaceous origin.

It is also reported that some species of *Gnetum* have pollen grains similar to those of the ancient (extinct) angiosperms. This similarity suggests that some of the present day (extant) species of *Gnetum* are the remnant of a gymnospermous clade – the sister group of the angiosperms.

It is presumed that the angiosperms originated in the tropics, especially when it is supported by the fact that many of the archaic families such as Hernandiaceae, Lactoridaceae, Chloranthaceae, Austrobaileyaceae and others are distributed in these localities. Because of the predominant occurrence of many Ranalean families in the southwestern Pacific and southeastern Asia, Bailey (1949), Axelrod (1952,1960), Takhtajan (1969) and others have suggested this area to be the probable place of origin of angiosperms. The pollen and spore analysis of different zones in Israel also gives the impression that the early angiosperms dominated a warm and humid climate (Brenner 1997).

Early angiosperms followed different ecological pathways during the initial diversification. Their initial competitors were ferns and sphenopsids — and together all these plants thrived near wet areas — in a fluvial ecosystem. All of them were perennial, rhizomatous, and herbaceous plants. The angiosperms of this period were growing rapidly but were small-sized at maturity. With their highly reduced and efficient reproductive system and better dispersal agents (seeds vs. spores), they were in an advantageous position to establish themselves on unstable or ephemeral sites like the fluvial ecosystem. They also occupied the levées[3], competing with other seed-bearing herbaceous, perennial plants. From here the angiosperms diversified and radiated to occupy various other niches like the aquatic ecosystems, margins of backswamps, and floodplain terraces.

The angiosperms with their newly acquired woody habit and ability to grow at a faster rate as seedlings (when compared with other tree members like the Bennettitaleans, *Caytonia* and other conifers) could easily compete with other members growing in these habitats. As reported by Rettalack and Dilcher (1981a) and Crane (1987), many of the probable competitors of flowering plants became extinct during the mid-Cretaceous period (97myr) and the angiosperms became more successful.

The Ancestral Angiosperms. To determine which angiospermous group are the least specialized, one must assume the features that characterized the ancestors of extant angiosperms. Reconstruction of the evolutionary pathway of early angiosperms is not yet complete. A poor fossil record and an apparent wide

[3]Levées – Natural embankment of alluvium built up by river on either side of the channel.

evolutionary gap between angiosperms and other seed plants have impeded progress. However, the presence of certain "living fossils[4] " (the non- missing links) like the relict *Saruma henryi* with numerous primitive (or unspecialized) features makes the situation more satisfactory. Studies carried out on some of these phylogenetic relicts (such as the families Winteraceae, Illiciaceae, Austrobaileyaceae, Hernandiaceae and others) bring out the probable ancestral characteristics and the evolutionary trends in many angiosperm organs and tissues (stamens, pollen grains, carpels, ovule, embryo sac characters as well as vegetative growth and habit, leaf character, and xylem and phloem characters). In addition, comparative studies of gymnosperms and ferns, and paleobotanical studies on fossils have added new data to our knowledge of the ancestral angiosperms. Even clasdistic[5] and molecular[6] investigations (Chase et al. 1993, Qiu et al. 1993, Sytsma and Baum 1997) have helped in drawing conclusions. The plesiomorphic (primitive) characters are most invaluable to get an idea about the ancestral angiosperms because these (primitive) features form the link amongst the least-specialized angiosperms, and also with their extinct ancestors. Thorne (1997) attempts to suggest these ancestral features, and trends of specialization amongst the primitive angiosperms. The four major groups of least specialized angiosperms are: Maglonianae, Nymphaeanae, Rosanae, and Lilianae. The primitive features characteristic of the archaic members of these four groups are:

(a) Woody habit, or perennial herbaceous habit.

(b) Simple, alternate, pinnately-veined, estipulate leaves; often with sheathing leaf base (e.g. Ranunculaceae).

(c) Vesselless or otherwise primitive xylem (long cambial initials and tracheids, bordered tracheary pittings, heterogamous rays and diffuse to tangentially-banded wood parenchyma as in *Drimys winteri).*

(d) Small, actinomorphic, bisexual flowers, often epigynous. Epigyny has existed for at least 130 million years but in phylogenetic terms has still not outnumbered hypogyny (Gustafsson and Albert 1999). Friis et al. (1994) described several epigynous flowers from the early Cretaceous of Portugal, some of which resemble extant members of the families Hernandiaceae and Chloranthaceae

(e) Perianth many-parted, undifferentiated, arranged spirally (absent in Chloranthaceae).

(f) Stamens basifixed, broad, unmodified into filament, anther and connective; microsporangia marginal.

(g) Pollen grains monosulcate, exineless.

(h) Carpels styleless with 'unsealed ' stigmatic crest as in Winteraceae.

(i) Ovules anatropous on marginal or submarginal placenta.

(j) Folliculate fruits with many seeds (Crane and Dilcher 1984).

(k) Seeds endospermous, with a rudimentary embryo (Martin 1946).

Recent developments in molecular and morphological based systematics and powerful programs for analysing large data sets in a phylogenetic context have greatly improved the potential for more stable results (Fig. 8.10).

---

[4]Living fossil – Present day members with all its ancient characters retained.

[5]Cladistic – A type of analysis of characters that attempts to summarise knowledge about the similarities among organisms in terms of a branching diagram called a Cladogram.

[6]Molecular data – Nuclear rRNA sequencing and chloroplast gene rbcL data.

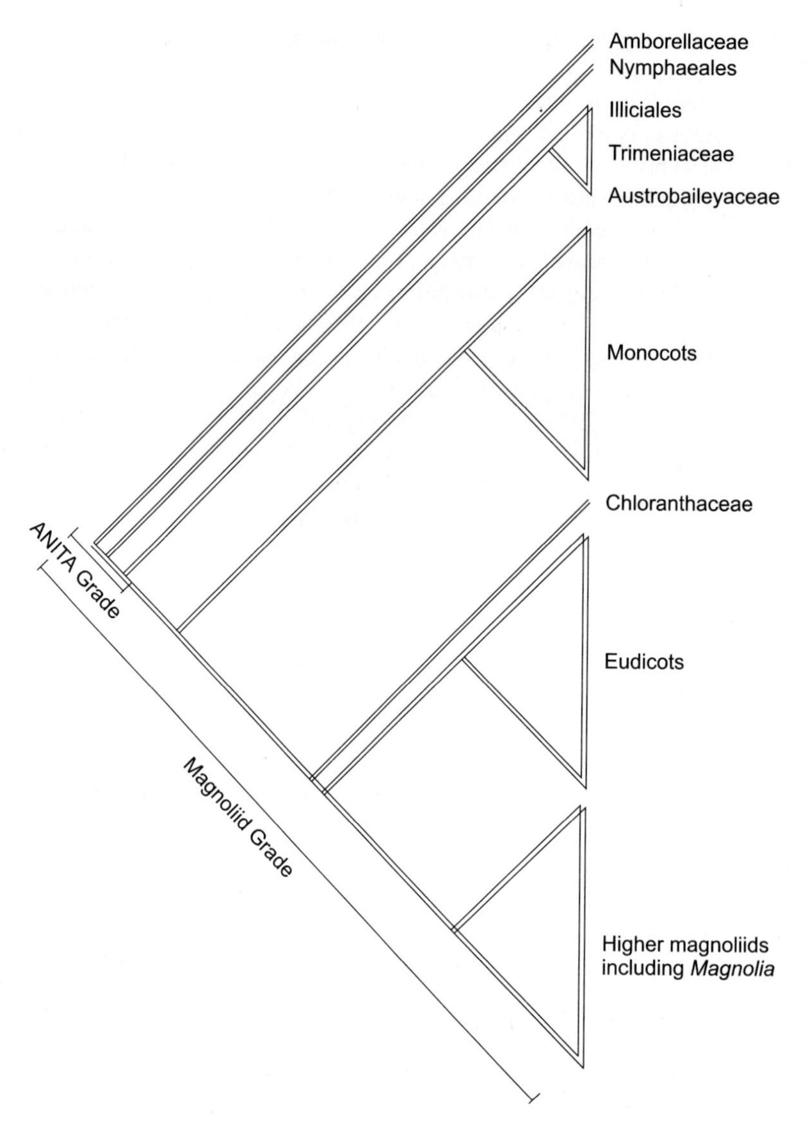

Fig. 8.10 Hypothetical diagram to depict the relationships among angiosperms based on molecular analyses.

Furthermore, the discovery of three-dimensionally preserved flowers, fruits, seeds and stamens from Cretaceous period have provided a new source for studying reproductive structures in early angiosperms.

The new phylogenetic analyses no longer support the traditional division of angiosperms into two distinct lineages, dicots and monocots. Instead, two major monophyletic groups—higher dicots (or eudicots) and monocots are now recognized. These are embedded in a grade of basal dicots referred to as the magnoliids. Eudicots are distinguished from the magnoliids and monocots by their pollen, which is three-aperturate or an aperture configuration derived from this basic type. Most magnoliids and monocots have pollen with a single aperture. The new analysis also split the paleoherbs — a group of herbaceous magnoliids and monocots – which according to earlier analyses were resolved as a monophyletic clade.

In many earlier classifications, *Magnolia* was a basal taxon, and its large, many-parted flowers (Fig. 8.5) were presumed to be primitive amongst the angiosperms. The fossil record, however, indicates that the ancestral flowers might have been smaller and simpler (Fig. 8.7). This is in line with the new phylogenetic analyses that identify a basal grade of magnoliids referred to as ANITA (from its constituents Amborellaceae, Nymphaeales, Illiciales, Trimeniaceae and Austrobaileyaceae), which includes several taxa with small simple flowers (Fig. 8.10).

Conclusions: When, where and how did the angiosperms originate is an "abominable mystery". The early Cretaceous floras from Portugal are extremely rich in angiosperm fossils. Currently about 140-150 taxa of flowers, fruits and seeds have been identified. A survey of pollen in flowers, in dispersed stamens, and on fruit surfaces shows that about 85% of pollen taxa have a single aperture indicating affinity to magnoliids and perhaps some basal monocots. Only 15% of the taxa have equatorial aperture characteristic of eudicots. The ratio of basal angiosperms to eudicots strongly contrasts with that of Late Cretaceous and Tertiary floras, where eudicots are clearly dominants and basal angiosperms typically constitute less than 5% of the taxa. Even in the present floras basal angiosperms constitute a minor fraction of the angiosperm diversity.

Most of the Early Cretaceous angiosperms from Portugal do not belong to any existing family or order and many of them are clearly of extinct lineages. The Chloranthaceae are currently the only extant family that has been documented among these fossils. This family is presumed to be close to the base of the angiosperm tree, and its early occurrence in angiosperm history is also supported by its pollen type. Other fossil taxa from Portugal have close similarity to extant members of the ANITA grade, particularly to the Amborellaceae, Nymphaeales and Illiciales. They, however, exhibit certain characters unknown in the extant taxa. There are also fossils exhibiting monocot affinity as well as some others showing a mixture of monocot and magnoliid features. But, so far, no extant monocot taxon has been identified from the Early Cretaceous. Eudicot fossils from this period (in Portugal) are all of simple structure and organization, and share features with taxa resolved at the base of eudicots.

In all Early Cretaceous angiosperms the reproductive organs are very small, ranging in length up to about 0.4mm. Flowers are uni- or bisexual, often with a few parts, simple, naked or with a simple perianth. The androecium often constitutes the most prominent part. Stamens usually have a massive sterile tissue—the connective and minute pollen sacs; filaments usually poorly developed; dehiscence by laterally hinged valves that open like doors or by longitudinal slits. The gynoecium consists of one to many free carpels, the number of seeds per carpel is always low and many fruiting structures are one-seeded.

Qiu et al. (1999) reported a phylogenetic analysis of DNA sequences of five mitochondrial plastids, and nuclear genes from all basal angiosperm and gymnosperm lineages. This study, too, demonstrated that *Amborella*, Nymphaeales and Illiciales — Trimeniaceae—*Austribaileya* represent the first stage of angiosperm evolution with *Amborellla* being sister to all other angiosperms. *Amborella* is a shrub of the monotypic New Caledonian family Amborellaceae.

Subsequent Evolution. Mid-Cretaceous angiosperm fossils (about 115-95 myr) exhibit an increased complexity in pollen features, leaf architecture and floral structure. Eudicot angiosperms became more frequent and the number of taxa assignable to modern taxa is higher both among eudicots as well as magnoliids. Magnoliids with simple but large floral structures are still predominant.

In the Late Cretaceous period, angiosperms passed through another major radiation phase. During this period eudicots expanded dramatically, and more specialized flowers with well-differentiated calyx and corolla, floral tubes and nectaries appeared for the first time. By the end of Cretaceous, all major eudicot groups were established and also several monocot lineages. Another major radiation took place in the Tertiary when more specialized groups such as the grasses and orchids proliferated

The angiosperms are basically tropical, i.e. grow in large numbers in the warm and humid regions of the world. Even today, these archaic groups with innumerable primitive (or least specialized) features are restricted to the tropical montane, i.e. summer-wet and warm-temperate forests. It is presumed, therefore, that the fossil remains of angiosperms would be available in areas, which were tropical or subtropical during earliest Cretaceous (140 myr) or late Jurassic (146 myr) time (Thorne 1997). A point to be remembered (here) is, that the equatorial zone could have shifted during the geological past, and it is not certain that the present-day tropics represent the tropics of the Cretaceous period. Moreover, the climate all over the earth during Cretaceous period was probably less varied than it is today (Beck 1976).

Many archaic living angiosperms with vesselless wood or with unspecialized vessels are mostly restricted to highly mesic sites such as the banks of streams, ponds, and rivers that are with minimal seasonal water stress (Carlquist 1975).

As plants in later stages had more highly evolved xylem with more efficient conductivity, they moved from mesic highlands to hot tropical lowlands with fluctuating soil moisture. With further evolution, these early (or ancient) angiosperms occupied areas with more stressful climates. As indicated by fossils, these plants adapted quite early to extreme habitats by undergoing various changes in their growth forms.

Comparative studies (of different groups of angiosperms) in anatomy, embryology, palynology, paleobotany, phytochemistry and molecular taxonomy, point towards unidirectional trends in specialization that is seldom reversible. Data from all these fields coupled with fossil evidences and those of relict flowering plants enable one to recognise the probable characteristics of ancestral angiosperms (Taylor and Hickey 1990, 1992,1997, Thorne 1976, 1992, Wolfe et al. 1976).

It is also heartening to note that although the ancestral features and trends of specialization amongst angiosperms were developed much earlier and, in fact, before the advent of "Molecular approach" to angiosperm phylogeny, the rbcL sequence variation (Chase et al. 1993, Qiu et al. 1993) and evidences from mitochondrial, plastid and nuclear genomes (Y-L Qiu 1999) support the present evolution and phylogeny of the angiosperms.

9

Theory of Evolution

Evolution is usually considered a series of processes involving descent of populations with modification marked by successive adaptations to environmental conditions, governed by competition and natural selection acting on variation. Within each population, there is always variation between individuals. The theory of natural selection asserts that the contribution of offspring to the next generation is not entirely at random but is correlated with this variability.

Phylogeny is origin and evolution of taxa or the study of genealogy (the study of ancestral relationship) and evolutionary history in a group of organisms. Relationship between different taxa can be depicted diagrammatically in the form of a **phylogram,** which is based on the level of advancement of the descendents.

Phylogenetic analysis or **cladistics** is yet another method for classifying organisms. The approach focuses on the branching of one lineage from another in the cause of evolution and identification of monophyletic groups or **clades** that possess ancestral character-states. The result of cladistic analysis is a **cladogram**—it provides a graphical representation of a working model of branching sequences.

All existing groups of organisms have evolved from the progenitors of the past. Logically, therefore, these ancestral groups might have evolved along certain definite pathways to give rise to the present day forms. According to their similarity pattern they fall into distinctive groups and the pathways of evolutionary changes among organisms are known as 'evolutionary trends'

Stebbins (1974) suggested different methods by which evolutionary trends can be deduced. The two important lines of evidences are (1) from fossils and (2) synthesis and interpretation of different lines of evidence from the extant or living groups. Fossil evidences are too meager to throw light on the intricate processes of evolutionary divergence. At the same time, data from various fields (other than morphological) for the majority of extant plants is insufficient. The working taxonomists therefore, depend mainly on comparative morphology of the existing (extant) taxa, to discern the evolutionary trend of the past. The process has been termed—"backward systematics" by Hughes (1976).

There are four evolutionary concepts and they are:

1. Primtive and Advanced forms
2. Homology and Analogy
3. Parallelism and Convergence
4. Monophyly and Polyphyly

Primitive (plesiomorphic) and advanced (apomorphic) forms

Determination of the primitive or plesiomorphic and advanced or apomorphic characters, is the first step in constructing phylogeny when gradational series of fossils is wanting (Van Valen 1978). Bessey (1915) enumerated many such trends that would determine the primitive or advanced condition of various taxa. Several taxonomists of a later date, like Cronquist, Dahlgren, Hutchinson, Takhtajan and Thorne are of the same Besseyan outlook though their classification systems vary to a large extent. Various methods have been suggested for the estimation of 'character polarity' i.e. determination of primitive and advanced features. Amongst these the only satisfactory one is that, "the oldest character-state is the most primitive".

However, as the information on true phylogeny is lacking, determination of the oldest character is also not possible (Crisci and Stuessy 1980). This, therefore, turns out to be that, "the earlier a character-state appears, the more primitive it is likely to be. And on the other hand, those character-states, which appear in more recent taxa, are the advanced ones."

According to Willis (1922) the character-states shared by the largest number of taxa in a group is likely to be more primitive in the group. Wagner (1961) and Taylor and Campbell (1969) held the same viewpoint. If in a group of 100 species of the same genus, 95 have character-state A, and the rest five have character-state B, it is likely that the ancestral taxon also possessed A, and hence, it is a primitive character-state. However, the result may be disputable depending upon whether we consider only the living members or both living as well as extinct members.

Stebbins (1974) modified Willis' concept to say that 'primitive' characters are the ones which are "most common in the members of the group that first achieved widespread success" because these taxa would display the ancestral (primitive) characters better than the modern forms which have diverged considerably from the ancestral ones.

In classification of flowering plants the floral characters have been given more importance than vegetative characters since long. It has been assumed that floral parts are less affected by environmental changes and therefore exhibit more primitive features, i.e. they are more conservative. Although the floral features are subject to lesser degree of evolutionary pressure, they are affected by strong selection pressure because of various aspects of reproductive biology such as pollinator relation (Clifford and Lavarack 1974). It is thus observed that vegetative and floral (or reproductive) characters are affected by different types of selection pressure to which they react in different ways. It is useless, therefore, to look for same type and degree of constancy and conservation in vegetative and reproductive characters of flowering plants.

To add to the complications, it has been noticed that amongst the flowering plants different organs of different taxa and often of the same taxon, do not evolve at the same rate. During the stages in their evolutionary history, the rate of evolution of a lineage may vary widely. For example, the genus *Tetracentron* (family Tetracentraceae) has advanced tricolpate pollen grains but primitive vesselless wood. The genus *Dicentra* (family Papaveraceae) has advanced zygomorphic flowers and primitive rootstock. In the taxon Amentiferae, highly specialized (reduced) flowers are associated with primitive tree-like habits. Order Alismatales includes members with primitive features like apocarpous ovary, conduplicate carpels and lateral- laminar placentation but advanced feature like non-endospermous seeds. Such an occurrence of both primitive (plesiomorphic) as well advanced (apomorphic) features in a taxon, at the same time, is know as **heterobathomy**. The phenomenon of unequal rate of evolution of different features within one lineage has been termed as "Mosaic evolution" (De Beer 1954). An organism might have a mosaic combination of various characters at different evolutionary levels due to heterobathomy. Because of strong expression of heterobathomy in certain taxa, taxonomic information provided by different sets of characters turns out to be contradictory. In such instances one has to look for reliable phyletic markers.

However, the concept of primitiveness and advanceness are only relative, and carry no meaning unless they are studied with particular reference to a group or taxon.

Due to extensive gaps in the knowledge of evolutionary history of different groups of organisms, the only positive way to ascertain primitive members of any taxonomic group is by their possession of a relatively large number of **primitive** or **plesiomorphic** characters and advanced members by comparatively few primitive characters, others being lost or replaced by **apomorphic (advanced)** characters. It has been observed that certain morphological characters are statistically correlated.

Sporne (1974) proposed a list of 24 features in Dicots (magnoloid) and 14 features in Monocots (amarylloid), which show positive correlation. Based on the distribution of these character- states **advancement index** for each family of the angiosperms can be calculated. These values when represented in the form of a circular diagram, the most primitive families are placed near the center and the most advanced ones along the periphery (Singh 2000). All living families are advanced in some respect at least.

Here are some examples where the primitive and advanced taxa have been distinguished from each other by way of comparison. When a taxon exhibits certain adaptation or specialization because of which it has restricted distribution and/or reproduction, it is an advanced one. If *Acacia auriculiformis* and *A. nilotica* are compared, the former is an advanced taxon due to the presence of phyllodes and its ability to live under xeric condition.

A taxon with multiple floral parts, e.g. *Magnolia* (Magnoliaceae) is comparatively primitive to *Salix* (Salicaceae) with reduced floral parts. *Magnolia* flowers with several spirally arranged perianth lobes are insect-pollinated—a primitive feature whereas highly reduced flowers of *Salix* are wind-pollinated—an advanced feature.

An organism with free floral parts is primitive and one with connate or fused parts is more advanced. *Ranunculus* with free petals is primitive to *Delphinium* with united petals.

Organs of angiospermous plants evolve in four different lines that are independent of each other: (a) Underground parts, (b) Aerial parts, (c) Flowers and (d) Fruits. The different evolutionary trends in gross morphology of angiosperms have evolved as a result of the following processes.

1. Reduction in structure, e.g. Amentiferae.
2. Fusion of structure, e.g. Syncarpous condition from apocarpous condition.
3. Change in symmetry e.g. Zygomorphy from actinomorphy.
4. Elaboration—Development and differentiation of vascular tissues.

A primitive or advanced group may be ancient or recent depending upon the time when the taxon got separated from the main evolutionary line. In figure 9.1 AA□ is the main evolutionary line and the taxon A□ is extinct. Each time a new branch arose from this primitive stock (indicated by an angle) it is assumed that advancement occurred. The taxa B and K are equally advanced but B is more recent (or younger) than K. L is more advanced than many other taxa in this group but yet ancient than B, C, and D. Both G, and H are primitive because they have diverged only once from the ancient stock; yet G is more ancient and H more recent as they got separated from the main evolutionary line at different time periods.

Very often ecological conditions have speeded up evolution, e.g. xerophytic conditions. Stebbins (1974) pointed out that numerous structures in plants have evolved as a result of adaptation towards xeric conditions. Annual habit, bulbous or rhizomatous rootstock, succulent habit, presence of hairs on leaves, formation of phyllodes and phylloclades are some such adaptations. The families Cactaceae, Casuarinaceae and Zygophyllaceae have evolved in response to xerophytic conditions.

All taxonomists who believe in evolution also believe in the existence of various evolutionary trends in nature. However, as there is no fossil evidence to support these trends, it is difficult to understand the direction of the trend. The problem is further accentuated, as there is frequent occurrence of **reversal of traits**. For example, the new-world *Cleome* species have perennial habit and old world *Cleome* species show annual habit. Phylogenetic studies have shown that the New World Cleomes have evolved from Old World Cleomes although perennial habit is primitive and annual habit is advanced. This is reversal of trait. Flowers of family Euphorbiaceae are unisexual and mostly wind-pollinated. Genus *Euphorbia* is the most highly evolved and advanced taxon but show insect pollination—a primitive feature. Insect pollination is also seen in the ancestral taxon Geraniales. This is also reversal of trait as wind pollination is advanced as compared to insect pollination.

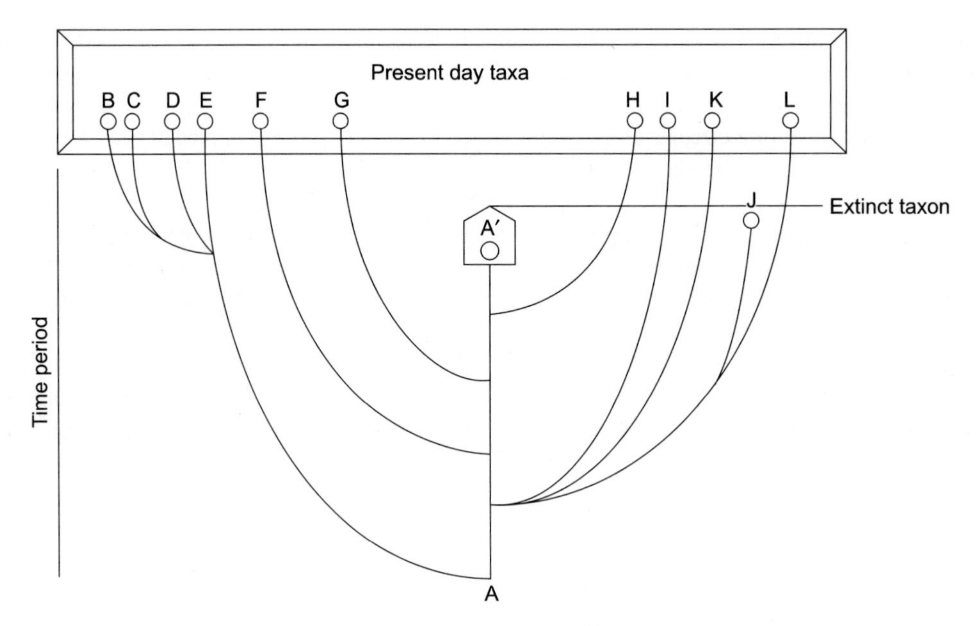

Fig. 9.1 Phylogenetic tree: advanced and primitive taxa. AA′ is the main evolutionary line of which A′ is an extinct taxon. Each time a new branch arose from this primitive stalk, it is assumed that an advancement occurred. **B–L** are present day taxa (except J which is an extinct taxon).

All parts of organisms at all stages of their development may produce evidence of evolutionary significance. Evolutionary trends progressive or retrogressive are generally consistent but occasionally reversible with changes in environmental conditions.

Homology and Analogy

Different plants resemble each other in certain characters. Various taxa of different ranks are constructed on the basis of overall resemblances, which may be due to **homology** or **analogy.**

Owen (1840) defined **homologous** organs as 'same organ' in different taxa with a variety of form and function, e.g. stem of *Solanum tuberosum*, potato (tuber), *Amorphophalus campanulatus*, elephant foot yam (corm), *Duranta repens*, golden dew drops (spine), cacti (phylloclade), and *Gloriosa superba*, glory lily (aerial climbing stem; these are all stems in different taxa but their form and function are different.

Analogous organs are part of a taxon, which have same function as some other part of another taxon, e.g. stem tuber of *Solanum tuberosum* and root tuber of *Ipomoea batatas*—both storage organs. Darwin (1859) used these two terms with reference to both animals and plants According to him homologous organs refer to organs similar in evolutionary and developmental origin regardless of function, such as wing of bird and forearm of mammals. Superficially similar structures that are not the result of common ancestry are analogous organs, e.g. phyllode of *Acacia auriculiformis* and cladode of *Ruscus*.

An organ or structure is homologous with another because of "what it is" and not "what it does".

If floral parts of different plants are considered, they are homologous. But 'petals' of *Anemone* are not homologous with those of *Ranunculus*. They are homologous with the sepals of *Ranunculus* as true petals are absent in *Anemone*.

Simpson (1961) defined homology as "the resemblance due to inheritance from a common ancestry." Analogy, likewise, exhibits "functional similarity and not due to inheritance from a common ancestry." Mayr (1969) held more or less similar viewpoint. According to him homology is "occurrence of similar features in two or more organisms which can be traced back to the same feature in the common ancestor of these organism". Analogy, on the other hand, is "the occurrence of similar features in different organisms which cannot be traced back to be common ancestor if any".

Wiley (1981) has given yet another interpretation of these two terms. Homology may be sought after between two characters or between two organisms, for a particular character. "Two characters are homologous if one is directly derived from the other". A series of character like this is termed as evolutionary **transformation series** or **morphoclines** or **phenoclines.** In this series the character in the beginning is termed **plesiomorphic,** i.e. primitive and the characters at later stages of the series are **apomorphic,** i.e. advanced; these are derived characters.

Three or more characters may be homologous if they belong to the same evolutionary transformation series (aestivation of sepals or petals—twisted ? imbricate ? valvate, or ovary ? superior ? half inferior inferior) As shown in Fig. 9.2 the two terms plesiomorphic and apomorphic are relative. In the evolutionary transformation series no. II, character B is apomorphic in relation to character A but plesiomorphic in relation to character C. The possession of plesiomorphic character-states in common by a group of taxa is known as **symplesiomorphy**, and possession of derived character-states in common is **synapomorphy**.

In Fig. 9.3, diagram I shows ancestral taxon X giving rise to three taxa A, B and C. Similarity between B and C is due to synapomorphy. In diagram II the similarity between A and B is due to symplesiomorphy.

Homologous features have a common origin whereas analogous features have a common function but different evolutionary origin. Superficial similarity is not a useful criterion for taxonomic decisions. Similarity of a certain feature may reflect a common ancestral inheritance or adaptation towards similar environmental conditions.

In practice, a few criteria may be useful to identify homologous and analogous characters:

(i) Similarity in position, e.g. a branch will be positioned always in the axil of a leaf no matter what its modification may be;

(ii) Similar ontogeny; and

(iii) Similarity in anatomical and histological characters, e.g. evolution of vessels from tracheids. Primitive vessels appear more like tracheids with their elongated form, and oblique end walls.

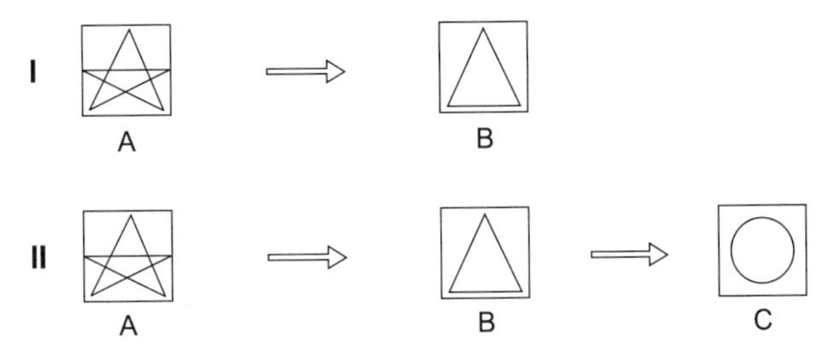

Fig. 9.2. Homology amongst characters: In I character **A** is plesiomorphic (primitive) and **B** is apomorphic (advanced). In II **A** is plesiomorphic, **B** is apomorphic when compared to **A** and plesiomorphic when compared to **C**—as **A** ▯ **B** ▯ **C** belong to the same evolutionary transformation series.

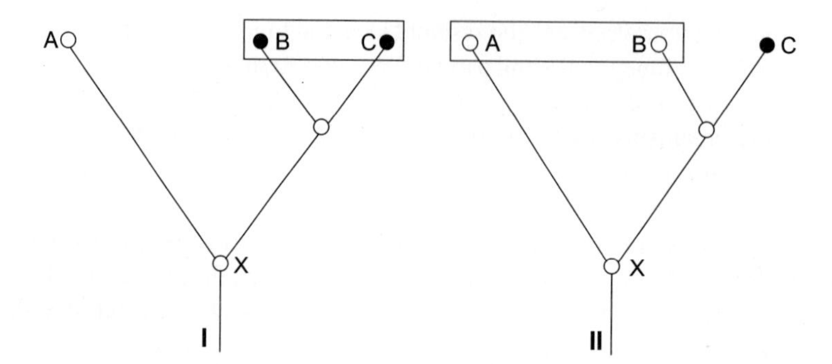

Fig. 9.3 Diagram I shows ancestral taxon giving rise to three taxa **A, B** and **C;** similarity between **B** and **C** is due to synapomorphy (both are with advanced features); in diagram II similarity between **A** and **B** is due to symplesiomorphy. [complete circles represent advanced features; hollow circles represent primitive features]

Although, there are limitations, these criteria have been useful particularly with regard to morphological characters.

In chemosystematic studies, however, not much importance had been attached to homology and analogy until recently. This is mainly because of the time and difficulties involved in establishing these facts. Taxonomists usually concern themselves with the presence or absence of certain chemicals, irrespective of the possibility of different biosynthetic pathways resulting in the same end product. Presence of the chemical betalains in various members of Centrospermae, such as Amaranthaceae, Basellaceae, Chenopodiaceae, Nyctaginaceae and others is due to homology. All these families originated from a common ancestral family Phytolaccaceae. Another chemical—amentoflavone occurs in Pinaceae and Casuarinaceae. As these families do not have a common ancestor, this similarity is due to analogy.

An interesting example can be that of Cyanogens. Hegnauer (1977) reported that if adequately and carefully used, cyanogenesis might prove to be of considerable value for a natural classification of plants. More than two thousand species of plants so far discovered contain cyanogens. There are at least five different groups of cyanogenic glycosides that are the precursors of prussic acid or hydrocyanic acid. The major groups of vascular plants are characterized by their own type of biogenesis of these compounds: pteridophytes have phenylalanine; gymnosperms, tyrosine; and angiosperms valine, leucine and cyclopentenyl glycene besides the former two as the precursors. Cyanogenesis can be used as a systematic character at lower levels also but the fact is that the end product of biogenesis—prussic acid—in itself is no indicator of relationship. It is the chemical characterization of the cyanogenic constituents, attributable to distinct biosynthetic pathways which is much more useful in plant systematics than cyanogenesis itself. This will, therefore, be known as homology of biosynthetic pathways (Sivarajan 1984).

Parallelism and Convergence

The concept of natural relationship between various taxa is based on phenotypic similarities The greater the phenotypic similarity between two groups, the higher is their potentiality to undergo parallel evolutionary change. Evolutionary parallelism tends to indicate relationship.

Parallelism is the possession of similar features by two or more taxa that have a common ancestor but those features are not present in the common ancestor. **Convergence,** on the other hand is the possession of similar characteristics by two or more groups without a common ancestor; presumably arising as a response to evolutionary pressure (Fig. 9.4).

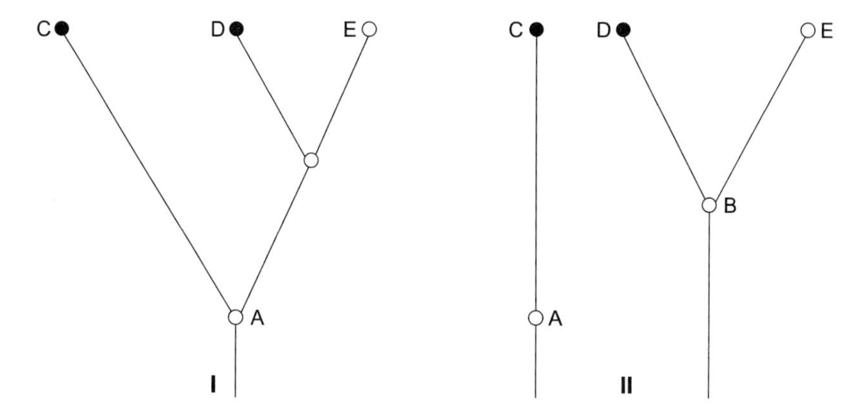

Fig. 9.4 Parallelism and convergence. In diagram I similarity between **C** and **D** is due to false synapomorphy caused by parallelism. In diagram II similarity between C and D is due to false synapomorphy caused by covergence. In both I and II **C** and **D** show derived or apomorphous characters.

Simpson (1961) defined parallelism and convergence as – **parallelism** is the independent occurrence of similar changes in different taxa with a common ancestry and **convergence** is similarity or occurrence of similar features in two or more different taxa belonging to distinct phyletic lines.

Various other authors have also defined the two terms. Parallelism: "development of similar features separately in two or more genetically similar, fairly closely related lineages ...." (Heywood 1967).

"When an essentially similar evolutionary trend has developed independently in two or more lineages where originally similar organs were involved to give rise to homologous, ecologically and typologically similar structures, the adequate term is parallelism." (Dahlgren 1970)

"................ development proceeding from similar forms to new similar forms ...." (Kuprainova 1974).

Convergence: "Development of similar features separately in two or more genetically diverse and not closely related lineages and not due to a common ancestry ...." (Heywood 1967).

"................ When more or less dissimilar organs become similar and/or ecologically equivalent along independent evolutionary lines ...." (Dahlgren 1970).

"................ Development proceeding from dissimilar forms to similar ones ...." (Kuprainova 1974).

"................ When two or more closely related groups tend to evolve by acquiring similar character-states, it is called parallelism and when distantly related or unrelated ones acquire similar character-states, it is convergence" (Sivarajan 1984).

There is a greater emphasis on the genotype when parallelism is concerned. Closely related groups of taxa show similar evolutionary pattern as they have similar evolutionary potentiality and are likely to produce similar mutations (Cronquist 1969). For unrelated taxa there are very few evolutionary strategies left and are easily affected by environmental influence and similar selection pressure. Parallel and convergent evolution both have brought about similarities among diverse plant taxa to reflect evolutionary relationship between different taxa, it is necessary to discriminate between these two evolutionary concepts.

These two frequently used terms in phylogenetic studies are relative and carry no meaning unless they are studied with reference to certain groups. Three species of *Ranunculus* subgenus *batrachium* viz *Ranunculus hederaceous, R. tripartites* and *R. fluitans* exhibit aquatic habit and presence of dissected leaves. This is an example of evolutionary parallelism. Genus *Utricularia* include species of three different habitats: aquatic, terrestrial and epiphytic. The aquatic species have floating stem and highly dissected submerged leaves, some of the segments of which are modified into bladders. The terrestrial species have stem-anchored, no leaves and bladders are seen on the lower side of some spathulate, foliaceous structures; and all the epiphytic species are characterized by long-petioled orbicular leaves. After the genus has split into three separate evolutionary lines, species belonging to each of these lines have acquired similar characteristics. These are some examples of **parallelism** or parallel evolution.

Convergence brings forth increasing similarity between genetically dissimilar groups. Presence of pollinia in Asclepiadaceac and Orchidaceac is a feature common to these two unrelated families. This similarity is due to convergence. Similarly, stem of *Equisetum, Ephedra* and *Casuarina* exhibit the common features—jointed stem, ridges and furrows in the internodal region and whorl of reduced scale leaves at the nodes, although they are unrelated.

Convergence is usually due to adaptation towards similar ecological niche or habitat, e.g. all aquatic plants have reduced root system, lack root hairs and root cap but all possess aerenchyma tissue to help them remain afloat; ephemeral annuals and succulents are characteristic of desert areas.

Convergence may also be due to adaptation towards similar mode of pollination as the members of several unrelated families like Salicaceae, Urticaceae and Poaceae are adapted towards wind pollination or due to adaptation towards similar mode of dispersal as the hairy seeds of several Malvaceae, Asclepiadaceae and Asteraceae.

Comparable selective forces may act on plants growing in similar habitats but in different parts of the world, and the appearance of totally unrelated plants may turn out to be similar, e.g. *Euphorbia* spp. (Euphorbiaceae), *Echinocereus* sp. (Cactaceae) and *Stapelia* sp. (Asclepiadaceae).

The concept of convergence can also be extended to certain metabolic activities. There are three different types of photosynthetic pathways in plants—Calvin cycle or C_3 pathway, dicarboxylic acid pathway or C_4 pathway and Crassulacean acid metabolism or CAM. Of these three, C_3 pathway is most common. C_4 and CAM pathways occur in diverse groups of plants that are adapted to xerophytic environment. They are not phylogenetically similar and have evolved by convergence (Sivarajan 1984).

Phytochemical taxonomy is mainly based on the presence or absence of certain chemical compounds. This may not always lead to the correct conclusion. Biosynthetic pathways should also be given equal importance. Homologous biosynthetic pathways contribute to parallelism and non-homologous ones to convergence (Tetenyi 1973).

Monophyly, Polyphyly and Paraphyly

In taxonomic and evolutionary literature very few terms have lead to so much confusion as monophyly, polyphyly and paraphyly. They have far-reaching consequences and have often been extensively debated. Broadly speaking, **monophyly** means origin of a taxon from a single ancestral form and **polyphyly** means origin of a taxon from more than one ancestral form (Fig. 9.5.).

Use of these terms is purely relative and depends upon how far back we are ready to trace the ancestry of a certain taxon. (Fig. 9.6).

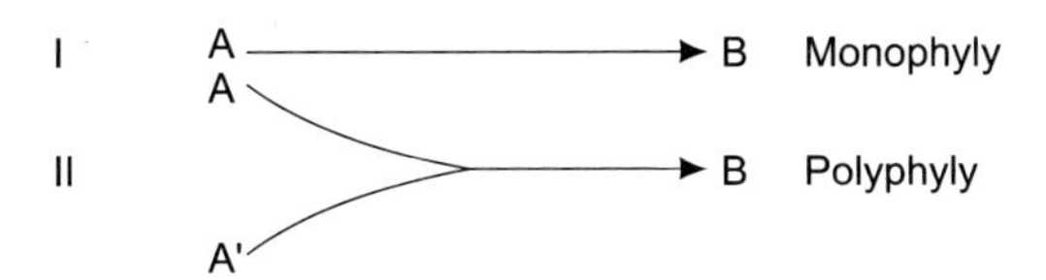

Fig. 9.5 Monophyly and polyphyly. In diagram I taxon **B** is derived from taxon **A**. In II taxon B is derived from two ancestral taxa **A** and **A**.

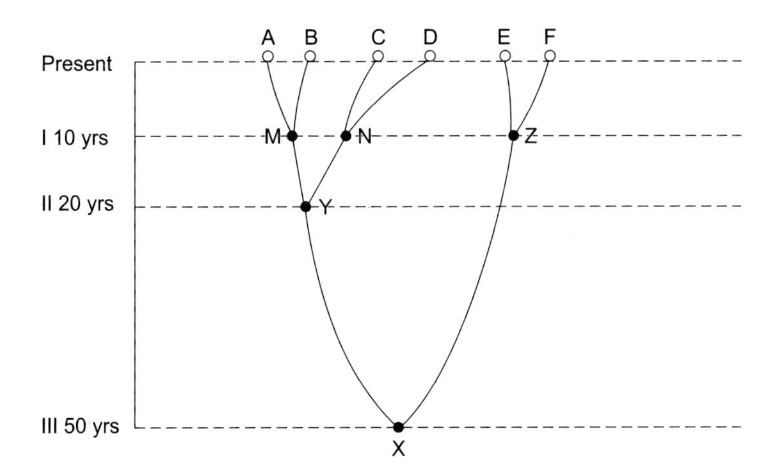

Fig. 9.6 Traced up to 10 yrs (level I) the taxa **AB, CD,** and **EF** are polyphyletic, each pair with one ancestral taxon **M, N, Z;** traced upto 20 yrs (level II) **AB** and **CD** are derived from a common ancestor **Y** and traced up to 50 yrs (level III) all the taxa are derived from a common ancestor **X**.

All organisms, plant or animal, can be traced back to an original, single ancestral form and hence, are monophyletic in origin. The concept of monophyly and polyphyly is a much-debated issue. Systematists with different approaches have provided different definitions for these terms and it is necessary to be precise and more meaningful about these terms.

Simpson (1961) defined monoplyly as "the derivation of a taxon through one or more lineages from one immediately ancestral taxon of the same or lower rank." This definition would mean that there are different degrees or levels of monophyly. If Genus X is the one that might have arisen from Genus Z through only one species then it is monoplyletic at the same rank (Genus) and also at a lower rank (species). But if two different species of Genus Z are involved in the origin of Genus X then it is monophyletic at genus level but polyphyletic at the species level (lower rank).

Heslop-Harrison (1958) defined monophyletic group as the one developed from a single ancestor whose characters are still represented (visible) in the present-day taxon. This is more strict an interpretation for monophyly. Hennig (1966) too supported this definition. Hence, it may be said there are two levels of monophyly: (a) **minimum monophyly**—where a taxon is derived from another of the same **rank** and (b) **strict monophyly**—where one higher taxon is derived from a single evolutionary species along with the retention of important characters from the ancestral species.

The terms monophyly and polyphyly with the definitions given by Simpson (1961) and Heslop-Harrison (1958) can be put to practice in the genera of family Scrophulariaceae. The two genera *Celsia* and *Verbascum* resemble each other closely except that *Celsia* has 4 stamens and *Verbascum* has 5 of them. They can form both intergeneric and interspecific hybrids. Cytological and genetical studies have shown that the genus *Verbascum* has given rise to *Celsia* through two different species.

According Simpson (1961) *Celsia* is monophyletic at the genus level and polyphyletic at species level as 2 different species are involved. According to Heslop-Harrison *Celsia* is polyphyletic in origin because it is derived from two different taxa whose important character (number of stamens) is not the same as in the present-day taxon *Celsia.*

In a classification scheme that reflects phylogeny in true sense, every taxon is ideally monophyletic i.e. members of any taxon (genus, species etc.) have originated from a single ancestor.

If the monophyletic requirement were interpreted loosely (minimum monophyly sensu Simpson 1961) rather than strictly (strict monophyly sensu Heslop-Harrison 1958 and Hennig 1966), most of the conflict between phylogeny and taxonomy would disappear.

According to Cronquist (1968) monophylesis and polyphylesis are not utterly distinct terms. There is a continuous gradation from the strictest monophylesis to the highest level of polyphylesis. If a taxonomic group has its position towards the monophyletic end of this gradation, it is natural and acceptable.

Mayr (1969) and Melville (1983) are the strong supporters of minimum monophyly. Heslop-Harrison (1958), Hennig (1966), Ashlok (1971) and Wiley (1981) are all in favor of strict monophyly. According to these authors all supraspecific taxa (above the species level) comprise independently-evolved individual lineages and cannot be ancestral to one another.

Hennig (1966) defined monophyletic group as one, which contains all the descendents of a group of individuals derived from the most recent common ancestor. Ashlok (1971) termed such a group of individuals as **holophytic**. When all the descendents of the most common ancestor are not included in one group – they are called **paraphyletic** (Fig. 9.8). In both these monophylesis the descendents must have belonged to a single species at one time.

A polyphyletic group according to Ashlok (1971) is the one for which the most recent ancestor is cladistically not a member of that group, i.e. it does not possess the ancestral character-states. However, Cronquist (1973) rejected the requirement of cladistic membership for the ancestral group as impractical. Can we argue that there is an original angiosperm, which is ancestral to all other angiosperms of today? Angiosperms cannot be traced back to a single common ancestor. Hutchinson (1959) has claimed that the angiosperms have evolved monophyletically from the hypothetical pro-Angiosperms. According Heslop-Harrison's definition it is polyphyletic origin because the unique characteristic features of pro-angiosperms are not found in the angiosperms, i.e. they do not belong to the same group.

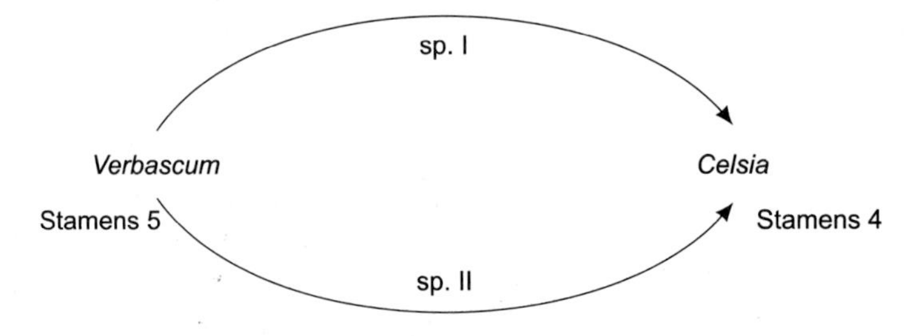

Fig. 9.7 Relationship between the two genera *Verbascum* and *Celsia*.

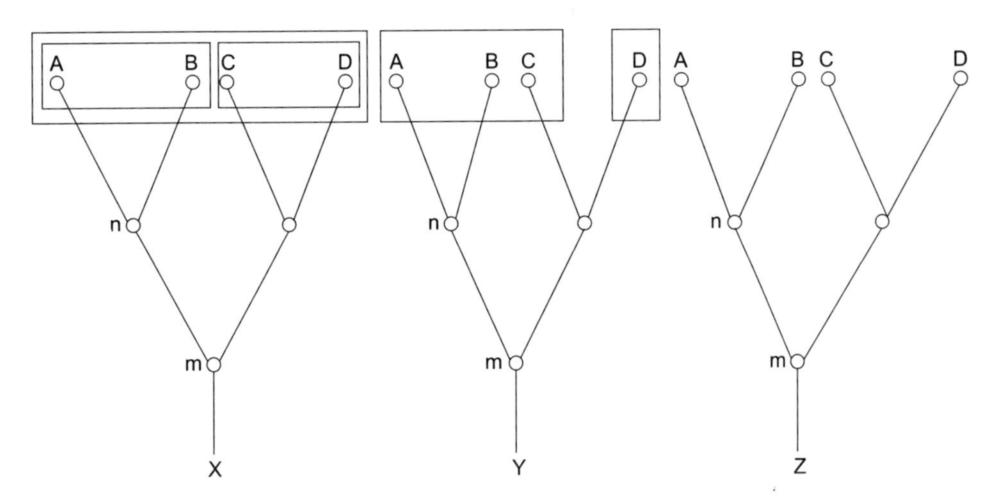

Fig. 9.8 Concepts of monophyly, paraphyly and polyphyly. In **X** the groups AB and CD are monophyletic as each of them have a common ancestor at level *n*; group ABCD is also monophyletic as their common ancestor is at level *m*; in **Y** ABC is paraphyletic because the descendent D from common ancestor at level *m* is outside the group; in **Z** the group BC is polyphyletic as they are arising from different ancestors at level *n* (after Singh 2000).

The general principle of any phylogenetic classification is that all the taxonomic groups above species level must be monophyletic but very often it is not possible to have the ideal condition. The family Ranunculaceae sensu lato originally included the genus *Paeonia* that differs conspicuously from other members of the family. As a result it has been segregated and now it belongs to the monotypic family Paeoniaceae. Another genus *Sphenoclea*, originally of Campanulaceae has been segregated to Sphenocleaceae. The segregation has made these families natural and monophyletic. But the families Euphorbiaceae or Rosaceae (although polyphyletic) cannot be segregated into monophyletic groups due to the presence of continuous variations amongst the members.

10

Evolutionary Trends in Angiosperm Flowers

The ultimate goal of taxonomy is to provide a classificatory system based on the similarities and dissimilarities among the organisms classified. With the publication of Darwin's *Origin of Species* in 1859, the concept of evolution was incorporated in the classificatory systems. According to Darwin's Evolutionary Theory, species represent lineages, and species within a genus have evolutionary affinities with each other. Genera and families are progressively higher orders of divergence in the lineages.

Some of the pre-Darwinian systems of classification did aim at providing a natural system by making use of as many characters as possible. However, in these early periods, many well-known techniques and sophisticated instruments of today were unknown and these systems were based on exomorphic characters only. Darwin's new concept led to the development of phylogenetic systems of classification, in which it is understood that the genetic relationship between the taxa is revealed. Although this concept did not affect the levels of classification for the species, genera and families etc., it did bring about changes in the arrangement of families from simple linear sequence to branching format which reflected evolutionary sequence.

For example, the probable relationship among the orders of Magnoliidae, according to Cronquist (1968), is shown below:

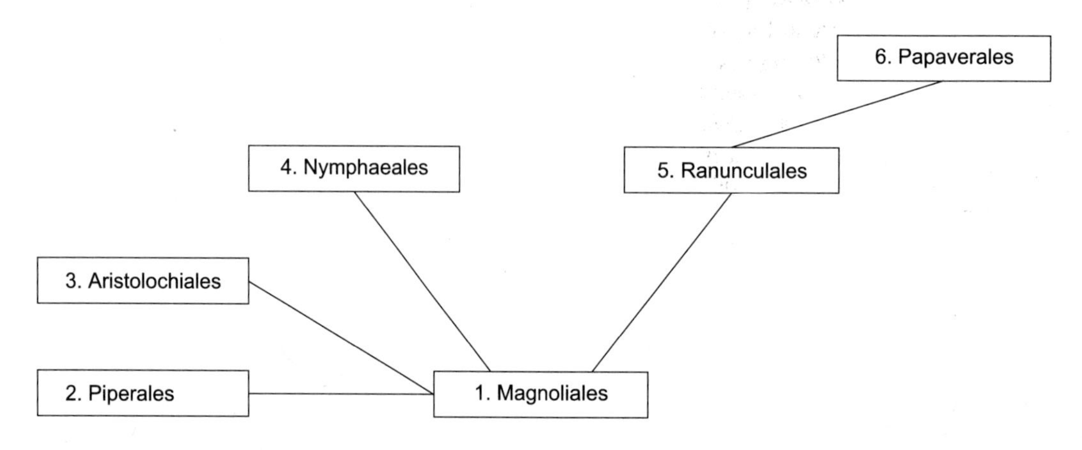

The present systems of classification are mostly phylogenetic, attempting to bring the related taxa together. Now, the importance lies not only in exomorphic features but also in internal structures as revealed by studies in anatomy, embryology, palynology and cytology, biochemistry and plant geography. A number of new and sophisticated techniques are available to determine the relationship between taxa, such as protein analysis, various chromatographic techniques, serological methods and DNA hybridization techniques.

New information and the reinterpretation of information already available time and again bring about changes in the existing classificatory systems.

One of the major difficulties faced by angiosperm phylogenist is the non-availability or inadequacy of the fossil record. There are other problems, too, such as convergent evolution and extreme structural modifications of many flowering plants (Thorne 1963, 1976). It has, however, been shown that the formation of a phylogenetic or evolutionary classification is difficult, but not impossible. An evolutionary systematist who claims that his system is phylogenetic, means that the hierarchial system of taxa is so arranged that it represents the sequence of repeated branching or "cladogenesis" and the degree and character of evolutionary modifications of branches (Takhtajan 1980).

Angiosperms are the dominating vegetation of most terrestrial ecosystems and consist of ca. 250,000–300,000 extant species. But this dominant plant group's origin has remained shrouded in mystery because of the apparently uninformed fossil record, uncertain relationships among the living members of this group and probably the inseparable morphological 'gaps' between the angiosperms and other seed plants, i.e. the gymnosperms.

The angiosperms could have originated variously—from some extinct gymnosperms, from the Bennettitales, from the Pteridosperms (seed-ferns) and more or less all the groups of extinct and extant gymnosperms could be the potential ancestors. Recent studies based on morphological and molecular evidences support monophyletic origin of the angiosperms and also the earlier ideas that the extinct Bennettitales and extant Gnetales are the two taxa most closely related to the angiosperms. This group (Bennettitales, Gnetales and Angiosperms) has been termed 'anthophyta', i.e., they possess flower-like reproductive structures. But even then, in many respects the morphological gap between the various members of anthophyta is rather wide (Crane et al. 1995).[7]

The earliest fossil angiosperms include relatives of the modem *Liriodendron* and *Magnolia,* as well as some members of the Amentiferae. This has led to controversy amongst the phylogenists: according to some, the Amentiferae should be considered as among the most primitive angiosperms, and according to others the Ranales should be treated as the basal group in the evolution of the angiosperms. According to Bessey (1915), within any one phylad (an evolutionary line), the evolution of different organs may proceed at different rates. At any one time, any particular group may present both relatively advanced as well as relatively primitive characters. As a result, no one family will have all primitive features; instead, they will have some comparatively advanced features in addition. On the other hand, it is also true that a family with highly primitive features will never show highly advanced, but only comparatively advanced, characters. For example, in the primitive subclass Magnoliidae (sensu Cronquist 1968, 1981), some members are primitive in certain features but advanced in certain others. The family Winteraceae is vesselless (primitive character) but has sieve tubes and companion cells (advanced character). Austrobaileyaceae have vessels but very primitive, gymnosperm-like phloem, without any companion cells (Cronquist 1968).

The hypothetical primitive angiosperm was an evergreen tree or large shrub with alternate, simple, entire, stipulate, pinnately net-veined leaves. The flowers were borne singly in the axils of leafy bracts, were bisexual, large and showy with numerous, spirally arranged perianth lobes which were not distinguishable as calyx and corolla, or only sepals were present and petals absent. Numerous stamens were spirally arranged and were laminar, i.e. without any differtiation into filament and anther. Carpels were numerous and free, usually stipitate and with unsealed margnis which were covered with glandular hairs, that formed the elongated stigma. Placentation in many taxa was laminar, i.e. unilocular ovary bearing ovules over the whole abaxial surface. Ovules were anatropous, bitegmic and crassinucellate, and seeds endospermous.

---

[7]An attempt has been made to unravel this mystery in Chapter 8 of this book.

Recent studies emphasize great diversity of floral forms in the Magnoliidae—variation in number and arrangement of floral parts is extreme and both large, multiparted, bisexual flowers as well as small, simple, often unisexual flowers quite common amongst the members. Current morphological and molecular evidences favour phylogenetic models with small, trimerous flowers (or even simpler) to represent the basal angiospermous form. The available fossil record (from Portugal and North America) also reveals that these flowers are generally few-parted and often with undifferentiated perianth; stamens with small pollen sacs, valvate dehiscence and apically extended connective; and carpels with poorly differentiated stigmatic surface (Crane et al. 1995).

The trends in evolution of the flower have taken place by (1) reduction in number, (2) fusion, (3) specialisation of parts, and (4) changes in symmetry.

Flower. The primitive angiosperm flower might be considered as having numerous, spirally arranged, prominent tepals, numerous, spirally arranged stamens and numerous unsealed carpels. All these characters occur in one or the other member of the present-day Magnoliales. Flowers of *Degeneria* and Winteraceae are of medium size, and with a moderately elongated receptacle. According to Stebbins (1974), the early angiosperms had flowers of moderate size, which supports the hypothesis that they were small woody plants, inhabiting pioneer habitats exposed to seasonal drought. "Under these ecological conditions, rapid development of flowers and seeds would have had an adaptive advantage and would be most easily acquired by reduction in size of the reproductive shoots" (Stebbins 1974). Hallier (1912) and Parkin (1914), however, had suggested large-sized flowers as the most primitive.

According to Takhtajan (1980), large flowers, as seen in many Magnoliaceae and Nymphaeaceae, as well as large flowers like *Rafflesia arnoldii* are of secondary origin, and evolved in response to selection pressure for different methods of pollination. Small flowers like those of Monimiaceae and Amborellaceae are also derived and may be correlated with specialisation of the inflorescence or reduction of the whole plant.

As evolution progressed, shortening of the receptacle brought the floral parts closer together, in a series of whorls. The number of floral parts in each whorl is reduced, and they become connate or adnate.

Other tendencies are towards elaboration and differentiation of parts. These two apparently opposing tendencies may sometimes be expressed in the same structure. Sympetalous corolla has evolved from polypetalous corolla through union and at the same time, many sympetalous flowers are also irregular, e.g. many members of the Scrophulariaceae (*Antirrhinum majus*).

However, these evolutionary tendencies are independent of each other and, as a result, a flower selected at random will show some relatively primitive and some relatively advanced features, e.g. sympetalous flowers may be polysepalous as in the Sapotaceae, or vice versa, as in some Caryophyllaceae; in Papilionaceae the calyx is synsepalous and regular but the corolla is polypetalous and irregular.

Pentamerous vs. Trimerous Flowers. Trimerous flowers are one of the unifying traits of the monocotyledons. In a few cases where they are absent (Pandanaceae and Sparganiaceae) in this group, some hypotheses relate the simpler flower structure to the trimerous ground plan. The same trimerous flowers are also seen in the Magnoliales, Laurales, Ranunculales and Piperales, and to some extent in the Polygonales.

The evolutionary origin and systematic significance of trimerous flowers is debatable. Based on the distribution of trimerous flowers and their correlation with other characters, Dahlgren (1983a) concludes that this trait is ancient and must have appeared in the angiosperms before differentiation of the monocotyledons, and that the trimerous flowers presumably occurred side by side with flowers with helically arranged parts. Referring to the frequent occurrence of trimerous flowers in the Ranunculales, Dahlgren (1983a) goes on to say that the pentamerous condition has most likely evolved out of a trimerous state in this order, and a similar situation is also observed in the Polygonales. In Nymphaeaceae, he presumes that the large polymerous flowers are derived from small or medium-sized trimerous flowers as observed in the present-day *Cabomba.* Dahlgren and Clifford (1982) also indicate the possible derivation of large, polymerous flowers with helical phyllotaxy (Nymphaeaceae, Magnoliaceae and Illiciaceae) from fairly small and oligomerous flowers in which there was "a tendency for the whorls to be trimerous."

This would mean a return from trimerous condition to pentamerous condition and spiral anthotaxy.[8]

Burger (1978) proposes the derivation of trimerous flowers from the phylogenetic fusion of three simple flowers by contraction of internodes. This concept—and claiming the phylogenetic primacy of the monocotyledons (Burger 1981)—would turn the situation upside down, because ultimately trimerous flowers would have given rise to pentamerous or even polymerous flowers.

A transition from spiral to trimerous anthotaxy has been demonstrated in the Magnoliaceae by Erber and Leins (1982, 1983). In *Magnolia stellata* the flowers exhibit fully spiral anthotaxy, whereas in *M. denudata* and *Liriodendron tulipifera* the floral envelope consists of three whorls, one considered as calyx and two as corolla. This condition is derived from two situations: (1) after the inception of three floral primordia, there is no space left for any further primordia in the same whorl, and (2) a pause after inception of every third element of the floral envelope leads to its trimerous arrangement. Tucker (1960) and Hiepko (1965) also presented similar findings in *Michelia* and *Magnolia*, respectively. It is a fact that the pentamerous condition is the most common amongst the dicotyledons. In the Ranalean complex, however, co-occurrence of the pentamerous, trimerous, as well as dimerous condition has been observed, and sometimes even in the same individual. For example, almost all genera of the Berberidaceae have a trimerous flower, but in *Berberis vulgaris* the terminal flower of the inflorescence is often pentamerous. Another genus, *Epimedium,* has a dimerous flower. This feature reflects a close association between trimery, pentamery and dimery. Similar examples are also reported amongst the members of Lauraceae (Mez 1889, Kasapligil 1951). Dimery frequently occurs with trimery in members of Annonaceae, Lauraceae, Berberidaceae and Papaveraceae (Kubitzki 1987). In Monimiaceae, many primitive genera like *Hortonia* (Endress 1980a), *Laurelia* (Sampson 1969), *Trimenia* (Endress and Sampson 1983) have spiral anthotaxy, while in the more advanced genera like *Wilkiea, Kibara, Tetrasynandra, Steganthera* and *Tambourissa* (Endress 1980b), floral envelopes are dimerous. The trimerous condition is absent altogether. This makes it clear that dimerous and trimerous conditions have both originated from spiral anthotaxy, and the pentamerous condition might be the evolutionary equivalent of two trimerous or dimerous whorls (Kubitzki 1987). The transition from spiral to whorled anthotaxy is documented in several extant families of the Ranalean complex and must have taken place in various parallel evolutionary lines. Except for its predominance amongst the monocots and frequent occurrence in the Ranalean complex, trimery is relatively rare in other dicots. Kubitzki (1987) proposes that trimery is a morphological constraint which offers only very limited possibilities for meristic variation, with no possible return to pentamery or spiral anthotaxy.

Unisexual vs. Bisexual Condition. The primitive flower was bisexual with both stamens and pistils, and unisexual flowers are derived from the bisexual ones. Sometimes, there is an intermediate stage with both types of sex organs well developed, but only one is functional, e.g. in Sapindaceae, some members have an apparently bisexual but functionally unisexual flower. Some members of Compositae exhibit a

[8]Anthotaxy—a term used as against "phyllotaxy" for the floral parts (Kubitzki 1987).

unique case of unisexuality where the disc florets are functionally staminate. In these florets, the ovary is phyletically lost, but pollen from the anther tube is still pushed out by the elongating style.

Monoecious and dioecious groups are also derived from bisexual ancestors through many intermediate stages.

Hypogynous vs. Epigynous Flower. The primitive flower was hypogynous and with free sepals, petals and stamens arising from below the ovary. The basal parts of the different floral whorls are frequently fused to form a structure—the hypanthium, which in some plants apparently resembles a calyx. In such plants, the petals and stamens appear to be inserted on the calyx tube, e.g. members of the Papilionaceae. In some others, however, such as Cucurbitaceae, the hypanthium is distinct, so that the calyx lobes, corolla and stamens appear to be arising from the apex of the hypanthium. Flowers with hypanthium and a superior ovary are perigynous, which is intermediate between hypogynous and epigynous conditions. The epigynous condition is derived from the perigynous when the hypanthium becomes adnate to the ovary, or the bases of the outer floral whorls may be fused with the ovary wall, as in the Rubiaceae, or the ovary itself may become submerged in the receptacle, as in the Santalaceae.

Perianth. Although the primitive angiospermous flowers had no corolla, and the perianth consisted of the sepals only, in modem angiosperms, the presence of a corolla is a primitive feature and its absence is derived. Petals (or perianth) might have originated either from bracts (as in the Magnoliales, Illiciales and Paeoniales) or from stamens as in the Nymphaeales, Ranunculales, Papaverales, Caryophyllales and Alismatales (Takhtajan 1980). The extra petals in "double" flowers of many species are staminodial in origin.

The usual tendency of the perianth to consist of an outer protective part, the calyx, and an inner attractive part, the corolla, is essentially functional. There is, however, much variation and deviation from this usual pattern, e.g. in *Delphinium* and *Aconitum* of Ranunculaceae, the sepals are large and showy like petals, and the petals function as nectaries. In another example, *Mirabilis jalapa* of Nyctaginaceae, the calyx is corolloid and is subtended by a calyx-like involucre of five basally united bracts. In the most advanced families, usually there is only one whorl each of calyx and corolla. Two sets of calyx and two sets of corolla are much less frequent (Cronquist 1968).

Androecium. Comparative studies of the stamens of living angiosperms show that the most primitive type of stamen is a broad, laminar, three-veined organ which is not differentiated into filament or anther. It develops two pairs of elongated microsporangia embedded in the abaxial or adaxial surface, between the midvein and lateral veins (Fig. 10.1). Canright (1962) regards the stamen of *Degeneria* as "the closest of all known types to primitive angiosperm stamen". In *Degeneria, Galbulimima, Lactoris, Beliolum* (Winteraceae), Annonaceae and *Liriodendron,* the microsporangia are situated on the abaxial surface and therefore the stamens are extrorse. On the other hand, in the Magnoliaceae (except *Liriodendron),* Austrobaileyaceae and Nymphaeaceae, the position of the microsporangia is on the adaxial surface, and the stamens are introrse (Takhtajan 1980). It is not certain, however, which of these two conditions is more primitive. According to Takhtajan (1980), both the abaxial and adaxial position of microsporangia have been derived from a common ancestral type, which could only have been marginal.

Many taxonomists, including Moseley (1958), Eames (1961), Canright (1962) and Cronquist (1968) consider that deeply sunken microsporangia in the staminal tissue, as in *Degeneria* and *Galbulimima,* is a primitive feature. Other evolutionary features observed in androecia include the reduction in the number of stamens from many to a few. This feature sometimes coexists within the same genus. In *Hibbertia*, of the family Dilleniaceae, all conditions from numerous free stamens, to five fascicles of stamens to five separate stamens and lastly to a single stamen are reported (Wilson 1964). In Guttiferae the genus *Hypericum* has a similar series from numerous separate stamens to five fascicles of three stamens each.

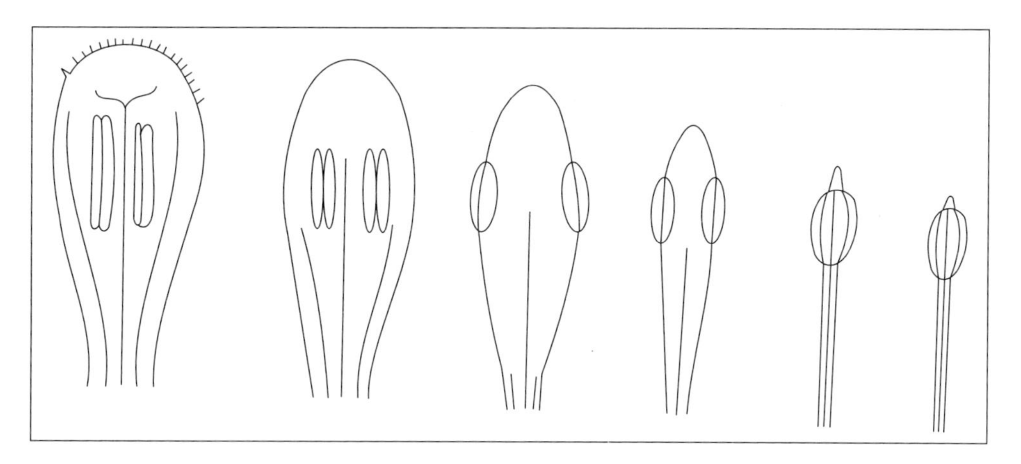

Fig. 10.1 Evolutionary Trends in Androecium.

Once the stamens have been reduced to a single whorl, the number of stamens usually does not increase. On the other hand, there is a possibility of further reduction in their number, as in Bignoniaceae, Labiatae and Scrophulariaceae. In these families, the number of corolla lobes is five, but the number of stamens is two or four, the missing stamens often represented by staminodia. However, according to some workers (see Cronquist 1968), in certain families of the Guttiferales and Malvales, progression has been from a few stamens in a single whorl to fascicled stamens and to numerous separate stamens. Eames (1961) interpreted the anatomical evidence in these families as indicating a reduction series. Cronquist (1968) also supports this view.

Pollen Grains. Most primitive angiosperm pollen is considered to have a single distal germinal furrow or colpus or sulcus in the sporoderm (Cronquist 1968, Sporne 1972, Stebbins 1974, Takhtajan 1980). Angiosperm pollen grains are mainly of two types: uniaperturate and triaperturate. The former has a single germinal pore or furrow usually on the surface opposite the contact point of the pollen grains in a tetrad. The latter has three germinal furrows which do not closely approach each other.

Uniaperturate grains are characteristic of the monocotyledons and some members of the Magnoliales and Ranales, and triaperturate grains are known in most of the dicotyledons. Apart from these two main types, multiaperturate and nonaperturate grains are also known. These are the derived ones and, hence, more advanced.

Pollination. Discoveries of the early- and mid-Cretaceous fossils of angiosperm flowers indicate that the early members of this group were insect-pollinated. Stamens in these fossil flowers have small anthers with valvate dehiscence, lesser amount of pollen grains; generally poorly developed and unelaborated stigmatic surface and smaller pollen grains (than those for wind dispersal). These early flowers were most probably pollinated by pollen-collecting or pollen-eating insects. Fossils of flowers pollinated by nectar-collecting insects appeared much later (Crane et al. 1995).

Gynoecium. The most primitive type of carpel is essentially a megasporophyll and this concept of megasporophyll-like carpel fits very well with the Ranalean theory of angiosperm evolution.

It was assumed for a long time that the primitive pre-angiospermous carpel had a row of ovules along each margin of the laminar structure, and with the closure of such a carpel, the two rows of ovules were brought together and formed a single row. Bailey and Swamy (1951) and Eames (1961) are of the opinion that the most primitive carpels are unsealed, conduplicate, and more or less stipitate structure with a large number of ovules scattered on the adaxial surface. This type of conduplicate carpel is seen in *Degeneria* and *Tasmannia* and also in some primitive monocotyledons. Another important characteristic of the primitive carpel is the absence of a true style, and stigmas occurring along the margin of the conduplicate carpels. These stigmatic margins are not fused at the time of pollination and, in the course of evolution, this primitive stigma was transformed into subapical and then apical stigmas (Fig. 10.2). All these transitional stages can be seen amongst the members of the Magnoliales and Ranales.

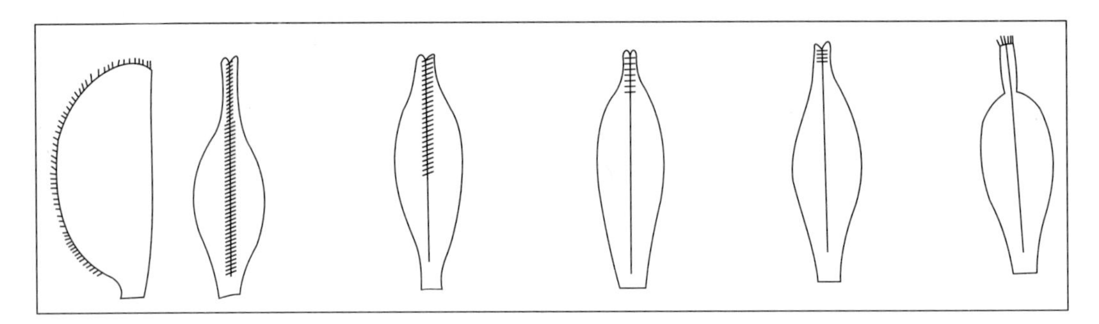

Fig. 10.2 Evolutionary Trends in Carpels.

Primitive angiosperm gynoecium consisted of a large number of carpels, arranged spirally on the surface of a more or less elongated receptacle. The carpels, each containing a large number of ovules, spread open when mature and let out the seeds. Amongst both primitive dicots and monocots, this type of carpel is common. Sometimes, these spirally arranged free carpels become fused to form a "pseudo-syncarpous" condition, as in some species of *Magnolia,* but this line of evolution did not progress any further.

In the normal course of evolution, the number of carpels became less and they became restricted to a single whorl. A syncarpous gynoecium emerged from an apocarpous gynoecium by lateral fusion of closely connivent carpels arranged in a whorl. The primitive form amongst these still shows free upper portions of the fertile region of the carpels but, eventually, a compound style with one compound stigma is achieved.

There may be further reduction in the number of carpels in some families, which is in accord with the usual trend of evolution. The genus *Linnaea* of the Caprifoliaceae has a trilocular ovary but only one of these locules contains a normal ovule which develops into the seed. The other two locules contain several abortive ovules. A similar feature is seen in *Valerianella* of the Valerianaceae, in which two locules of the trilocular fruit are sterile. In another member of this family *Valeriana,* the ovary is tricarpellary with a trilobed stigma but two of the carpels are vestigial. In addition, there are several other families (Leitneriaceae, Krameriaceae) with "pseudomonomerous" gynoecium, i.e. an apparently monocarpellary ovary which is phyletically derived from a many-carpelled ovary.

Placentation. According to Takhtajan (1980), "the main directions of evolution of the gynoecium determine the main trends of evolution of placentation". The most primitive type of placentation is the laminar-superficial type, i.e. ovules scattered over the adaxial surface. Ovule position restricted only to the margin of these laminar structures was an early evolutionary step, and is termed laminar-marginal. From this stage with complete closure, marginal plancentation was derived. When several carpels in a whorl, each with marginal placentation, fuse at their margins, axile placentation is derived. If, however, there is incomplete closer, fusion of such carpels in a whorl, will give rise to parietal placentation (Fig. 10.3). From axile placentation, through the process of reduction, both apical and basal placentation are derived.

In many Apocynaceae *(Catharanthus roseus)* and Asclepiadaceae *(Calotropis procera),* there are two separate ovaries but style and/or stigma are fused. This condition is secondary in origin and not a primitive feature. There is no evidence to show that carpellary fusion has proceeded from the stigmatic end. Whatever the reason of the secondary splitting of the ovaries may be, the style and stigma have remained fused so that both the carpels may be pollinated by the same agent from a common pollinating surface and this is, no doubt, advantageous.

Most of the advanced families show an axile, parietal, free-central, basal and apical type of placentation.

Ovules. Anatropous, orthotropous, amphitropous and campylotropous ovules are known amongst the angiosperms. Orthotropous is the most primitive type and is typical of the living gymnosperms. Most angiosperms have anatropous ovules, which means that this type of ovule must have formed very early in the evolution of the angiosperms. Campylotropous and amphitropous conditions are derived from the anatropous type. Although orthotropous ovules are known amongst the gymnosperms, in angiosperms they are believed to have been derived from anatropous ovules (Eames 1961, Cronquist 1968, Corner 1976, Takhtajan 1980).

The importance of the number of integuments in an ovule has long been recognized. The bitegmic ovules are considered to be more primitive as compared to the unitegmic ones. In particular, the multilayered integuments in the orders Magnoliales, Laurales, Aristolochiales, Piperales and Illiciales are generally regarded as the most primitive (Dahlgren 1975b). Bitegmic ovules are predominant amongst the monocotyledons and in most of the Polypetalae of the dicotyledons. Ovules are exclusively unitegmic in most of the gamopetalous orders of the dicotyledons. Unitegmic ovules sometimes occur in a number of isolated families of the orders with otherwise predominantly bitegmic ovules, proving that there are independent lines of evolution.

A unitegmic condition is a derived one and is formed either by fusion of the two integuments or by reduction of either one of them.

An interesting point of coincidence is that iridoids are almost 100% restricted to groups with unitegmic (and generally tenuinucellate) ovules (Jensen et al. 1975, Dahlgren 1975a). Usually, the primitive families amongst the angiosperms have crassinucellate ovules, and more advanced ones have tenuinucellate ones. In many groups of angiosperms, the development of the nucellus is directly related to the number of integuments and the type of endosperm formation (Dahlgren 1975b). Thus, the most primitive ovules are bitegmic and crassinucellate, and the most advanced ovules are unitegmic and tenuinucellate. The studies of Philipson (1974, 1975) have, however, shown that there are a large number of intermediate types, e.g. the bitegmic-tenuinucellate condition in Theaceae and Primulaceae, and the unitegmic-crassinucellate condition in Cornaceae and Araliaceae.

Seed. Seeds of primitive angiosperms are medium-sized, with abundant endosperm and a minute embryo which is often undifferentiated (Takhtajan 1980). In advanced seeds, however, the embryo is large but the endosperm is highly reduced and often absent. Both small and large seeds are derived.

Chemistry. Ellagitannins, and leucoanthocyanins, also called the proanthocyanins, are usually present in the primitive angiosperms. On the other hand, iridoids and glucosides are more common amongst the more advanced members, e.g. iridoids are distributed in most of the gamopetalous families, such as Gentianaceae, Apocynaceae and others.

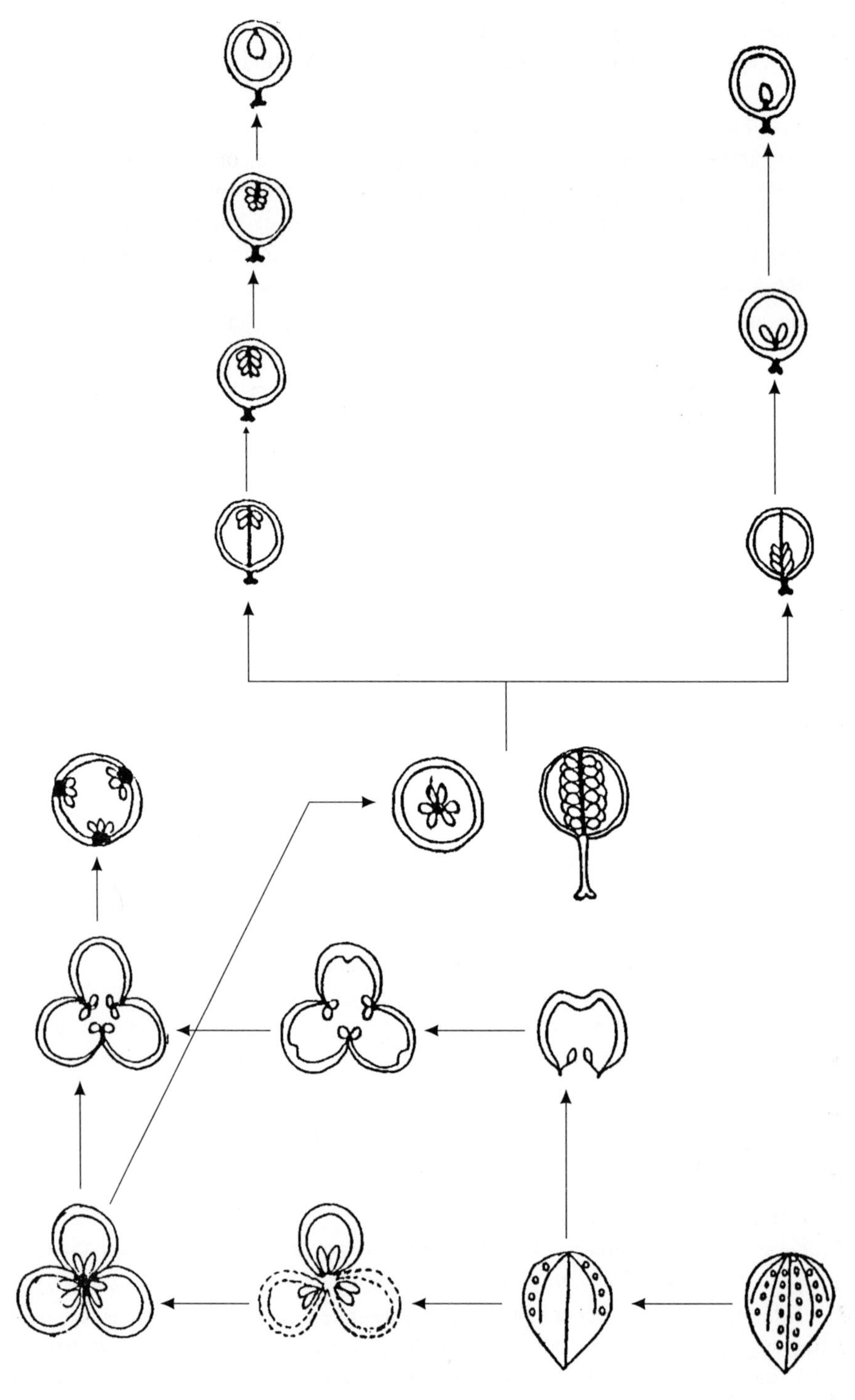

Fig. 10.3 Evolutionary Trends in Placentation.

To summarise, the trends of floral evolution are reduction in number, adnation or fusion of the floral parts, specialisation of these parts, and changes in symmetry. It has also been observed that the different parts of a plant do not evolve at the same rate, and as a result certain taxa show both primitive and advanced features. This phenomenon of unequal rate of evolution of different features within one lineage is known under various names: "chevauchement des specialisations" (Dollo 1893) and "mosaic evolution" (De Beer 1958). Different evolutionary stages or grades of different characters of the given taxon are the result of such mosaic evolution. Takhtajan (1959, 1966, 1980) named this difference in grades as "heterobathomy" (Greek bathmos-step, grade).

Because of heterobathomy, an organism may present a mosaic combination of characters of quite different evolutionary levels. For example, the genera *Trochodendron* and *Tetracentron* have primitive vesselless wood and, at the same time, rather specialised flowers. On the other hand, the genus *Magnolia,* with primitive flowers, has rather advanced wood anatomy. Heterobathomy can also be expressed within the flower, e.g. in *Delphinium* the flower is zygomorphic and therefore advanced, but the stamens are arranged spirally, which is a primitive feature.

The more strongly heterobathomy is expressed, the more contradictory is the taxonomic information provided by different sets of characters. The "more heterobathomic a taxon, more complete and alround must be its study" (Takhtajan 1980).

11

Some Selected Families of Angiosperms

DICOTYLEDONS—ARCHICHLAMYDEAE

Order Urticales

The order Urticales comprises five families—a monotypic family Rhoipteleaceae, Ulmaceae, unigeneric Eucommiaceae, Moraceae and Urticaceae (Engler and Diels 1936). Rendle (1925) split Moraceae into Cannabinaceae and Moraceae.

Plants with simple stipulate leaves, generally unisexual flowers, perianth inconspicuous, in one whorl, stamens equal to the number and opposite the perianth lobes; carpels unilocular with one ovule.

Moraceae

Moraceae is a large family of 53 genera and 1400 species distributed in the tropical and subtropical zones of the world.

Vegetative Features. Mostly trees and shrubs, a few are herbs, e.g. *Dorstenia,* deciduous or evergreen, mono- or dioecious. Leaves entire or deeply cleft or sometimes palmately lobed, usually alternate (Fig. 11.1 A), rarely opposite, stipulate, stipules 2, lateral or each pair forms a cap over the bud and leaves a cylindrical scar on the stem. All the plant parts usually contain latex (except *Morus).*

Floral Features. Flowers unisexual and inflorescence varied. In *Morus,* both male and female flowers in pendulous catkins (Fig. 11.1A, B), in *Maclura, Cudrania* and *Broussonetia,* female inflorescences are condensed into globose heads. In *Dorstenia,* the pedicels and peduncles are coalesced and dorsiventrally flattened into a laminate structure, the minute sessile flowers are borne on the ventral surface. In *Ficus*, the receptacle has developed into a hollow structure (syconium. Fig. 11.1 I), the flowers are borne on the inner surface. At the apex of the syconium is an ostiole with inwardly directed bracts on the inside (Fig. 11.1 I). On the inner fleshy wall of the syconium, pedicellate male flowers are borne around the closed ostiole; each flower has 3 hooded perianth lobes and a stamen (Fig. 11.1 J). Pistillate flowers are of two types: sessile or almost sessile, long-styled seed flowers and pedicellate, short-styled gall flowers. Each pedicellate flower has five perianth lobes, and an uniovulate ovary with a laterally attached style and papillate stigma (Fig. 11.1 K, L).

In other members, flowers are minute, perianth of 4 tepals in 2 whorls, free or slightly connate at the base (Fig. 11.1 C). Number of stamens in male flowers (Fig. 11.1 C), as many as the tepals or reduced, opposite the perianth lobes, filaments short, thick and prominent, incurved in bud; anthers versatile and longitudinally dehiscent. The female flowers with or without a perianth, when present, tepals 4; gynoecium syncarpouns, bicarpellary, one carpel often abortive; ovary superior or inferior, unilocular and 1-ovuled (Fig. 11.1 E). Styles 2, filiform, stigmas 2 (Fig. 11.1 D, F). Fruit usually compound or aggregate. Numerous small achenes or drupes are adnate to the perianth and the receptacles of the other adjacent flowers, forming an aggregate fruit. Seeds with curved embryo and fleshy endosperm. Floral diagrams for staminate and pistillate flowers depicted in Fig. 11.1 G, H.

Anatomy. Wood occasionally ring-porous, vessels usually medium-sized to large and predominantly solitary; spirally thickened, perforation simple, intervascular pitting minute to large. Parenchyma paratracheal, rays 1 to 15 cells wide, slightly heterogeneous to homogeneous, with a few uniseriate rays.

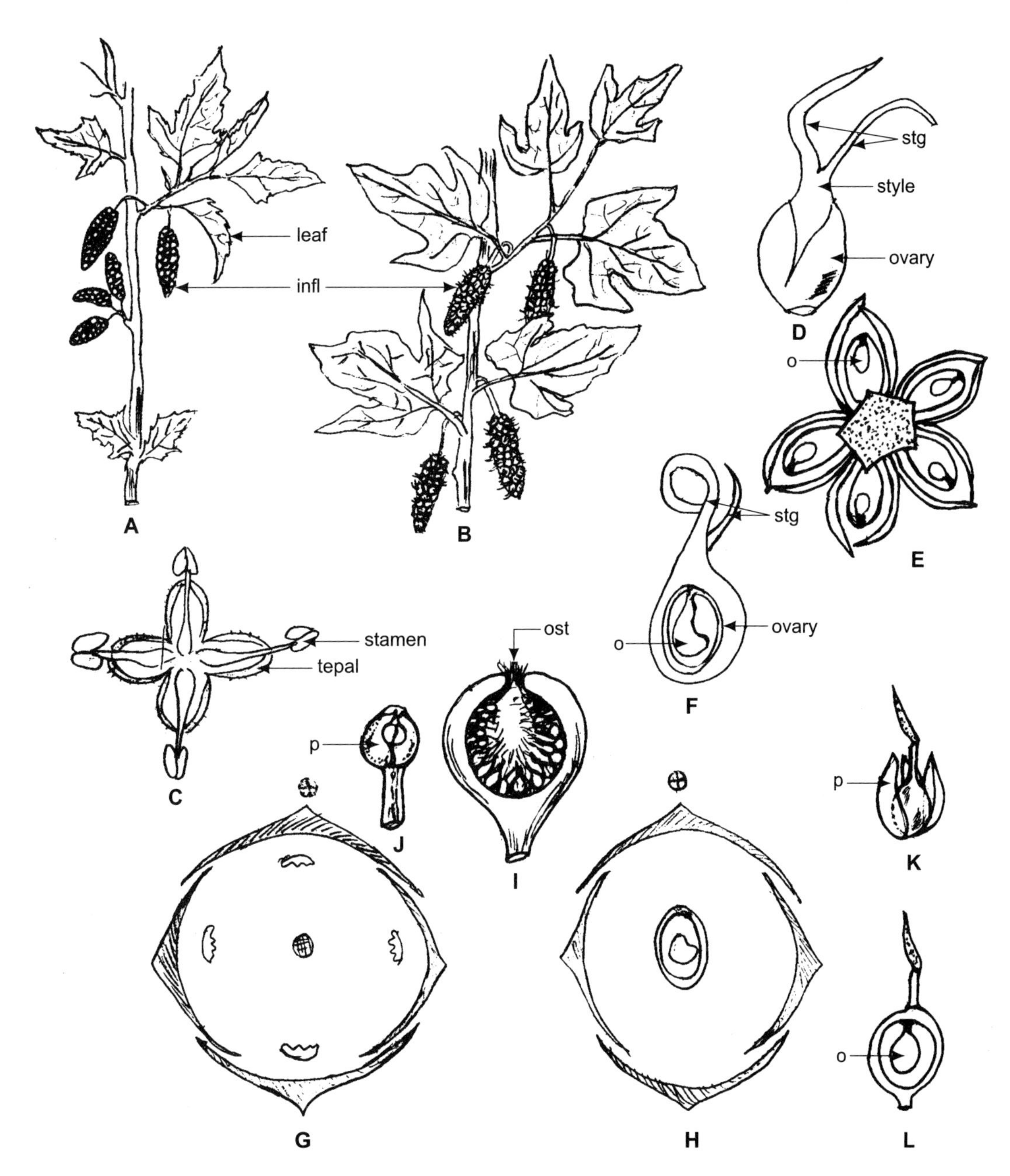

Fig 11.1 Moraceae: **A–H** *Morus alba*; **I–L** *Ficus glomerata* **A**, **B** Flowering twigs: male (**A**), femal (**B**). **C**, **D** Flowers: staminate (**C**) with four tepals and four stamens and pistillate (**D**) with a bicarpellary gynoecium, tepals removed. **E** Pistillate inflorescence, cross section. **F** Ovary, vertical section. **G**, **H** Floral diagrams for staminate (**G**) and pistillate (**H**) flowers. **I**. Inflorescence, vertical section. **J** Staminate flower. **K** Pistillate flower. **L** Same, vertical section. (*infl*, inflorescence, *o*, ovule, *ost*, ostiole, *p*, perianth *stg*, stigma).

Latex tubes present in rays of many genera. Leaves usually dorsiventral, hairs both glandular and non-glandular, stomata distribution variable, raised above or depressed below the leaf surface. Cruciferous type, as in *Conocephalus,* or Ranunculaceous type, or surrounded by a rosette of subsidiary cells, as in some species of *Ficus.*

Embryology. Pollen grains elliptic and biporate as in *Ficus or* spheroidal as in *Morus,* exine smooth, shed at 2-celled stage. Ovules hemianatropous *(Ficus) or* anatropous *(Morus, Dorstenia),* bitegmic, crassinucellate. Polygonum type of embryo sac, 8-nucleate at maturity. Endosperm formation of the Nuclear type. Seed coat is formed by both the integuments.

In different species of *Ficus,* the most interesting feature is the study of pollination by different fig pollinators. These insects have specific organs for collecting pollen, located on various parts of their body. *Ceratosolen arabicus*-pollinator of *F. sycomorus,* and *Blastophaga quadraticeps*—pollinator of *F. religiosa*–have thoracic pollen pockets closed from outside by a movable lid. Some other types of wasps have coxal corbiculae in the form of longitudinal depressions along the coxae, i.e. proximal joints of forelegs. In *B. psenes* pollen is stored in the intersegmental and pleural invaginations on the body of the wasp. Wasps that have no pollen pockets usually pollinate such species of *Ficus* as produce abundant pollen grains.

Growth of the syconium is dependent on pollination. In both gall and seed flowers, endosperm and embryo develop and in gall flowers, wasp larva grows concurrently. The growing larva brings about abortion of the young embryo and consumes the endosperm, ultimately ruling the gall completely (Johri et al. 1992).

Chromosome Number. Basic chromosome numbers are x = 7, 12, 13.

Chemical Features. Phytoalexin benzofurans and stilbenes are produced by members of Moraceae (Harborne and Turner 1984).

Important Genera and Economic Importance. Moraceae is a large family with many economically important taxa. *Maclura pomifera,* commonly called osage orange or bow wood, is an ornamental tree from the central United States. The round catkins of the female plants develop into globular fruits that look like large-sized lemons and consist of innumerable tightly packed wedge-shaped fruits. The wood was used by native Americans to make clubs, and it also yields a yellow dye. *Chlorophora tinctoria* (a tree from tropical America) yields another natural dye called fustic. Inner bark of *Broussonetia papyrifera* stem is the source of a good fibre used in the manufacture of paper. It is widely grown in Central Asia for making tapa cloth. Most spocies of the genus *Dorstenia* (from tropical South America and Africa) are highly poisonous. Various species of *Morus (M. alba, M. nigra)* are source of the mulberry fruit. Leaves of these plants are also used as feed for rearing silk moths. Other important fruit trees are *Artocarpus communis* (bread fruit), *A. heterophyllus* (jackfruit) and *Ficus carica* (fig). African bread fruit is the fruit of *Treculia africana. Castilla elastica* from Mexico and Central America gives Panama rubber. *Ficus elastica* from India is another source of rubber. The genus *Ficus* has more than 1000 species. Some of these, like *F. benghalensis, F. religiosa* and *F. krishnae* are shade trees and also have religious importance. *F. aurea* is an epiphyte, *F. pumila* and *F. radicans* are both climbers with mottled leaves from Japan. These three species are grown as indoor plants. *F. carica* and *F. glomerata* are grown for their edible fruits. *Ficus benghalensis* is a tree with large canopy and numerous aerial roots produced from the branches grow downwards. These roots reach the ground and help in supporting the tree. *F. pumila,* known as Indian ivy, is a root climber and the climbing roots secrete a gummy substance by which the roots become attached to the support.

Taxonomic Considerations. Bentham and Hooker (I 965c) treated this family as a highly advanced one and placed it in series Unisexuales of the Monochlamydeae along with other families like Euphorbiaceae, Urticaceae, etc. Hallier (1912), however, kept all its members in the family Urticaceae. Bessey (1915) treated it as a member of the Malvales. Cronquist (1968, 1981) also treats this family as a member of the Urticales and derives it from the Hamamelidales. Benson (1970) places it in the Urticales of his Thalamiflorae. Dahlgren (1983a) treats the Urticales (including the family Moraceae) as a member of the superoder Dillenianae. Takhtajan (1980, 1987) includes Moraceae in the suborder Urticinae of order Urticales. The two herbaceous genera *Cannabis* and *Humulus* (containing only watery sap and no latex) are kept separately in the family Cannabaceae by Hutchinson (1948).

Cannabinaceae (=Cannabaceae)

This family is represented by only two genera, *Cannabis* and *Humulus*, and 4 species, distributed in temperate Eurasia and North America. This family is often treated as a constituent part of Moraceae, although both genera are herbaceous and without any latex.

Vegetative Features. Annual or perennial herbs and climbers. Leaves simple and palmately lobed (Fig. 11.2 A) or compound, palmately veined, alternate, stipulate, stipules persistent.

Floral Features. Flowers unisexual (Fig. 11.2 B, C) borne on dioecious plants. Male flowers in loose racemes or panicles with 5 polysepalous sepals, corolla absent. Stamens 5, alternate with sepals, filaments distinct, anhters bicelled, basifixed, longitudinally dehiscent. Female flowers in dense clusters, (Fig. 11.2 A) sepals 5, completely fused to form a cup-like structure enclosing the ovary. Gynoecium syncarpous, bicarpellary, one carpel abortive so that the ovary becomes one-loculed with one ovule (Fig. 11.2 D, E). Fruit achene enclosed in persistent calyx.

Anatomy. Vessels medium-sized, solitary or in small irregular clusters, pits simple, perforation plates simple. Leaves dorsiventral, crystals and cystoliths present, stomata Ranunculaceous type.

Embryology. Pollen grains shed at 2-celled stage. Ovules anatropous, bitegmic, crassinucellate; nucellar cap present; Polygonum type of embryo sac, 8-nucleate at maturity. Endosperm formation of the Nuclear type with chalazal, free-nuclear haustorium, polyembryony reported. Inner integument collapses before maturity of seed. Mature seed with a rudimentary seed coat.

Chromosome Number. Basic chromosome number is $x = 8$ in *Humulus* and $x = 10$ in *Cannabis.*

Chemical Features. The resinous exudate contains D-tetrahydrocannabinol, an active hallucinogenic compound. Other chemical constituents are cannabidiolic acid, tetrahydrocannabinol-carboxylic acid, cannabigerol and cannabichromen (Kochhar 1981).

Important Genera and Economic Importance. There are only two genera, *Cannabis* and *Humulus,* and both are economically important. *Cannabis* is a multipurpose plant. It gives hemp fibre, hemp seed oil and narcotics. It is also used in indigenous medicines. The narcotic constituents are mainly concentrated in the resin produced by the glandular hairs, more abundant on the bracts surrounding the female flowers. Three forms of intoxicants available from this plant are bhang, ganja, and charas. Of these, charas or hashish is the strongest form. *Humulus lupulus* is used as a flavouring material in beer. Here also, the glands occurring on the female inflorescences are important.

Taxonomic Considerations. Earlier workers like Bentham and Hooker (1965c), Hallier (1912) and Bessey (1915) included Cannabinaceae in Moraceae. Lawrence (1951) placed the genera *Cannabis* and *Humulus* in the subfamily Cannaboideae of the Moraceae. Rendle (1925) and Hutchinson (1948) raised this subfamily to family rank and designated it as Cannabinaceae, This view is supported by most of the recent workers like Dahlgren (1983a), Cronquist (1981) and Takhtajan (1987).

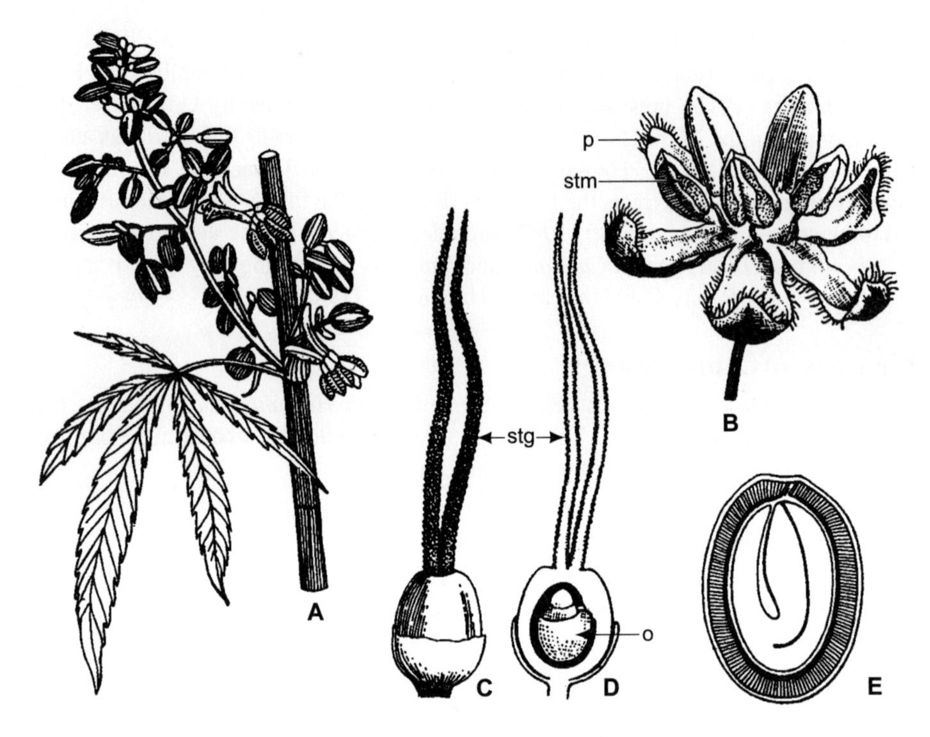

Fig. 11.2 Cannabinaceae: **A–E** *Cannabis sativa*. **A** Twig with female flowers. **B** Staminate Flower. **C** Pistillate flower **D** Same, longisection **E** Seed, vertical section. (*o* ovule, *p* perianth, *stg* stigma, *stm* stamen).

The family name has, however, been changed from Cannabinaceae to Cannabaceae, on the basis of the type genus *Cannabis*. It is distinct from other members of the family Moraceae as it is herbaceous, contains watery sap instead of latex, differs also in chemical constituents and embryological features.

Order Centrospermae (=Caryophyllales)

The order Centrospermae comprises 4 suborders and 13 families.

Suborder Phytolaccineae: Phytolaccaceae, Gyrostemonaceae, Achatocarpaceae, Nyctaginaceae, Molluginaceae and Aizoaceae;

Suborder Portulacineae: Portulacaceae and Basellaceae;

Suborder Caryophyllineae: Caryophyllaceae;

Suborder Chenopodiineae: Dysphaniaceae, Chenopodiaceae, Amaranthaceae and Didiereaceae.

The plants are herbs, shrubs, trees or vines. The leaves alternate (Chenopodiaceae) or opposite-decussate (Caryophyllaceae), usually estipulate, frequently show anomalous secondary growth in both stem and root. P-type plastids have been reported. Perianth typically biseriate; pollen grains 3-celled; ovules campylo- or amphitropous, bitegmic, crassinucellate with nucellar cap. Embryo generally coiled or curved. Most families (excluding Gyrostemonaceae, Caryophyllaceae and Molluginaceae) contain betalain pigments. The two families Caryophyllaceae and Molluginaceae contain anthocyanin pigments, and Gyrostemonaceae lacks both anthocyanin and betalain.

Nyctaginaceae

A family of 30 genera and 390 species, distributed in the temperate and tropical regions of the New World. Three of the genera are indigenous to the Old World—*Boerhavia, Oxybaphus* and *Pisonia.*

Vegetative Features. Herbs *(Boerhavia),* shrubs or trees *(Pisonia),* sometimes scandent *(Bougainvillea).* Leaves usually opposite, simple, entire, estipulate (Fig. 11.3 A). *Tricycla* is a spiny shrub and lateral shoots are spine-tipped in *Phaeoptilum.*

Floral Features. Inflorescence cymose; flowers bracteate, ebracteolate, bisexual or sometimes unisexual *(Phaeoptilum),* actinomorphic, hypogynous, incomplete. Bracts 2 to 5, foliaceous, resembling a calyx or coloured and subtend a flower as in *Bougainvillea* (Fig. 11.3 B). Perianth uniseriate, calyx of 5 sepals, connate to form a tube which resembles a sympetalous corolla (Fig. 11.3 A, C), plicate or contorted in bud; true corolla absent. Stamens vary from 1 to 30, usual number 5, filaments connate basally in a tube, unequal; anthers bicelled, longitudinally dehiscent (Fig. 11.3 D). Gynoecium monocarpellary, ovary unilocular with 1 basal erect ovule (Fig. 11.3 C), superior, style 1, slender (Fig. 11.3 E), stigma 1. Fruit an achene (Fig. 11.3 F), often enveloped by the persistent calyx which may be variously modified to facilitate dissemination. Seeds endospermous, embryo straight or curved.

Anatomy. Vessels small, typically in radial groups and clusters behind the phloem strands, perforations simple. Parenchyma scanty and limited to a few cells around the vessels, sometimes storied; rays small and 1 or 2 cells wide and hetero- or homogeneous or absent. Included phloem always present; anomalous growth in thickness of the axis, by the development of successive rings of collateral vascular bundles. Leaves dorsiventral or isobilateral, hairs of various types occur, stomata confined to the lower surface in most species, Ranunculaceous or Rubiaceous type.

Embryology. Pollen grains 3- to 4- colpate with coarse, reticulate or spinulose exine; polyporate (Nowicke 1970, Behnke 1976b). Ovules anatropous *(Pisonia)* or campylotropous, bitegmic, crassinucellate. Embryo sac of Polygonum type, 8-nucleate at maturity. Endosperm formation of the Nuclear type. Perisperm present.

Chromosome Number. Basic chromosome number is x = (13, 17) 20–29 (Behnke 1976b) or x = 10, 13, 17, 29 and 33 (Takhtajan 1987).

Chemical Features. The pigment betalain is present in members of Nyctaginaceae.

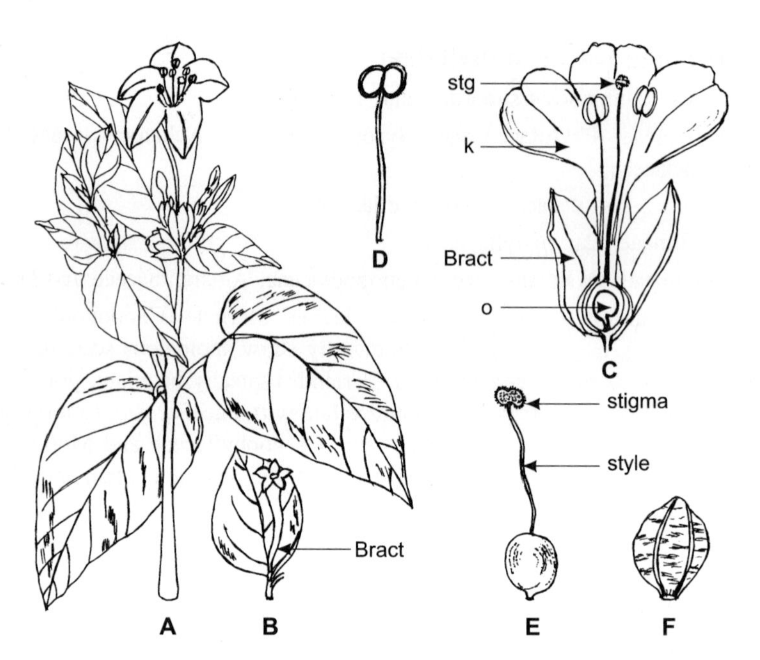

Fig 11.3 Nyctaginaceae: **A, C–F** *Mirabilis jalapa*. **B** *Bougainvillea spectabilis*. **A** Flowering twig. **B** Flower of *B. spectabilis*, subtended by a leafy bract. **C** Flower, vertical section. **D** Stamen. **E** Carpel. **F** Seed. (*k calyx*, *o* ovule, *stg* stigma). (Original).

Important Genera and Economic Importance. *Mirabilis multiflora* is an interesting genus in which the apparent calyx is only an involucre of bracts and the actual calyx is modified into an attractive and coloured structure that resembles a sympetalous corolla (Fig. 11.3 A, B). Species of the genus *Neea* often show cauliflory. Bracts petaloid and calyx less conspicuous in *Bougainvillea* and *Coligonia. Okenia* species produce geocarpic fruits. In *O. hypogaea* the peduncles elongate up to as much as 11 inches and dig deep into the sand, where the fruits mature. Although the plant appears to be perennial, it is actually annual. Every season the fruits buried deep down into the sand give rise to new plants. *Okenia* also shows anisophylly. *Pisonia alba,* popularly called "Tree Lettuee" is a perennial soft-wooded highly branched small tree. *P. alba* has a scanty distribution in small remote islands in Pacific and Indian Ocean and in the beach forest of the Andamans. It is a threatened medicinal plant highly valued for its use in rheumatic and pulmonary disorders (Jagadishchandra et al. 1999). *Pisonia grandis,* a large tree, is widely distributed by various sea birds as the fruits are covered by sticky glands and adhere to the feathers of the birds. Similar fruits are also seen in *Boerhavia diffusa,* a small prostrate herb with pink flowers. In *Ramisia brasiliensis,* the persistent calyx spreads out like wings in mature fruits to help in dissemination.

A few taxa are ornamentals like *Bougainvillea, Mirabilis jalapa* (the 4 o'clock plant) and *Abronia* (the sand verbenas).

Taxonomic considerations. The presence of betalain pigments and basal placentation do not leave any doubt about the inclusion of this family in Centrospermae. Ehrendorfer (1976a) considers it to be a derivative of Phytolaccaceae, because of the presence of "showy bracts and fused tepals (= sepals)".

Most taxonomists include this family in the Centrospermae; Hutchinson (1969, 1973) includes it in the Thymelaeales. Mabry et al. (1963) and Mabry (1976) include it in the suborder Chenopodiineae of the order Caryophyllales, as it contains betalain.

Cronquist (1981) and Takhtajan (1987) also place Nyctaginaceae in the Caryophyllales; Cronquist derives this family from the Phytolaccaceae.

Caryophyllaceae

A medium-sized family of 80 genera and 2000 species distributed primarily in the north temperate regions, a few genera in the south temperate regions and higher altitude areas of the tropics. The Mediterranean area is the centre of distribution. Two genera—*Colobanthus* and *Lyallia*—occur in Antarctica, and many genera such as *Silene, Lychnis, Spergula. Spergularia, Sagina, Arenaria, Paronychia* and *Scleranthus* grow in the cold Arctica.

Vegetative Features. Annual (11.4.2 A) or perennial herbs, sometimes suffrutescent shrubs. Stem herbaceous with swollen nodes, leaves opposite, rarely alternate, simple, mostly linear to lanceolate, often sessile and basally connected by a transverse line or by a shortly connate-perfoliate base as in *Dianthus*, stipulate with scarious stipules or estipulate.

Floral Features. Inflorescence simple or complex dischasial cyme (Fig. 11.4.1 A) or solitary terminal flowers, globose spiny heads in *Sphaerocoma.* Flowers ebracteate, often bracteolate (Fig. 11.4.1 H) as in *Dianthus*, bisexual, actinomorphic, hypogynous dimorphic in some *Stellaria* species; sterile flowers petalliferous and fertile ones apetalous. Calyx of 5 sepals, free as in *Cerastium, Stellaria* (11.4.2 B) *Spergularia*, (Fig. 11.4.1 B) and *Polycarpaea* or united as in *Dianthus* (Fig. 11.4.1 H), *Silene* and *Lychnis.* Corolla with 5 petals, polypetalous, with a distinct limb and claw (Fig. 11.4.1 I), often deeply bifid, e.g. in *Silene* and *Stellaria* (Fig. 11.4.1 B, C) or limb multifid as in *Dianthus* (Fig. 11.4.1 I). Corolla absent in *Cerdia, Colobanthus* and *Microphyes.* Stamens in 1 or 2 whorls, same as or double the number of petals, filaments free, anthers bicelled, longitudinally dehiscent (Fig. 11.4.1 B, H) petaloid staminodes sometimes present. Stamens alternate with sepals in apetaious flowers of *Colobanthus.* Gynoecium syncarpous, 2- to 5-carpellary (Fig. 11.4.1 D). Ovary unilocular with free-central placentation (Fig. 11.4.1 E, F, J, 11.4.2 E) or basally 3-5-loculed with axile placentation in the lower part and unilocular with free-central condition above as in some Silenoideae; or unilocular ovary with solitary, basal ovule as in *Acanthophyllum* and *Drypis*; ovules 1 to numerous. Ovary usually on a gynophore which is a stipe-like torus. Styles and stigmas 2 to 5 or as many as the carpels; fruit a capsule dehiscing apically by valves or circumscissilely or indehiscent utricle or achene. Seed usually with a hard endosperm (soft in *Agrostemma*) and a curved embryo (straight in, *Dianthus*), often winged as in *Spergula arvensis* (Fig. 11.4.1 G). Fig. 11.4.2 G is the floral diagram for *Stellaria media.*

Anatomy. In transverse section of stem, xylem and phloem appear to be in the form of a continuous cylinder or as distinct bundles separated by broad rays. Vessels are with simple perforations. Anomalous secondary growth occurs more frequently in roots of certain genera. Leaves dorsiventral or centric; central tissue serves as water storage tissue in *Sphaerocarpos.* Various types of hairs—both uniseriate and multiseriate and glandular—occur on leaf surface. Wax is thickly deposited on the leaf surface of certain species of *Cometes, Dicheranthus* and *Pteranthus;* stomata generally of the Caryophyllaceous type.

Embryology. Pollen grains 3-colpate, pantoporate, exine spinulose and tubuliferous-punctate or finely reticulate, shed at 3-celled stage. Ovules campylotropous, bitegmic, crassinucellate. Embryo sac of Polygonum type, 8-nucleate at maturity. Endosperm formation of the Nuclear type. Seed exarillate and perispermous.

Chromosome Number. Basic chromosome number is x = (5-) 9–15 (–19).

Chemical Features. Rich in anthocyanin pigments, some taxa also contain saponin.

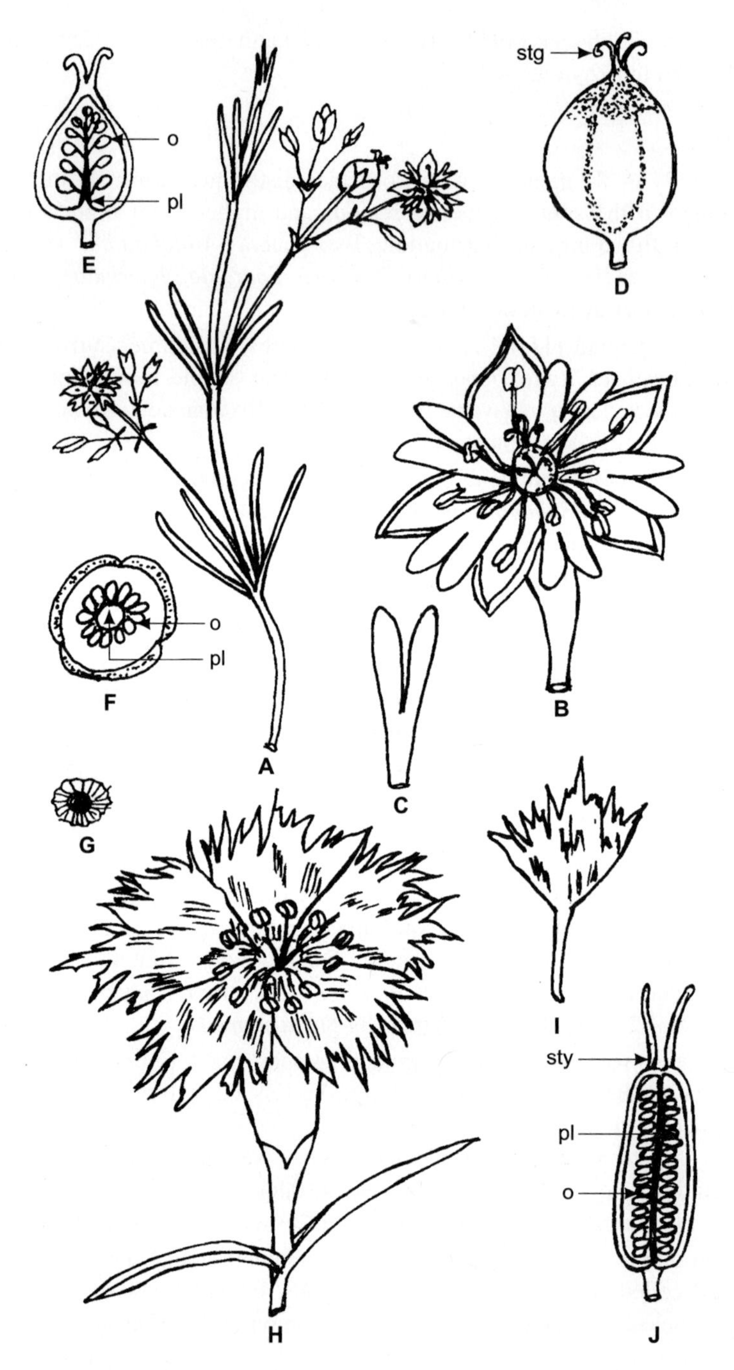

Fig 11.4.1 Caryophyllaceae: **A–G** *Spergula arvensis*, **H–J** *Dianthus caryophyllus*. **A** Flowering branch. **B** Flower. **C** Petal. **D** Carpel. **E** vertical section of carpel. **F** Ovary, cross-section. **G** Seed. **H** Flower of *D. caryophyllus*. **I** Petal with limb and claw. **J** Vertical section of carpel. (*o* ovule *pl* placenta, *stg*, stigma, *sty* style). (Original).

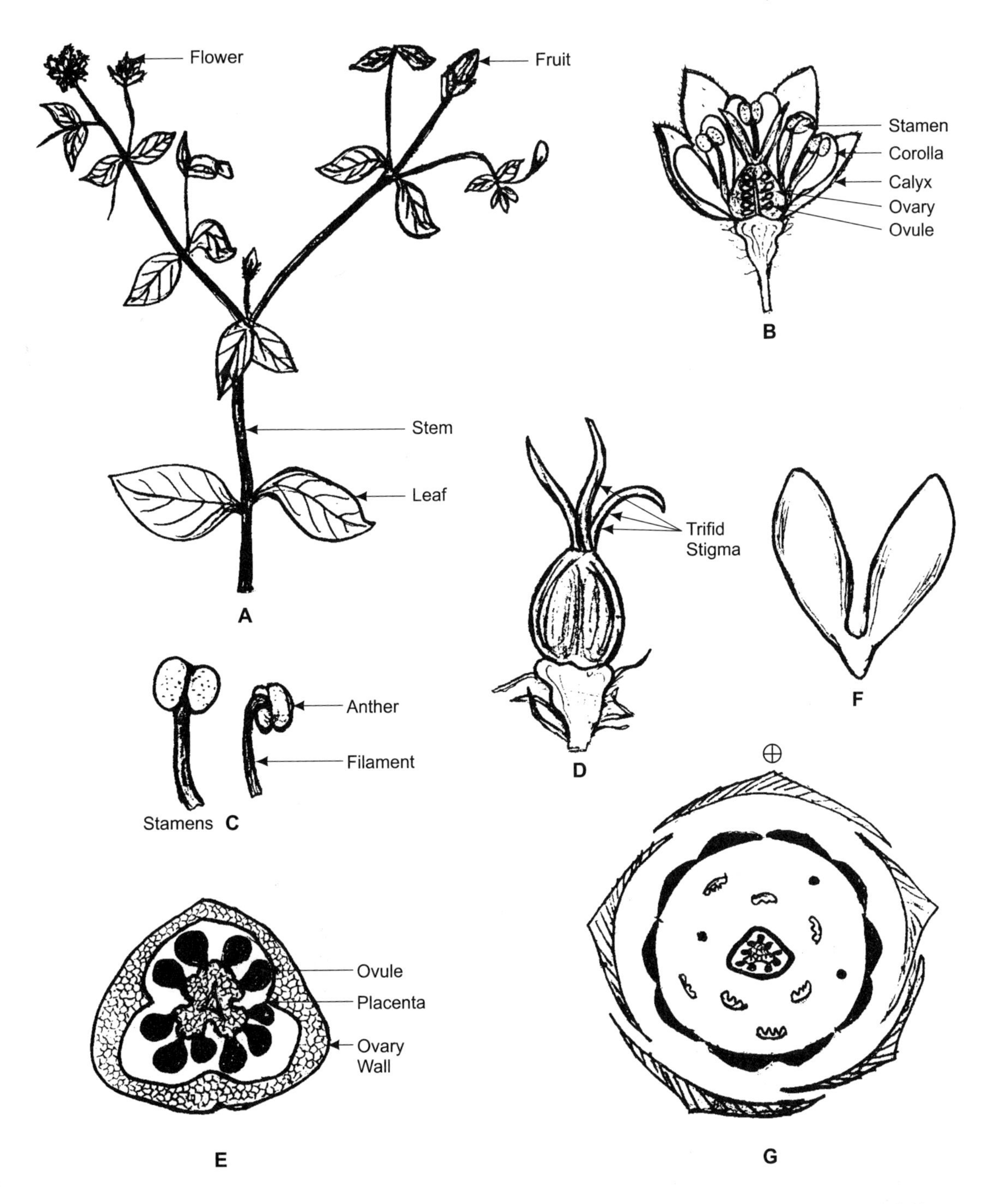

Fig 11.4.2 Caryophyllaceae: **A–G** *Stellaria media* **A** Flowering twig. **B** Flower, longisection. **C** Stamens. **D, E** Ovary **(D)** and transverse section of the same **(E). F** Bifid petal. **G** Floral diagram.

Important Genera and Economic Importance. A large number of genera have rather unusual distribution from South America to Africa to Australia and New Zealand. Many genera grow at an altitude of 2000 to 4000 ft in the Himalayas.

This family is economically important for the ornamentals, e.g. *Dianthus, Gypsophila, Saponaria Silene, Lychnis, Arenaria* and *Cerastium.* The roots of *Vaccaria pyramidata* yield saponin, which forms lather with water.

Taxonomic Considerations. The origin of Caryophyllaceae is disputed. One view is that it originated from the Phytolaccaceae, where the outer whorl of stamens becomes converted to petals, and the outer whorl of carpels to stamens. This view is supported by Eichler (1875; see Lawrence 1951), Pax (1927), Rendle (1925) and Wettstein (1935). The second view is that it originated from the Ranalean ancestors and is the source of origin for Amaranthaceae, Chenopodiaceae and the Primulales. This was proposed by Wernham (1911) and is supported by Bessey (1915), Lawrence (1951), Hutchinson (1973), Cronquist (1981), Dahlgren (1983a) and Takhtajan (1987). A third view, suggested by Dickson (1936), is that it has originated from the Geraniales. The present knowledge about Caryophyllaceae supports the view of Wernham.

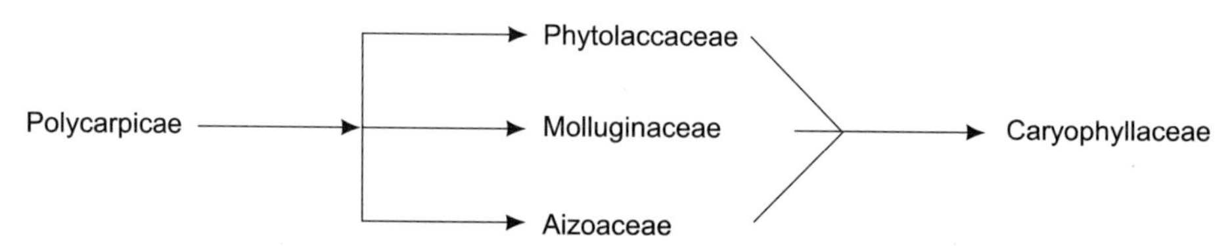

According to Ehrendorfer (1976a), Caryophyllaceae has originated from the woody Polycarpicae (with showy perianth, primary polyandry and anthocyanin pigments) through Phytoaccaceae-Molluginaceae-Aizoaceae although it has diverged from this group in course of evolution. They have retained the anthocyanin pigments from the ancestral "Polycarpicae" although anthocyanins have been replaced by betalain pigments in the immediate ancestors or the core families. Mabry (1976), on the basis of DNA-RNA hybridisation data, is also of the opinion that the Caryophyllaceae is derived from a common "Centrospermous" ancestor.

Chenopodiaceae

A family with 105 genera and 1600 species, more or less cosmopolitan in distribution, especially in the xeric environment and halophytic areas. The family is well represented on the prairies and plains of North America, the pampas of South America, the shores of the Red, Caspian and Mediterranean seas, the Central Asiatic region, the South African karroo and the salt plains of Australia (Lawrence 1951). Many genera are indicators of saline habitats.

Vegetative Features. Annual or perennial herbs (Fig. 11.5 A) or shrubs, rarely trees, e.g. *Haloxylon* of Central Asiatic Steppes. Plants mostly halophytes, *Salsola kali* grows near seashore and *Salicornia herbacea* in salt marshes, or in steppes and deserts which at one time were covered with sea water but are dry and supersaturated with salt at present. These plants exhibit typical xerophytic characters to reduce the rate of transpiration. Stems sometimes fleshy, jointed and nearly leafless. Leaves alternate (Fig. 11.5 A), rarely opposite as in *Nitrophila* and *Salicornia,* simple, estipulate, fleshy and terete in some and reduced to scales in others.

Floral Features. Inflorescence dichasial or unilateral cymes (Fig. 11.5 A). Flowers bracteate, ebracteolate, bi-or unisexual and the plants are dioecious (as in *Grayia)* or monoecious (as in *Sarcobatus),* actinomorphic, mostly hypogynous, epigynous in *Beta,* minute and greenish. Perianth uniseriate, pentamerous, sepaloid, basally connate (Fig. 11.5 C), persistent (sometimes absent in male flowers, only 2 in *Atriplex,* and 3 or 4 in *Salicornia),* imbricate. Stamens as many as perianth lobes and opposite the tepals, inserted on a staminal disc or a hypogynous disc. Filaments distinct, incurved in bud, anthers bicelled, introrse, dehisce longitudinally (Fig. 11.5 D). Gynoecium syncarpous, 2- or 3-carpellary, ovary unilocular with a solitary basal ovule (Fig. 11.5 E, F), superior or inferior, styles and stigmas 1 to 3. Fruit an indehiscent nut, nutlet or achene, enclosed in a persistent perianth; sometimes a large number aggregate together by connation of fleshy perianth. Seeds small with endosperm surrounded by a peripheral or coiled embryo. Endosperm scanty or none in *Salsola, Sarcobatus* and *Suaeda* (Lawrence 1951). Fig. 11.5 F represents the floral diagram of *Chenopodium album.*

Anatomy. Vessels small, typically in clusters on the inner side of the phloem strands, sometimes tending to be ring-porous; with spiral thickening, perforations simple; intervascular pitting alternate. Parenchyma conjunctive, linking the strands of phloem in broad irregular bands and scattered round and among the vessel groups; rays absent. Typical flattened leaves develop in only a few genera such as *Atriplex, Beta, Chenopodium, Hablitzia, Obione* and *Rhagodia.* Leaf highly reduced in size in other genera. Various types of trichomes occur (Carolin 1982). Epicuticular wax of platelet type is reported in Chenopodiaceae members (Engel and Barthlott 1988). Stomata occur on all parts of the surface of both cylindrical and flattened leaves; generally Ranunculaceous type, sometimes Rubiaceous, as in some species of *Camphorosma, Salicornia, Salsola* and *Suaeda.* Anomalous secondary growth recorded in stem and root. Mature stems contain numerous vascular bundles, laid down together with the conjunctive tissue around them, by a succession of rings or arcs of cambium. Although, usually situated in the pericycle, these may also originate in phloem (Metcalfe and Chalk 1972).

Embryology. Pollen grains polyporate with spinulose or tubuliferous punctate exine and shed at the 3-celled stage. Ovules campylotropous, bitegmic, crassinucellate. Polygonum type of embryo sac, 8-nucleate at maturity. Endosperm formation of the Nuclear type. Embryo annular or conduplicate in Cyclobeae and spirally coiled in Spirolobeae.

Chromosome Number. Basic chromosome number is x = (6–) 9.

Chemical Features. Chenopodiaceae is a betalain-containing family. The flavonoid chemistry (Young 1981) of this family is allied to Dilleniiflorae and Malviflorae than to Magnolliiflorae. According to Hartley and Harris (1981), ferulic acid is present in the cell walls of Chenopodiaceae.

Important Genera and Economic Importance. Some of the important genera are *Beta, Chenopodium* and *Spinacia. Beta* is the only genus (in this family) with inferior ovary. *B. vulgaris,* commonly called garden beet or sugar beet, is largely cultivated for its roots. The white variety is the source of sucrose or beet sugar and the red variety is used as a vegetable. *Chenopodium anthelminticum* yields an essential oil, "oil of wormwood", used as a vermifuge. C. *album* and C. *murale* are often used as pot herb. The seeds of C. *quinoa* are boiled and eaten like rice. *Spinacia oleracea* is used as a green vegetable. *Kochia indica* is an ornamental plant. *Suaeda, Salsola* and *Salicornia* are plant indicators for saline soil. *Salsola* is a good fodder for camels.

Taxonomic Considerations. The Chenopodiaceae is a member of the Centrospermae, according to Engler and Diels (1936), Wettstein (1935) and Melchior (1964). Bentham and Hooker (1965c) treated it as a member of the Monochlamydeae (because of uniseriate perianth), Hutchinson (1973) recognises a separate order Chenopodiales—advanced over the Caryophyllales; Cronquist (1981) includes it in his Caryophyllales (= Centrospermae). Takhtajan (1987) also includes all the betalain-containing families in the order Caryophyllales. However, he includes the two allied families—Amaranthaceae and Chenopodiaceae—in the same suborder Chenopodiineae of the order Caryophyllales.

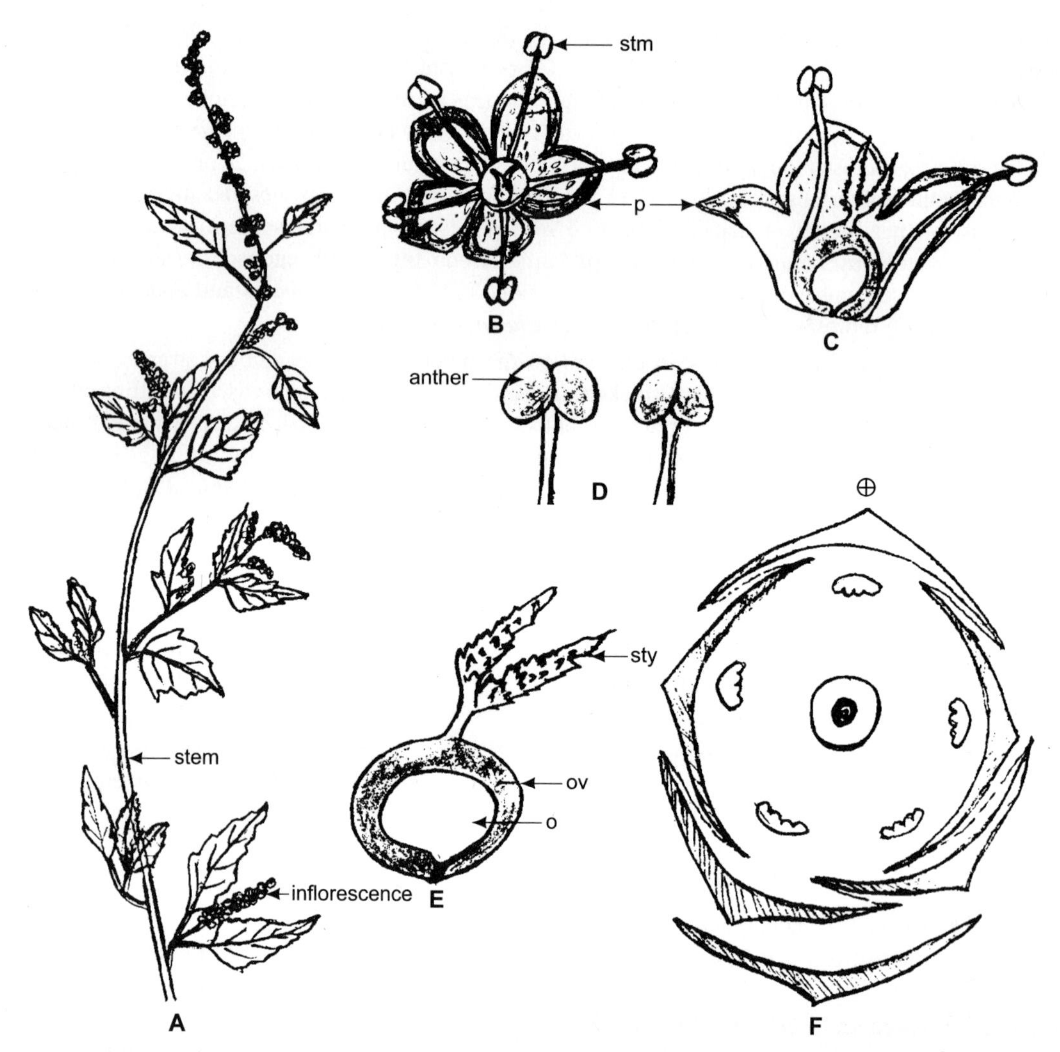

Fig 11.5 Chenopodiaceae: **A–F** *Chenopodium album* **A** Twig with leaves and inflorescences. **B** Flower, top view. **C** Flower, longisection **D** Stamens. **E** Ovary, longisection. **F** Floral diagram. (*o* ovule, *ov* ovary wall, *p* perianth, *stm* stamen, *sty* style). Sketched by Meenakshi Dua.

Amaranthaceae

A family of 65 genera and 900 species distributed widely, but abundant in tropical America and tropical Africa. Nearly one-third of the genera are monotypic.

Vegetative Features. Annual or perennial herbs (Fig. 11.6 A), rarely shrubs or trees; *Alternanthera aquatica* is an aquatic herb. Stem often angular or ridged, green or sometimes reddish, as in some species of *Amaranthus.* Leaves alternate or opposite, simple, entire, estipulate, often covered with adpressed hairs as in *Aerva tomentosa* and *Achyranthes aspera.*

Floral Features. Inflorescence a simple or branched spike or raceme, the ultimate branches often dichasial cymes. Flowers minute and often densely crowded in the inflorescence to give an attractive appearance, e.g. *Amaranthus* and *Celosia;* bracteate, bracteolate, bisexual, less commonly unisexual, actinomorphic, hypogynous. Perianth uniseriate, usually of 4 or 5 perianth lobes (Fig. 11.6 B), free or basally connate, dry, membranous, white or coloured, often hairy. Stamens same as the number and opposite the tepals, filaments usually partially connate along their entire length into a membranous tube (Fig. 11.6 C); lobed or fringed petaloid outgrowths may alternate with the anthers (Fig. 11.6 C); anthers 4-celled at anthesis in Amaranthoideae and bicelled in Gomphrenoideae, dehiscence longitudinal (Fig. 11.6 D). Gynoecium syncarpous, 2- or 3-carpellary (Fig. 11.6 E); ovary unilocular with a solitary basal ovule, superior; ovules several on a seemingly single basal funicle in Celosieae. Styles and stigmas 1 to 3. Fruit a circumscissile capsule as in *Celosia* or a utricle or nutlet, rarely a drupe or berry. Seed with embryo enveloping the mealy endosperm, usually disc-shaped and with a shiny testa.

Anatomy. Vessels small to medium-sized. perforations simple, intervascular pitting alternate and moderately large. Parenchyma paratracheal, scanty to vasicentric, sometimes storied. Rays absent and replaced by radial sheets of conjunctive parenchyma. Included phloem common. Stem frequently angular with collenchyma well-developed in the ribs. Leaves dorsiventral (isobilateral in *Celosia argentea).* Woolly or silky covering of uniseriate hairs is common in addition to various special types. Stomata

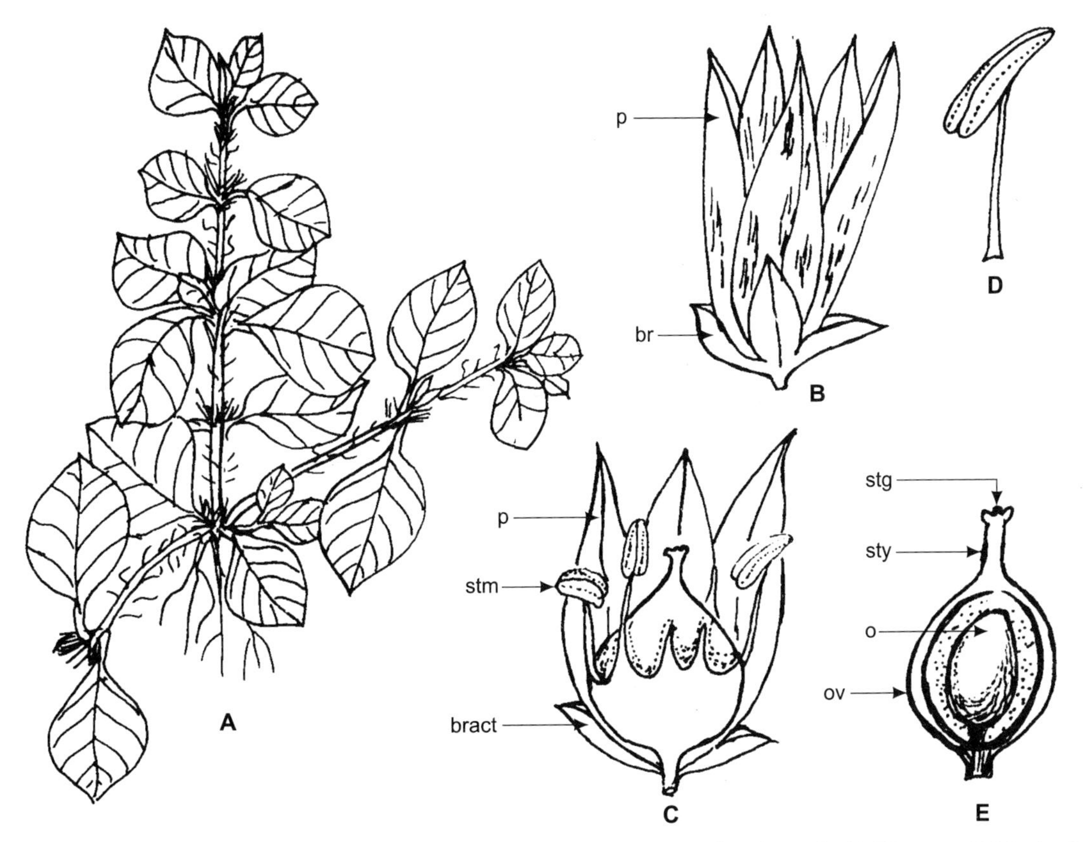

Fig. 11.6 Amaranthaceae: **A–E** *Alternathera pungens*. **A** Flowering branch. **B** Flower. **C** Vertical section of flower. **D** Stamen. **E** Carpel, vertical section. (*br* bract, *o* ovule, *ov* ovary, *p* perianth, *stg* stigma, *stm* stamen, *sty* style). (Original).

present on both surfaces but usually more numerous on the lower than the upper surface in *Achyranthes, Aerva*, *Allmania*, *Amaranthus*, *Celosia*, *Gomphrena* and *Pupalia.* Vascular bundles of both large and small veins are surrounded by sheaths of usually green, almost cubical, parenchymatous cells. Cluster crystals of very large size recorded in *Iresine* (Schinz 1934). Crystal-sand reported in the axis as well as leaf of *Acnida. Allmania. Amaranthus, Celosia, Cyathula, Deeringia* and *Pupalia* (Metcalfe and Chalk 1972).

Embryology. Pollen grains polyporate with spinulose or tubuliferous-punctate exine, shed at the 3-celled stage. Ovules campylotropous, bitegmic, crassinucellate. Endosperm formation of the Nuclear type.

Chromosome Number. Basic chromosome number is x = 7 – 9 (–13).

Chemical Features. Amaranthaceae members contain betalain pigments.

Important Genera and Economic Importance. The family is not very important except for a few species of *Amaranthus*, whose leaves and seeds are edible. A few species of *Amaranthus, Celosia* and *Gomphrena* are grown as ornamentals. A number of genera like *Achyranthes, Alternanthera, Aerva, Digera* and *Pupalia* grow as weeds.

Taxonomic Considerations. Anatomically, Amaranthaceae is affiliated to Nyctaginaceae but it does not contain any raphides or styloids. It is also close to Chenopodiaceae because of the similar type of anomalous secondary growth, and trichomes (Carolin 1982).

According to Lawrence (1951), Amaranthaceae was earlier presumed to be a primitive family but recent studies of the bracts and bractlets provide evidence that the basic inflorescence is a dichasium of 3 flowers, of which 2 have been lost, and only the bractlets remain. Each flower represents an ancestral dichasium and this is certainly an advanced feature. Hutchinson (1969) relates Amaranthaceae to the tribe Polycarpeae of the Caryophyllaceae in which the calyx is similarly dry and scarious. Cronquist (1981) and Takhtajan (1987) include Amaranthaceae in Caryophyllales and refer to its alliance with Chenopodiaceae.

Order Magnoliales

A large order of 22 primitive families, mostly distributed in the tropics and subtropics of the southern hemisphere. Some of the small, isolated relic families of this order are common in the tropical and subtropical rain forests in the Western Pacific region: Austrobaileyaceae, Eupomatiaceae, Himantandraceae, Idiospermaceae, Calycanthaceae are confined to Eastern Australia and Eastern Malaysia (Endress 1983).

Plants evergreen or deciduous, large to medium-sized trees or shrubs. Leaves simple, mostly alternate. Flowers usually solitary, large and showy, sometimes in inflorescence as in Trochodendraceae. Numerous whorls of sepals and petals and indefinite stamens are arranged spirally; anthers introrse. Gynoecium monocarpellary or many-carpelled and apocarpous. Wood anatomy shows primitive vessels with scalariform end-walls. Pollen grains are shed at 2-celled stage; shedding in permanent tetrads in Winteraceae, Monimiaceae and rarely in Annonaceae and Magnoliaceae. Ovules anatropous, bitegmic, crassinucellate. Many of the families are rich in alkaloids.

Magnoliaceae

A small family of about 14 genera and 240 species, distributed mainly in the tropics and subtropics of both the New and the Old World. Some members occur in North temperate zones also. The geological record shows that the family was at one time much more widely distributed in the Northern hemisphere.

Vegetative Features. Large to medium-sized trees or shrubs; evergreen or deciduous. Leaves simple, alternate, entire, stipulate, sometimes thick, coriaceous and shiny as in *Magnolia;* stipules often large and protective to the young buds, deciduous leaving a circular scar around the node as the leaves expand.

Floral Features. Flowers usually solitary, terminal or axillary; hermaphrodite (unisexual in *Kmeria),* actinomorphic, bracteate, ebracteolate, mostly very large and ornamental, often with fragrance as in *Michelia champaca* (Fig. 11.7 A, B). Calyx distinct or indistinct; when distinct, sepals 3, cyclic, green. Corolla of 6 petals or more, spirally arranged, often around the base of an elongated receptacle as in *Magnolia.* Stamens numerous, hypogynous, distinct, arranged spirally; anthers bicelled, introrse, dehiscence longitudinal (Fig. 11.7 C). Gynoecium sessile or borne on an elongated axis or gynophore; carpels numerous, free, arranged spirally on the axis (Fig. 11.7 E). Ovary monocarpellary, unilocular, placentation parietal, ovules 1 to numerous (Fig. 11.7 D), style 1, stigma 1. Fruit a follicle, berry or samara; an etaerio of follicles in *Magnolia.*

Anatomy. Vessels usually medium-sized but small (less than 100 mm*)* in *Alcimandra, Liriodendron, Magnolia, Michelia* and a few other taxa; solitary or in small groups, sometimes with spiral thickening; perforation plates typically scalariform with a few, widely spaced bars; but perforations simple in *Magnolia acuminata.* Intervascular pitting scalariform to opposite; parenchyma terminal; rays usually up to 3 or 4 cells wide, hetero- or homogeneous. Leaves dorsiventral, stomata mostly confined to the lower surface; usually Rubiaceous sometimes Ranunculaceous type. Calcium oxalate crystals not very common.

Embryology. Pollen grains uni- or triaperturate, monocolpate; mature grains oval-shaped with thick walls; shed at 2-celled stage in *Magnolia,* 3-celled in *Liriodendron tulipifera.* Ovules anatropous, bitegmic, crassinucellate; Polygonum type of embryo sac, 8-nucleate at maturity. Endosperm formation of the Cellular type.

Chromosome Number. Basic chromosome number is x = 19.

Chemical Features. D-pinitol, a methyl ether of inositol, has been isolated from *Magnolia* spp. A dimethyl ether of myo-inositol, liriodendritol, occurs in only two Species of *Liriodendron—L. chinense* and *L. tulipifera* (Plouvier 1963). An alkaloid, liriodendrine, occurs in the heartwood of *L. tulipifera*

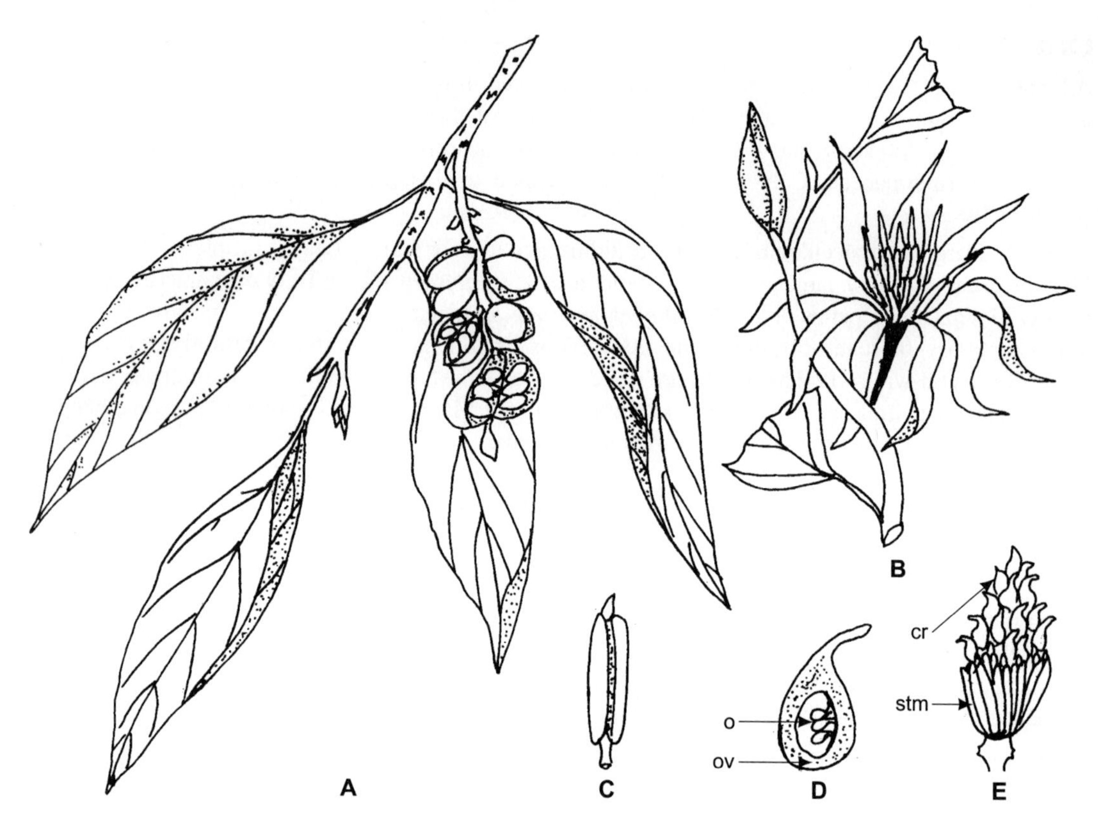

Fig. 11.7 Magnoliaceae: **A–C, E** *Michelia champaca*, **D** *Magnolia grandiflora.* **A** Fruiting twig. **B** Flower. **C** Stamen. **D** Carpel, vertical section. E Stamens and carpels arranged spirally. (*cr* carpel, *o* ovule, *ov* ovary, *stm* stamen). (**A, B** after DeWit 1963).

(Hegnauer 1963). Essential oils of different types have been reported in various species of *Magnolia* and *Michelia* (Thien et al. 1975). Sesquiterpene lactones have been recorded in *Liriodendron* and *Michelia.*

Important Genera and Economic Importance. A large number of trees are ornamental. *Michelia champaca* is a large tree and bears solitary axillary flowers with an elongated cone-like torus or gynophore; perianth petaloid in many whorls of 3 each, followed by spirally arranged free stamens and carpels. It is cultivated for its fragrant flowers. The timber of *M. nilagirica* is very handsome and used for furniture, railway sleepers; bark yields essential oil. Various species of *Magnolia* are large or medium-sized trees which bear beautiful, large, fragrant flowers and timber that is used for cabinet work. *Liriodendron tulipifera* is a cultivted deciduous tree of the eastern United States of America. It yields a commercial timber called Canary white wood. Timber of *Magnolia acuminata* is similar to that of *Liriodendron* and is soft and easy to work with. *Aromadendron* is another interesting genus in which the carpels are concrescent (fused), fleshy and indehiscent, and the ovules in each ovary are reduced to 2. In *Pachylarnax* also the carpels are fused but they open completely along their abaxial suture and partially along their line of junction. Therefore, the fruit in more or less like a woody loculicidal capsule (Hutchinson 1969). *Magnolia acuminata, Manglietia hookeri, Michelia baillonii, M. deltsopa* and *Pachylarnax pleiocarpa* produce valuable timber.

Taxonomic Considerations. The family Magnoliaceae is considered to be the most primitive family amongst the dicots (Bentham and Hooker 1965a, Hallier 1905, Hutchinson 1973, Stebbins 1974). Hallier (1905) even compared the elongated floral axis (torus) bearing the spirally arranged floral whorls with the sporophyll-bearing axis of the Bennettitales. Some members of this family show very primitive features, e.g. *Magnolia stellata* and *Aromadendron* have tepals, stamens and carpels, all spirally arranged, with large, more or less petaloid tepals that are not distinguishable into sepals and petals. On the other hand, *Liriodendron* has three series of three tepals each, the outermost series is comparable to sepals. Anatomically, too, Magnoliaceae members show primitive structures such as xylem vessels with scalariform perforation plates and spiral thickening. The tree-like habit with large, showy, solitary flowers is also a primitive feature.

Magnoliaceae is indeed a primitive family, but certainly not the most primitive. Melchior (1964), Benson (1970), Cronquist (1981), Dahlgren (1980a, 1983a) and Takhtajan (1987) agree to this suggestion.

Annonaceae

The Annonaceae are a common pantropical woody family comprising about 130 genera and 2300 species (Okada and Ueda 1984). A large family of aromatic trees, shrubs or climbers, which occur in tropical and subtropical regions. In the tropics of the Old World, they are usually of climbing or straggling habit, and occur in lowland dense evergreen forest. In tropical America, they are nearly all shrubby or arboreal, and grow mostly in the open grassy plains (Hutchinson 1964, Leboeuf et al. 1982). The only genus extending into the temperate zone is *Asimina,* which occurs in North America as far north as the Great Lakes. According to Takhtajan (1969), 51 genera and ca. 950 species are confined to Asia and Australasia, 40 genera and ca. 450 species in Africa and Madagascar and 38 genera and 740 species on the American continent. Thus Asia, together with Australasia, is the basic centre of distribution of the Annonaceae. However, on the basis of phytogeographical and palynological data, Le Thomas (1981) hypothesizes South America or Africa to be the centre of origin of this family.

Vegetative Features. The members of Annonaceae are trees (*Polyalthia longifolia*), shrubs (*Annona squamosa, A. reticulata*) or vines such as *Artabotrys odoratissima* (Fig. 11.8 A), *Uvaria* spp. with aromatic wood and foliage. Leaves simple, alternate, entire, estipulate; deciduous or persistent, gland-dotted.

Floral Features. Flowers usually solitary (Fig. 11.8 A), rarely inflorescent, bisexual (unisexual and plants dioecious in *Ephedranthus*, *Stelechocarpus*, and *Thonnera*), actinomorphic, hypogynous, spirocyclic, trimerous, usually scented. Perianth triseriate, the outer whorl, a calyx of 3 basally connate or free, valvate, persistent sepals, and two inner whorls, a corolla of usually 6 distinct, similar or dissimilar petals, imbricate or valvate. The axis usually extends and enlarges beyond the point of perianth attachment. Corolla sympetalous in *Asteranthe*, *Disepalum*, *Enneastemon*; outer petals with long dorsal appendages in *Rollinia*; petals 6 in *Haplostichanthus*, *Monanthotaxis* and *Monocyclanthus* and only 3 in *Xylopia.* Stamens numerous, free, spirally arranged, only 9 to 12 arranged in whorls in *Mezzettia*; filaments short and thick, anthers 4-celled at anthesis (Fig. 11.8 B), extrorse, but introrse in *Mezzettia parviflora*; anther lobes linear with a prolonged connective; anthers transversely locellate in *Cardiopetalum*, *Hornschuchia* and *Porcelia*, and connective is not elongated in *Alphonsea.* Gynoecium of a few to numerous carpels (basally connate in *Monodora*), arranged spirally; carpel solitary in *Kingstonia*, *Mezzettia*, *Monocarpia* and *Tridimeris.* Ovary superior, unilocular (Fig. 11.8 C), but multilocular in *Pachypodanthium*; ovules 1 to many, typically parietal but often appear to be basal; rarely all the carpels are united to form a unilocular ovary with parietal placentation as in *Monodora.* Style 1, very short or absent, stigma 1. Fruit a berry or etaerio of berries (Fig. 11.8 D) or the mature pistils are connate and adnate to the floral axis to form a fleshy, aggregate fruit as in *Annona squamosa* (Fig. 11.8 F). Seed large (Fig. 11.8 E) with a small embryo and copious ruminate endosperm, often arillate.

Anatomy. Vessels very small (less than 50 mm mean tangential diameter), as in some species of *Malmea, Orophea, Oxandra* and *Popowia*, or large (more than 200 mm, diameter), as in some species of *Cananga, Cleistopholis, Guatteria, Rollinia* and *Unona*; usually a few, with simple perforations. Parenchyma apotracheal, in numerous fine lines, often storied. Rays typically wide and high, 3–16-celled wide, slightly heterogeneous to homogeneous. Leaves generally dorsiventral; stomata Rubiaceous type, confined to the lower surface. Simple, stellate and peltate hairs on both surfaces. Both solitary and clustered crystals occur in epidermal cells.

Embryology. Pollen grains either inaperturate or monocolpate and provided with a proximal germ pore. Generally mature pollen grains show a tendency to remain in groups in tetrads in *Annona*; in pairs in *Cananga*, and are shed as tetragonal or rhomboidal tetrads (Parulekar 1970). Exine sculpturing varies

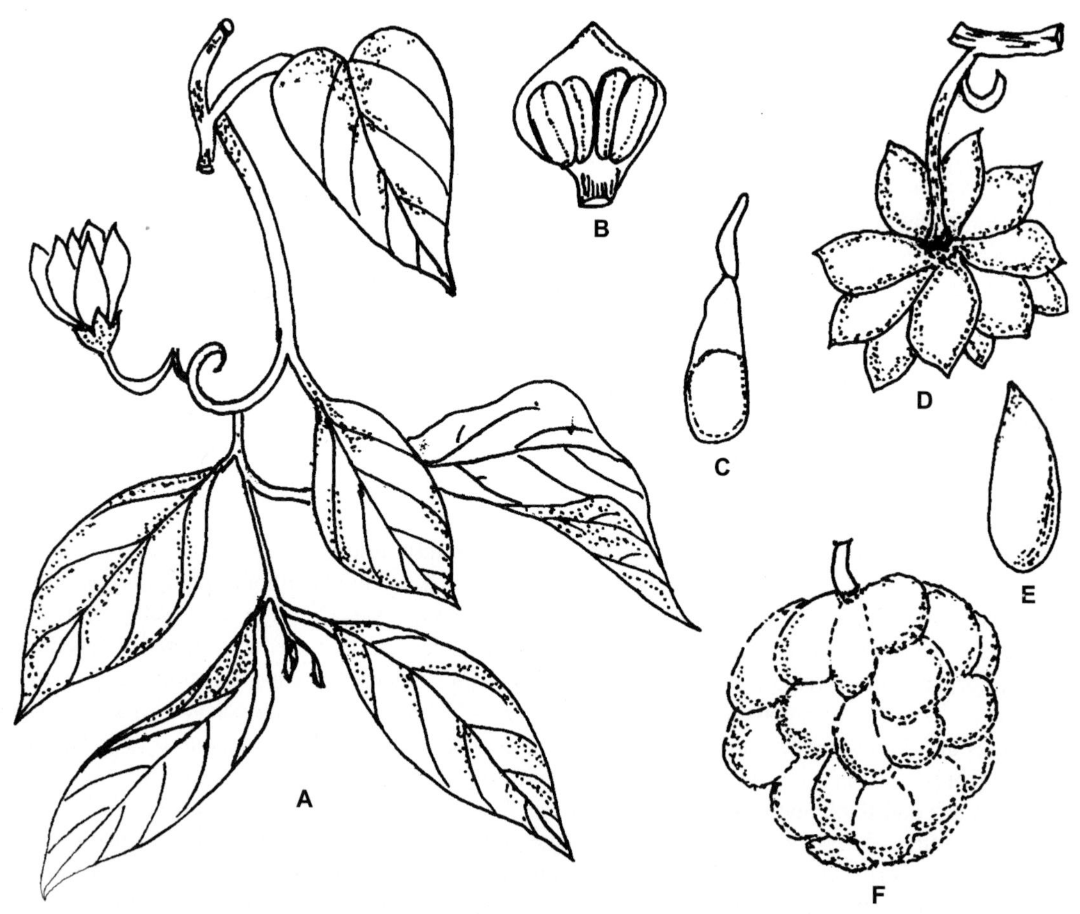

Fig. 11.8 Annonaceae **A–E** *Artabotrys odoratissima*, **F** *Annona squamosa*. **A** Twig with flower. **B** Stamen. **C** Carpel. **D** Fruit. **E** Seed. **F** Aggregate fruit of *A. squamosa*. (**A–E** adapted from DeWit (1963)

from psilate, verrucate, reticulate to echinate (Canright 1962); shed at 2-celled stage. Ovules anatropous, bi- or tritegmic, crassinucellate. Embryo sac of Polygonum type, 8-nucleate at maturity. Endosperm formation of the Cellular type and becomes horny and ruminate in the mature seed.

Chromosome Number. Basic chromosome numbers are x = 7, 8, 9, 14 (Takhtajan 1987). Okada and Ueda (1984) report that among Asian genera of Annonaceae, 2n = 14 occurs only in *Mezzettia*; 2n = 16 is widespread and is also reported in *Anaxagorea* (with some primitive characters). 2n = 18 is reported for 11 genera, and tetraploidy (2n = 36) has been observed in *Polyalthia* species. Basic no. x = 8 or 9 has been suggested for the Asian genera.

Chemical Features. Annonaceae members produce a wide range of non-alkaloidal compounds including carbohydrates, lipids, amino acids and proteins, polyphenols, essential oils, terpenes, aromatic compounds, and a host of other substances in addition to a large number of alkaloids (Leboeuf et al. 1984). Among the alkaloidal components, it is interesting that isoquinolines, which have benzylisoquinolines as in vivo precursors (Shamma and Moniot 1978), are the main alkaloidal constituents of the members of Annonaceae. Isolation of isoquinolines is significant from the chemotaxonomic point of view. The occurrence of the alkaloids—anonaine, xylopine, liriodenine and lanuginosine—in *Xylopia brasiliensis* and *Annona squamosa* suggests a chemotaxonomic relationship between these two taxa (Bhaumik et al. 1979).

Important Genera and Economic Importance. The climber *Artabotrys odoratissima* climbs with the help of strong hooks borne on its flower stalks. *Polyalthia longifolia* is an ornamental tree with long, lanceolate, shiny waxy leaves and yellowish-green flowers in dense umbels. *Monodora myristica* of tropical Africa is the only genus of this family with polycarpellary, syncarpous ovary. Its flowers are also zygomorphic. The habit of the genus *Geanthemum* is remarkable; the flowers are borne on subterranean sucker-like shoots.

Economically, the fleshy fruits of various species of *Annona* are edible. The white custard-like pulp of custard apple (*Annona squamosa*) is sweet. The fruits of *A. reticulata* (bullock's heart) are so named because of their shape like a heart; these are sour to taste. The fruit of *A. muricata* is also sour to taste. *Artabotrys odoratissima*, *Unona discolor* and *Cananga odorata* are cultivated for their fragrant flowers. The essential oil, "ylang-ylang", is obtained from the flowers of *Cananga odorata*, and has widespread use in perfumery. Oils and fatty acids have also been isolated from fruits and/or seeds of various species of *Annona*, *Asimina triloba*, *Dennettia tripetala*, *Xylopia aethiopica*, *X. brasiliensis*, *X. longifolia* and from the leaves of *Annona muricata* and *A. senegalensis* (Hegnauer 1964, Leboeuf et al. 1984). These could possibly be used as edible oils after refining. Some products are used as spices, e.g. seeds of *Monodora myristica* as nutmegs, and *Xylopia aethiopica* as the source of Ethiopian pepper. The beautiful leaves of *Polyalthia longifolia* are often used for festive decorations.

Taxonomic Considerations. The family Annonaceae is closely related to the Magnoliaceae; it differs from Magnoliaceae in having estipulate leaves, valvate corolla, and ruminate endosperm. There is general agreement that the Annonaceae is a derivative from Magnoliaceous stock. Hutchinson (1973) erected a separate order Annonales and included in it Annonaceae and Eupomatiaceae; he also pointed out that Annonales was related to, but more advanced than, the Magnoliales.

According to Cronquist (1968), within the Magnoliales, 5 families—Magnoliaceae, Winteraceae, Degeneriaceae, Himantandraceae and Annonaceae—form a cluster and share some common primitive characters. Later, in 1981, Cronquist changed the position of Annonaceae and placed it with the Myristicaceae and Canellaceae. Dahlgren (1980a) treats Annonales including Annonaceae, Myristicaceae, Eupomatiaceae and Canellaceae as the most primitive dicot order. Takhtajan (1969, 1980) at first placed Annonaceae in the Magnoliales, but in 1987 he also erected a separate order Annonales, comprising Annonaceae, Canellaceae and Myristicaceae. Takhtajan treats Annonales as an advanced group over the Magnoliales and Eupomatiales.

The families Magnoliaceae and Annonaceae have many common features: (a) tree habit (a few exceptions in Annonaceae), (b) multilacunar nodes, (c) Rubiaceous type of stomata (Ranunculaceous type in some Magnoliaceae), (d) bisexual, trimerous flowers, (e) embedded microsporangia, (f) monocolpate pollen grains (acolpate in some Annonaceae), (g) free carpels arranged spirally (exceptions are *Monodora* and *Isolona* of Annonaceae where gynoecium is syncarpous), (h) anatropous, bitegmic, crassinucellate ovules, (i) Polygonum type of embryo sac, (j) Cellular type of endosperm, and (k) follicular fruit. All these features strongly support the close relationship of Magnoliaceae and Annonaceae. At the same time, the differences between them are no less significant (see Table 11.1).

Table. 11.1 Comparative Data for Magnoliaceae and Annonaceae

Magnoliaceae	*Annonaceae*
1. Stamens with 3-7 vascular trace	Stamens with 1 vascular trace
2. Pollen grains released as monads	Pollen grains released in permanent tetrads
3. Endosperm not ruminate	Endosperm ruminate
4. Seeds not arillate	Seeds arillate
5. Basic chromosome number x = 19	Basic chromosome number x = 7, 8, 9, 14
6. Leaves stipulate	Leaves estipulate
7. Fruits follicular	Fruit a fleshy aggregate in which the follicles fuse with the floral axis
8. Vessels with scalariform perforation plates	Vessels with simple perforation plates

There is no doubt that the Annonaceae belong to the Magnolian stock, but as the Annonaceae have several features advanced over the Magnoliaceae, the erection of a separate order Annonales is justified.

Order Ranunculales

The order Ranunculales comprises 2 suborders and 7 families.

The suborder Ranunculineae includes Ranunculaceae, which is the largest and most primitive. Berberidaceae, Sargentodoxaceae, Lardizabalaceae and Menispermaceae are the other families. All the families have herbaceous members or soft woody climbers or climbing shrubs.

The two families of the suborder Nymphaeineae—Nymphaeaceae and Ceratophyllaceae—are both aquatic herbs with submerged rhizomes and long-petioled, floating or submerged leaves.

Flowers inflorescent in Ranunculineae but solitary axillary in Nymphaeineae. Flowers of both suborders have spiral floral organs. Sepals, petals and stamens are all spiral. Anthers adnate, bicelled, longitudinally dehiscent. Carpels few to many, free or fused; when free, spirally arranged. Ovaries uni- or multiovulate. Nectaries at the base of the petals occur in Ranunculaceae, Berberidaceae and Lardizabalaceae. The members of this order are characterised by triaperturate pollen, anatropous, uni- or bitegmic ovules.

The isoquinoline alkaloid, berberine, is reported in a few families of this order.

Ranunculaceae

A moderately large family with ca. 35 genera and 2000 species, chiefly distributed in the cooler temperate zones of the northern hemisphere. In the subtropics and tropics, they are fewer and occur at higher altitudes.

Vegetative Features. Annual or perennial herbs, a few are woody climbers such as *Clematis;* some aquatic, e.g. *Ranunculus aquatilis:* some are amphibious, e.g. *R. sceleratus* (Fig. 11.9.1 A). *Naravalia zeylanica* is a common climbing shrub in the tropical forests at the foothills of the Himalayas. Stem usually herbaceous, rarely woody at the base. The perennial herbs persist by their rhizomes.

Leaves usually alternate, petiolate, estipulate, opposite in *Clematis;* compound [entire in *Caltha* (Fig. 11.9.1 B), *Coptis*], pinnately compound in *Xanthorrhiza* and *Actaea,* palmately compound in *Nigella* and *Delphinium.* In *Clematis aphylla* the entire leaf is modified into tendril. In *Naravelia* leaves trifoliate with the terminal leaf modified to tendril. Leaves radical in *Anemone* and *Callianthemum.* Leaf base broad and sheathing. Heterophylly is seen in *Ranunculus aquatilis,* the submerged leaves much dissected and the floating ones lobed.

Floral Features. Inflorescence varied: dichasial cyme in *Ranunculus* (Fig. 11.9.1 A), raceme in *Delphinium,* (Fig. 11.9.2 A) solitary axillary in *Clematis cadmia,* solitary terminal in *Trollius, Nigella* and a panicle in *Clematis nutans.* Flowers bracteate, bracteolate, bisexual (unisexual in some species of *Thalictrum),* actinomorphic or zygomorphic as in *Delphinium* (11.9.2 B) Calyx of 5 sepals, polysepalous, green or petaloid. imbricate or quincuncial; induplicate-valvate in *Clematis* and *Naravelia,* more or less half-imbricate and half-valvate in *Clematopsis;* spurred or saccate in *Myosurus.* Corolla of 5 or more petals, usually polypetalous, imbricate (Fig. 11.9.1 C), pocket-like nectaries at the base of each petal; *R. pinguis* of New Zealand has 2 or 3 nectaries on the petals, petals represented by petaloid staminodes in some species of *Clematis;* totally absent in *Anemone* and most species of *Clematis.* Stamens indefinite, polyandrous, spirally arranged, hypogynous, distinct (Fig. 11.9.1 C); anthers bicelled, basifixed or adnate (11.9.2 C, D), dehiscence longitudinal (Fig. 11.9.1 D). Gynoecium apocarpous or syncarpous, 3- to many-carpellary, very rarely monocarpellary, e.g. *Delphinium,* (11.9.2 E, F), when polycarpellary and apocarpous, arranged spirally on the receptacle (Fig. 11.9.1 C), ovaries superior, unilocular with one basal ovule in *Ranunculus;* in monocarpellary gynoecium of *Delphinium,* and each carpel of polycarpellary gynoecium in *Caltha* (Fig. 11.9.1 E, F), ovules 1 to many, placentation parietal along the ventral suture (Fig. 11.9.2 F, G); in polycarpellary *Nigella* ovary as many loculed as the number of carpels, ovules many, placentation axile. Style and stigma 1 (many in *Nigella).* Fruit typically a follicle. Sometimes achene as in *Ranunculus,* or berry as in *Actaea,* or a capsule as in *Nigella.* Seed with minute embryo and copious endosperm. Fig. 11.9.2 G is floral diagram for *Delphinium ajacis.*

Anatomy. The vascular bundles in transverse section of stem appear to be widely spaced. Vessels in tangential or irregular groups, ring-porous, with spiral thickening, perforation plates simple, intervascular pittings alternate; parenchyma paratracheal, storied. Rays large, up to 12 or more cells wide. Leaves generally dorsiventral, hairs both glandular and non-glandular. Stomata Ranunculaceous type confined to the lower surface or on both surfaces.

Embryology. Pollen grains tricolporate with smooth exine; 2-celled at shedding stage. Ovules anatropous, unitegmic as in *Anemone, Clematis* or bitegmic as in *Adonis.* Both functional and non-functional ovules occur. Non-functional ovules ategmic and lack micropyle as well as vascular supply; Polygonum or Allium type of embryo sac, 8-nucleate at maturity. Endosperm formation of Nuclear type, later becomes cellular, persistent.

Chromosome Number. Ranunculaceae members show three different basic numbers: x = 7, 8 or 9. Most of the members, including *Anemone, Clematis, Adonis, Ranunculus,* etc. show x = 8; *Coptis* is the only genus which has x = 9.

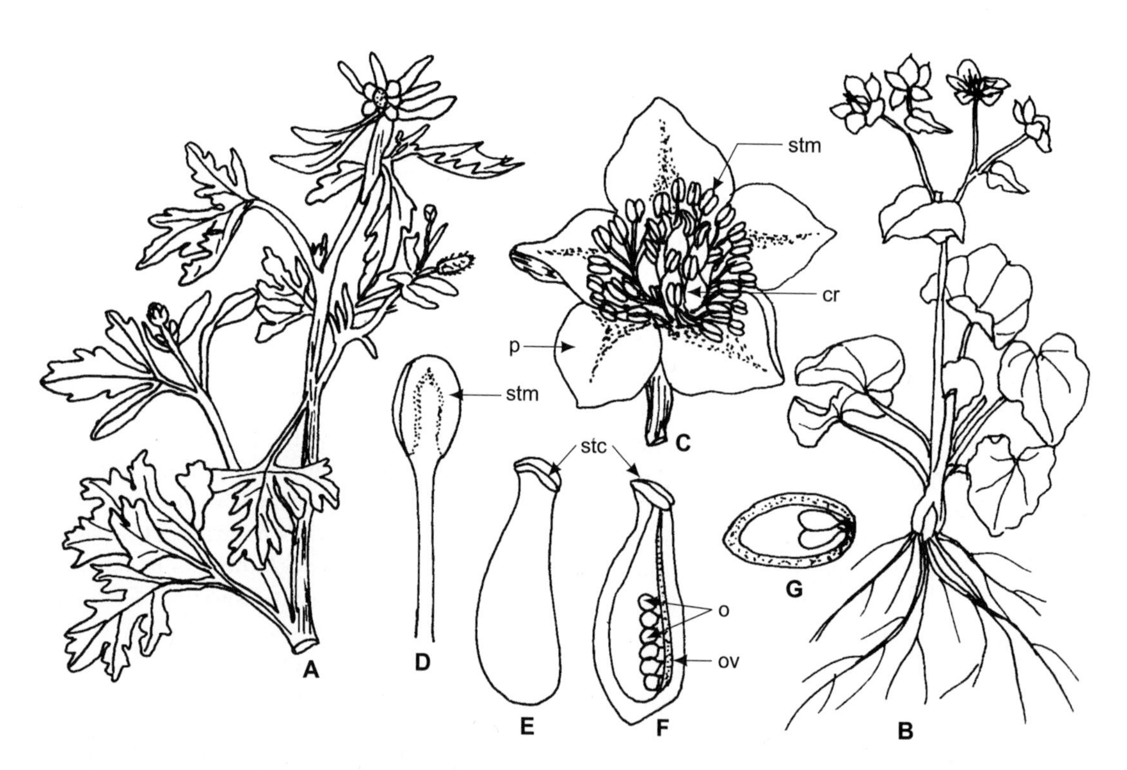

Fig. 11.9.1 Ranunculaceae: **A** *Ranunculus sceleratus.* **B-G** *Caltha palustris.* **A** Flowering twig. **B** Habit of *C. palustris.* **C** *Flower.* **D** *Stamen.* **E** *Carpel.* **F, G** *Carpel,* longisection and transection. (*cr* carpel, *o* ovule, *ov* ovary, *p* perianth, *stc* stigmatic crest, *stm* stamen). (**A** original. **B-G** adapted from Radford 1987).

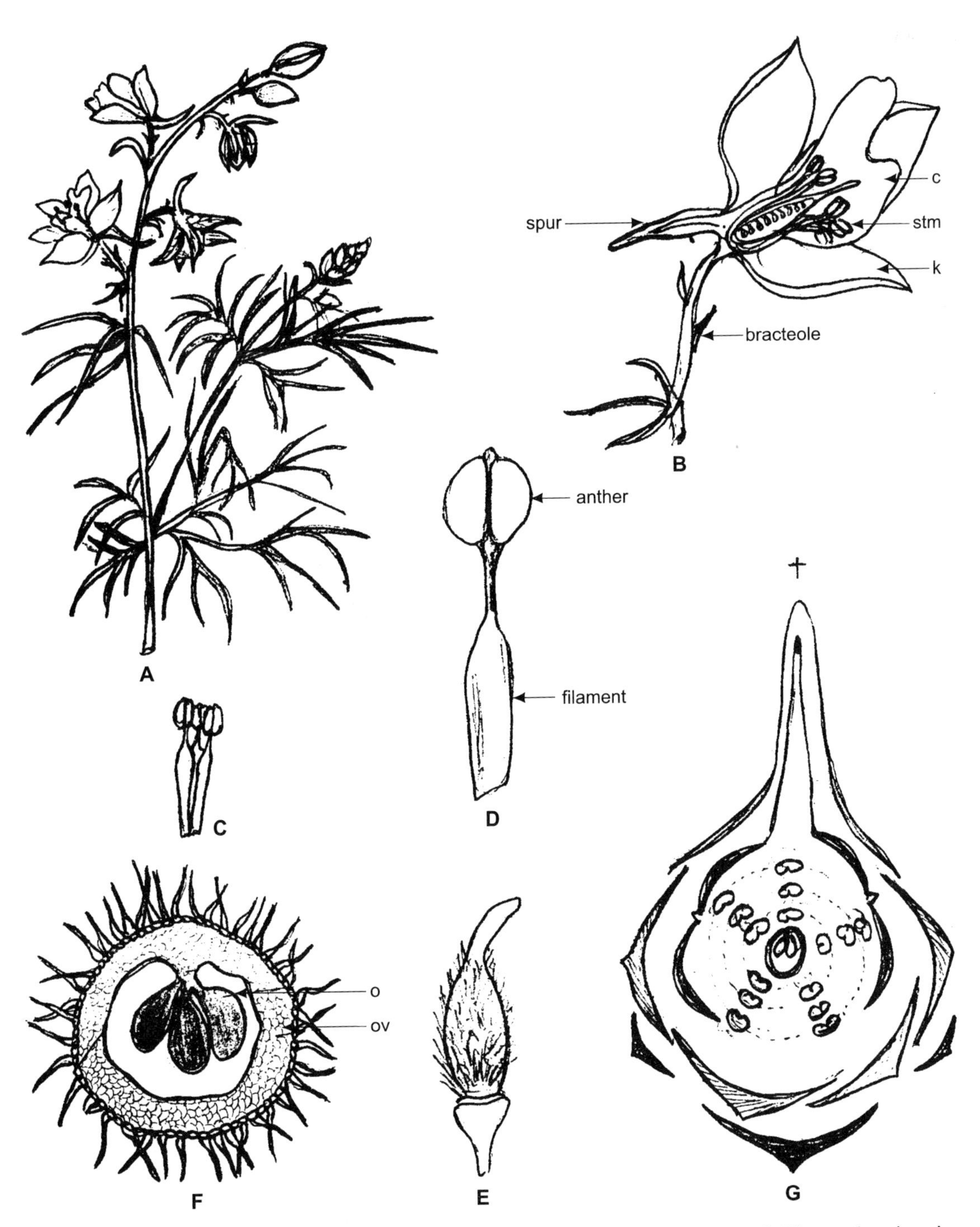

Fig. 11.9.2 Ranunculaceae: **A–G** *Delphinium ajacis* **A** Twig with inflorescence. **B** Flower, longisection **C, D** A bunch of **(C)** and a single stamen **(D)**. **E** Gynoecium. **F** Ovary transverse section. **G** Floral diagram (*c* corolla, *k* calyx, *o* ovule, *ov* ovary, *stm* stamen).

Chemical Features. Isoquinoline alkaloid ranunculin is reported from many members (Ruijgrok 1968). *Aconitum napellus* is known to contain another alkaloid, aconitine, which is used medicinally. The two alkaloids, magnoflorine and berberine, have been isolated from some species (Hegnauer 1963). Ecdysones, an insect-moulting hormone, has been detected in *Helleborus* sp., along with saponins and bufadienolides (Hardman and Benjamin 1976).

Important Genera and Economic Importance. *Ranunculus* is the largest genus, with about 250 species, amongst which hydrophytes, amphibians, and mesophytes are known. Often it is a troublesome weed in the cropfields. *Clematis* is another large genus of woody climbers with about 200 species. Many species are cultivated as ornamentals because of the clusters of fragrant flowers. *Delphinium* with 250 species and *Anemone* with 100 species are other large genera. *Aquilegia, Thalictrum, Myosurus, Cimicifuga* and *Coptis* are sometimes grown as ornamentals. Various members are used medicinally. Dried tuberous roots of *Aconitum napellus, A. heterophyllum, A. chasmanthum* and *A. deinorrhizum* contain some very toxic alkaloids like aconitine, aconine and benzoylaconine. These alkaloids are used externally for neuralgia and rheumatism, and internally to relieve pain and fever (Kochhar 1981). *Cimicifuga* or black snakeroot, also used in medicine, consists of dried rhizomes and roots of *C. racemosa* (Metcalfe and Chalk 1972). Seeds of *Nigella damascena* are used as spice and also medicinally.

Taxonomic Considerations. As early as 1783, De Jussieu (see Lawrence 1951) concluded that the Ranunculaceae was the most primitive amongst the dicot families. This fact has been accepted by most later workers like Bentham and Hooker (1965a), Hallier (1912), Bessey (1915) and Hutchinson (1948). Engler and Diels (1936) and later Lawrence (1951), Melchior (1964) and Stebbins (1974) did not agree with this assignment. On the basis of wood anatomy of Ranunculaceae members, and those of some monocots like Alismataceae, Metcalfe and Chalk (1972) conclude it to be the most primitive dicot family.

According to Cronquist (1968, 1981), the order Ranunculales, is the herbaceous equivalent of the Magnoliales. Ranunculaceae is ancestral to all other families in this order but is itself a derivative of the Magnoliales. Dahlgren (1983a) and Takhtajan (1987) also accept this view.

It is evident that, although anatomically Ranunculaceae is primitive, in floral features, it is advanced. There is no vesselless genus in this family and members with zygomorphic flowers, a derivative feature, are not known in more than one genus. Ranunculaceae should be treated as advanced over Magnoliales.

Glaucidium, a monotypic genus, is usually included in Ranunculaceae (Melchior 1964, Cronquist 1981). Hutchinson (1973) includes this genus in Helleboraceae, a family separated from Ranunculaceae. Langlet (1928) and Miyaji (1930) suggested its inclusion in Berberidaceae. Tamura (1963, 1972), Takhtajan (1966), Dahlgren (1975a, 1983a), Tobe (1981) and Thorne (1983) treat it as a member of an independent family, Glaucidiaceae. Tamura (1972) includes it in the Hypericales. Tobe (1981) places it together with Paeoniaceae. Dahlgren (1983a) and Thorne (1983) are also of the same opinion. Takhtajan (1987), however, erected a separate order, Glaucidiales, and placed it between Ranunculales and Paeoniales. Embryologically, *Glaucidium* is so distinct that it does deserve a family rank (Tobe 1981). It is so distinct from other nearby families morphologically, anatomically, chemically and in chromosome numbers, that it deserves an ordinal rank too.

Order Papaverales

The order Papaverales comprises 6 (or 7) families under 4 suborders:

The suborder

(a) Papaverineae includes Papaveraceae,

(b) Capparineae includes Capparaceae, Cruciferae (= Brassicaceae) and Tovariaceae,

(c) Resedineae includes Resedaceae, and

(d) Moringineae includes monotypic Moringaceae (Takhtajan 1987).

The members are mostly herbaceous, except *Moringa* of Moringaceae. Flowers bisexual, actino- or zygomorphic, hypogynous; number of stamens variable—numerous in Papaveraceae, to 6 (in most Brassicaceae) or even 2, as in *Coronopus didymus* of Brassicaceae. Gynoecium syncarpous with parietal placentation (axile in Tovariaceae). Pollen grains shed at 2-celled stage, ovules anatropous, bitegmic, crassinucellate, endosperm mostly absent (except in Tovariaceae). Most members have benzylisoquinoline (Papaverineae) or methylglucosinolates.

Capparaceae

Capparaceae, a medium-sized family of 42 to 45 genera and 850 species (Takhtajan 1987) is distributed palaeotropically in both hemispheres.

Vegetative Features. Annual or perennial herbs *(Cleome),* shrubs *(Capparis)* or trees *(Crataeva),* sometimes lianas *(Maerua arenaria),* often xerophytic such as *Capparis decidua.* Stem herbaceous or woody, without latex. Leaves alternate, simple *(Maerua)* or palmately compound, pentafoliate, trifoliate or unifoliate *(Steriphoma peruviana);* stipulate or estipulate, stipules minute, glandular or spinose *(Capparis decidua,* C. *sepiaria).*

Floral Features. Inflorescence terminal (Fig. 11.10 A, E) or axillary racemes or cymes; flowers solitary in *Niebuhria.* Flowers bisexual, complete, sometimes unisexual — the plants are then monoecious, e.g. *Podandrogyne;* actinomorphic (Fig. 11.10 B) or zygomorphic *(Capparis decidua,* Fig. 11.10 G), bracteate, ebracteolate. Flowers borne on old wood in *Bachmannia.* Calyx of 4 sepals, polysepalous, valvate, outer surface often covered with glandular hairs as in *Cleome viscosa;* sepals unequal in *Capparis decidua,* the posterior one forms a hood-like structure; corolla of 4 petals, polypetalous, sometimes connate, the two posterior ones form a large hood-like structure in *Emblingia,* imbricate, sometimes absent; all equal or 2 posterior ones larger, clawed or sessile. Stamens 4 to many, never tetradynamous, but are derived from tetrandrous condition by splitting of the 4 primordia and then many filaments often lack anthers; anthers bi- or tetracelled, dehisce longitudinally (Fig. 11.10 C). In some genera the androecium and gynoecium are borne on androgynophore as in *Cleome gynandra* (Fig. 11.10 B). Gynoecium syncarpous, bicarpellary, ovary usually borne on a long or short gynophore (Fig. 11.10 C), superior, unilocular with parietal placentation (Fig. 11.10 D) and numerous ovules; style 1, short or filiform and elongate, stigma bilobed or capitate (Fig. 11.10 B, C). Fruit a capsule dehiscing by valves as in *Cleome* or elongated or torulose berry as in *Capparis* (Fig. 11.10 H) and with transverse constrictions, as in *Maerua* (Fig. 11.10 F), sometimes indehiscent nuts as in *Emblingia.* Seeds reniform with a curved, folded embryo and fleshy endosperm.

Anatomy. Vessels very small to medium-sized, often in clusters or long, radial multiples, perforations simple, intervascular pitting alternate, small. Parenchyma paratracheal, sparsely vasicentric, sometimes storied, rays up to 2 to 5 cells wide, homogeneous. Included phloem occurs in some genera like *Boscia, Cadaba, Forchhammeria, Maerua* and *Stixis.* Leaves dorsiventral, isobilateral or centric, hairs of various

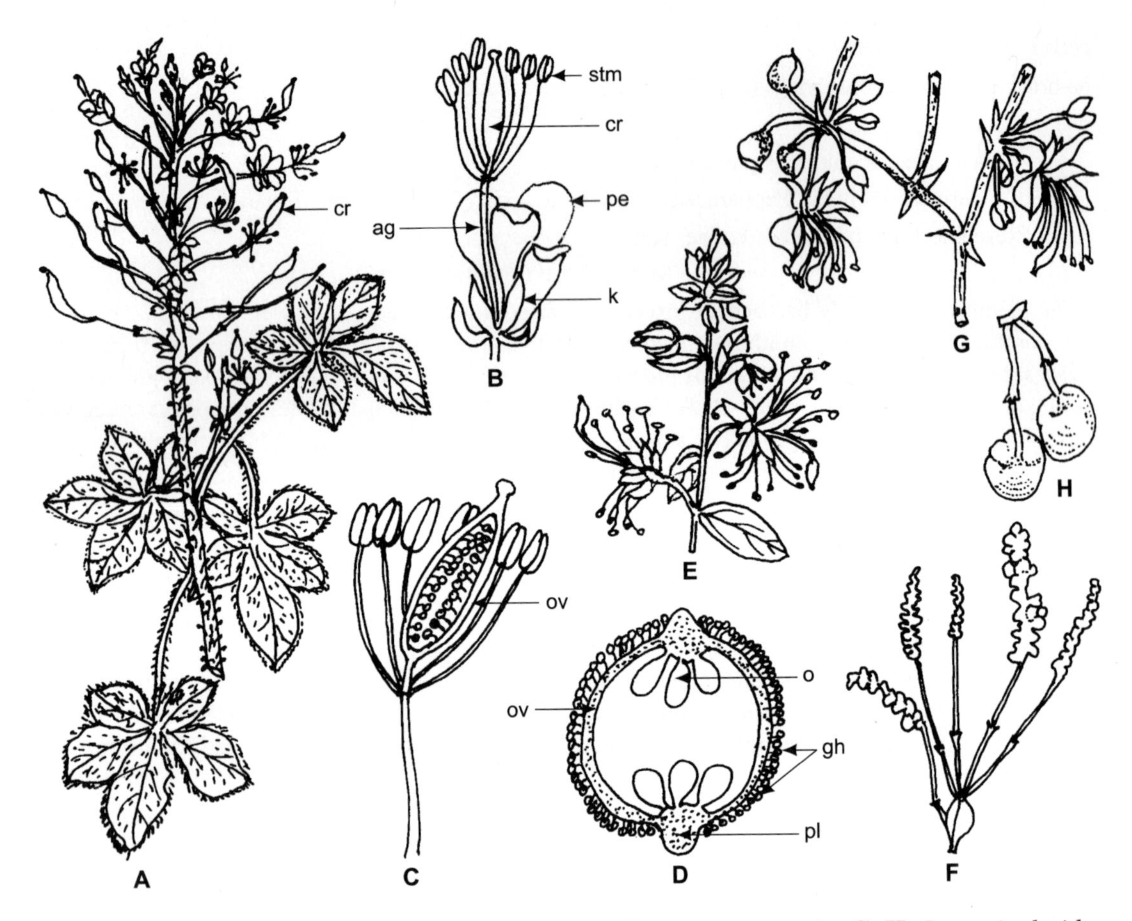

Fig. 11.10 Capparaceae: **A–D** *Cleome gynandra*, **E, F** *Maerua arenaria*, **G, H** *Capparis decidua*. **A** Flowering twig. **B** Flower. **C** Longisection, sepals and petals removed. **D** Ovary, cross section. **E** Flowering twig of *M. arenaria*. **F** Fruits. **G** Flowering twig of *C. decidua*. **H** Fruits. (*ag* androgynophore, *cr* carpel, *gh* glandular hairs, *k* calyx, *o* ovule, *ov* ovary, *pe* petal, *pl* placenta, *stm* stamen). (Original).

types; stomata Ranunculaceous type, on both the surfaces in isobilateral and centric leaves and confined to lower surface in dorsiventral leaves.

Embryology. Pollen grains spherical, tricolpate, with small spine-like structures on the thick exine; shed at 2- or 3-celled stage. Ovules campylotropous (anatropous in *Crataeva*), bitegmic, crassinucellate. Polygonum type of embryo sac, 8-nucleate at maturity. Endosperm formation of the Nuclear type.

Chromosome Number. Basic chromosome numbers are x = 8 to 17. As both *Podandrogyne* and *Cleome* have a very high basic chromosome number x = 29, Cochrane (1978) suggested phyletic relationship amongst the species of these two genera.

Chemical Features. Methylglucosinolates prominent (Ettlinger and Kjaer 1968). Ellagitannins usually absent. Common flavonols and their O-methyl derivatives also occur (Gornall et al. 1979).

Important Genera and Economic Importance. Amongst the important genera are *Cleome* (150 spp.), *Capparis* (250 spp.), *Crataeva* (9 spp.), *Cadaba* (30 spp.), *Maerua* (100 spp.) and *Roydsia.* Economically, the Capparaceae is not so important. *Capparis spinosa* is grown in the Mediterranean area for the unopened flower buds called capers which are useful in seasoning foods. Fruits and flower buds of *C. decidua* are edible either raw or pickled. The bark of *Crataeva nurvala* finds application as a remedy for bladder stones; the fruit is edible and the rind is used as a mordant in dyeing. *Cleome gynandra, C. brachycarpa* and *C. viscosa* are noxious weeds. *Capparis sepiaria* is a hedge plant and climbs by hooked spines. *Oceanopapaver* is a new monotypic genus from New Caledonia (Schmid et al. 1984).

Taxonomic Considerations. The family Capparaceae has been placed in Rhoeadales by the earlier authors (Lawrence 1951), in Papaverales by Melchior (1964) and in Capparales by Cronquist (1968, 1981), Dahlgren (1980a, 1983a) and Takhtajan (1980, 1987). It is intermediate between the families Papaveraceae and Cruciferae, but more closely allied to the Cruciferae. A comparison between the two families is given in Table 11.2.

Table 11.2 Comparative Data for Capparaceae and Cruciferae

Capparaceae	*Cruciferae*
Mostly shrubs and trees; very few herbs	Mostly herbs
Leaves simple or palmately compound, not auriculate	Leaves simple, lobed or pinnatifid; auriculate
Stamens 6-8-numerous, never tetradynamous	Stamens 6, tetradynamous
Androphore and/or gynophore present	Androphore and gynophore absent
Fruit capsule or berry, rarely siliqua	Fruit siliqua/silicula, never a capsule

Capparaceae is a natural taxon and well placed as the most primitive family amongst the Capparales. That it is a primitive taxon is further proved by the fact that Capparaceae is distinct, mainly due to the presence of alanine-derived methylglucosinolates. The other families of the Capparales have more complex glucosinolates which are derived from it (Harborne and Turner 1984).

Its placement is more desirable in the Capparales than in the Papaverales.

Aleykutty and Inamdar (1978) considered it unnecessary to separate the tribe Cleomoideae (from the Capparaceae) as a distinct family Cleomaceae.

Cruciferae (= Brassicaceae)

A large family of 350 genera and 2500 species distributed primarily in northern hemisphere and in colder alpine regions of the tropics. Some of the genera, such as *Cardamine, Lepidium, Sisymbrium* are cosmopolitan.

Vegetative Features. Annual, biennial or perennial herbs, rarely undershrubs. Stem herbaceous, soft and green, sometimes modified for storage of food, e.g. *Brassica caulorapa.* Roots of some genera are also modified for food storage, e.g. *Armoracia rusticana, Brassica rapa* and *Raphanus sativus.* Leaves alternate, radical or cauline, simple, auriculate, often lyrate as in *R. sativus,* and *Sisymbrium irio* (Fig. 11.11 A); estipulate.

Floral Features. Inflorescence racemes, spikes or corymbs. Flowers bisexual, actinomorphic, zygomorphic in *Iberis* and *Teesdalia,* owing to the enlargement of two outer petals, hypogynous, ebracteate, ebracteolate. Calyx of 4 sepals, polysepalous (Fig. 11.11 B, H), in 2 whorls of 2 each; corolla of 4 petals, polypetalous (Fig. 11.11 J), with a distinct limb and claw, cruciform; often petals very small as in *Coronopus didymus* (Fig. 11.11 J) or absent as in *Lepidium.* Stamens 6 in 2 whorls (Fig. 11.11 C), often reduced to 4 *(Cardamine hirsuta)* or 2 *(Coronopus didymus,* Fig. 11.11 G, I) or sometimes numerous as in *Megacarpaea* (16 stamens); usually tetradynamous, filaments of each inner pair sometimes connate, those of outer pair may be winged or toothed. Nectariferous glands present in between the filaments, anthers bicelled, dorsifixed, dehiscence longitudinal. Gynoecium syncarpous, bicarpellary, ovary unilocular but bilocular later due to the formation of a false septum called 'replum' *(r* in Fig. 11.11 D, K), superior, placentation parietal, ovules numerous or only a few; style one or obsolete, stigmas 2, capitate. Fruit a siliqua or a silicula. Generally, the ovary is sessile on the receptacle but there is a stipe between the ovary and receptacle in *Stanleya.* Fruit a 1-seeded indehiscent nut in *Bunias* and *Isatis.* Seeds with scanty or no endospern and a large embryo.

Anatomy. Vessels small, perforations simple, intervascular pitting alternate with horizontal apertures. Parenchyma paratracheal, extremely sparse, rays up to 2 to 4 cells wide, heterogeneous. Leaves

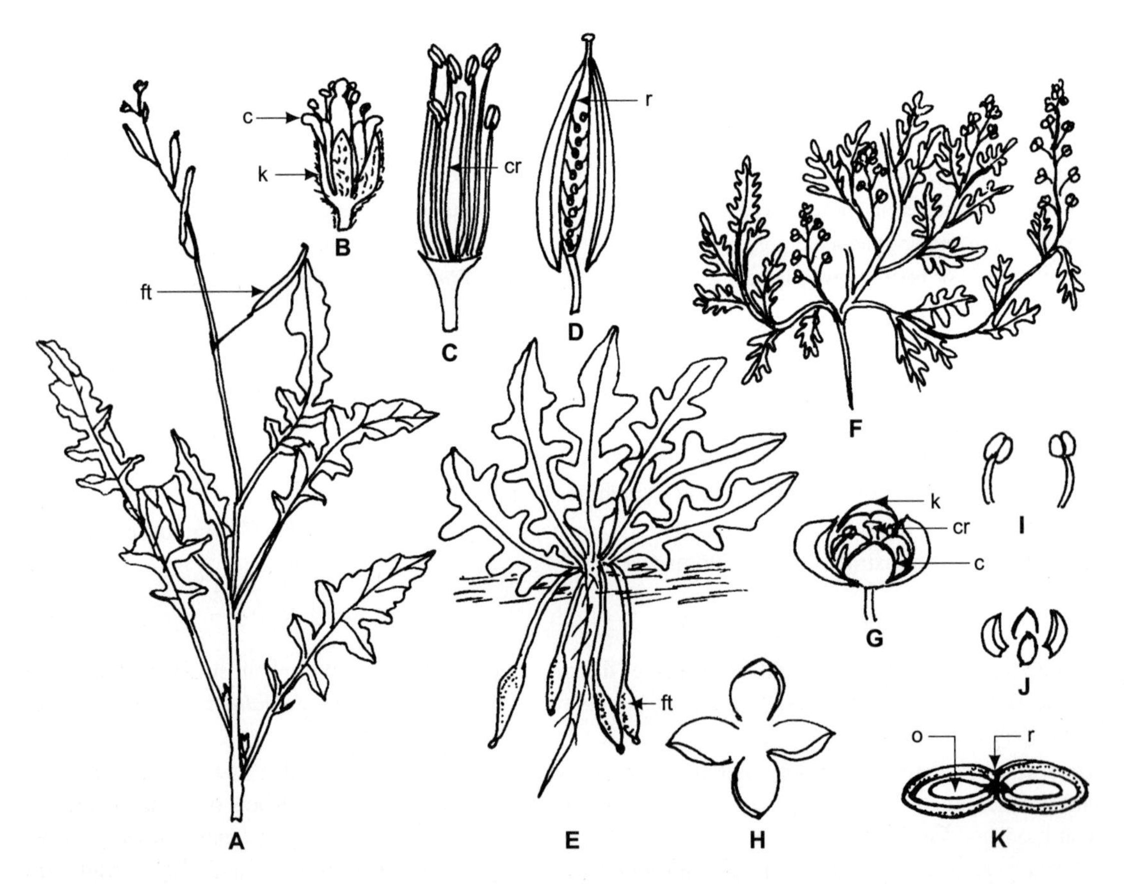

Fig. 11.11 Brassicaceae: **A–D** *Sisymbrium irio,* **E** *Geococcus pusillus,* **F-K** *Coronopus didymus.* **A** Flowering twig. **B** Flower. **C** Tetradynamous stamens and carpel. **D** Replum with seeds on it. **E** Fruiting plant of *G. pusillus.* **F** Flowering plant of *C. didymus.* **G** Flower. **H** Dissected calyx lobes. **I** Stamens. **J** Corolla. **K** Ovary, cross section. *c* corolla, *cr* carpel, *ft* fruit, *k* calyx, *o* ovule, *r* replum. (**E** adapted from Hutchinson, 1969, **A–D** and **F–K** Original)

dorsiventral or isobilateral, stomata Cruciferous type, hairs variable. Myrosin cells distributed throughout leaf parenchyma.

Embryology. Pollen grains mostly 3-colpate but inaperturate in *Matthiola* (P.K.K. Nair 1970); shed at 3-celled stage. Ovules bitegmic, tenuinucellate. Polygonum type of embryo sac, 8-nucleate at maturity. Endospenn formation of Nuclear type (Dahlgren 1980a).

Chromosome Numbers. Basic chromosome numbers are x = 3–13.

Chemical Features. The seeds of Brassicaceae are rich in fatty acids. Erucic acid, a rare fatty acid, occurs in higher quantities in the seed oil from *Brassica napus.* They are also rich in isothiocyanates and glucosinolate sinigrins (Ettlinger and Kjaer 1968).

Important Genera and Economic Importance. Important genera include *Brassica* with 50 species and a number of varieties, common in Europe, Asia and Mediterranean region, 8 species are cultivated in warmer and temperate regions. *Raphanus* is another important genus with modified roots for storage and are edible. *R. sativus* has numerous strains that are cultivated. Economically also this family is very important. A number of winter vegetables such as cabbage, *Brassica oleracea* var. *capitata,* cauliflower, *B. oleracea* var. *botrytis,* Brussels sprouts, *B. oleracea* var. *gemmifera,* Kohl rabi, *B. caulorapa,* radish, *Raphanus sativus,* cress, *Lepidium sativum* are obtained from this family. Fatty oils used for cooking purposes are obtained from *Brassica juncea* var. *sarson.* Fatty oil from *Eruca sativa* is used for adulteration of mustard oil, has unpleasant taste and aroma. Table mustard is made from powdered seeds of *B. nigra* and *B. alba.* In *Geococcus pusillus* (Fig 11.10 E) the fruits mature underground.

lsatis tinctoria is the source of a blue dye 'woad'. Dame's violet—*Hesperis matronalis,* sweet alyssum—*Lobularia maritima* (= *Alyssum maritimum*), wallflowers—*Matthiola incana* and *M. tristis,* candytuft— *lberis amara,* and *stock—Cheiranthus* are some of the cultivated ornamentals. A few, like *Capsella bursa- pastoris, Coronopus didymus* (Fig. 11.10 F), *Nasturtium* sp. and *Sisymbrium irio,* are weeds in waste places.

Taxonomic Considerations. Cruciferae (= Brassicaceae) was placed in the order Parietales (Bentham and Hooker 1965a), Rhoeadales (Rendle 1925, Lawrence 1951, Melchior 1964), Cruciales (Hutchinson 1969), Brassicales (Hutchinson 1973) and Capparales (Cronquist 1981, Dahlgren 1983a, Takhtajan 1987). It is a natural taxon. The only controversy about its origin is whether it is derived from Papaveraceae or Capparaceae. On the basis of androecial and gynoecial morphology and anatomy, it is derived from Capparaceous ancestors (Lawrence 1951). Chemical features also support this view; isothiocyanates are reported in both Capparaceae and Cruciferae (Harborne and Turner 1984). From the Papaveraceae it differs chemically and in endospermous seeds, although there are a few resemblances in androecial and gynoecial features and the tetramerous perianth.

On the basis of studies of different taxonomists, it is apparent that the Cruciferae has originated from Capparaceous ancestors and are better placed in the order Capparales.

Order Rosales

A large order comprising 19 families distributed in four suborders:

1. Suborder Hamamelidineae with unigeneric Platanaceae, Hamamelidaceae, and unigeneric Myrothamnaceae.
2. Suborder Saxifragineae includes Crassulaceae, unigeneric Cephalotaxaceae, Saxifragaceae, monogeneric Brunelliaceae, Cunoniaceae, unigeneric Davidsoniaceae, Pittosporaceae, unigeneric Byblidaceae, and Roridulaceae, and Bruniaceae.
3. Suborder Rosineae includes Rosaceae, Neuradaceae, and Chrysobalanaceae.
4. Suborder Leguminosineae includes Connaraceae, Leguminosae, and Krameriaceae.

The members are trees, shrubs and herbs, of cosmopolitan distribution. Flowers generally cyclic, typically pentamerous, hypo-, peri- or epigynous, stamens mostly in many whorls; gynoecium apo- to syncarpous, style and stigma distinct; ovules bitegmic, seeds endospermous. The taxa of this order are characterised by the presence of common flavonols and, to a lesser extent, by flavones together with their O-methyl derivatives.

Leguminosae

A large taxon, mostly treated as distinct order comprising 3 families: Papilionaceae (Fabaceae), Caesalpiniaceae and Mimosaceae. Of all these families, Papilionaceae is predominantly herbaceous with a few shrubs and trees, but both the Caesalpiniaceae and Mimosaceae are chiefly arborescent. The flowers in racemose inflorescence, bisexual, actino-or zygomorphic, usually highly ornamental. Stamens few to numerous, basifixed, mostly dehisce longitudinally. Ovary monocarpellary, superior, unilocular, placentaion marginal. Fruit a dehiscent or indehiscent legume.

Mimosaceae

A family of ca. 56 genera and 2800 species, more or less confined to the tropics and subtropics of both the hemispheres.

Vegetative Features. Mostly trees and shrubs, often with spiny outgrowths on stem, xerophytes common, hydrophytes also reported *(Neptunia).* Leaves bipinnate, unipinnate in *Affonsea* and *Inga,* sometimes reduced to phyllodia (Fig. 11.12.1 A*);* rachis pulvinate, generally gland-bearing (Fig. 11.21 A), stipulate, stipules spiny; leaves show sleeping movement. Leaves of *Mimosa pudica* are sensitive to touch.

Floral Features. Flowers in spike or head (condensed racemes) inflorescences, involucre common. In *Dichrostachys* upper part of the spike is bisexual, the lower neutral with long staminodes. Flowers bisexual, actinomorphic, hypogynous. Sepals 4 or 5, inconspicuous, cup-shaped (Fig. 11.12.1 B, C), valvate, odd sepal anterior. Petals 4 or 5, polypetalous, sympetalous in *Acacia* (Fig. 11.12.1 D) and *Albizia,* valvate. Stamens 10 or equal to the number of petals, free or monadelphous (Fig. 11.12.1 E) as in *Inga,* all fertile. Anthers gland-tipped in some genera—*Acacia* (Fig. 11.12.1 D, E), *Adenanthera, Parkia* and *Prosopis;* bithecous and longitudinally dehiscent; pollen granular or agglutinated into tetrads or polyads (Fig. 11.12.1 F). Gynoecium monocarpellary, ovary unilocular, superior, placentation marginal (Fig. 11.12.1 I); ovules numerous; style long, filiform, coiled in bud (Fig. 11.12.1G), stigma truncate (Fig 11.12.1 H). Fruit a legume or lomentum as in *Entada, Pseudoentada* and *Plathymania.* Seeds dorsiventrally flattened, funicle long and coiled, pleurogram present (absent in *Pithecellobium);* embryo straight, endosperm present.

Fig. 11.12.1 Mimosaceae: **A–I** *Acacia glaucescens.* **A** Flowering twig, note adpressed hairs all over the plant body. **B** Flower bud. **C** Calyx cup. **D** Flower with sympetalous corolla and gland-tipped stamens. **E** Staminal tube. **F** Anther. **G** Pistil from flower bud, style coiled. **H** From open flower, style uncoiled. **I** Ovary, cross section. (*k* calyx, *gl* gland, *o* ovule). (After Rangaswamy and Chakrabarty 1966).

Anatomy. Wood diffuse-porous, vessels medium-sized to large, typically solitary, spiral thickenings absent; parenchyma abundant, paratracheal, rays 2 to 5 cells wide, homogeneous, mostly of small cells. Fibres with few, small, simple pits. Anomalous structure rarely seen. Leaves dorsiventral, isobilateral or centric. Hairs of both glandular and non-glandular types occur. Stomata Rubiaceous type, confined to the lower surface in Adenanthereae, Ingeae and Parkieae and uniformly distributed on both sides of leaf in Acaciae, Eumimoseae and in *Dichrostachys* and *Neptunia.*

Embryology. Pollen simple, granular in *Neptunia, Leucaena, Prosopis* and *Desmanthus;* shed as tetrads or polyads in other genera. Ovules anatropous, campylotropous or amphitropous, bitegmic, crassinucellate. Polygonum type of embryo sac, 8-nucleate at maturity. Endosperm formation of the Nuclear type; chalazal haustorium common.

Chromosome Number. Diploid chromosome numbers for various genera are 2n = 16, 22, 24, 26, 28, 36, 44, 52, 56 and 104 (Kumar and Subramanian 1987).

Chemical Features. Rich in tannins; a glycoside, dihydroacacipetalin reported in *Acacia* sp. Non-protein amino acid albizzine occurs in seeds of *Albizia julibrissin, Acacia* (except series Gummiferae; Seneviratne and Fowden 1968), and *Mimosa.* Carotenoids present in yellow-flowered *Acacia decurrrens* var. *mollis, A. discolor* and *A. linifolia.* Cyanogenic glucosides reported in some *Acacia* spp. (Secor et al. 1976, Seigler et al. 1978). Gum containing sugars arabinose, rhamnose, galactose and glucoronic acid present in *Acacia.*

Important Genera and Economic Importance. The genera *Acacia, Albizia, Adenanthera, Inga, Entada, Enterolobium, Mimosa* and *Pithecellobium* are important. Leaves of *M. pudica,* commonly called touch-me-not or the sensitive plant are sensitive to touch. Leaves of two aquatic plants *Neptunia oleracea* and *N. plena* also show similar characteristics. *Entada scandens* is an immensely woody climber with almost 1-meter-long fruits, and round seeds that are 2" in diameter.

Many genera are economically important. *Acacia nilotica* var. *gangeticus* yields fuel and gum. The wood is also used for making agricultural implements, tent pegs, etc. *A. senegal* is the source of gum arabic. The heartwood of *A. catechu* (on boiling) gives a tannin known as katha and is commonly used with betel leaf. A yellowish dye obtained from this heartwood is also used in dyeing khaki cloth. Pods and bark of *A. farnesiana* are used medicinally and for tanning; the flowers yield a perfume. *Xylia dolabriformis* or iron-wood tree from Burma yields valuable timber.

Acacia auriculiformis, the phyllode-bearing Australian *Acacia* tree, *Albizia lebbeck, A. procera, Enterolobium,* etc. are grown as avenue trees. *Calliandra haematocephala* is an ornamental shrub.

Taxonomic Considerations. The family Mimosaceae is allied to the Rosaceae and shares common characters: actinomorphic flowers and numerous stamens. However, it can readily be distinguished because of the hypogynous stamens, anteriorly-placed odd sepal, superior ovary and the fruit a legume or lomentum. Occurrence of trees and shrubs and rarely herbs, actinomorphic flowers and numerous stamens are primitive characters of this family. On the other hand, predominantly bipinnate leaves and the existence of xerophytes and hydrophytes are the advanced features.

It is considered to be the most primitive family/subfamily of the Leguminosae.

Caesalpiniaceae

A large family of ca. 152 genera and over 2800 species (Kumar and Subramanian 1987), distributed in the tropics and subtropics of both hemispheres, abundant in America.

Vegetative Features. Predominantly arborescent, xerophytes less common (*Parkinsonia aculeata;* Fig. 11.12.2A), hydrophytes not known. Stem mostly glabrous. Leaves unipinnate (*Cassia*), or bipimnate

(*Delonix*) or rarely simple (*Bauhinia*); rachis pulvinate, rarely gland-bearing (*Cassia*), stipulate, stipules sometimes foliaceous, e.g. C. *auriculata.* Leaflets exhibit sleeping movement; sometimes with mucronate tip.

Floral Features. Flowers in corymb or simpte raceme (Fig. 11.12.2 A), sometimes pendulous as in *Cassia fistula;* bracteate, ebracteolate, pedicellate, bisexual, zygomorphic. Sepals 5 (4 in *Amherstia*) polysepalous, rarely gamosepalous as in *Bauhinia;* descendingly imbricate, odd sepal anterior and outermost. Petals 5 (Fig. 11.12.2 B, D), rarely fewer (3 in *Tamarindus* and *Amherstia*; 1 in *Afzelia*), and absent in *Saraca*; petals dissimilar, often clawed. Stamens 10 or less (3 in *Tamarindus*, 3 to 8 in *Saraca*), free as in *Cassia* and *Caesalpinia* or monadelphous as in *Tamarindus* or diadelphous as in *Amherstia.* In *Caesalpinia, Delonix*, and *Parkinsonia* (Fig. 11.12.2 D), all stamens fertile. Anthers basifixed, bithecous, longitudinally dehiscent or poricidal; filaments free massive, dilated at the base (Fig. 11.12.2 C). Heteroanthery that is large, medium and small types of stamens have been observed in many species of *Cassia.* These are arranged in two whorls and either poricidal or longitudinally dehiscent (Chauhan et al. 2003). In *C. occidentalis* and *C. siamea*, three small stamens are reduced to staminodes. Gynoecium monocarpellary, ovary superior, unilocular, placentation marginal (Fig. 11.12.2 E, F), ovules numerous; style massive, often slightly recurved as in *Cassia*, stigma capitate. Fruit a legume or large woody pod, dehiscent or indehiscent, cylindrical (*Cassia fistula*), or flattened (*Delonix regia*). Seeds dorsiventrally flattened, funicle longer than seed; endosperm present.

Anatomy. Wood diffuse-porous, vessels typically medium-sized, solitary, sometimes with spiral thickenings; parenchyma moderately abundant, paratracheal; rays 1 to 3 cells wide, heterogeneous. Leaves generally dorsiventral except in certain species of *Hoffmanseggia* and *Hymenaea.* Hairs glandular or nonglandular; stomata mostly Rubiaceous or Ranunculaceous type, variations common. Abnormal anatomy reported in many species of the genus *Bauhinia.*

Embryology. Pollen grains basically 3-colpate in this highly multipalynous family; shed singly or as tetrads as in *Afzelia*, at 2-celled stage. Ovules anatropous, campylotropous, or amphitropous, bitegmic, crassinucellate. Polygonum type of embryo sac, 8-nucleate at maturity. Endosperm formation of the Nuclear type; chalazal haustorium present.

Chromosome Number. Diploid chromosome numbers are 2n = 16, 18, 22, 24, 26, 28, 42, 48, 52 and 56.

Chemical Features. Plants rich in tannins; carotenoids in *Delonix regia* flowers and anthraquinones in *Cassia* are reported. Non-protein amino acids occur in the seeds of some species of *Caesalpinia* (Evans and Bell 1978).

Important Genera and Economic Importance. Important genera of this family are *Bauhinia, Cassia, Caesalpinia, Delonix, Amherstia, Haematoxylon, Hardwickia, Humboldtia, Parkinsonia* (Fig. 11.12.2 A), *Saraca,* and *Tamarindus.* Economically, the family Caesalpiniaceae is quite important. Fruits of *Cassia fistula* are used medicinally, and those of *Tamarindus indica* have carminative and laxative properties. The seeds of *Cassia occidentalis* are powdered and mixed with coffee powder as an adulterant. The heartwood of *Haematoxylon campechianum* is the source of the dye heamatoxylin, used as nuclear stain in biological sciences. *Hardwickia binata* is the source of valuable timber. The wood of *Parkinsonia aculeata* is good for making charcoal. Many genera are grown as ornamentals (*Bauhinia racemosa, B. purpurea, B. variegata*) or as avenue trees (*Tamarindus indica*).

Bauhinia vahlii is a large woody climber with stem tendrils. The simple apically notched leaves are almost 30 cm in diameter and are used as substitute for plates; ropes made from the bark of this plant are very tough and used for making suspension bridges over small rivers and rivulets in the Himalayan region. *B. anguinia* is another such climber with flat, ribbon-like and twisted stem giving the appearence of a snake, and the common name is 'nagpat'.

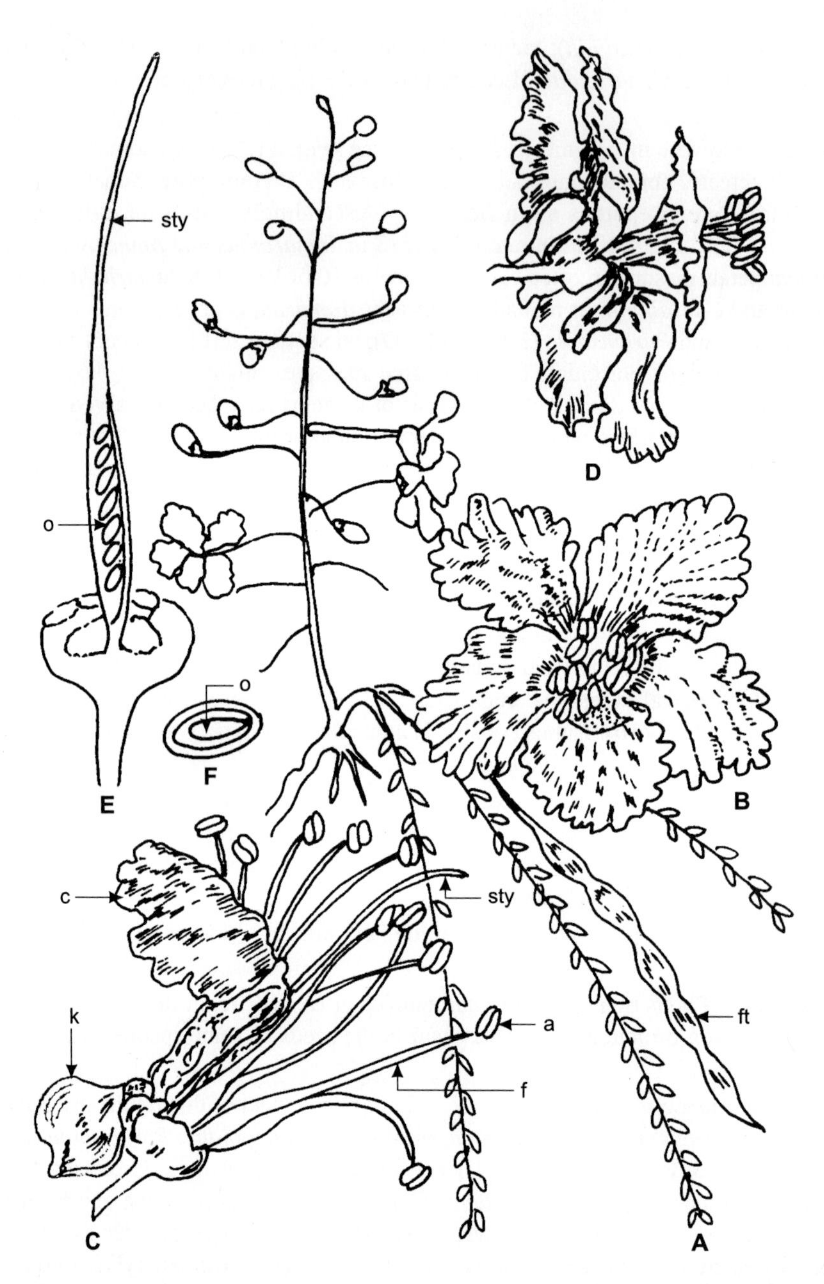

Fig. 11.12.2 Caesalpiniaceae: **A–F** *Parkinsonia aculeata.* **A** Branch with bipinnate leaf and inflorescence. **B** Flower, top view, showing corolla and stamens. **C** Flower with sepals and petals partly removed. **D** Flower, side view. **E** Pistil, longisection. **F** Cross section. *a* anther, *c* corolla, *f* filament, *ft* fruit, *k* calyx, *o* ovule, *sty* style. (Adapted from Benson 1970)

The flower buds of *Bauhinia variegata* are used as a vegetable. *Caesalpinia bonducella* or fever nut tree is also used medicinally. The wood of C. *sappan* yields a red dye used for dyeing wool and silk. The red colour mixed with starch powder is used during the Holi festival in India.

Taxonomic Considerations. According to Hutchinson (1969, 1973), Caesalpiniaceae is the most primitive amongst the members of the Leguminosae, and is therefore closest to the Rosales, from which it has been derived. The vertical sections of the flowers of *Parinari* of Rosaceae and *Bauhinia* of Caesalpiniaceae resemble each other. The position of this family should be between the Mimosaceae and Papilionaceae. The family Caesalpiniaceae has been retained as a subfamily Caesalpinioideae in the family Leguminosae by Stebbins (1974) and Takhtajan (1980, 1987), as a family Caesalpiniaceae in the order Fabales by Dahlgren (1977a, 1980a, 1983a); in the order Leguminales by GN Jones (1955), Hutchinson (1973) and Rangaswamy and Chakrabarty (1966).

Papilionaceae (Fabaceae)

A very large family of about 482 genera and 12000 species, cosmopolitan, abundant in tropics and subtropics and some in temperate zones.

Vegetative Features. Chiefly herbs and climbers, some are shrubs (*Sesbania sesban*), trees (*Pongamia pinnata*, *Sophora* sp.) and woody climbers (*Abrus precatorius*); xerophytes (*Alhagi pseudalhagi*) and hydrophytes (*Aeschynomene aspera*) rare. Bacterial root nodules are commonly present. The stem surface mostly hairy except in woody species. Leaves simple (*Indigofera cordifolia*) or unipinnate, often trifoliate as in *Cajanus cajan*, *Rhynchosia minima*; rachis pulvinate, grooved, lamina gland-dotted in *Rhynchosia minima*; stipulate, stipules spiny in *Robinia*; sleeping movement common in many genera. Leaflets often modified to tendrils, as in *Pisum sativum* (Fig. 11.12.3. A), *Vicia hirsuta* and others.

Floral Features. Flowers in a raceme or spike, sometimes highly condensed to form heads, e.g. *Medicago lupulina, Trifolium pratense;* bracteate, ebracteolate, pedicellate, bisexual, hypogynous, zygomorphic (Fig. 11.12.3 B, C). Sepals 5, connate, campanulate (Fig. 11.12.3 C), odd sepal anterior and inferior, often coloured, as in *Pongamia.* Corolla papilionaceous, petals 5, the posterior odd petal outermost and is called standard, two lateral ones the wings, and the two anterior ones fused to form a keel or carina (a boat-shaped structure (Fig. 11.12.3 D, E). Stamens 10, diadelphous (Fig. 11.12.3 G), rarely 9 and monadelphous, as in *Abrus*; in *Erythrina* 10 monadelphous stamens, all fertile, filaments fused to form staminal column but free near the apex. Anthers basifixed, bithecous, longitudinally dehiscent, pollen granular. Gynoecium monocarpellary, superior, unilocular, ovary with marginal placentation and one row of ovules (Fig. 11.12.3 I, J); style thick and curved or reflexed at base, stigma brushy (Fig. 11.12.3 F) or capitate. Fruit a dehiscent (Fig. 11.12.3 H) or indehiscent legume. Seeds reniform or rounded (Fig. 11.12.3 I, J); funicle shorter than seed. Embryo pleurorhizal, endosperm present.

Anatomy. Wood ring-porous, vessels medium-sized to very small, spiral thickenings occasional; parenchyma moderate to abundant, paratracheal. Rays as in Caesalpiniaceae. Anomalous growth occurs in many genera, climbers in particular. Leaves usually dorsiventral, less frequently isobilateral. Hairs both glandular and non-glandular; glandular leaf-teeth in *Myroxylon pubescens* and extrafloral nectaries are present on the stipules of *Canavalia, Dolichos, Erythrina* and *Vicia*. Stomata variable in structure and distribution. Stomata are present on both surfaces of the leaf in many species of *Alysicarpus, Arachis, Argyrolobium, Canavalia, Crotalaria, Rhynchosia, Smithia* and many others; confined to upper surface in *Coelidium, Dillwynia, Eutaxia, Geoffraea, Ormocarpum,* and *Pultenaea;* confined to lower surface in *Aeschynomene, Chadsia, Clitoria, Derris, Desmodium, Dioclea, Dumasia, Millettia, Mucuna, Strongylodon* and some others. The stomata may be Rubiaceous type as in *Alysicarpus, Arachis, Bowdichia, Cicer* and many others, Rubiaceous type but with two pairs of subsidiary cells parallel to the

Fig. 11.12.3 Papilionaceae : **A–J** *Pisum sativum.* **A** Flowering twig. **B** Flower. **C** Flower bud. **D** Papilionaceous aestivation of petals. **E** Keel. **F** Pistil. **G** Stamens and pistil. **H, I** Pod (**H**) l.s (**I**). **J** Seed and placenta. (*c* corolla, *car* carina, *k* calyx, *o* ovule, *ov* ovary, *st* standard, *stg* stigma, *stp* stipule, *w* wing). (Original)

pore, e.g. species of *Aotus, Brachysema, Dillwynia* and *Oxylobium;* surrounded by 3 or more subsidiary cells as in most Galegeae, Hedysareae, Podalyrieae and Sophoreae; approximating to the Cruciferous type in species of *Crotalaria, Lebeckia, Lotononis, Rafnia* and *Viborgia*; surrounded by a rosette of cells in *Anarthrophyllum, Genista, Lebeckia* and *Templetonia;* Ranunculaceous type in most Loteae and Vicieae.

Embryology. Highly multipalynous family with the fundamental form as 3-colpate; pollen grains shed singly at 2-celled stage; usually smooth, spinuliferous in *Dolichos* (Dnyansagar 1970). Ovules anatropous, campylotropous or amphitropous, bitegmic, crassinucellate, Polygonum type of embryo sac most common, Allium type and Oenothera type are also known, 8- or 4-nucleate at maturity. Endosperm formation of the Nuclear type; chalazal part forms a haustorium.

Chromosome Number. Haploid chromosome numbers are n = 7, 8, 10, 11, 12 and 13 of which 7 and 8 are more common. Natural hybridization occurs amongst various species of *Baptisia* (Alston and Turner 1963).

Chemical Features. Many non-protein amino acids are present, e.g. canavanine in *Canavalia ensiformis* (Jackbean), dopa or tyrosine in *Mucuna prurita* and *Vicia faba* seeds, lathyrine in seeds of *Lathyrus tingitanus,* and pipecolic acid in the seeds of *Phaseolus vulgaris.* Isoflavones occur profusely only in the Papilionaceae (amongst all the plant groups): in the flowers, leaves, seeds, roots and heartwood of the genera *Cytisus, Ulex, Trifolium* and *Lathyrus* (Harborne and Turner 1984). *Baptisia,* a genus with 18 species of perennial herbs from North America, contains 9 flavones, 16 flavonols and 18 isoflavone glycosides (Markham et al. 1970). Another related genus *Thermopsis* is also rich in flavonoids (Dement and Mabry 1972). Glycosides are present in the flower pigments of the tribe Vicieae of Papilionaceae (Harborne and Turner 1984). Quinolizidine in lupins (*Lupinus*) deters the feeding of herbivores and inhibits the growth and development of bacteria and fungi and also inhibits the germination of grass seeds. Alkaloid-free lupins have higher incidence of herbivory and disease (Wink 1985).

Important Genera and Economic Importance. Papilionaceae includes numerous important genera. Many are ornamental trees such as *Butea monosperma, Erythrina* spp., *Sophora* sp., *Pongamia pinnata, Sesbania grandiflora* and others; some are climbers, viz. *Derris elliptica, Lathyrus odoratus* and *Wisteria chinensis.* Various members of Papilionaceae provide many essential commodities. Food from *Pisum sativum, Dolichos lablab, Cajanus cajan, Lens esculenta, Phaseolus* spp., *Vigna* spp., *Canavalia* sp., *Cyamopsis tetragonoloba,* etc., fatty oil from *Arachis hypogaea;* fodder from *Trifolium* and *Trigonella* spp., dye from *Indigofera tinctoria* and *Butea monosperma;* timber from *Dalbergia sissoo. D. latifolia* and *Pterocarpus santalinus; P. santalinus* or Red Sanders, an endemic taxon of India, is renowned for its characteristic timber of exquisite odour and beauty and rank among the finest luxury woods in the world with an export potential to various countries. It also yields a natural dye, 'santalin'. *P. marsupium,* Indian Kino or Malabar Kino tree is known for its excellent timber next only to teak and rosewood in South India (Anuradha and Pullaiah 1999). Medicines from *Glycyrrhiza glabra* is used for sore throat and cough; and an insecticide is obtained from *Derris elliptica.* Dried flowers of *Butea monosperma* yield yellow colour, used for dyeing, and also during the festival of colour—"Holi". Cowage or *Mucuna prurita* is a climber whose pods are covered with stinging hairs. It is useful as green manure and cover crop. Seeds have medicinal value. Some are ornamental herbs like *Lupinus* and climbers like *Lathyrus odoratus* and *Clitoria ternatea.* There are many herbs of wild growth: *Alysicarpus, Indigofera. Heylandia, Medicago, Melilotus, Tephrosia, Vicia* and *Zornia.* The seeds of the stout liana *Abrus precatorius* are the source of the protein, abrin. The seeds of this plant are bright red with a black spot and oval-shaped. Each seed has such accurate weight that they are used for weighing gold and silver. *Adenanthera pavonia* (Mimosaceae) is a tree with similar bright red disc-shaped seeds that are used as curios after scooping out the cotyledons and filling the empty space with miniature animals made of ivory. Another interesting plant is *Aschynomene aspera,* the light spongy wood is of ivory colour and used for making decorative articles. The decorations for the foreheads of brides and bridegrooms in Bengal are made of this wood. Guar, *Cyamopsis tetragonoloba* is an useful source of industrial gum. It is also being used as a substitute for agar in laboratories (Koul 2003).

Taxonomic Considerations. This family is the most advanced of the three families—Mimosaceae, Caesalpiniaceae and Papilionaceae—with predominantly herbaceous members, some xerophytes and hydrophytes, zygomorphic flowers and fewer and diadelphous stamens. The status of Papilionaceae has changed repeatedly according to the changes in the status of the order Leguminales (= Leguminosae). The presence of phenylated flavones, flavonoids with a methylenedioxy group and 5- and 7-deoxy-flavonoids in Fabaceae (= Papilionaceae) and Rutaceae indicate their close association (Wollenweber 1982). On the other hand, cyanogenic glycosides are reported to occur (Seigler 1977) in both Fabaceae and Rosaceae, supporting their alliance with Rosaceae.

Taxonomic Considerations of the Order Leguminales

The order Leguminales/Fabales comprises three families: Mimosaceae, Caesalpiniaceae and Fabaceae or Papilionaceae. Whether the taxon Leguminales (= Leguminosae) represents a family of 3 subfamilies or an order embracing 3 families is still disputed (see Table 11.3). Bentham and Hooker (1965a), Rendle (1925), Wilber (1963), Cronquist (1968) Benson (1970) and Takhtajan (1980, 1987) consider the Leguminosae as a family of 3 subfamilies.

That the 3 subfamilies enjoy the rank of individual families and, therefore, the Leguminosae constitute an order is not, however, the latest view. As early as 1814, Brown reported that: "this extensive tribe, i.e. Leguminosae, may be considered as a class (that is, an order in present-day terminology) divisible into at least 3 orders (that is families) namely Mimosae, Lomentaceae or Caesalpiniaceae, and Papilionaceae". Hutchinson (1926, 1969, 1973), Stebbins (1974) and Dahlgren (1975a, 1977a, 1980a, 1983a) also treat the Leguminosae as an order. Following the International Code of Botanical Nomenclature, GN Jones (1955) proposed the ordinal name Leguminales. Stebbins (1974) and Dahlgren (1980a, 1983a) have further changed the name to Fabales based on the type Family Fabaceae (= Papilionaceae).

Table 11.3 Taxonomic Status of Leguminosae

Leguminosae as a family	*Leguminosae as an order*	*Leguminales*	*Fabales*
Bentham and Hooker (1965a)	Brown (1814)	GN Jones (1955)	Stebbins (1974)
Rendle (1925)	Hutchinson (1926)	Hutchinson	Dahlgren (1980a,
Wilber (1963)	Hallier (cf.	(1959, 1969)	1983a)
Cronquist (1968)	Lawrence 1951)		
Benson (1970)	Dahlgren (1975a, 1977a)		
Takhtajan (1980, 1987)	Stebbins (1974)		

On the basis of the above discussion, the three taxa should be treated as distinct families: Mimosaceae, Caesalpiniaceae and Papilionaceae, and included in a distinct order, Leguminales, independent of the Rosales, but next to and derived from it. As Fabaceae is the alternate name for Papilionaceae, according to ICBN, the term Fabales may also be used alternatively for the Leguminales.

Order Geraniales

The order Geraniales comprises three suborders and nine families: Limnanthaceae, Oxalidaceae, Geraniaceae, Tropaeolaceae, Zygophyllaceae, Linaceae, Erythroxylaceae, Euphorbiaceae and Daphniphyllaceae.

Most families are predominantly herbaceous except some members of Erythroxylaceae, Euphorbiaceae and Daphniphyllaceae. The members of Geraniales are characterised by obdiplostemonous stamens, pendulous ovules with a ventral raphe, and the micropyle pointing upwards or erect ovules with a dorsal raphe and the micropyle pointing downwards. Ovary syncarpous, the number of carpels vary, styles often persistent and seeds normally without endosperm.

Euphorbiaceae

A large family of 300 genera and 7500 species of cosmopolitan distribution except in the arctic and antarctic regions.

Vegetative Features. Annual or perennial herbs, shrubs or trees, sometimes xerophytic. The genus *Euphorbia* has prostrate herbs (*E. thymifolia, E. hirta*), shrubs (*E. tirucalli, E. neriifolia*) and trees (*E. nivulia*). Other tree members are *Bischofia javanica, Mallotus philippensis, Putranjiva roxburghii, Emblica officinalis* and *Cicca acida.* Shrubs include *Jatropha gossypifolia, Poinsettia pulcherrima* and *Antidesma ghesaembilla*, and herbs are *Phyllanthus fraternus, P. simplex, Croton bonplandianum* (Fig. 11.13A), *Acalypha indica* and others.

Many genera are latex-bearing (except *Bridelia, Phyllanthus, Baccaurea* and *Poranthera*); stem usually soft, herbaceous and green, sometimes modified to phyllodes as in *Xylophylla*; woody in tree members. Leaves mostly alternate, sometimes opposite, as in *E. hirta*, or whorled as in *Acalypha indica*, entire or lobed, as in *Ricinus communis* and *Jatropha* spp., stipulate, stipules modified to glandular hairs (*J. gossypifolia*) or spines (*E. milli*; Fig. 11.13 B).

Floral Features. Inflorescence shows variations. Usually the first branching of racemose type is followed by cymose types. Catkins or pendulous racemes are seen in *Acalypha indica*, 1 or 2 axillary flowers have been observed in *Emblica officinalis* and *Phyllanthus fraternus* (Fig. 11.13 F). An erect raceme is known in *Ricinus communis*; terminal dichasial cymes in *Jatropha* (Fig. 11.13 N); simple or compound racemes in *Manihot* and in *Trewia*, the male flowers in drooping catkins and a large female flower is solitary on a long peduncle. In *Euphorbia* spp. the inflorescence gives the appearance of a single flower, cyathium. In a cyathium, a centrally situated, highly reduced female flower is surrounded by a large number of male flowers, each represented by a single stalked stamen subtended by a bract. All these flowers are enclosed within an involucre formed of 5 bracts alternating with 5 nectaries and there are 1 or 2 large bracts forming the outermost layer (Fig. 11.13 J – L). The reduced male flowers are in 2 – 5 groups and arranged in scorpioid cymes, the oldest is nearest to the female flower.

Flowers unisexual (Fig. 11.13 G, H, K, L), plants may be mono- or dioecious, male and female flowers may be borne on the same inflorescence or on separate ones, complete or incomplete, zygo- or actinomorphic, hypogynous, bracteate, ebracteolate. Cauliforal inflorescences seen in *Baccaurea* sp. Both calyx and corolla are present in *Jatropha* (Fig. 11.13 N). Sepals 5, polysepalous, imbricate; petals 5, polypetalous, valvate or contorted. Stamens usually 10, in two whorls of 5 each, filaments basally connate. In female flowers the gynoecium is syncarpous, 3-carpellary, ovary superior, trilocular with axile placentation and one ovule per locule (Fig. 11.13 M). In *Croton*, the female flowers may be with or without a conspicuous corolla (Fig. 11.13 D, E). In *Phyllanthus* and *Ricinus*, both male and female flowers are apetalous and only a sepaloid perianth is present (Fig. 11.13 G, H). In *Manihot*, the calyx is

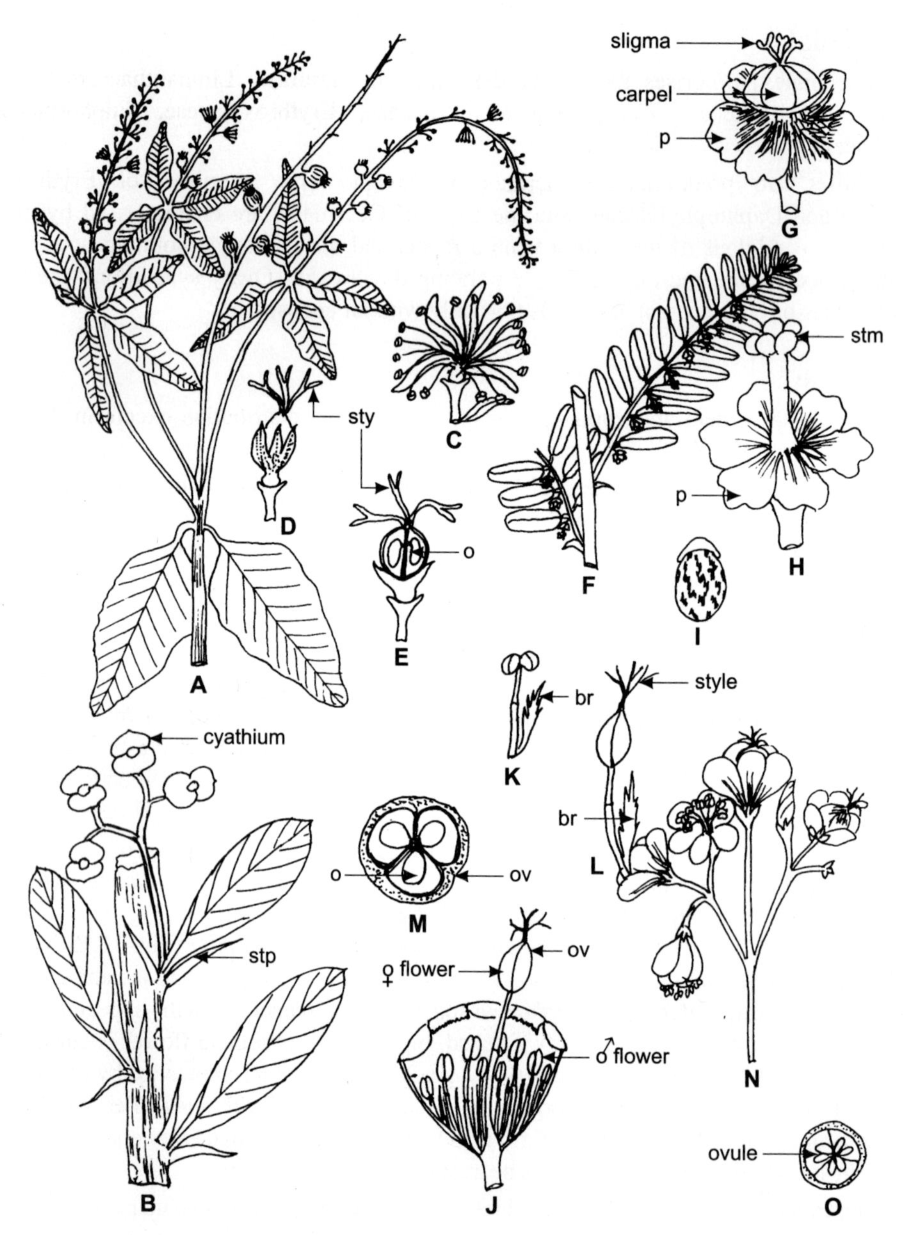

Fig. 11.13 Euphorbiaceae: **A**, **C–E** *Croton bonplandianum,* **F–H** *Phyllanthus fraternus*, **B**, **J–M** *Euphorbia milli*, **I** *Ricinus communis*, **N, O** *Jatropha* sp. **A** Flowering twig of *C. bonplandianum.* **B** Of *Euphorbia milli.* **C, D** male (**C**) and female (**D**) flowers of *C. bonplandianum.* **E** Female flower, longisection. **F** Flowering twig of *P. fraternus.* **G** Pistillate flower. **H** Staminate flower. **I** Seed of *R. communis.* **J** Cyathium cut open of *E. milli.* **K, L** Staminate and pistillate flower. **M** Ovary, cross section. **N** Inflorescence of *Jatropha* sp. **O** Ovary, cross section. *br* bract, *o* ovule, *ov* ovary, *p* perianth, *stm* stamen, *stp* stipule, *sty* style (Sketched by Arindam Bhattacharyya)

petaloid and in *Euphorbia* the flowers are without any calyx and corolla (Fig. 11.13 J – L). Stamens may be 1 to numerous in male flowers (Fig. 11.13 C), free or variously branched. In *Euphorbia* spp., each male flower is represented by one stalked stamen (Fig. 11.13 K) which is bracteate. In *Ricinus*, there are 5 stamens that are profusely branched and each branch terminates in an anther. In *Jatropha*, *Crozophora* and *Phyllanthus*, the stamens are basally connate or monadelphous (Fig. 11.13 H). Pistillodes are sometimes present. Anthers bicelled, dehiscence longitudinal, transverse or poricidal.

Gynoecium syncarpous, tricarpellary, ovary superior, trilocular with 1 or 2 ovules per locule (Fig. 11.13 M, O), placentation axile; styles three, free or basally connate (Fig. 11.13 D, E, G, L), each often bilobed, stigmas 3 or 6, linear or broadened, often papillate or dissected into filiform segments. Fruits usually 3-valved schizocarpic capsule, splitting into three l-seeded cocci that dehisce ventrally; seeds with straight or curved embryo and fleshy endosperm. Seeds with a caruncle in some members, e.g. *Jatropha*, *Ricinus* (Fig. 11.13 I).

Anatomy. The anatomical structure exhibits a wide range of variation with the diversity of habit, and there is no important character throughout the numerous tribes into which this family is divided (Metcalfe and Chalk 1972). Vessels variable in size, sometimes even within the same genus, with simple (Crotonoideae and Glochidion type of Phyllanthoideae) or scalariform perforations (Aporosa type of Phyllanthoideae). Parenchyma abundant, apotracheal in Crotonoideae, diffuse in Phyllanthoideae Aporosa type); absent or as rare cells about the vessels in Phyllanthoideae, (Glochidion type). Rays mostly of two distinct sizes; typically 2 to 3 cells wide or exclusively uniseriate as in Crotonoideae. Leaves may be of ordinary laminate types with a distinct dorsiventral mesophyll, rolled and furrowed forms which are often centric or sometimes much reduced as in succulent species. Hairs of glandular, non-glandular and stinging types known; extrafloral nectaries also common. Stomata usually Rubiaceous type, generally Cruciferous type in *Andrachne*, *Aporusa*, *Baccaurea* and *Richeria*, predominantly of Ranunculaceous type in European species of *Euphorbia*; usually confined to the lower surface, rarely on both the surfaces. Latex tubes are sometimes present in the rays.

Embryology. Pollen grains 2- or 3-celled at the time of shedding, have smooth or reticulate exine; generally triporate, non-aperturate in *Baliospermum montanum* and *Croton bonplandianum*; quadriporate in *Acalypha indica*, *A. alnifolia* and *Micrococca mercurialis*; 10- to 12-porate in *Melanthesa rhamnoides.* Ovules mostly ana-, hemiana- (*Chrozophora*) or orthotropous (*Breynia patens*), bitegmic, crassinucellate and have an obturator, prominent nucellar beak, hypostase and vascular supply in the integuments of some species. Embryo sac of Polygonum type, 8-nucleate at maturity. Endosperm formation of the Nuclear type and eventually becomes cellular. Seeds usually with a caruncle. Polyembryony occurs in *Alchornea* and *Euphorbia dulcis.*

Chromosome Number. Basic chromosome number is $x = 6 - 12$.

Chemical Features. The seeds of many Euphorbiaceae are rich in linolenic, linoleic and oleic acids: *Antidesma diandrum, Bischofia javanica, Euphorbia heterophylla, E. marginata, Mercurialis annua* and others (Shorland 1963). Natural polyols have been reported in *E. pilulifera* and in the latex of *Hevea* (Plouvier 1963). Coumarin glycoside with aglycone aesculetin occurs in *E. lathyris* (Paris 1963); triterpenes in resins and bark of trees, and in the latex of *Euphorbia* and *Hevea.* The acrid, milky or colourless juice of most members contain triterpenoids, flavonoids and alkaloids, coumarins, cyanogenic compounds and tannins (Rizk 1987).

Important Genera and Economic Importance. *Euphorbia* is the most important genus of this family, showing much variation in habit and habitat. Many of the xerophytic euphorbias bear so much resemblance to cacti that they can be separated only by the presence of latex and pairs of stipulary spines.

The genus has a remarkable floral structure not shared by many genera. There are other important genera like *Ricinus communis, Hevea brasiliensis, Emblica officinalis, Croton, Jatropha, Phyllanthus* and others.

Hevea brasiliensis is an important tree from the Amazon River valley, and its latex is the source of rubber. Its mature fruit is a hard, woody, trilobed capsule that dehisces violently into 3 pieces (when dry) throwing the seeds to a distance away from the mother tree. The seeds are recalcitrant due to their high moisture content and lose viability if stored under open-air condition (Thomas et al. 1996). *Croton tiglium* and *Ricinus communis* are two other important plants which yield croton oil and castor oil, respectively. Both the oils have many commercial uses. *Hura crepitans,* commonly called the sandbox tree, is a large or medium-sized tree. Its trunk is usually covered with short, sharp spines. The male flowers are in a dense spike, and the female flower is a solitary one borne on the side of the stalk of the male inflorescence. The capsule looks like a small pumpkin and consists of about fifteen-seeded woody chambers; when ripe, it explodes with a loud report. The milky sap of the plant is poisonous and is often mixed with meal to stupefy fish. Another poisonous plant is *Hippomane mancinella* from Panama, Venezuela, the West Indies and South Florida.

Aleurites fordii of China is the source of tung oil used in varnishes; *A. moluccana* yields candlenut oil used as a preservative for the hulls of vessels. The bark of *Bischofia javanica* and *Bridelia retusa* are useful in tanning. Candelilla wax is extracted from the stems of *Euphorbia antisyphylitica* and *Pedilanthus pavonis,* both from Mexico and Texas. The oil from the seeds of *Givotia rottleriformis* is used as a lubricant. Ink is prepared from the ripe fruits of *Kirganelia reticulata.* Chinese tallow tree or *Sapium sebiferum* is a native of subtropical China and has been cultivated for at least 14 centuries as a seed oil crop (Seibert et al. 1986).

Some plants yield edible fruits and roots. Fruits of *Aleurites moluccana, Baccaurea sapida, Bridelia squamosa, Hemicyclis andamanica, H. sepiaria* and *Trewia nudiflora* are used for culinary purposes. The fruits of *Cicca acida* and *Emblica officinalis* are rich source of vitamin C. Starchy roots of *Manihot esculenta* are the commercial tapioca.

Some taxa are cultivated as ornamentals: *Acalypha, Codiaeum, Dalechampia, Jatropha, Euphorbia, Poinsettia* and others. *Baccaurea* sp. with its colourful caulifloral inflorescences is a spectacular tree from Kerala and some parts of Western Ghats in South India. .

Taxonomic Considerations. There is much controversy regarding the systematic position of Euphorbiaceae. Hallier (1912) regarded it as a member of his Passionales; Lawrence (1951), Melchior (1964) in Geraniales; Rendle (1925) and Wettstein (1935) in Tricoccae; and Bentham and Hooker (1965c) placed it in the Unisexuales under Monochlamydeae, on the basis of its floral structures. Hutchinson (1969, 1973) placed Euphorbiaceae in a separate order of its own Euphorbiales, next to the Malpighiales. According to him, it is a highly evolved family, almost comparable to the Asteraceae because of so much reduction in the floral structure. Also, the family Euphorbiaceae comprises a group of genera which have been derived from different stocks—like Tiliaceae, Sterculiaceae, Malvaceae and Celastraceae. It also bears a relationship with the Geraniales and Sapindales, on account of the nature of ovules. There is no doubt that the Ephorbiaceae has a polyphyletic origin.

Cronquist (1981) includes Euphorbiaceae in the Euphorbiales along with four other families—Buxaceae, Daphniphyllaceae, Aextoxicaceae and Pandaceae. All these families have unisexual, mostly monochlamydeous flowers, 1 or 2 bitegmic ovules per locule and copious endosperm. Takhtajan (1987) includes Dichapetalaceae in Euphorbiales, in addition to Pandaceae and Aextoxicaceae, and removes Buxaceae and Daphniphyllaceae to separate orders. The families Buxaceae, Dichapetalaceae and Daphniphyllaceae have earlier been treated as different tribes of Euphorbiaceae. They are all embryologically distinct from Euphorbiaceae, but they show common features such as: monochlamydeous, unisexual flowers, bitegmic ovules and copious endosperm. Three genera of the 4 members of the Pandaceae were earlier included in the Euphorbiaceae and the Pandaceae is a monotypic family. Forman (1966) moved these three genera back to Pandaceae. The two families are closely related. Webster (1967) agreed to this assignment and expressed similar views about the families Aextoxicaceae, Buxaceae and Daphniphyllaceae.

Hence, it appears that there should be a separate order, Euphorbiales to include all these 6 families—Buxaceae, Euphorbiaceae, Dichapetalaceae, Daphniphyllaceae, Pandaceae, and Aextoxicaceae.

Order Rutales

The order Rutales comprises 3 suborders:

(a) Suborder Rutineae includes 6 families,

(b) suborder Malpighiineae 3 families, and

(c) suborder Polygalineae 2 families.

Members of the Rutales are mostly trees and shrubs, a few climbers and a few herbs. Leaves simple or pinnately or palmately compound, mostly estipulate, often glandular-punctate (as in Rutaceae). Inflorescence much variable, flowers usually actinomorphic, scented and with a nectariferous interstaminal disc (absent in Polygalineae). Embryologically and anatomically, these families are allied. Pollen grains are monads and usually shed at 2-celled stage. Ovules mostly bitegmic, crassinucellate, and anatropous, hemianatropous or epitropous.

Rutaceae

A large family comprising 150 genera and 1500-1600 species, widely distributed in both temperate and tropical zones of the New as well as the Old World.

Vegetative Features. Mostly shrubs (*Citrus*) and trees (*Aegle marmelos, Feronia elephantum*), sometimes climbers, e.g. *Paramignya scandens*, a woody climber with strong axillary, recurved spines; herbs rare, e.g. *Ruta graveolens*, (Fig. 11.14 A), *Boenninghausenia* sp. Stem herbaceous in *Boenninghausenia, Monnieria* and *Dictamnus*; woody below and herbaceous above in trees and shrubs. Leaves alternate or opposite, simple (*Boronia, Pitavia*), or palmately (*Citrus*) or pinnately (*Ruta*) compound (Fig. 11.14A), sometimes reduced to spine, estipulate. In *Citrus* spp. the petiole is winged and separated from the lamina by a distinct joint—often considered as a unifoliate, palmately compound leaf. Leaves glandular-punctate (Fig. 11.14 B).

Floral Features. Inflorescence usually cymose, e.g. *Toddalia,* sometimes racemose—a raceme or corymb (*Murraya paniculata*), or sometimes solitary axillary as in *Triphasia aurantiola.* In the Australian genus *Diplolaena*, the flowers are densely grouped into a head with a 3- or 4-seriate involucre of bracts, the inner is petaloid (Fig. 11.14 D, E). Epiphyllous flowers are borne in *Erythrochiton hypophyllanthus* from South America.

Flowers mostly bisexual (Fig. 11.14 C, E, G), rarely unisexual as in *Evodia* and *Zanthoxylum,* actinomorphic (zygomorphic in *Dictamnus*), usually pentamerous, may be tetra- or trimerous also, e.g. in *Ruta,* the terminal flower is pentamerous and the lateral ones tetramerous. Calyx mostly of 3 to 5 sepals. poly- or gamosepalous (*Citrus*), imbricate or quincuncial; corolla of 3 to 5 petals, polypetalous (gamopetalous and campanulate in *Correa speciosa*), imbricate. Stamens 3 to 10 or more, obdiplostemonous or in 2 whorls—outer ones opposite the petals; a disc present between stamens and ovary. In *Citrus* stamens numerous and polyadelphous (Fig. 11.14 G, H); in *Aegle marmelos* the number of stamens is 50 or more, in *Murraya* 10, in *Skimmia* 5, and in *Zanthoxylum* 3 to 5. All stamens attached at base or rim of the nectariferous disc (Fig. 11.14 C, I), some occasionally reduced to staminodia, free or basally connate, rarely adnate to petals, usually straight and unequal. Anthers bicelled, introrse, dehiscence longitudinal (Fig. 11.14 C, E, G, H), connective often with glandular apex. Gynoecium syncarpous, 5- or 4-carpellary, or carpels weakly connate or only basally or apically connate (Lawrence 1951); ovary superior, usually deeply lobed, typically 4- to 5-loculed with axile placentation (Fig. 11.14 F) (unilocular with parietal placentation in *Feronia limonia*); ovules 1, 2 or more in each locule; styles as many as carpels and free or only 1, stigma 1, capitate. Fruit various—a valvate capsule or

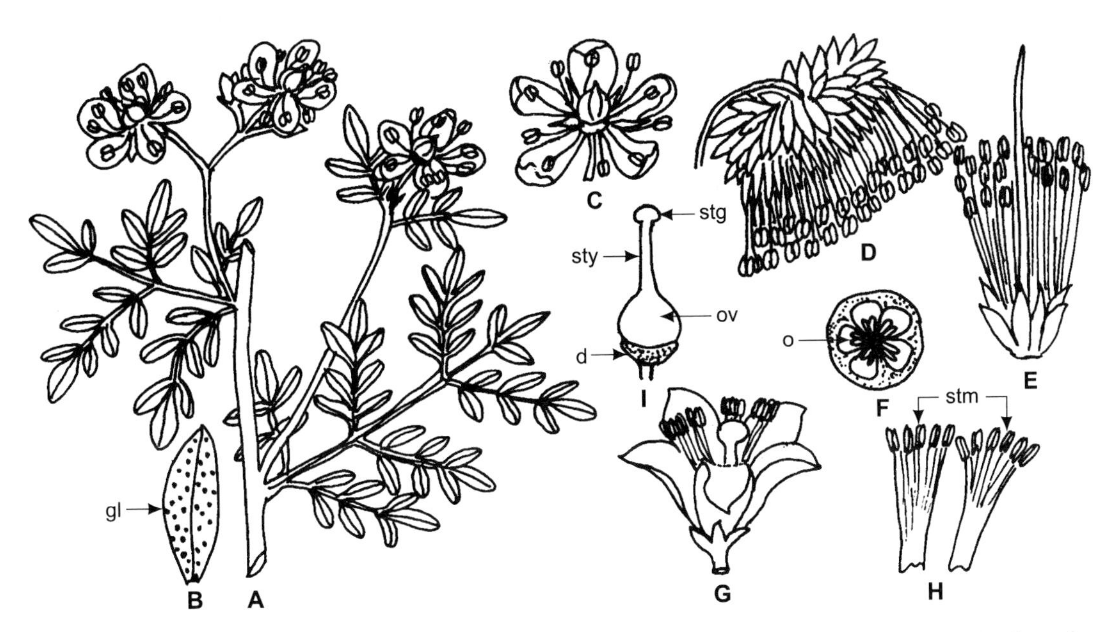

Fig. 11.14 Rutaceae: **A-C** *Ruta graveolens*, **D-F** *Diplolaena dampieri*, **G-I** *Citrus limon.* **A** Flowering twig. **B** Glandular-punctate leaflet. **C** Flower, face view. **D** Inflorescence of *D. dampieri.* **E** Flower. **F** Ovary, cross section. **G** Flower of *Citrus limon.* **H** Polyadelphous stamens. **I** Gynoecium. (*d* disc, *gl* gland, *o* ovule, *ov* ovary, *stg* stigma, *stm* stamen, *sty* style). (**A-F** adapted from Hutchinson 1969, **G-I** original)

a hesperidium, or separating into mericarps or a winged berry or drupe or samara. Seed with a large, straight or curved embryo and fleshy endosperm or endosperm absent.

Anatomy. Wood ring-porous or semi-ring-porous; vessels small to medium-sized, typically in multiples and sometimes with a distinct radial or oblique pattern. Perforation mostly simple, intervascular pitting alternate. Parenchyma terminal, paratracheal, usually vasicentric, often include crystal cells. Rays uniseriate, up to 2 to 4 cells wide; homogeneous or weakly heterogeneous. Leaves generally dorsiventral, sometimes centric. Stomata of various types occur on both the surfaces in *Cneoridium dumosum* and *Ruta graveolens*, generally only on the lower surface. Secretory cavities appear as transparent dots.

Embryology. Pollen grains 2- to 8-colporate with reticulate exine and shed at 2-celled stage. Ovules anatropous, bitegmic and crassinucellate. Embryo sac of Polygonum type, 8- nucleate at maturity. Endosperm formation of the Nuclear type (Johri et al. 1992)

Chromosome Number. Basic chromosome numbers are x = 7–11, 13, 17, 19.

Chemical Features. Flavanones and alkaloids are widespread. A large number of genera are relatively rich in essential oils. The Rutaceae is particularly rich in coumarins (Price 1963).

Important Genera and Economic Importance. *Citrus* with about 16 species grows throughout the temperate and tropical regions of the world. These are mostly shrubs or small trees with palmately unipinnate leaves, spiny branches and white fragrant flowers. The genus *Citrus* is difficult to classify as there is a tendency to form natural hybrids and mutants. Some of the important species are *C. reticulata* (sweet orange), *C. sinensis* (tight-skinned orange or mousambi), *C. aurantiifolia* (lime), *C. limellioides* (sweet lime), *C. maxima* (shaddock), *C. medica* (citron) and *C. limon* (lemon). Apart from being used as fruits, they produce a large number of commercial products such as essential and fixed oils, citric acid and pectin. Fruits also find use in preparation of juices, squashes, marmalades and jellies. The juice of *C. limon* is rich in vitamin C. Seeds contain varying amount of fixed oil, protein and limonin; the oil is used in soap industry. The waste pulp finds use in the production of food yeast, industrial alcohol and ascorbic acid.

Aegle marmelos is important for its medicinal use. It is a medium-sized spiny tree with tripinnate leaves and fragrant flowers. Pulp of the ripe fruit is a good laxative; unripe fruits after boiling or roasting is often used against diarrhoea and dysentry. The mucilage around the seeds is used as an adhesive. The leaves are beleived to be sacred in Hindu mythology and are offered in prayers to Lord Shiva. *Ruta graveolens* is a strongly smelling ornamental herb with yellow flowers; 'oil of rue' is distilled from its leaves. Trees of *Chloroxylon swietenia* and *Zanthoxylum flavum* yield useful timbers, commonly called satinwoods. Timber from *Flindersia brayleana* is an important hard wood from Australia, used for cabinet work, veneers, aeroplane construction and rifle stocks. Various plants have medicinal value—*Cusparia febrifuga* gives cusparia bark used as a substitute for quinine; the dried leaves of *Agathosma (=Barosma) betulina, A. crenulata* and *A. serratifolia* form the drug buchu, used as diurectic.

The violet-scented 'oil of boronia' used in perfumery is derived from *Boronia megastigma.* 'Mexican elemi', an oleoresin, is obtained from *Amyris balsamifera* and *A. elemifera.* The twigs of *Glycosmis pentaphylla* and *Zanthoxylum alatum* are used as chewsticks or toothbrushes. 'Chinese box', *Murraya paniculata* is a garden ornamental, and *M. koenigii* is grown extensively for its leaves, that are used as a condiment. The roots of *Toddalia asiatica*, a spiny shrub, are the source of a yellow dye.

Taxonomic Considerations. Bentham and Hooker (1965a) included Rutaceae in Geraniales and so did Engler and Diels (1936). But Melchior (1964) regarded it as a member of a separate order Rutales and placed it next to the Geraniales. Rendle (1925), and Hutchinson (1973) also place Rutaceae in Rutales, but the circumscription of the order varied.

Cronquist (1981) treats it as a member of the Sapindales although Takhtajan (1980, 1987) agrees with the views of earlier workers. The family Rutaceae is more allied to Meliaceae, Sapindaceae and Anacardiaceae in their exomorphic and anatomical features, and it appears that this family is better placed in an order distinct from the Geraniales. Rutales could be an appropriate order for this family.

Tetradiclis. This genus has been included in various families like Crassulaceae, Elatinaceae, Zygophyllaceae and Rutaceae. Fenzl (1841), Hallier (1908, 1912), and Takhtajan (1966) placed it in the Rutaceae, but many taxonomists place it in Zygophyllaceae. According to Takhtajan (1987), *Tetradiclis* differs sufficiently from both Rutaceae and Zygophyllaceae to deserve the rank of an independent family, Tetradiclidaceae. This family is somewhat closer to the Rutaceae.

Order Malvales

The Malvales is a large order comprising 4 suborders and 7 families; Elaeocarpaceae, Sarcolaenaceae, Tiliaceae, Malvaceae, Bombacaeae, Sterculiaceae and Scytopetalaceae.

The taxa are predominantly woody plants of the tropics and subtropics of both the northern and southern hemisphere. Vegetative parts often stellate-pubescent and mucilage-producing. Flowers bisexual, actinomorphic, hypogynous and mostly pentamerous. Calyx valvate, corolla valvate or contorted. Stamens numerous, in one or more than one whorl, often monadelphous. Ovary multicarpellary, multilocular and usually with axile placentation. Nectary glands are characteristic multicelluar hairs packed close together to form cushion-like growths.

Tiliaceae

A medium-sized family of 46 genera and 450 species (Takhtajan 1987), mostly restricted to tropical regions with a few members distributed in temperate zones. The genus *Carpodiptera* has 3 of its 5 species in the West Indies, one in the coastal zone of east tropical Africa, and the fifth in the Comoro Islands.

Vegetative Features. Trees or shrubs, rarely herbs, e.g. *Corchorus* (Fig. 11.15 A), *Triumfetta*; with mucilage. Stem mostly woody, leaves usually alternate, rarely opposite as in *Plagiopteron,* simple, entire, stipulate, stipules deciduous; margin dentate, serrate or lobed, often oblique at the base.

Floral Features. Axillary or terminal cymes, often highly reduced so as to appear as a solitary flower, e.g. *Corchorus*; often peduncles winged up to the middle with the adnate, foliaceous or coloured bract. Flowers ebracteate, ebracteolate, actinomorphic, bisexual or rarely unisexual as in *Vasivaea* and *Carpodiptera* (Fig. 11.15 G, K), hypogynous. Calyx of 4 or 5 sepals, free or slightly united at base, usually valvate. Corolla of 4 or 5 petals (Fig. 11.15 B) or absent as in *Prockia*, sometimes sepaloid, polypetalous, valvate or imbricate. Stamens 10 or more, free or basally connate or polyadelphous as in *Grewia* (Fig. 11.15 G); anthers introrse, bicelled, dehisce longitudinally (Fig. 11.15 C) or by apical pores. Gynoecium syncarpous, 2- to 10-carpellary, ovary superior, 2- to 10-loculed, placentation axile (Fig. 11.15 D), ovules 1 to numerous in each locule; style 1, simple, stigmas usually as many as locules. Fruit fleshy, berry-like or drupaceous (*Grewia*, Fig. 11.15 H) or dry, capsule (*Corchorus*, Fig. 11.15 E, F), dehiscent or indehiscent. Seeds small, usually endospermic, with straight embryo; sometimes covered with stellate hairs as in *Triumfetta* (Fig. 11.15 I).

Anatomy. Vessels small to medium-sized with radial multiples of 4 or more cells in some genera, semi-ring-porous, perforations simple; parenchyma predominantly apotracheal but paratracheal and intermediate types also reported. Rays uni- or multiseriate, 2–3 cells or 4–15 cells wide. Leaves generally dorsiventral but consist wholly of palisade tissue in some species of *Apeiba, Berrya, Corchorus, Diplodiscus* and *Grewia.* Hairs unicellular, uniseriate, stellate, tufted; stomata generally confined to the lower surface, Ranunculaceous type. Mucilage cells and cavities observed in cortex and pith.

Embryology. Pollen grains shed at 2-celled stage (Sharma 1969). Ovules anatropous, bitegmic, crassinucellate. Polygonum type of embryo sac, 8-nucleate at maturity. Endosperm formation of the Nuclear type.

Chromosome Number. Basic chromosome number is x = 7–10, 41.

Chemical Features. Phytoalexin sesquiterpene occurs in *Tilia* and C-glycoflavones in members of the Tiliaceae (Gornall et al. 1979).

Important Genera and Economic Importance. *Berrya cordata*, *B. cordifolia*, *Erinocarpus nimmonii*, *Tilia americana* and *T. vulgaris* are important timber-yielding trees. Linden or basswood is

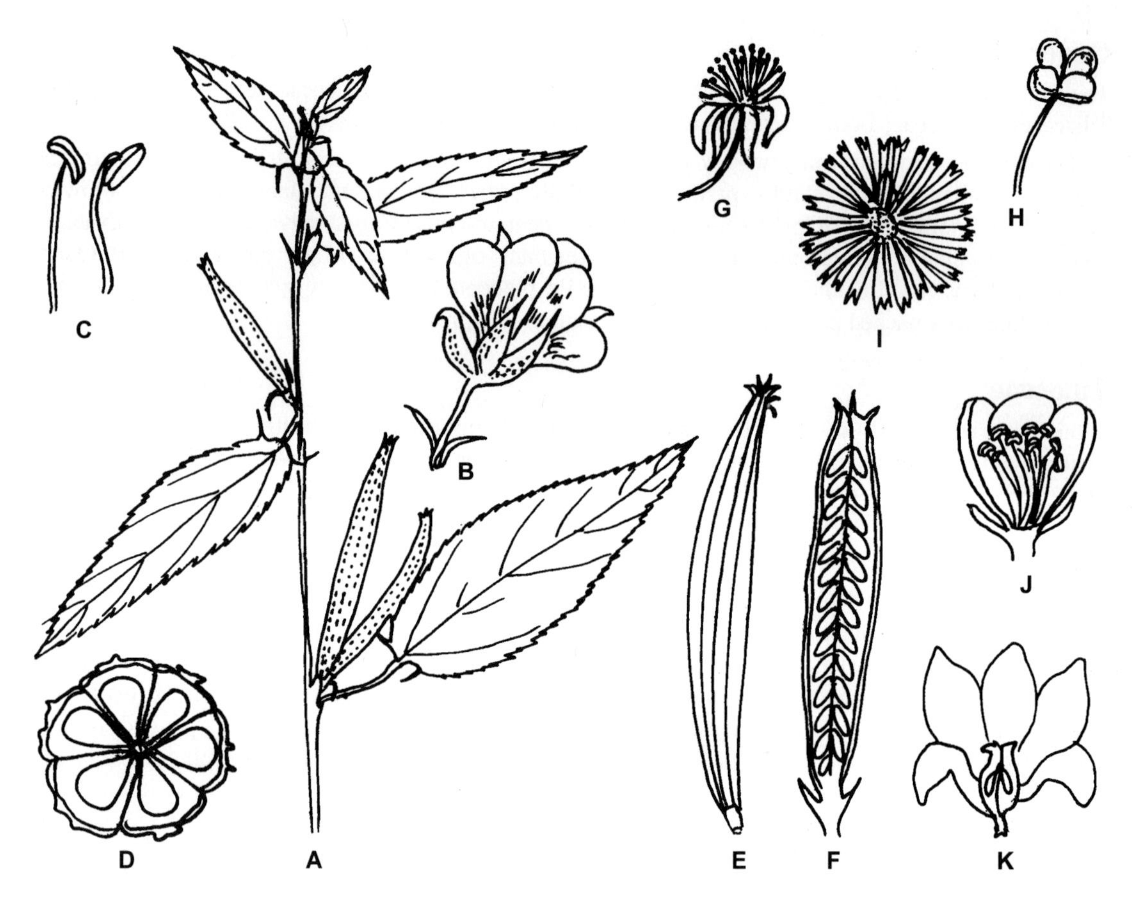

Fig. 11.15 Tiliaceae : **A–F** *Corchorus olitorius*, **G, H** *Grewia tenax*, **I** *Triumfetta setulosa*, **J, K** *Carpodiptera ameliae.* **A** Flowering and fruiting twig. **B** Flower. **C** Stamens. **D** Ovary, cross section. **E** Fruit. **F** Longisection. **G–H** Flower and fruit of *G. tenax.* **I** Fruit of *T. setulosa*, with spiny projections. **J, K** Staminate and pistillate flower of *Carpodiptera ameliae.* (**A-H** Original, **I–K** adapted from Hutchinson, 1969)

obtained from *Tilia americana. Grewia tiliafolia* is another tree, its berries are edible. *Corchorus* is yet another important genus, two species—*C. capsularis* and *C olitorius*—yield the bast fibre, jute. *C. aestuans* and *Triumfetta rhomboidea* are weeds of wastelands.

Taxonomic Considerations. Most taxonomists treat Tiliaceae as a member of the Malvales. Hutchinson (1973) includes Tiliaceae in an independent order, Tiliales. *Pakaraimaea* is a disputed genus. Although Maguire et al. (1977) placed it in Dipterocarpaceae, according to Kostermans (1978, 1985) it belongs to the Tiliaceae. The anatomical structure of *Pakaraimaea* indicates that it should be retained in the order Theales of the Dilleniidae; Sarcolaenaceae and Dipterocarpaceae are its closest allies. Along with two African genera, *Marquesia* and *Monotes, Pakaraimaea* easily forms a group separate from the true Dipterocarps. The numerous stamens on androgynophore, tricolpate pollen grain, gum sacs instead of resin canals, uniseriate wood rays, and absence of glandular hairs bring this group closer to the Tiliaceae. Takhtajan (1987) treats this group as a separate family, Monotaceae, of the Malvales and places it next to the family Tiliaceae. Takhtajan's Malvales also include Dipterocarpaceae.

Pakaraimaea, *Marquesia* and *Monotes* should be included in a distinct family, Monotaceae, and placed near the Tiliaceae.

Malvaceae

A medium-sized family with 75 to 85 genera and 1500-1600 species, distributed predominantly in the tropics.

Vegetative Features. Trees (*Thespesia populnea, Kydia calycina*), *shrubs* (*Hibiscus mutabilis, H. rosa-sinensis*) or herbs (*Malvastrum, Sida, Urena, Malva*—Fig. 11.16 A), the entire plant contains mucilaginous sap. Leaves alternate, simple, petiolate, stipulate, stipules free-lateral; margin dentate, crenate, entire or deeply lobed as in *Abelmoschus esculentus*, usually palmately veined.

Floral Features. Inflorescence basically cyme (Fig. 11.16 A) but often solitary, axillary as in *H. rosa-sinensis* (Fig. 11.16 D). Flowers bracteate, bracteolate, bracteoles usually form a whorl of epicalyx (exceptions *Sida, Abutilon*), pedicellate, actinomorphic, hypogynous, bisexual, pentamerous. Epicalyx of 7 to 10 lobes, free, green, hairy. Calyx of 5 sepals, gamosepalous, valvate, outer surface hairy; corolla of 5 petals, polypetalous (Fig. 11.16 B), often basally adnate to the staminal column, convolute or twisted, contain mucilage. Stamens numerous, monadelphous, form a staminal column around the style, fused with the basal parts of the corolla. Upper part of the branched filament is free, each bears a monothecous, reniform, half-anther which opens by a transverse slit; extrorse, pollen formation profuse. All the stamens are derived by copious branching of 5 antipetalous stamens; the outer antisepalous whorl of stamens is lost, though in *Hibiscus* it is represented by 5 teeth of the staminodes on the summit of the staminal tube. Gynoecium. syncarpous, 2- to many-carpellary, ovary superior, 2- to many-loculed (Fig. 11.16 C), usually in a ring or infrequently superposed as in *Malope* (Fig. 11.16 E); placentation axile, ovules 1 to many in each locule, style 1 and apically branched or as many as the carpels; stigmas as many or twice as many as the carpels (Fig. 11.16 D, E), capitate or discoid. Fruit typically a loculicidal capsule (Fig. 11.16 F) or schizocarpic or carcerulus (Fig. 11.16 H), or rarely a berry as in *Malvaviscus* or a samara. Seeds mostly reniform (Fig. 11.16 I); often pubescent or comose as in *Gossypium* (Fig. 11.16 G).

Anatomy. Vessels small to medium-sized, semi-ring-porous; perforations simple, intervascular pitting small and alternate. Parenchyma predominantly paratracheal in tribes Malveae and Ureneae, and predominantly apotracheal in Hibisceae. Rays may be of 2 types: (a) multiseriate rays usually up to 4–9 and rarely up to 23 cells wide, low to high, and (b) uniseriate; typically heterogeneous. Leaves dorsiventral, centric in *Malva parviflora,* covered with both glandular and non-glandular hairs. Extrafloral nectaries also occur in some members. Stomata of the Ranunculaceous type, always on the lower surface. Mucilage cells and secretory cavities common.

Embryology. Pollen grains spherical, pantoporate or multiporate (Venkata Rao 1954a), spinous and very large. Pollen sterility is reported in *Thespesia populnea.* Ovules campylotropous, bitegmic, crassinucellate; outer integument longer than the inner one. Polygonum type of embryo sac, 8-nucleate at maturity. Endosperm formation of the Nuclear type.

Chromosome Number. Basic chromosome number is variable—$x = 5–13, 15–17, 19–23, 29, 33, 39$.

Chemical Features. In the members of Malvaceae, the seed fats contain palmitic, oleic and linoleic acids as major components. Unusual fatty acids like malvalic and sterculic acid occur in many members. Aglycone anthocyanin pigments such as cyanidin and malvidin are present in many flowers. Cotton fibres are rich in cellulose.

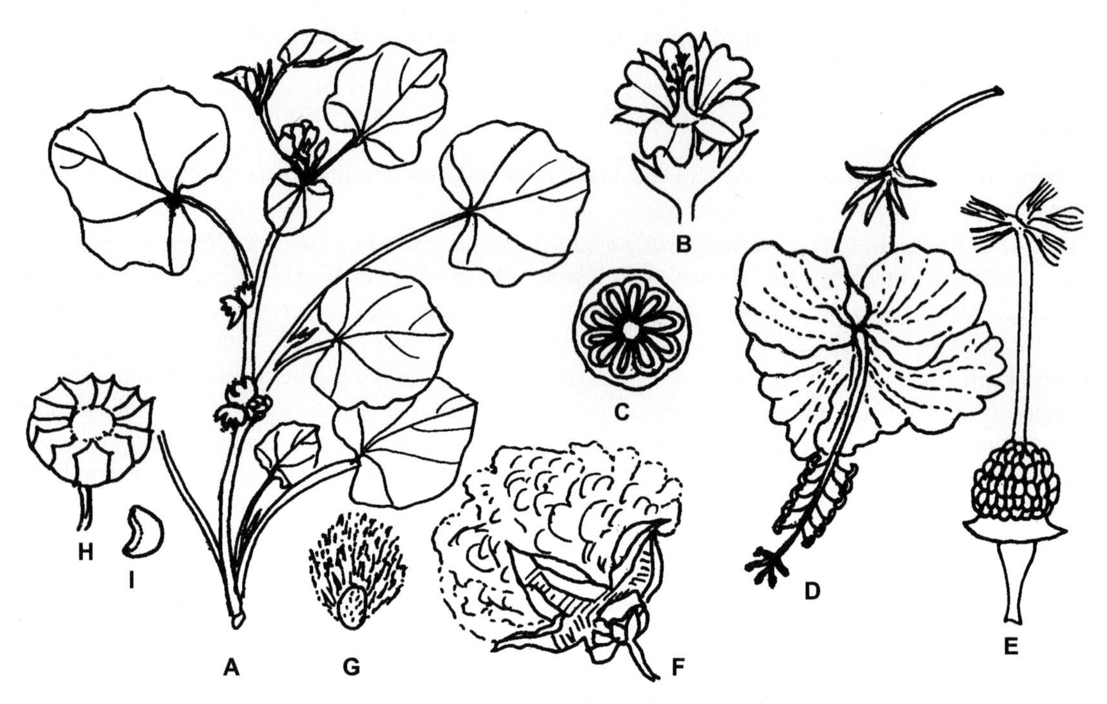

Fig. 11.16 Malvaceae: **A–C** *Malva parviflora*, **D** *Hibiscus rosa-sinensis*, **E** *Malope trifida*. **F, G** *Gossypium hirsutum*, **H, I** *Malvastrum coromandelianum*. **A** Twig bearing flowers and fruits. **B** Flower. **C** Ovary, cross section. **D** Flower of *H. rosa-sinensis*. **E** Gynoecium of *Malope trifida*. **F, G** Boll and seed of *G. hirsutum* **H, I** Fruit and seed of *Malvastrum coromandelianum*. (**A, B, H, I** after J K Maheshwari 1965, **E** after Hutchinson 1969).

Important Genera and Economic Importance. *Hibiscus* with 300 species and *Pavonia* with 200 species each are the largest genera. *Hibiscus* species, mostly with the large showy flowers, are distributed in the tropical and subtropical regions of both the New and the Old World. *Pavonia* and *Sida* also occur in the tropics and subtropics. Economically, the Malvaceae is very important. The commercial cotton is derived from the densely hairy seeds of *Gossypium*. Cotton fibres have been traced to the Indus Valley civilization (ca. 3000 BC). The Greeks found this crop when they invaded India with Alexander the Great and they called it 'lamb on the tree'. *Gossypium hirsutum*, cultivated mainly in the USA, yields long staple cotton, *G. arboreum* and *G. herbaceum* which yield short staple cotton, are cultivated in Asia. Cotton seeds are rich in fat and yield edible oil called the cotton seed oil. It is also used in manufacturing soaps and lubricants; the oil cake is used as a cattle feed. *Hibiscus cannabinus* is another plant which yields fibre, used widely for cordage, ropes, etc.; fatty oil used in the manufacture of linoleum, paints and varnishes and also as edible oil and oil cake is used as a cattle feed. Other fibre-yielding plants of minor importance are *Abutilon indicum, A. persicum* and *A. theophrastii* (China jute), *Sida cordifolia, S. acuta, Urena lobata* and many others. *Hibiscus sabdariffa* (commonly called roselle) is a shrub, native to the West Indies; its epicalyx and calyx are fleshy, rich in acids and pectin and used in preparation of jellies and confectionery. Many species of *Hibiscus* such as *H. mutabilis, H. rosa-sinensis, H. schizopetalous* are cultivated as ornamentals; *H. elatus* (blue mahoe) is the national flower of Jamaica, *H. syriacus* growing in eastern Mediterranean has large flowers of pinkish colour and is commonly called 'rose of Sharon'.

The fruits of *Abelmoschus esculentus* or lady's finger are used as a vegetable. Roots of *Althea officinalis* or marshmallow are used medicinally, considered to be a very useful herbal remedy for cough. *A. rosea* or hollyhock is a cultivated garden ornamental. The mucilaginous root of *Pavonia hirsuta* from Zimbabwe is added to milk to hasten butter production (Dutta 1988). The mucilaginous substance obtained from the stem of *Kydia calycina* is used for clarifying sugar; its wood is used for matches, packing cases, pencils, shoe heels, picture frames, veneers, plywood, paper and rayon-grade pulp (Singh et al. 1983). The Portia tree or *Thespesia populnea* is an avenue tree. *Anoda hastata, Pavonia odorata,* various species of *Malva, Malvastrum, Decaschistia, Sida* and *Urena* are weeds of waste-lands. *Sidalcea nelsoniana* is an endemic species in Oregon, USA and is a threatened species. Its seed coat is very hard, and must be softened to such an extent that water and oxygen enter and the embryo is allowed to expand and break out of the seed coat (Halse and Mishaga 1988).

Taxonomic Considerations. There is no controversy about the assignment of the family Malvaceae in the Malvales. The distinguishing features include presence of mucilage cells, monadelphous stamens and monothecous anthers, spinous and polyporate pollen grains, and pentacarpellary ovary with the style passing through the staminal tube.

Order Cucurbitales

The order Cucurbitales includes only one family—Cucurbitaceae. The plants are easily recognisable by their prostrate or climbing habit, often with the help of tendrils. Stem and branches herbaceous, covered with hairs; leaves simple, lobed or divided, hairy on both surfaces. Flowers unisexual, epigynous and ovary with parietal placentation. Many fruits are edible.

In transverse section of stem, the vascular bundles are bicollateral, and the pith is hollow. Seeds of many species are rich in amino acids, oils and storage proteins (Jeffrey 1980). Seed oils contain mostly linoleic, oleic, linolenic or conjugated polyethenoid acids as major components (Shorland 1963).

Cucurbitaceae

A large family with ca. 90 genera and 700 species distributed in the tropics and subtropics of both the Old and the New World.

Vegetative Features. Annual or sometimes perennial climbers or prostrate, monoecious or dioecious herbs, rarely trees, e.g. *Dendrosicyos*, a small tree on Socotra Islands. Stems cylindrical or more often pentangular, e.g. *Cucurbita, Luffa;* hairy, sometimes rooting at nodes, usually fistular. Leaves estipulate or stipulate, alternate, hairy, palmately 3- to 5-lobed, often with extrafloral nectaries embedded in leaf lamina, tendrils simple or multifid, usually in extra-axillary position.

Floral Features. Inflorescence axillary racemes or cymes of male flowers, female flowers solitary (Fig. 11.17 A); unisexual (hermaphrodite in *Schizopepon*), actinomorphic, epigynous (Fig. 11.17 B, C), incomplete, usually yellow or white. Calyx lobes 5, gamosepalous, outer surface hairy, imbricate, often with extrafloral nectaries on outer surface as in *Coccinia, Luffa* (Fig. 11.17 A, G). Corolla of 5 petals, gamopetalous (Fig. 11.17 B, C, F, H) (polypetalous in *Fevillea*), salver-form as in *Momordica* or campanulate as in *Cucurbita* (Fig. 11.17 F, H), margin of upper parts of the petals often fringed as in *Trichosanthes*; valvate, imbricate or quincuncial. Staminate flowers show much variation in the form and structure of the 5 stamens, due to cohesion and fusion. Stamens 5, free in *Fevillea*; in *Thladiantha*, of the 5 stamens, 4 are in 2 pairs by slight cohesion of the basal portion of the filaments, and the 5th is free. In *Bryonia* and *Momordica*, apparently there are 3 stamens—2 by the fusion of 2 each and the 3rd is the free 5th stamen. In *Cucurbita*, the situation is more complex; anthers of all the 5 stamens are spirally twisted into a central column (Fig. 11.17 F, G) and the filaments are connate except at the extreme base (Fig. 11.17 F). In *Cyclanthera*, the stamens are monadelphous with the thecae of the anthers in 2 horizontal link-like rings around the edge of a peltate mass of connective and filament tissue, the thecal links dehisce (seemingly) transversely by a single suture. Rudimentary ovary may be present in a staminate flower.

In pistillate flowers staminodes often present, pistil single; 5-carpellary in *Fevillea*, mostly 3-carpellary, syncarpous, ovary unilocular with parietal placentation (Fig. 11.17 D, E) (rarely trilocular with axile placentation), ovules usually numerous (unilocular with 1 large apically parietal ovule in *Sechium edule*); style 1, rarely 3 with as many branches as of stigma (Fig. 11.17 D, H). Fruit a berry with soft or hard pericarp and numerous seeds (Fig 11.17 J), also called a pepo. Seeds many, germinate within the fruit in *Sechium edule;* embryo straight with large cotyledons, endosperm absent.

Anatomy. Predominantly bicollateral vascular bundles separated from one another by broad strips of ground tissue and frequently arranged in two rings. Xylem vessels in older stems with wide lumina and simple perforations. Sieve tubes of the phloem are very large and conspicuous with transverse sieve plates. Leaves usually dorsiventral; various types of hairs recorded on leaf surfaces. Stomata on both surfaces of leaves or confined only to the lower surface; Ranunculaceous type.

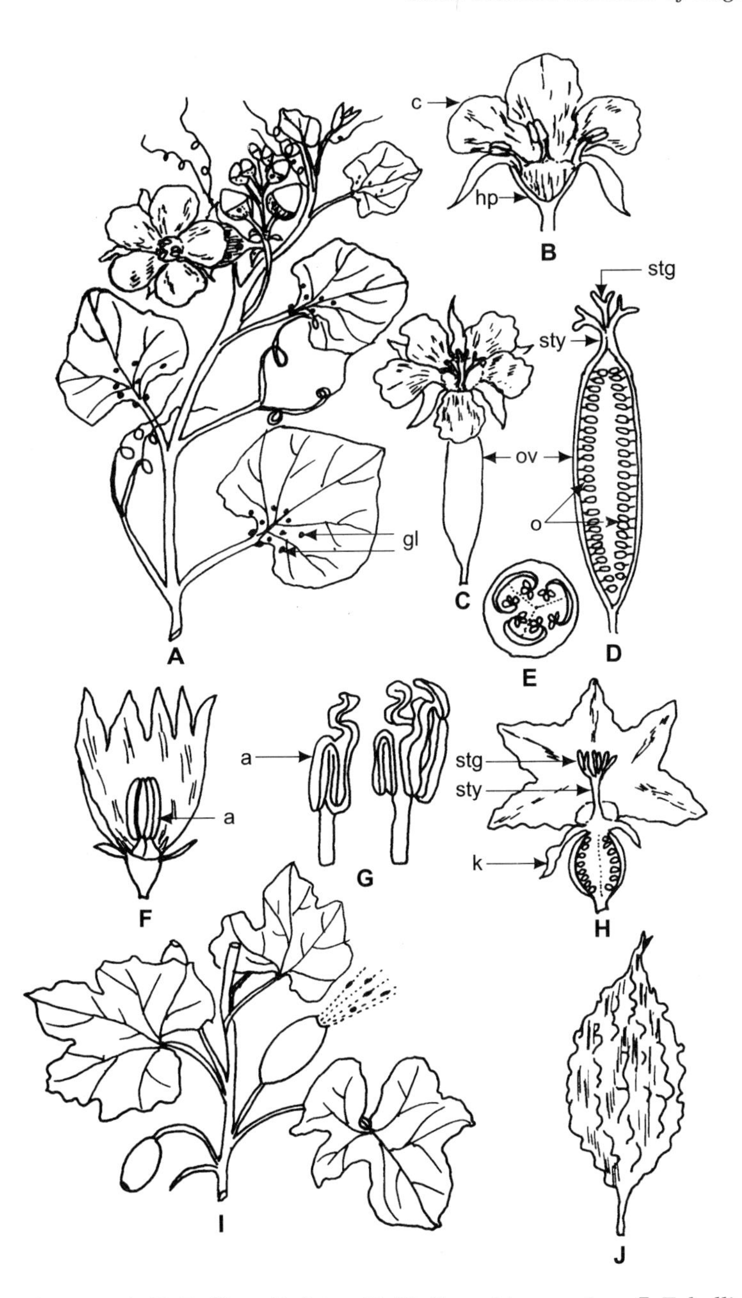

Fig. 11.17 Cucurbitaceae: **A–E** *Luffa cylindrica*, **F–H** *Cucurbita maxima*, **I** *Ecballium elaterium*, **J** *Momordica charantia.* **A** Plant with staminate inflorescence. **B** Staminate flower, vertical section. **C** Pistillate flower. **D** Carpel, vertical section. **E** Cross section. **F** Staminate flower of *C. maxima*, corolla cut open. **G** Anthers. **H** Pistillate flower, corolla cut open. **I** Plant of *E. elaterium* with squirting fruits. **J** Fruit of *Momordica charantia.* (*a* anther, *c* corolla, *gl* gland, *hp* hypanthium, *k* calyx, *o* ovule, *ov* ovary, *stg* stigma, *sty* style). (**I** adapted from Hutchinson 1969, **A–H, J** Original)

Embryology. Pollen grains 2- or 3-celled at shedding stage, the family is eurypalynous, 3- to 10-colporate, 3- to 5-porate or 3-aperturate. Exine smooth or variously sculptured. Ovules anatropous, bitegmic, crassinucellate. Polygonum type of embryo sac, 8-nucleate at maturity. Seed coat is formed by the outer integument only. Endosperm formation of the Nuclear type; chalazal haustorium common. Endosperm haustoria of varied forms occur in a number of genera (Johri et al. 1992).

Chromosome Number. The basic chromosome numbers are x = 7–14.

Chemical Features. A group of bitter triterpenes, the cucurbitacins are almost characteristic components of the vegetative parts and fruits of Cucurbitaceae. *Lagenaria* fruits are the rich source of the enzyme catalase and the seeds of water melon (*Citrullus lanatus*) are rich in the enzyme urease. The seeds of Cucurbitaceous plants are rich in fatty acids of various kinds.

Important Genera and Economic Importance. A large number of genera of this family are grown for their fruits that are consumed as vegetables, e.g. *Benincasa hispida* (white gourd), *Citrullus lanatus* (water melon), *Coccinia grandis* (scarlet gourd), *Cucumis melo* (melon), *C. sativus* (cucumber), *Cucurbita maxima* (pumpkin), *Lagenaria siceraria* (= *L. vulgaris)* (bottle gourd), *Luffa acutangula* (ribbed gourd), *L. cylindrica* (vegetable bath sponge), *Momordica charantia* (bitter gourd), *Sechium edule* (squash), *Trichosanthes anguina* (snake gourd) and *T. dioica* (parwal).

Some plants of this family are of medicinal value, e.g. bryony, used against common cold, is obtained from *Bryonia dioica.* A purgative called elaterium is produced from the fruits of *Ecballium elaterium* and a powerful laxative colocynth is obtained from dried fruit pulp of *Citrullus colocynthis.* The seed (testa) extract of *Momordica charantia* is useful for diabetic patients. A few taxa are also grown as ornamentals, e.g. *Coccinia cordifolia, Ibervillea sonorae, Xerosicyos* sp., etc.

Interesting plants of this family are: *Acanthosicyos horridus*, an erect, shrubby and very spiny genus from Africa; *Dendrosicyos,* a tree genus from Socotra Islands; *Hodgsonia macrocarpa,* a gigantic climber attaining a length of ca. 25 m; *Ecballium elaterium*, the squirting cucumber (Fig. 11.17) with explosively dehiscent fruits; *Alsomitra macrocarpa*, with winged seeds; and the Indomalayan *Hodgsonia* with ovules and seeds in pairs.

Taxonomic Considerations. The systematic position of the Cucurbitaceae is controversial. Most phylogenists include this family amongst those with trilocular ovary and parietal placentation. The stamen and fruit morphology are unique to agree to this association. The family Cucurbitaceae is distinguished by the structure of its anthers; the anther cells in most cases are very long, winding up and down on the outer surface of the connective. The stamens may be monadelphous, or syngenesious and variously disposed upon the filaments, usually vertical but sometimes horizontal or inclined. The anther cells show various type of configurations: simple to curved, variously convoluted or with annular patterns. The structure of stamens, and the ovary, habit of the plants, and presence of tendrils together make this family unique amongst the dicot families.

On the basis of gamopetalous corolla, unusual stamen structure and inferior ovary, Wettstein (1935) and Eichler (1878; see Lawrence 1951) allied it to the Campanulales. Engler and Diels (1936) constructed a separate family, Cucurbitaceae, under Cucurbitales, and placed it next to the Campanulales.

According to Bentham and Hooker (1965a), the presence of unisexual flowers, inferior unilocular ovary with parietal placentation and stem tendrils, make it the nearest relative of the Passifloraceae. Hutchinson (1973) included this family in his Cucurbitales along with Begoniaceae, Datiscaceae and Caricaceae. Cronquist (1968, 1981) placed Cucurbitaceae in Violales along with all those families with unilocular ovary and parietal placentation. Mez (1926), by his, serological studies, showed that the Cucurbitaceae are akin to the Campanulaceae and not to the Passifloraceae. Many embryological features also favour its inclusion in the Sympetalae but the constantly unisexual flowers, inferior ovary and presence of tendrils bring the family closer to the Passifloraceae (Dutta 1988).

Chakravarty (1966) observed that the Cucurbitaceae are highly evolved. There has been progression together with consequent metamorphosis of the structures associated with the reduction. amalgamation, connation, adnation and sterilization of the sex organs. It resembles the Passifloraceae in: placenta retained on the wall, single-chambered ovary, fleshy fruit, tendril structure and extrafloral nectaries. However, in spite of these resemblances, Cucurbitaceae deserves a separate rank by itself because of some of its typical features, especially of the male organs and the typical gourd-type fruit.

Takhtajan (1987) presumed its affinity with Begoniaceae and Datiscaceae, and therefore places the Cucurbitales near the Begoniales. Serologically, the Cucurbitaceae are related to Datiscaceae and Begoniaceae and do not show any affinity with the Loasaceae and Caricaceae of the Violales (Jeffrey 1980, Gershenzon and Mabry 1983).

The ovary structure, unisexual flowers, predominantly herbaceous nature and serological affinity indicate that the Cucurbitaceae is closely allied to some of the most advanced members of the Violales.

The family Cucurbitaceae is divided into 2 subfamilies and 9 tribes (Takhtajan 1987, Dutta I988):

Subfamily 1. Cucurbitoideae—tendrils unbranched or branched (2–7) from lower part, spiralling above the point of branching. Pollen of various types; style 1, seeds not winged.

Tribe (i) Benincaseae—pollen sacs convolute and pollen grains with reticulate exine. Hypanthium in pistillate flowers short; ovules horizontal. Fruit usually smooth and indehiscent. e.g. *Benincasa*, *Citrullus* and others.

Tribe (ii) Cucurbiteae—pollen grains large, spiny, with many pores. Ovules erect or horizontal. Fruit fleshy, indehiscent and 1- to many-seeded, e.g. *Cucurbita.*

Tribe (iii) Cyclanthereae—stamen filaments united into a single column; pollen grains punctate. Ovules erect or ascending in a unilocular ovary. Fruit often spiny, usually dehiscent, often explosive, e.g. *Cyclanthera, Marah.*

Tribe (iv) Joliffieae—pollen grains reticulate. Hypanthium short. Petals fimbriate or with basal scales. Ovules usually horizontal, e.g. *Momordica, Telfairia.*

Tribe (v) Melothrieae—pollen-sacs straight or almost so and pollen grains reticulate. Hypanthium campanulate or cylindrical and alike in both sexes. Ovules horizontal, e.g. *Cucumis, Zehneria.*

Tribe (vi) Schizopeponeae—stamens 3, free, pollen grains reticuloid. Ovules pendulous in a trilocular ovary. Fruit dehisce explosively into three valves, e.g. *Schizopepon.*

Tribe (vii) Sicyoeae—filaments united into a central column; pollen grains spiny. Ovules single, pendulous in a unilocular ovary. Fruit 1-seeded, indehiscent, usually hard or leathery or fleshy, e.g. *Sechium.*

Tribe (viii) Trichosanthieae—pollen grains striate, smooth or knobbed, never spiny. Hypanthium long and tubular in both sexes. Petals entire or with fimbriate margin. Ovules horizontal. Fruit fleshy or dry, dehiscing by three valves, e.g. *Trichosanthes.*

Subfamily II. Zanonioideae—Tendrils apically bifid, mostly perennial and dioecious; stamens 5, all alike.

Tribe (ix) Zanonieae—tendrils twice-branched from near the apex, spiralling above or below the point of branching. Pollen grains small, striate, uniform. Styles three. Seeds often winged, e.g. *Alsomitra, Zanonia.*

On the basis of vegetative and floral morphology, palynology and seed morphology, the erection of two subfamilies and nine tribes is fully justified.

Order Myrtiflorae

The order Myrtiflorae comprises 17 families in 3 suborders:

1. Suborder Myrtineae includes Lythraceae, unigeneric Trapaceae, Crypteroniaceae, Myrtaceae, unigeneric Dialypetalanthaceae, Sonneratiaceae, Punicaceae, Lecythidaceae, Melastomataceae, Rhizophoraceae, Combretaceae, Onagraceae, monotypic Oliniaceae and Haloragaceae.
2. Suborder Hippuridineae includes unigeneric Theligonaceae and Hippuridaceae.
3. Suborder Cynomoriineae comprises only Cynomoriaceae.

Mostly trees and shrubs, some herbs and a few aquatic herbs are included in this order. Members of Myrtales usually have opposite, simple, commonly entire and mostly estipulate (or with caducous stipules) leaves. Flowers cyclic, mostly tetramerous, usually with a hypanthium which may in some families be wholly adnate to the ovary; and shows a transition from perigyny to epigyny. Stamens often numerous and develop in a centripetal sequence. Gynoecium 2- to 5-carpellary, ovary with as many locules and axile or apical, rarely parietal placentation. Seeds with scanty or no endosperm.

Anatomically Myrtales is characterised by the presence of intraxylary phloem.

Myrtaceae

A large family of 145 genera and 3650–4000 species distributed in tropical and subtropical regions of the world. The two centres of distribution are Australia and America.

Vegetative Features. Shrubs or trees, stems when sufficiently young have wings. Leaves usually opposite (Fig. 11.18.1 A), whorled in *Callistemon lanceolatus* (Fig. 11.18.1 E), simple, entire, coriaceous, gland-dotted, with intra-marginal venation (11.18.2 A)

Floral Features. Inflorescence cymose (*Eucalyptus*, 11.18.2 A), racemose (*Callistemon*) or paniculate (*Syzygium aromaticum*), rarely solitary axillary flowers as in *Psidium guajava* (Fig. 11.18.1 A). Flowers bisexual, actinomorphic, mostly epigynous (Fig. 11.18.1 B). Calyx usually 4 or 5, mostly inconspicuous or thrown off as a calyptra as in *Eucalyptus* (Fig. 11.18.1 F), free or basally connate. Petals 4 or 5, free, or connate forming an operculum, as in *Eucalyptus*, imbricate. Stamens numerous, bent inwards in bud (Fig. 11.18.1 F, 11.18.2 B) free or polyadelphous; anthers dorsifixed or versatile (Fig. 11.18.1 C, 11.18.2 C), rarely basifixed as in *Calothamnus*, bicelled, introrse, dehiscing longitudinally or apically by 2 pores as in *Actinodium*, *Darwinia*, *Homoranthus* and many others; the connective often conspicuous and gland-tipped. Gynoecium syncarpous, 2- to 5-carpellary; ovary inferior or half-inferior, uni-, bi-, tri-, or tetralocular with parietal (when unilocular) or axile placentation (Fig. 11.18.1 B, D; 11.18.2 E, F); ovules 2 to many on each placenta, obliquely pendulous; style and stigma 1 (Fig. 11.18.2 D). Fruit a berry (*Psidium*), or loculicidal capsule, rarely drupaceous or nutlike. Seeds usually a few, embryo variously shaped, endosperm scanty or absent. Fig. 11.18.2 represents floral diagram for *Eucalyptus camaldulensis*.

Anatomy. Vessels small, numerous, solitary and without any definite arrangement in most genera (exception *Eucalyptus*), perforations simple, intervascular pitting alternate. Parenchyma diffuse or in uniseriate bands in the wood with solitary vessels, predominantly paratracheal in wood with numerous multiple and some intermediate forms with both diffuse and paratracheal types. Rays exclusively uniseriate or up to 2 to 6 cells wide. Intraxylary phloem is common to all the members; intercellular canals and cavities containing oil are also present. Leaves isobilateral or centric as in *Callistemon linearis*; hairs mostly unicellular, stomata usually Ranunculaceous type and occur on both surfaces of vertically placed leaves.

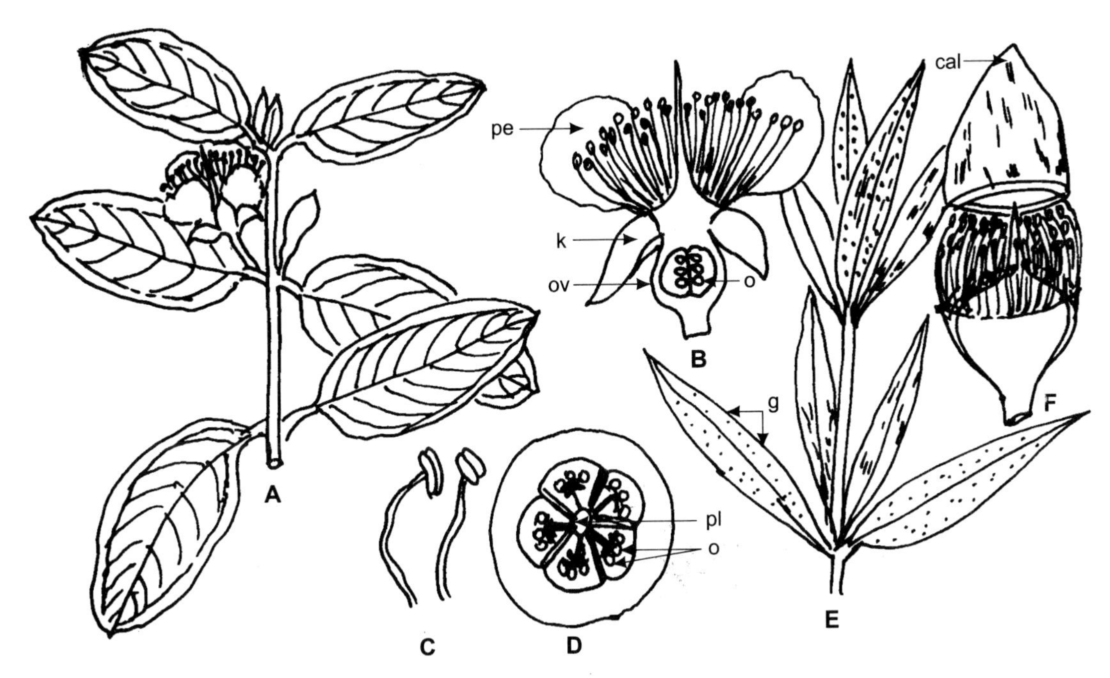

Fig. 11.18.1 Myrtaceae: **A–D** *Psidium guajava*, **E** *Callistemon lanceolatus*, **F** *Eucalyptus citriodora*. **A** Flowering twig. **B** Flower. **C** Stamens. **D** Ovary, cross section. **E** Twig of *Callistemon lanceolatus*, leaves gland-dotted. **F** Flower of *Eucalyptus citriodora*. (*cal* calyptra, *g* glands, *k* calyx, *o* ovule, *ov* ovary, *pe* petal, *pl* placenta). (Original)

Embryology. Pollen grains shed at 2-celled stage. Ovules ana- to campylotropous, bitegmic (unitegmic in *Eugenia*, *Syzygium*), crassinucellate; micropyle formed by both integuments in *Wehlia*. Polygonum type of embryo sac, 8-nucleate at maturity. Endosperm formation of the Nuclear type.

Chromosome Number. Basic chromosome numbers are $x = 6–11$.

Chemical Features. Presence of ellagitannins is characteristic. Anthocyanins of 3-glucoside type and cyanidin and delphinidin are present in *Eucalyptus* and malvidin and delphinidin in *Metrosideros*. The glycoside prunasin occurs in *E. corynocalyx*; D-quercitol in fruits of *Syzygium cumini* and L-quercitol in *Eucalyptus populnea*. Triterpenoid saponins are also present in some Myrtaceae members.

Important Genera and Economic Importance. One important genus is *Eucalyptus*, some species are tall handsome trees bearing white or red flowers. *E. regnans* from Australia is considered as one of the tallest angiosperms, second only to redwoods (gymnosperm) of California (MP Nayar 1984). Another interesting genus is *Callistemon*—commonly called bottle-brush tree because of the resemblance of its inflorescence, and it is cultivated on a large scale as an avenue tree. *Acmena acuminatissima* is a tree from the Andamans. *Decaspermum fruticosum* is a slender tree of Khasi Hills in Assam. *Psidium guajava* or guava, *Syzygium cumini*, black plum or Indian blackberry, *S. jambos*, and *S. malaccense* or Malaya apple are all cultivated as well-known fruit trees.

Economically, Myrtaceae is of great importance for fruits, oils, spices, timbers and ornamentals. Various species of *Eucalyptus* yield valuable timber, e.g. *E. diversifolia*, *E. leucoxylon*, *E. marginata* and *E. robusta*. The leaves of *E. globulus* on distillation yield eucalyptus oil, *E. maculata* and *E. rostrata* yield

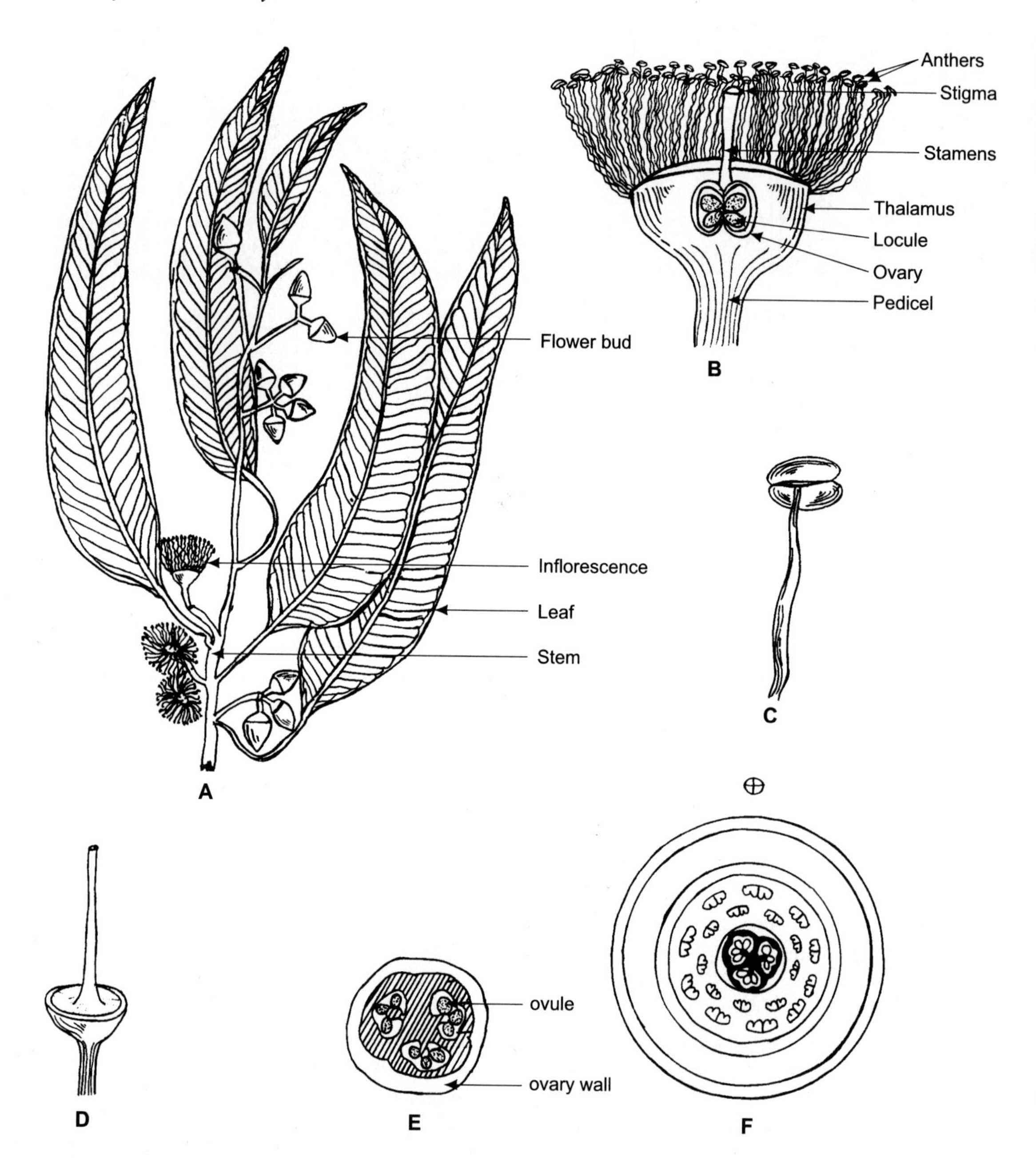

Fig. 11.18.2 Myrtaceae: **A–F** *Eucalyptus camaldulensis*. **A** Flowering twig. **B** Flower longisection. **C** Stamen. **D** Gynoecium. **E** Ovary, transverse section. **F** Floral diagram (sketched by Uma Joshi).

citron gum and red gum, respectively, and are used in medicine. *E. occidentalis* supplies mallet bark, rich in tannin. *Leptospermum laevigatum* is planted extensively in Australia for reclamation of moving sand. Cajuput oil is obtained from *Melaleuca cajuputi* (Burma to Australia). Allspice and cloves are the spices derived from *Pimenta dioica* and *Syzygium aromaticum*, respectively.

Taxonomic Considerations. The Myrtaceae is divided into 2 subfamilies and 3 tribes:

Subfamily I. Leptospermoideae: leaves opposite or alternate; fruit a capsule or nut-like, dry.

Tribe (i) Chamaelancieae: ovary unilocular, e.g. *Calytrix, Verticordia.*

Tribe (ii) Leptospermae: ovary multilocular, e.g. *Callistemon*, *Eucalyptus*, *Melaleuca.*

Subfamily II. Myrtoideae: leaves always opposite, fruit fleshy, typically a berry, rarely a drupe.

Tribe (iii) Myrteae: ovary 2- or 5-locular, e.g. *Myrtus*, *Psidium*, *Syzygium.*

The family Myrtaceae has been included in the Myrtales by all taxonomists and there is no dispute about its position.

Order Umbellales

The order Umbellales comprises 7 families: Alangiaceae, Nyssaceae, monotypic Davidiaceae, Cornaceae, unigeneric Garryaceae, Araliaceae and Umbelliferae.

Members of this order are mostly trees and shrubs, herbs in Umbelliferae, and climbing vines in Araliaceae. Most of them are tropical and subtropical in distribution except the members of the Umbelliferae, which are cosmopolitan but prefer cold climate. Leaves are simple in Alangiaceae, Nyssaceae, Davidiaceae, Cornaceae and Garryaceae but pinnately compound or highly dissected in Araliaceae and Umbelliferae. The inflorescence mostly umbellate, capitate, spicate or racemose. Inflorescence of *Davidia* is different in the sense that the capitulum or head has numerous male flowers but only one perfect flower. Garryaceae and Nyssaceae have separate staminate and pistillate inflorescences. Flowers pentamerous, epigynous, often bracteate, complete or incomplete, uni- or bisexual. Calyx mostly adnate to the inferior ovary and represented by small teeth. Petals distinct, valvate or imbricate, sometimes absent. Stamens diplostemonous or obdiplostemonous or the number is same as the number of petals. Pollen grains 2- or 3-celled. Ovary inferior with varied number of carpels and locules, placentation axile; locules uniovulate, ovules pendulous, anatropous, usually unitegmic, rarely bitegmic.

Anatomically there are two groups:

1. Wood parenchyma diffuse, vessels with scalariform perforations, nodes trilacunar, secretory canals absent.
2. Wood parenchyma vasicentric, vessels with simple perforations, nodes multilacunar, schizogenous secretory canals present.

There are two distinct groups with respect to their chemical constituents. Nyssaceae-Cornaceae alliance contains iridoids and ellagitannins. Umbelliferae-Araliaceae group contains aromatic oils, unsaturated petroselinic acid in seed fats, and falcarinone alkaloids.

Umbelliferae

A very large family with 200 genera and 2900 species distributed widely all over the world, chiefly in the temperate zones but also reported from the tropics and subtropics. One species of *Azorella* grows in the Antarctica.

Vegetative Features. Mostly normal mesophytic plants except a few taxa that grow on wet soil, e.g. *Hydrocotyle* or water pennywort; *Actinotus bellidioides* grows in peaty soil at high altitudes in Tasmania. *Eryngium campestre* is a xerophyte.

Mostly biennial or perennial, aromatic herbs. Stem herbaceous, rarely woody, with prominent, swollen, solid nodes and hollow internodes, surface ridged, hairs when present, non-glandular, sometimes dendroid, e.g. *Xanthosia pilosa* and stellate, e.g. *Bowlesia tropaeolifolia.* Pith large, shrinks or dries at maturity.

Leaves alternate, rarely opposite as in *Apiastrum*, often basal, sometimes heteromorphic, e.g. *Coriandrum sativum* (Fig. 11.19 A); usually pinnately, or palmately compound or decompound, e.g. *Daucus carota*; sometimes simple and reniform as in *Centella asiatica* (Fig. 11.19 H), petiole sheathing, estipulate.

Floral Features. Inflorescence simple or compound umbel (Fig. 11.19 A), in some genera, the pedicels are lost by reduction and the inflorescence is capitate, e.g. *Eryngium.* When compound, each primary ray or peduncle is terminated by an umbelet, whose pedicels are termed secondary rays. An umbel is often subtended by an involucre of distinct, simple or compound bracts. Similarly, an umbelet is subtended by an involucel of bractlets. The bracts may be caducous or persistent in fruit. Flowers generally bisexual but unisexual in *Echinophora*; in *Petagnia saniculifolia* (Fig. 11.19 J, K), a central female flower is surrounded by male umbels (Dutta 1988), plants dioecious in *Arctopus*, polygamous flowers in *Torilis anthriscus*. Polygamomonoecism exists when a few longer-pedicelled staminate flowers occur with a majority of bisexual ones as in *Astrantia*. Flowers actinomorphic/zygomorphic, pedicellate, bracteate, ebracteolate, epigynous (fig. 11.19 B, I).

Sepals 5, adnate to the ovary, persistent and mostly small, teeth-like, valvate (Fig. 11.19 B, I); in *Pentapeltis peltigera* calyx lobes disc-like. Corolla of 5 distinct petals, polypetalous, inflexed in bud, often notched at the apex, more or less bifid in *Coriandrum sativum* (Fig. 11.19 B, C) and *Foeniculm vulgare*, imbricate. Stamens 5, inflexed in bud but spreading later, alternipetalous, arising from an epigynous disc. Anthers bicelled, basi- or dorsifixed, longitudinal dehiscence (Fig. 11.19 B, D). Gynoecium syncarpous, bicarpellary, ovary inferior, bilocular, placentation axile, ovules pendulous, 1 in each locule (Fig. 11.19 E, F). Styles 2, with a swollen base called stylopodium (Fig. 11.19 F, I), stigmatic tip scarcely differentiated. Fruit a cremocarp composed of 2 mericarps (Fig. 11.19 G), coherent and dehisce by their faces (commissure), flattened dorsally, i.e. parallel to the commissural face or laterally, i.e. at right angles to the commissural face. Each mericarp has 5 ribs distinguished as lateral, dorsal and intermediate; the ribs thin or corky and filiform or winged (*Thapsia villosa*) or spinescent as in *Daucus carota*. Oil ducts or vittae are present in the spaces between two ribs or under the ribs (Fig. 11.19 F). Each mericarp is l-seeded and usually suspended after dehiscence by a slender wiry stalk or carpophore. Seed with a minute embryo, endosperm abundant.

Anatomy. Vessels very small to medium-sized, often in clusters, perforations usually simple, rarely scalariform as in *Heracleum sphondylium*. Parenchyma paratracheal, rays heterogeneous to almost homogeneous. Trichomes of various types common. Leaves dorsiventral, more stomata on lower surface, of Rubiaceous or Ranunculaceous type. Crystals absent, secretory ducts present.

Embryology. Bilateral, 3-colporate and 3-celled pollen grains. Ovules pendulous, anatropous, unitegmic, tenuinucellate. Polygonum type of embryo sac most common; Allium, Drusa and Penaea types also recorded (Johri et al. 1992); 8- or 16-nucleate at maturity. Abortive ovules reported. Endosperm formation of the Nuclear type.

Chromosome Number. The basic chromosome numbers are x = 3–12, most commonly 8 in Hydrocotyloideae and Saniculoideae, and 11 in Apioideae.

Chemical Features. Umbelliferae seeds are rich in the unsaturated fatty acid, petroselinic acid (Hegnauer 1973a). Monoterpenoids—carvone, limonene and fenchone—also occur. Polyacetelene falcarinose is common in most Umbelliferae members (Hegnauer 1971). The seeds contain an oligosaccaride, umbelliferose, in addition to various volatile oils. The bitter flavour of *Foeniculum vulgare* is due to fenchone and the sweet flavour is due to anethole. The pleasant delicate aroma of the fruits of *Coriandrum sativum* is due to coriandrol. The characteristic aroma and sharp taste of *Carum carvi* is due to the presence of caraway oil containing carvone, a ketonic substance (50–60%) and d-limonene. The main constituent of the volatile oil of cumin from *Cuminum cyminum* is cuminaldehyde (Kochhar 1981). Some of the aroma constituents of root, leaf or fruit are of phenylpropanoid type. Two specific constituents are parsleyapiole from *Petroselinium crispum* and dillapiole from *Anethum graveolens*. Another substance is myristicin, typical for Umbelliferae. This chemical is reported to have hallucinogenic effect in man (Shulgin 1966). It is reported in 14 genera of the subfamily Apioideae. Breeding and cultivation affects its concentration, and for taxonomic purposes the surveys should be

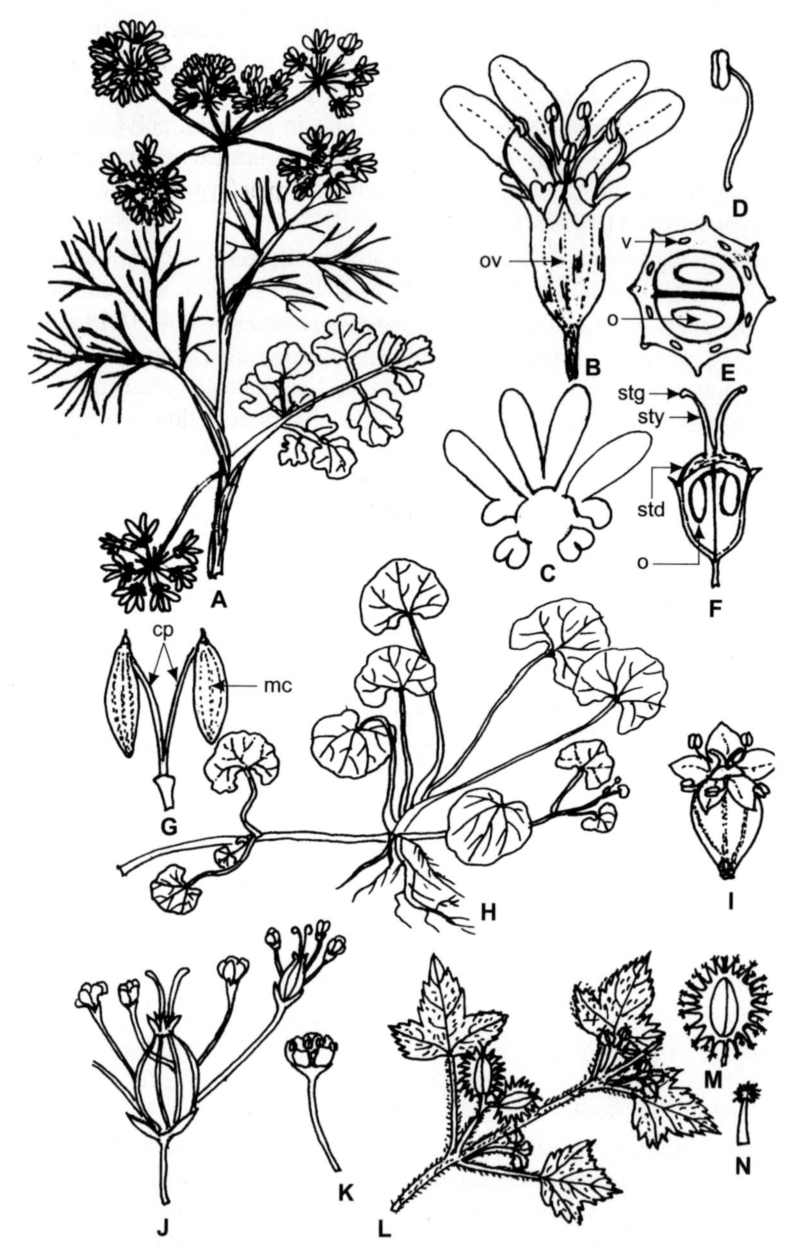

Fig. 11.19 Umbelliferae : **A–G** *Coriandrum sativum,* **H, I** *Centella asiatica,* **J, K** *Petagnia saniculifolia,* **L-N,** *Drusa oppositifolia.* **A** Plant with inflorescence. **B** Flower. **C** Petals, dissected. **D** Stamen. **E** Ovary cross section. **F** carpel, vertical section. **G** Cremocarp. **H** Plant of *C. asiatica.* **I** Flower. **J, K** Inflorescence and staminate flower of *P. saniculifolia.* **L** Flowering twig of *D. oppositifolia* **M** Fruit. **N** Hair, from the surface. (*cp* carpophore, *mc* mericarp, *o* ovule, *ov* ovary, *std* stylopodium, *stg* stigma, *sty* style, *v* vittae). (**A-G** Original, **H–N** adapted from Hutchinson 1969)

limited to wild species only. Myristicin acts as a very interesting taxonomic marker in the tribe Caucalideae. It occurs in closely related genera *Daucus* and *Pseudorlaya.* In the latter, it occurs in both species, but in *Daucus* it occurs in only the two polyploid members, *D. glochidiatus* and *D. montanus.* Another noteworthy aromatic volatile oil of this family is coumarin.

Important Genera and Economic Importance. The family Umbelliferae is very important economically as they produce food, condiments, spices, and ornamentals. *Daucus*, *Pastinaca*, *Apium* and *Petroselinium* are cultivated for their food value. Fruits of *Coriandrum*, *Foeniculum*, *Carum*, *Anethum*, *Cuminum*, *Pimpinella* and *Anthriscus* are used as spices and condiments. A number of plants are grown as ornamentals such as *Ammi majus*, *Trachymene spp.*, *Aegopodium* sp., *Angelica*, *Eryngium* and *Heracleum.*

Some taxa are poisonous, e.g. *Conium maculatum*[9] (poison hemlock), *Cicuta* (water hemlock) and *Aethusa* (fool's parsley). The aromatic oil from *Foeniculum* and *Coriandrum* finds use in medicine as a carminative. In India the seeds of *Carum carvi* and *Cuminum cyminum* have long been used as stomachic and carminative. The residual mass after extraction of oil is a valuable cattle feed.

Chinese folk medicine have been making use of a crude extract of *Angelica acutiloba* var. *acutiloba-kitagawa*, due to its analgesic, sedative and antibacterial effects. Chemical examination proved the anaesthetic nature of the components.

Asafetida, *Ferula assafoetida*, is another member in which the flavouring constituents are present in the gummy, resinous exudate that collects on the surface of chopped roots or rhizomes. In western countries, liquors, particularly gin, are often flavoured with *Coriandrum* seeds. Apart from numerous economically important genera, this family also includes a few morphologically interesting plants. *Actinotus bellidioides* is a diminutive taxon which grows on wet peaty soil at high altitudes in Tasmania. *A. helianthi* of South Australia is so named because of its inflorescence resembling a sunflower (*Helianthus*). *Drusa oppositifolia*, an endemic of the Canary Islands, has opposite leaves, plant body covered with dense hairs and fruits with peculiar anchor-like processes on the margin (Fig. 11.19 L). Another remarkable plant is *Petagnia saniculifolia* from Sicily; the flowers are unisexual; the female flower has attached to it 2 or 3 male flowers; their pedicels are partially adnate to a rib of the calyx (Fig. 11.19 J). *Platysace deflexa* from Western Australia bears edible underground tubers. The leaves are terete and jointed in *Ottoa oenanthoides*, which grows from Central America to Ecuador.

Taxonomic Considerations. There are no two opinions about the placement of the family Umbelliferae. The Araliaceae and Umbelliferae have always been retained in the same order i.e. Umbelliflorae of Bentham and Hooker (1965a); Umbellales of Engler and Diels (1936) and Melchior (1964); and Araliales of Hutchinson (1959), Dahlgren (1975a) and Takhtajan (1980). Hutchinson (1969) treats this family as the only member of his Umbellales. However, Thorne (1973, 1983) and Roth (1977) include Umbelliferae (= Apiaceae) in Araliaceae. Cronquist (1981) and Takhtajan (1987) use the alternate terminology, Apiales, for the order.

---

[9]The hemlock poison was served to Socrates who was condemned to death (for political reasons).

SUBCLASS SYMPETALAE

Order Primulales

The order Primulales comprises three families—Theophrastaceae, Myrsinaceae, and Primulaceae. According to Takhtajan (1987), there is a fourth family, Aegicerataceae, which Willis (1973) includes in the Myrsinaceae.

Both Theophrastaceae and Myrsinaceae are predominantly arborescent. The Primulaceae are mostly perennial herbs, and most advanced amongst the three. All three families have sympetalous, pentamerous flowers with or without staminodes. Ovary unilocular, superior with free-central placentation; ovules mostly anatropous and bitegmic. None of the members appear to be adapted to any particular ecological niche.

Primulaceae

A family of 23-27 genera and about 1000 species (28 genera and 800 species according to Lawrence 1951), widely distributed but more common in north temperate regions.

Vegetative Features. Mostly mesophytes, but a hydrophyte *Hottonia*, and a halophyte *Glaux maritima* are also reported. Mostly perennial herbs, perennate by rhizomes as in *Primula*, or by tuber as in *Cyclamen*; sometimes annual, e.g. *Anagallis*. The aerial stem internodes are often suppressed so that the leaves appear to be in the form of dense radical rosettes, e.g. *Primula*, *Dodecatheon* (Fig. 11.20A, E) and others; in *Lysimachia vulgaris* the stem is erect and well developed; stem creeping in *L. nummularia* and winged in *Anagallis*. Leaves usually simple, estipulate, often with toothed margin, alternate, opposite or whorled, obovate-spathulate; gland-dotted or farinose (*Primula* sp). In *Hottonia*, an aquatic genus, the leaves are submerged and finely dissected.

Floral Features. Inflorescence variable—racemose such as an umbel, e.g. *Dodecatheon* (Fig. 11.20 A), *Cyclamen*, *Primula* (Fig. 11.20 E) or racemes or spikes, e.g. *Lysimachia*; flowers solitary, axillary in *Anagallis* and *Trientalis borealis*. Flowers bracteate, ebracteolate, hermaphrodite, actinomorphic (zygomorphic in *Coris*), hypogynous, pentamerous, often show heterostyly. Calyx of 5 sepals, polysepalous, often foliaceous, persistent. Corolla mostly of 5 petals (rarely 4–9), gamopetalous (polypetalous in *Pelletiera* and totally absent in *Glaux*,), corolla shape varies—a longer tube with spreading limb in *Primula* (Fig. 11.20 F), a shorter tube with spreading limb in *Anagallis*. In *Soldanella* it is campanulate and in *Dodecatheon* drooping (Fig. 11.20 A, B). *Pelletiera*, an exceptional genus, has 3-merous flowers. Stamens as many as corolla lobes, epipetalous and oppositipetalous in one whorl. The missing outer whorl sometimes represented by scale-like staminodes as in *Soldanella* and *Samolus*. Anthers bicelled, introrse, dehiscence longitudinal (Fig. 11.20 C, D). Gynoecium syncarpous, 5-carpellary, ovary superior (half-inferior in *Samolus*), unilocular, ovules few to numerous, placentation free-central (Fig. 11.20 H); style 1, heterostyly common, stigma capitate (Fig. 11.20 G). Fruit usually a 5-valved or 5-toothed (sometimes 10-toothed) capsule or a pyxis (as in *Anagallis* and *Centunculus*); seeds small with transparent endosperm, and a small straight embryo.

Anatomy. Vascular bundles usually widely spaced and arranged in a circle around the central pith, e.g. in *Anagallis arvensis* with one bundle opposite each of the four angles of stem; in *Lysimachia vulgaris* with a large number of bundles and in *Samolus valerandi* with a few very small bundles. In mature xylem of *L. hillebrandii* the vessels are arranged in long radial multiples and provided with simple perforations. Intervascular pits small, alternate; xylem parenchyma and rays absent. Leaves usually dorsiventral; centric in xeromorphic forms. Stomata of Ranunculaceous type, confined to either lower or upper surface

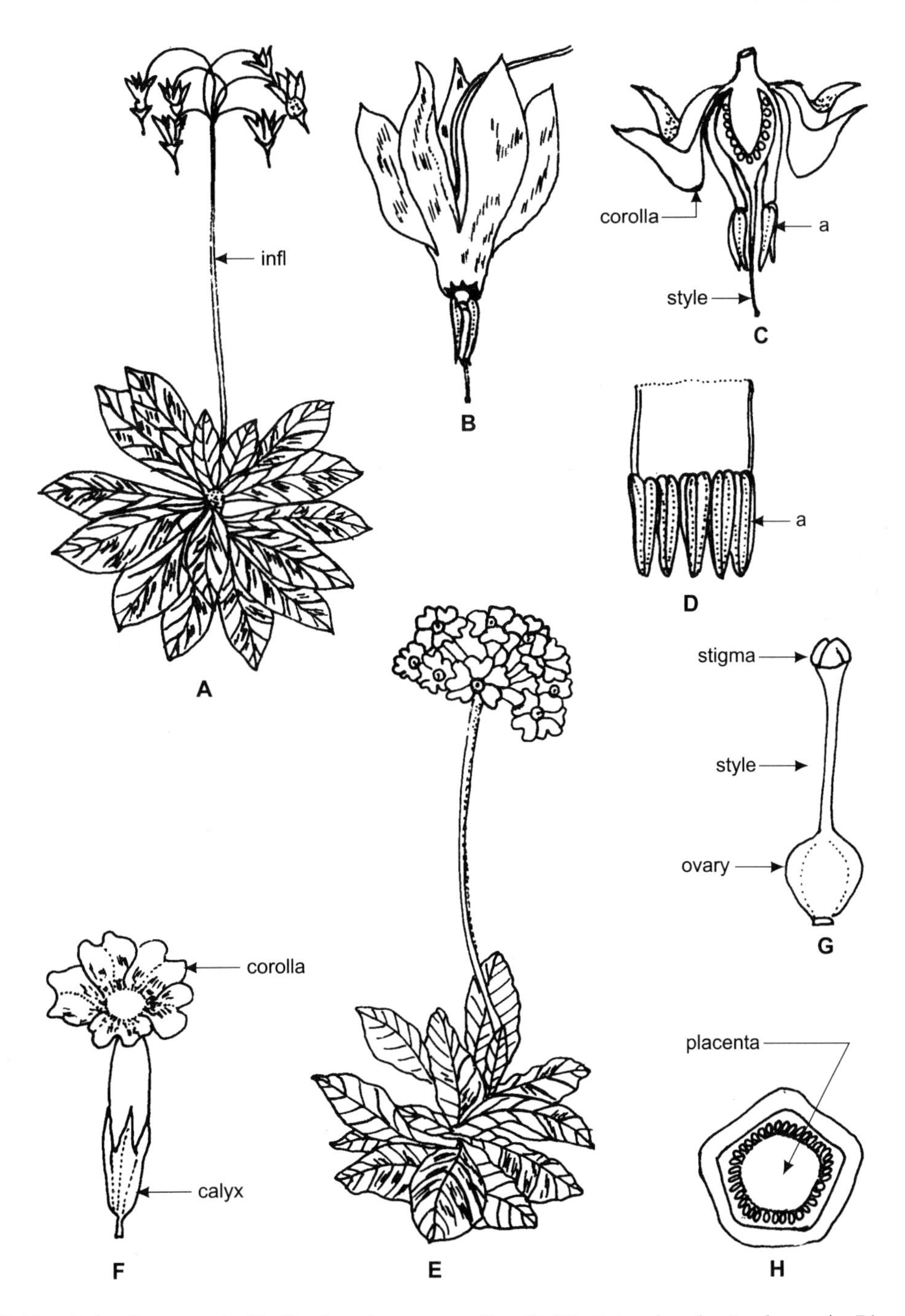

Fig. 11.20 Primulaceae: **A–D** *Dodecatheon meadia*, **E–H** *Primula denticulata.* **A** Plant with inflorescence. **B** Flower. **C** Same, vertical section. **D** Stamens (spread out). **E** Plant of *P. denticulata* with basal rosette of leaves and inflorescence. **F** Flower. **G** Gynoecium. **H** Ovary, transection. (*a* anther, *infl* inflorescence). (Adapted from Lawrence 1951).

or present on both surfaces, in different species. Hydathodes frequent, usually at the end of the midveins. Secretory cells with reddish-brown contents appear in the leaves of *Anagallis* and *Centunculus.*

Embryology. Pollen grains 3-colporate, 2-celled at the dispersal stage. Ovules hemiana- or anatropous as in *Hottonia* and *Samolus*, bitegmic, tenuinucellate. Polygonum type of embryo sac, 8-nucleate at maturity. Endosperm formation of the Nuclear type.

Chromosome Number. Basic chromosome number is variable x = 5, 8–15, 19, 22.

Chemical Features. Anthocyanin pigments hirsutidin and rosidin occur in *Primula* and *Dionysia* (Harborne 1968). Although herbaceous, the family is rich in leuco-anthocyanins like leucodelphinidin and leucocyanidin (Smith 1976). .

Important Genera and Economic Importance. Amongst the well-known genera is *Primula*, with about 400 species, which mostly occur in the temperate regions of the northern hemisphere and a few in the southern hemisphere; many taxa are grown as ornamentals in greenhouses and as borders and rockeries; commonly called primroses. *Lysimachia* or loosestrife is another large genus with about 100 species which are ornamental. *Androsace* with 85 species and *Cyclamen* with 20 species are also important genera. Economically, the family is not so important except that some taxa are grown as ornamentals, such as *Primula*, *Cyclamen* and others.

Taxonomic Considerations. The family Primulaceae is the most advanced family amongst the Primulales which are predominantly herbaceous. Hutchinson (1959, 1969, 1973) disagreed with this view and placed Primulaceae and Plumbaginaceae together in the order Primulales, and derived them from the Caryophyllales. However, most of the recent authors (Cronquist 1968, 1981, Dahlgren 1980a, 1983a, Takhtajan 1980, 1987) treat this family as the most advanced amongst the three families of the order Primulales and presume its alliance with the Ebenales.

Order Gentianales

The order Gentianales (sensu Melchior 1964) includes 7 families—Loganiaceae, unigeneric Desfontainiaceae, Gentianaceae, Menyanthaceae, Apocynaceae, Asclepiadaceae and Rubiaceae. In addition, Takhtajan (1987) considers Saccifoliaceae, Carlemanniaceae, Dialypetalanthaceae, Theligonaceae, Spigeliaceae and Plocospermataceae also as members of the order Gentianales.

The plants have opposite, simple or pinnately compound, estipulate leaves; internal phloem; hypogynous, actinomorphic flowers with contorted aestivation; 1 whorl of alternipetalous stamens; unitegmic, tenuinucellate ovules, mostly Nuclear type of endosperm, and many genera contain alkaloids or glucosides. The family Menyanthaceae, however, differs from the other families as they are aquatic plants, have alternate leaves, integumentary tapetum and Cellular type of endosperm. They also lack internal phloem.

Many of the Gentianalian genera are medicinally important as they contain alkaloids, glycosides and other chemicals.

Apocynaceae

A large family with 300 genera and about 1300 species; cosmopolitan in distribution, but more common in the tropics.

Vegetative Features. Members are herbs (*Catharanthus roseus*, 11.21.1 A), shrubs (*Nerium indicum*), or trees (*Alstonia scholaris*); *Beaumontia grandiflora* and *Trachelospermum jasminoides* are large, stout, woody climbers. Stem woody in shrubs and trees; soft in herbs; usually contain milky latex (sap colourless in (*Catharanthus*). Leaves opposite, decussate, sometimes alternate, as in *Thevetia* or whorled, as in *Alstonia*, *Rauwolfia*; simple, entire, petiolate, estipulate.

Floral Features. Flowers either solitary or in axillary pairs (*Catharanthus*, 11.21.1 A) or in cymose inflorescence; axillary cymes in *Allemanda,* terminal cymes in *Beaumontia*, axillary or terminal cymes in *Mandevilla* and *Melodinus.* In *Carissa* the flowers are in corymbose cymes; in *Alstonia* and *Rauwolfia* in umbellate cymes.

Flowers bracteate, ebracteolate, pedicellate, hermaphrodite, actinomorphic, complete, hypogynous, typically pentamerous. Calyx of 5 sepals, gamosepalous but deeply lobed, aestivation quincuncial, often with a hairy outer surface. Corolla of 5 petals, gamopetalous, contorted in bud (11.21.2 A), salverform or infundibuliform; throat of the corolla tube hairy or appendaged foming a corona-like structurre (11.21.2 B). Stamens 5, free, alternate with the petals, epipetalous (11.21.1 B; 11.21.2 B); anthers introrse, often sagittate, bicelled (11.21.1 C; 11.21.2 C), free or adherent (by viscid exudates) to the stigma or 'clavuncle'; filaments short. Pollen grains separate or rarely in tetrads as in *Condylocarpon.* Gynoecium composed of 2 carpels, which may be either fully syncarpous as in *Carissa* or may remain free below with 2 distinct ovaries as in *Catharanthus*, but with a single style and stigma (11.21.1 B, D). Each ovary supeior to half-inferior, unilocular with parietal placentation (11.21.1 E, F); when syncarpous, bilocular with axile placentation (11.21.2 E, F). Style simple, a thick or dumble-shaped stigma (Fig. 11.21.1 B, C); ovules few to many. Fruit a follicle (*Plumeria*, *Nerium*), capsule (*Allemanda*), drupe (*Thevetia*) or a berry (*Carissa*). Seeds winged in *Plumeria.* with a tuft of hairs in *Alstonia.* Seeds with fleshy endosperm and a straight embryo.

Anatomy. Vessels usually small; medium-sized in some species of *Alstonia, Aspidosperma, Kibatalia Rauwolfia, Vallaris* and some others; and sometimes large, as in species of *Anodendron* and *Landolphia*; perforations simple. Parenchyma most commonly apotracheal, as scattered cells or narrow bands; sometimes paratracheal. Rays typically 2 or 3 cells wide or exclusively uniseriate, heterogeneous. Leaves

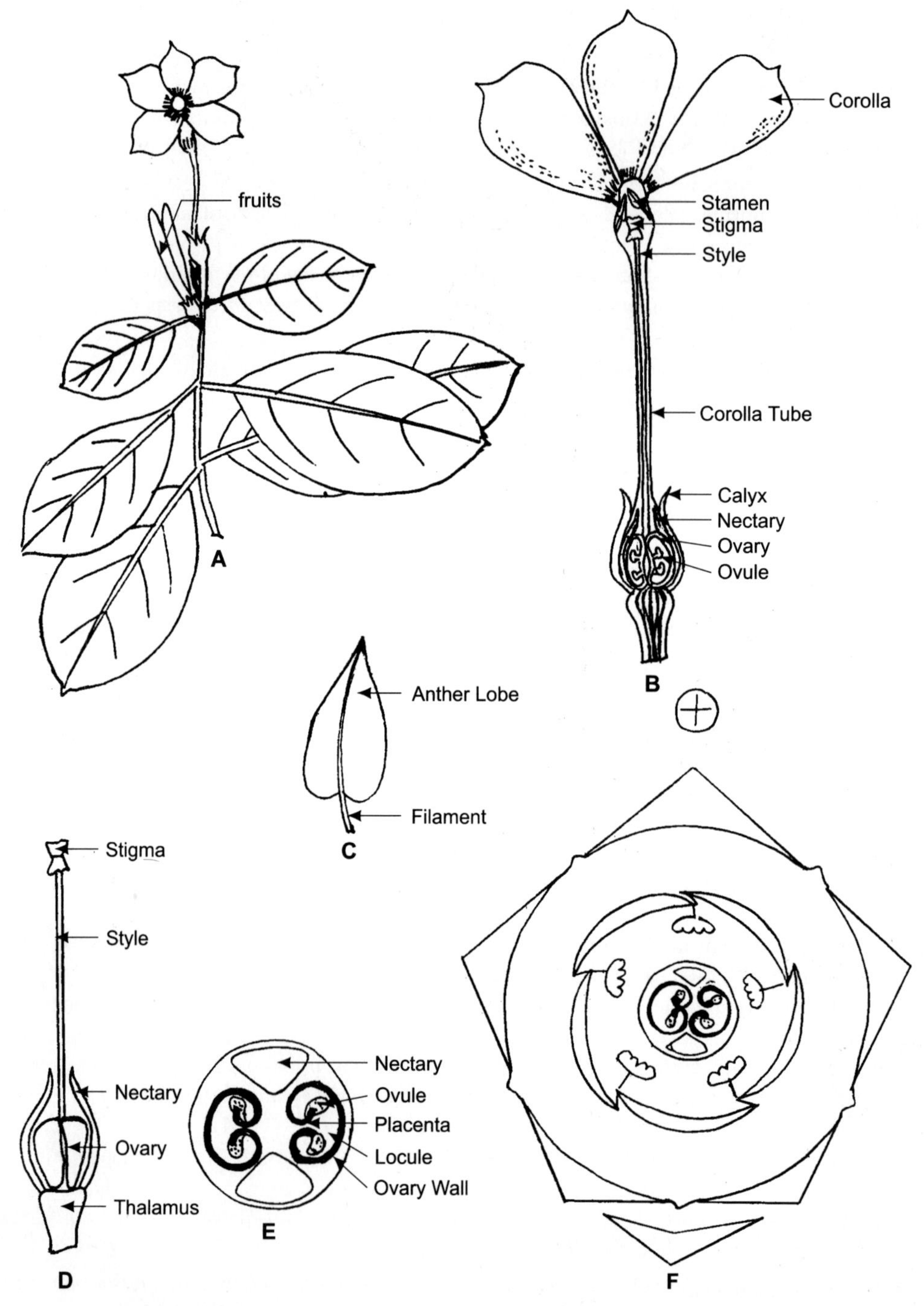

Fig. 11.21.1 Apocynaceae: **A–F** *Catharanthus roseus* **A** Twig bearing flower and fruits. **B** Flower, longisection. **C** Stamen. **D** Gynoecium. **E** Ovaries, transection. **F** Floral diagram (note semiapocarpous ovaries).

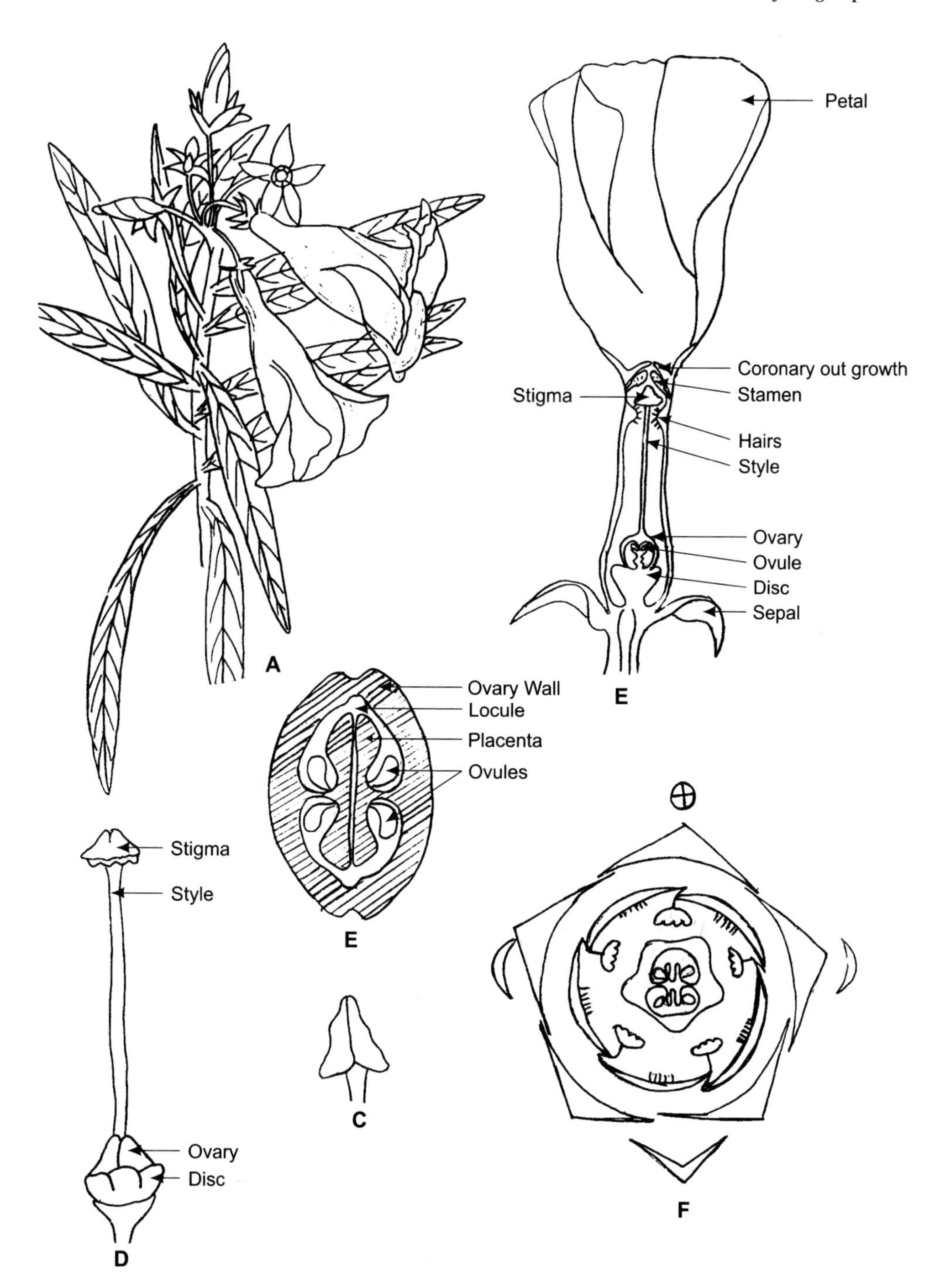

Fig. 11.21.2 Apocynaceae: **A–F** *Thevatia peruviana* **A** Flowering twig. **B** Flower, longisection. **C** Stamen. **D** Gynoecium. **E** Ovary, transverse section. **F** Floral diagram (sketched by Jaishree Chaudhary).

dorsiventral, occasionally isobilateral as in *Nerium*. Laticiferous canals and intraxylary phloem are reported. Stomata mostly Ranunculaceous type but Rubiaceous type in some taxa. Twenty three different types of stomata have been reported in *Catharanthus roseus* (Baruah and Nath 1996).

Embryology. Pollen grains triporate and 3-celled at shedding stage; often 2-celled as in *Catharanthus, Holarrhena* and *Plumeria*. Ovule hemianatropous or anatropous, rarely orthotropous (*Rauwolfia sumatrana*) (Johri et al. 1992). Embryo sac of the Polygonum type, 8-nucleate at maturity. Occasionally twin and abnormal embryo sacs with fewer or more than 8 nuclei occur. Degeneration of the embryo sacs common (Johri et al. 1992). Endosperm formation of the Nuclear type.

Chromosome Number. Basic chromosome number is variable: x = 8–12, 18 or 23.

Chemical Features. Apocynaceae members are rich in indole alkaloids. Cyclitols are reported in *Nerium* and *Catharanthus*.

Important Genera and Economic Importance. *Apocynum* is the type genus with all the usual features and is an erect herb or undershrub. The type species *A. androsaemifolium* is interesting. Its flowers emit a sweet honey-like fragrance which attracts insects. These insects, *Musca pipiens* in particular, enter the flower, are caught in the honey-like substance, and die. Some of the genera exhibit xerophytic features. *Adenium* and *Pachypodium* have fleshy stems and *Carissa* has spiny stems. Other interesting genera are: *Lepinia, Pleiocarpa* and *Notonerium* with 3- to 5-carpellary ovary and *Epigynum* and *Ichnocarpus* with half-inferior ovary. The fruits of *Alyxia concatenata* is a pair of moniliform follicles. In *Allemanda* an unusual type of gynoecial deveopment has been observed. The two carpels are free at initiation but fuse completely during development resulting in a unilocular ovary with parietal placentation (Fallen 1985).

Economically, the family has limited importance. *Rauwolfia serpentina* roots are used medicinally as a cure for epilepsy, high blood pressure, insanity and cardiac diseases. A number of alkaloids such as reserpine, reserpinine, serpentine, serpentinine, ajmalinine, isoajmaline and a few more alkaloids are reported from this plant and reserpine is the most important alkaloid amongst these. *R. tetraphylla* and *R. vomitoria* from America and Africa are also useful sources of reserpine. For several centuries *R. serpentina* has been extensively used as a source of drug. So much so that the species is now endangered (in India) due to its limited cultivation and indiscriminate collection from the wild. Recently the export of this crude drug has been restricted by Government of India. Keeping this in view, in vitro culture of shoot apices of the introduced species *R. tetraphylla* has been done and well-rooted plantlets were established in the field by Ghosh and Banerjee (2003). *Catharanthus roseus* is the source of anticancer drugs—Vincristin and Vinblastin. .

Many plants such as *Plumeria* spp., *Nerium odorum, Thevetia peruviana, Tabernaemontana divaricata* and *Allemanda* sp. are grown as ornamentals. The wood of *Wrightia tinctoria* and *Alstonia scholaris* is soft and used for wood carvings. The fruits of *Carissa carandas* are sour and edible. *Landolphia comoriensis, L. kirkii* and *Funtumia elastica,* three vines from Africa, yield rubber from their latex.

Taxonomic Considerations. The family Apocynaceae is closely related to the more advanced family Asclepiadaceae. It differs from Asclepiadaceae in: (a) single style, (b) absence of a corona, (c) mostly free pollen grains (never in pollinia), (d) stamens free from the stigma, and (e) in the absence of translators connecting the pollinia from the adjacent anther sacs.

Bentham and Hooker (1965b) treated this family as a member of the order Gentianales. Hallier (1912) combined the two families—Apocynaceae and Asclepiadaceae—and derived them from the Linaceae. Bessey (1915) and Rendle (1925) included these two families in the Gentianales. Engler and Diels (1936) placed it in the order Contortae. Hutchinson (1948) included the Apocynaceae alone in his Apocynales. In this order, Hutchinson (1969, 1973) also included three other families—Plocospermataceae, Periplocaceae and Asclepiadaceae—and derived Apocynales from the Loganiales. Cronquist (1968, 1981), Dahlgren (1980a, 1983a) and Takhtajan (1980, 1987) treat Apocynaceae as a member of the order Gentianales.

In the advanced family Apocynaceae, about 70% genera are apocarpous; the gynoecium consists of two carpels at the time of initiation. These are free above and usually united just at the base. During floral development, the uppermost part of the carpels undergoes a temporary postgenial fusion along their ventral flanks. According to Fallen (1986), the presence of the two fusion zones provide evidence for interpreting the apocarpous condition as phylogenetically secondary.

The family Apocynaceae is a natural taxon and is correctly placed in the order Gentianales.

Asclepiadaceae

A large family of 250 genera and about 3000 species distributed mainly in the Old World tropics and some in tropical America.

Vegetative Features. The members are mostly herbs or climbers and a few shrubs (*Calotropis gigantea*); perennial (*Asclepias*, *Gymnema*, *Stapelia*) or annual (*Pergularia daemia*); the vegetative parts yield latex. Stem herbaceous, climbing or twinning, sometimes covered with dense tomentum (*Calotropis*; 11.22 A) or with wax (*Hoya parasitica*). The stem of *Stapelia* is succulent, green and the leaves are scale-like. Leaves opposite, simple, petiolate, estipulate, entire. In *Dischidia* the leaves are modified into pitchers which collect water and adventitious roots from the next upper node grow in this pitcher.

Floral Features. Inflorescence di- or monochasial cyme or umbellate (11.22 A) or racemose. Flowers bracteate, ebracteolate, hermaphrodite, actinomorphic, hypogynous, typically pentamerous except the carpels. Calyx of 5 free or basally connate sepals, quincuncial or open; corolla of 5 petals, gamopetalous, or only basally fused, contorted or valvate, salver-shaped or funnel-form, corona of 5 or more scales or appendages (11.22 B, C) attached to the corolla tube or staminal tube. Stamens 5, usually adnate to the gynoecium to form a gynostegium; rarely free as in *Cryptostegia;* filaments flat and united to from a fleshy staminal tube around the ovary. The apical part is adnate to the pentangular stigmatic disc (11.22 B). Anthers bicelled, the pollen of each cell agglutinated in granular masses of tetrads as in *Cryptostegia* or waxy pollen masses or pollinia as in *Asclepias*, *Pergularia*, *Calotropis* (11.22 D) and others. Pollinia of the adjoining cells of the contiguous anthers are united in pairs, either directly or by appendages (caudicles) to black dot-like glands (corpusculum) which lie at the angles of the pentangular stigmatic disc (11.22 F). The 10 pollen masses are united in 5 pairs forming five translators. Gynoecium apocarpous, bicapellary, ovary unilocular, placentation marginal (11.22 B, E, F), superior, ovules numerous; styles 2, distinct, stigmas fused to form a pentangular stgmatic disc (11.22 B). Fruit an etaerio of follicles, seeds many, usually flattened, crowned with a tuft of long silky hairs; endosperm thin and scanty, embryo large.

Anatomy. Vessels small to medium-sized, with numerous radial multiples, sometimes ring-porous; perforations simple, intervascular pitting rather large and alternate. Parenchyma as scattered cells and irregular, uniseriate lines or scanty, paratracheal. Rays narrow and sometimes storied. Leaves usually dorsiventral; isobilateral in fleshy leaves of *Ceropegia* and *Hoya.* Hairs unicellular or uniseriate; stomata mostly Rubiaceous type but Cruciferous type in *Hoya* and *Stapelia* and Ranunculaceous type in *Sarcostemma, Solenostemma* and *Vincetoxicum.* Anomalous structure reported in climbing species.

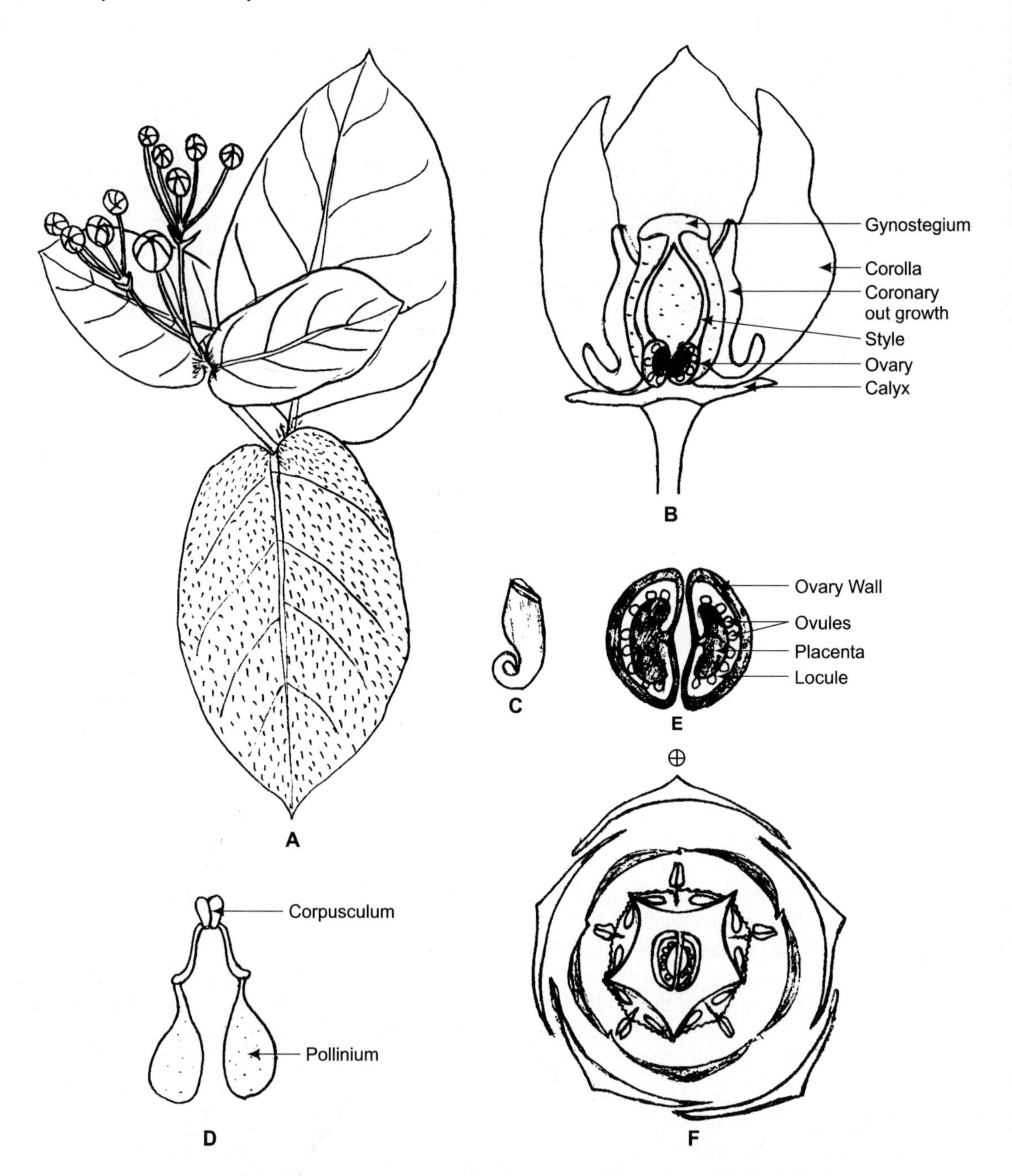

Fig. 11.22 Aslepiadaceae: **A–F** *Calotropis procera* **A** Twig with inflorescence. **B** Flower, longisection. **C** Corona. **D** Pollinia. **E** Ovaries, transverse section. **F** Floral diagram (sketched by Jaishree Chaudhari).

Embryology. Pollen grains are 3-celled and remain aggregated in pollinia. Ovules anatopous, unitegmic or ategmic and tenuinucellate. Ategmic ovules are reported in *Araujia, Asclepias, Gomphocarpus* and *Marsdenia* (Johri et al. 1992). Polygonum type of embryo sac, 8-nucleate at matuirty. Endospem formation of the Nuclear type.

Chromosome Number. Basic chromosome number is x = 11.

Chemical Features. Iridoids are totally absent. Seeds of various species of *Asclepias* are rich in linoleic acid (Shorland 1963).

Important Genera and Economic Importance. *Asclepias* is the largest genus and includes some ornamentals like *A. curassavica* or 'blood' flower and *A. tuberosa* or 'butterfly' weed. *Oxypetalum caeruleum* (blue milkweed), *Hoya carnosa* (wax plant), and *Stapelia* spp. (carrion flower) are some other ornamentals. *Dischidia rafflesiana* is an interesting plant from the tropical rain forests of Malayan region, with its leaves modified into pitchers that collect rain water and adventitious roots from adjacent nodes grow into it. Economically, the members of this family have limited importance. The seed hairs of *Asclepias* are used as substitute for 'kapok'—another seed fiber from *Salmalia malabarica* (Bombacaceae). The hairs are fine, silky and light. The plants of *Tylophora indica* are used medicinally against asthma. Latex from *Cryptostegia grandiflora* is a source of rubber. Latex of *Gymnema laticiferum* and *Oxystelma secamore* can be consumed as 'milk'. Latex of *Matalea*, on the other hand, is used as an arrow poison. *Calotropis procera* and *Pergularia daemia* are some of the weeds.

Taxonomic Considerations. The family Asclepiadaceae is closely allied to the family Apocynaceae, the differences between the two families are: (1) Stamens modified to pollinia and accompanied by translators, and (2) presence of a gynostegium.

All taxonomists retain this family next to the Apocynaceae, in the order Gentianales. This is fully justified.

Rubiaceae

A large family of about 500 genera and 6000 species, distributed mostly in the tropics and a few in temperate and arctic regions.

Vegetative Features. Trees (*Mitragyna parviflora, Anthocephalus cadamba*), shrubs (*Gardenia, Ixora, Hamelia*; 11.23.2 A) or herbs (*Oldenlandia*); sometimes lianas (*Galium, Rubia*); mostly perennial, rarely annual (*Oldenlandia*). Leaves opposite, decussate (*Ixora*) or whorled (*Gardenia*), simple, subsessile or petiolate, entire or rarely toothed, stipulate, Stipules inter- or intrapetiolar, frequently united to one another and to the petioles forming a sheath around the stem. The two stipules—one from each leaf standing next to each other—are usually united in the tribe Rubieae; often leaf-like and as large as normal leaf. The entire structure appears as a whorl of leaves in the tribe Galieae; or stipules reduced to glandular setae, as in *Pentas.*

Floral Features. Inflorescence basically axillary or terminal dichasial cyme—reduced to single flowers in *Gardenia;* sometimes aggregated into globose heads (Fig. 11.23.1 A) with the flowers basally adnate, e.g. *Morinda, Sarcocephalus,* (= *Nauclea*), *Anthocephalus, Uncaria* and others. Flowers hermaphrodite, actinomorphic, epigynous, 4- or 5-merous (Fig. 11.23.1 B). Calyx 4- or 5-lobed, aestivation open, persistent, enlarged in fruit in *Nematostylis.* Sometimes one sepal larger than others (Fig. 11.23.1 D-I) and brightly coloured, as in *Mussaenda.* Corolla of 4 or 5 petals, rarely 8–10 petals, gamopetalous, valvate, convolute or imbricate, rotate, salverform or funnel-form. Stamens 4 or 5, alternipetalous, epipetalous; anthers bicelled, dehiscence longitudinal (11.23.2 C), introrse. Gynoecium syncarpous, usually bicarpellary (pentacarpellary in *Hamelia patens* (11.23.2 E, F); monocarpellary with parietal placentation

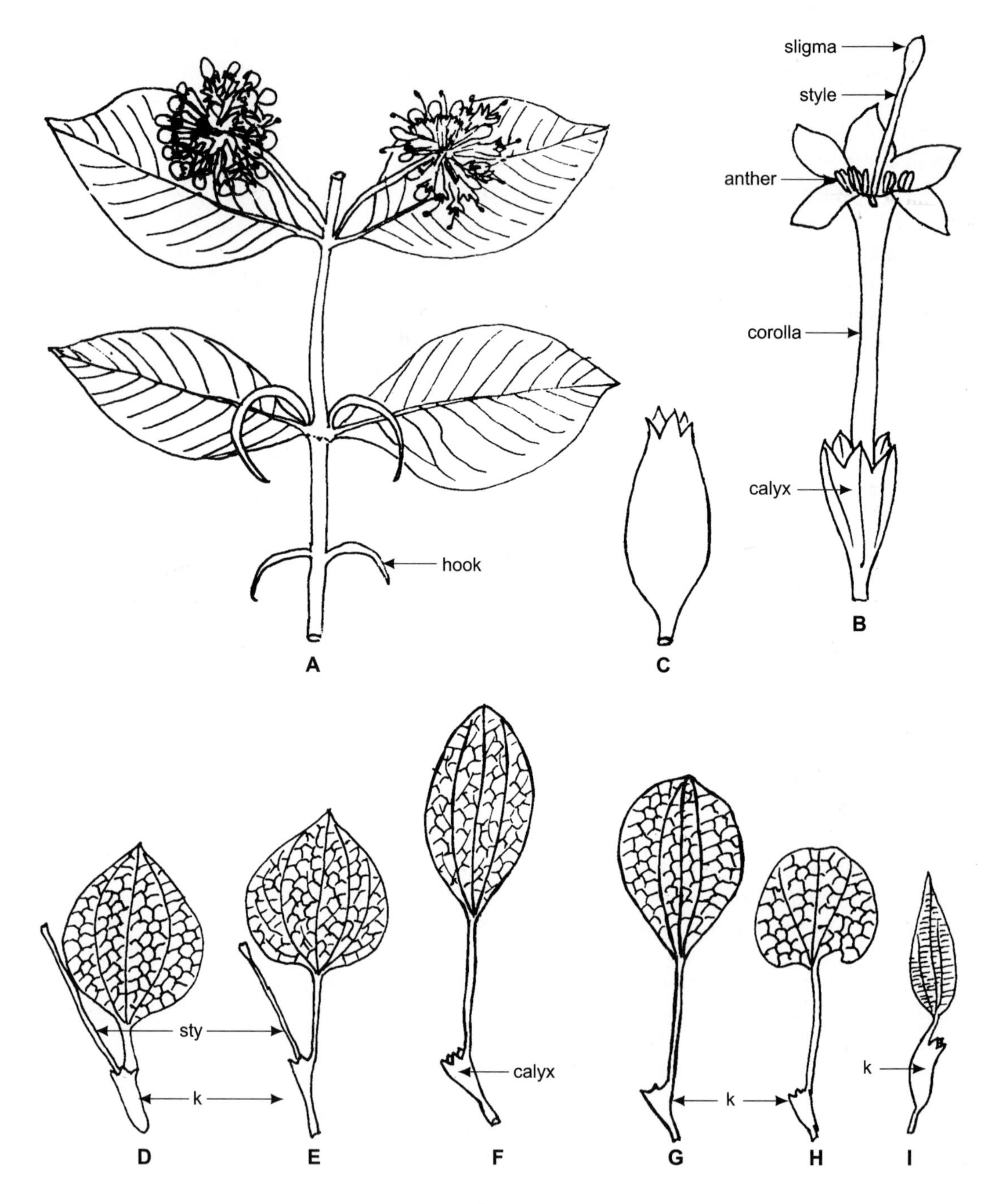

Fig. 11.23.1 Rubiaceae: **A-C** *Uncaria gambir*, **D** *Pinckneya pubescence*, **E** *Pogonopus tubulosus*, **F** *Capirona wurdeckii*, **G** *Calycophyllum spectabile*, **H** *C. candidissium*, **I** *Cosmocalyx spectabilis.* **A** Twig with inflorescence. **B** Flower. **C** Fruit. **D-I** Genera with one enlarged and coloured sepal. (*k* calyx, *sty* style). (Adapted from Hutchinson 1969).

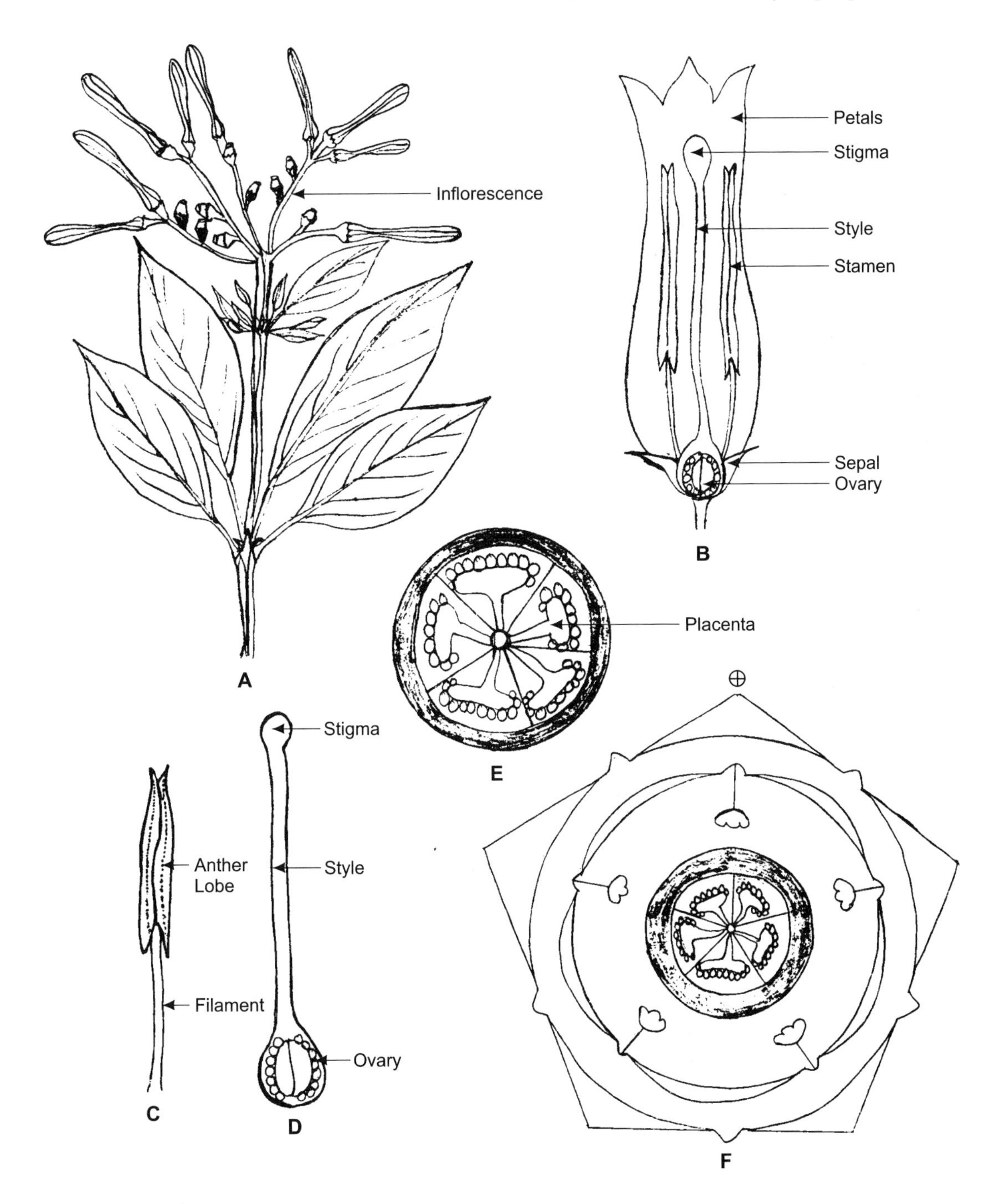

Fig. 11.23.2 Rubiaceae: **A–F** *Hamelia patens*. **A** Twig with inflorescence. **B** Flower. **C** Stamen. **D** Gynoecium, ovary in longisection. **E** Ovary, transverse section, note: T-shaped placenta. **F** Floral diagram (sketched by Jaishree Chaudhary).

in *Gardenia*). Ovary bilocular (pentalocular in *Hamelia patens*), axile placentation, inferior, rarely superior as in *Gaertnera* and *Pagamea;* ovary half-inferior in *Synaptantha.* Ovules usually numerous per locule; uniovulate in *Pavetta* with the ovule sunken in fleshy funiculus. Style 1 and slender, often 2-branched, stigma linear, capitate (11.23.2 B, D) or lobed. Distyly is sometimes reported, as in *Hedyotis salzmannii* (Riveros et al. 1995). Fruit a septi- or loculicidal capsule or indehiscent and separating into 1 seeded segments (*Galium, Rubia, Oldenlandia*); a fleshy berry in some, e.g. *Uncaria* (Fig. 11.23.1 C), *Coffea* and *Mitchella.* Seeds sometimes winged; embryo small in rich endosperm.

Anatomy. Vessels typically small or medium-sized, solitary or rarely in multiples of four, numerous, rarely ring-porous; perforation plates simple, parenchyma apotracheal in species with non-septate fibers and absent in species with septate fibres. Rays mostly narrow, up to 2-3 cells wide. Leaves dorsiventral (centric in *Asperula cynanchica).* Hairs unicellular, uniseriate, tufted or rarely peltate; stomata nearly always on lower surface, Rubiaceous type. Raphides and crystal sand are reported.

Embryology. The pollen grains are 2- or 3-colporate or 3-pororate with smooth exine. Pollen tetrad reported in *Gardenia.* Pollen trimorphism occurs in some genera (Mathew and Philip 1993). They are 2- or 3-celled at the dispersal stage. Ovules anatropous, unitegmic, tenuinucellate. Polygonum type of embryo sac, 8-nucleate at maturity. Twin embryo sacs reported in *Borreria hispida, B. stricta* and *Galium asperifolium* (Johri et aI. 1992). Allium type of embryo sac in *Scyphiphora hydrophyllacea* and Peperomia type in *Crucianella* spp. (Davis 1966). Endosperm formation is of the Cellular type.

Chomosome Number. Basic chromosome numbers are: x = 6–15, 17, x = 9 and 11 being more common.

Chemical Features. Iridoids and complex alkaloids such as quinine and quinidine in *Cinchona* and caffeine in *Coffea* have been reported. Iridoid glucosides occur in *Gardenia jasminoides* fruits (Inouye et al. 1974); loganin in *Mitragyna* (Cordell 1974).

Important Genera and Economic Importance. An interesting family, both morphologically as well as economically. Bacterial nodules develop on the leaves of various species of *Pavetta, Chomelia asiatica* and *Psychotria bacteriophylla.* These bacteria have been shown to fix atmospheric nitrogen (Boodle 1923).

Enlarged and petaloid calyx lobes are reported in many genera (Fig. 49.7 D-I). In *Uncaria gambir* of Malaya the lower inflorescences are barren and transformed into hooks (Fig. 49.7 A) by which the plant climbs. Many genera are spinescent, e.g. *Catesbaea spinosa* of the West Indies; leaves modified to spines in *Phyllacantha grisbachiana.*

Myrmecodia beccarii of Australia is a myrmecophilous plant with a large tuberous stem which houses ants.

Economically, the family is important. The bark of *Cinchona calisaya, C. ledgeriana, C. officinalis* and *C. succirubra* yields the antimalarial drug quinine. *Cephaelis ipecacuanha* is another medicinal plant yielding the well-known drug vinum ipecac, used against dysentry and liver diseases. The family yields another important product, coffee, from the dried and powdered seeds of *Coffea arabica, C. liberica* and *C. robusta.* The fruits of *Anthocephalas cadamba,* and *Randia dumetorum* are edible. *Uncaria gambir,* an Indonesian plant, is the source of a resin called gambier, used in medicine. Some plants yield timber: *Calycophyllum candidissimum* and *Ixora ferrea.* From the roots of *Rubia cordifolia* and the bark of *Morinda angustifolia*, important dyes are extracted. *Morinda umbellata*, a medicinal woody climber from Khasi Hills, Bihar and Deccan Peninsula is a rich source of anthraquinone derivatives. Nair and Seeni (2002) reports rapid multiplication through nodal and shoot tip explants as the natural wild populations are severely depleted owing to injudicious exploitation.

Many Rubiaceae are grown as ornamentals: *Asperula, Bouvardia, Gardenia, Hamelia, Ixora, Mussaenda, Nertera* and *Pentas. Oldenlandia corymbosa* is a common garden weed.

Taxonomic Considerations. Bessey (1915) placed the Rubiaceae in the order Rubiales along with Caprifoliaceae, Adoxaceae, Valerianaceae and Dipsacaceae. Bentham and Hooker (1965b) also placed it in the Rubiales, but only with Adoxaceae and Caprifoliaceae, and transfer the other families to the order Asterales. Wagenitz (1959) did not recognise the Rubiales as a natural taxon. He commented that the similarities between Rubiaceae and the members of the order Contortae (= Gentianales) are more important, as compared to the similarities between Rubiaceae and other families of the order Rubiales (sensu Bentham and Hooker 1965b). Wagenitz (1964) included the Rubiaceae in the order Gentianales along with Loganiaceae, Apocynaceae, Asclepiadaceae, Gentianaceae and Menyanthaceae. Takhtajan (1969) and Thorne (1968) also support this view. Based on chemical data, Dahlgren (1975a) supports the inclusion of the Rubiaceae in the order Gentianales. Lee and Fairbrothers (1978), based on serological evidences on Rubiaceae and other related families, indicate that the Rubiaceae has maximal affinity to the Cornaceae and Nyssaceae; next to the Gentianaceae, Caprifoliaceae and Asclepiadaceae, and least to the Apocynaceae and Dipsacaceae. Cytologically also, the Rubiaceae is more akin to the Gentianales. The basic chromosome number for both the Rubiaceae and the Gentianales is x = 11. In both taxa, the pollen grains are eurypalynous, ovules mostly anatropous, unitegmic, tenuinucellate and endospern formation of the Nuclear type. Therefore, Wagenitz's placement of the Rubiaceae in the order Gentianales is justified. According to Bremekamp (1966), however, the absence of intraxylary phloem is very much against placing the Rubiaceae amongst the Gentianales. Also, Rubiaceae differs distinctly from the Gentianales in having inferior ovary. Bremekamp (1966) and Cronquist (1968, 1981), therefore, treat it as a monotypic order—Rubiales with more affinities to the Gentianales. Cronquist also regards Rubiales to be the connecting link between the Gentianales and the Dipsacales.

This assignment is supported because: (1) the basic chomosome number x = 11 is not so frequent in the Gentianales as in the Rubiaceae, (2) the extent of eurypalyny in the Rubiaceae is much more in contrast to that in the Gentianales which are moderately eurypalynous, (3) ovary inferior in Rubiaceae and superior in all Gentianales members, and (4) intraxylary phloem is absent in the Rubiaceae.

The last two characters bring this family close to the order Dipsacales, and to the family Caprifoliaceae, in particular. Taxonomists have expressed doubt that the two families, Rubiaceae and Caprifoliaceae, should be treated separately (see Cronquist 1968). However, the Dipsacales differ from the Rubiaceae in Cellular type of endosperm, near-absence of the alkaloids and absence of the special glandular trichomes, colleters on the inner surface of the stipules.

Rubiaceae is a large family (ca. 500 genera and 6000 species) and would be too dominating if included either in the order Gentianales or Dipsacales. It will be best to treat this family as a distinct order, the Rubiales.

Order Tubiflorae

The Tubiflorae is a large order with 6 suborders and 26 families. The suborder Solanineae comprises 15 families, and the suborders Myoporineae and Phrymineae, 1 family each. The members are predominantly herbaceous. A few members of Bignoniaceae and Verbenaceae are trees and shrubs. All members of Lennoaceae and Orobanchaceae are parasites. Some members of Lentibulariaceae and the genus *Trapella* of Pedaliaceae are aquatic plants. Fouquieriaceae includes a xerophytic member, *Fouquieria.*

Inflorescences are of varied types. Flowers bisexual, mostly zygomorphic, bilipped, calyx persistent, corolla gamopetalous, floral parts in four whorls, stamens epipetalous. Ovary mostly superior (inferior in Columelliaceae), bicarpellary (tricarpellary in Polemoniaceae, pentacarpellary in Nolanaceae), syncarpous, bilocular (trilocular in Polemoniaceae, unilocular in Phrymaceae), placentation axile. Ovules numerous (only one in Phrymaceae), anatropous, unitegmic (bitegmic in Fouquieriaceae), tenuinucellate (rarely crassinucellate). Fruit mostly capsule, sometimes nutlets, drupe or berry.

Labiatae (= Lamiaceae)

A large family of about 200 genera and 3200 species, cosmopolitan in distribution; especially abundant in the Mediterranean region, the Old World and the mountains of subtropics.

Vegetative Features. Predominantly annual or perennial herbs, sometimes shrubs (*Lavandula dentata*), rarely trees (*Hyptis* and *Leucosceptrum*), or erect, perennial marsh plant (*Scutellaria galericulata*). Stem herbaceous or woody as in tree members, glandular-pubescent, erect, sometimes prostrate and with suckers as in *Mentha viridis,* quadrangular. Leaves simple, opposite-decussate, estipulate, petiolate, with varied type of margins, surface glandular hairy. These glands contain volatile oil which makes the leaves aromatic.

Floral Features. Inflorescence is typical of the family. It is a spike or raceme of pairs of dichasial cymes at each node (Fig. 11.24 A), rarely solitary, axillary as in *Scutellaria.* Sometimes the primary axis has short internodes and the pairs of biparous cymes are condensed together to form a globular head, as in *Lamium, Hyptis, Prunella* and *Monarda*. Flowers scattered in a raceme in a few species of *Scutellaria* and *Teucrium.* . Each cyme at the axil is subtended by a foliaceous bract which may be longer than the cyme and coloured (*Salvia*). Flowers hermaphrodite, bracteolate or ebracteolate, zygomorphic (Fig. 11.24 B) (actinomorphic in *Mentha* and *Elsholtzia*), pedicellate, complete, hypogynous, often showy. Calyx of 5 sepals, gamosepalous, sometimes bilobed (Fig. 11.24 C), as in *Alvesia, Ocimum;* sometimes the lobes are absent or appear to be only 2 with 5, 10, or 15 conspicuous ridges or ribs. Corolla of 5 petals, gamopetalous, imbricate, mostly bilabiate, the lower lip of 3 petals and upper lip of 2 petals. In *Ocimum* a single petal forms the lower lip and 4 petals together form the upper lip; the upper lip is rudimentary in *Teucrium.* Stamens 2 as in *Salvia* (Fig. 11.24 D, E, F) or 4 as in *Ocimum,* didynamous, monadelphous in *Coleus;* staminodes rarely present, as in *Orthodon* and *Hypogomphia,* epipetalous. Anthers bicelled, longitudinal dehiscence, connective elongated with the two lobes at two ends as in *Salvia lyrata* (Fig. 11.24 F), one lobe rudimentary in *S. urticifolia* (Fig. 11.24 D, E). Hypogynous, nectariferous disc often present between the stamens and the ovary (Fig. 11.24 C, G). Gynoecium syncarpous, bicarpellary; ovary superior, bilocular but tetralocular at maturity due to the intrusion of the ovary wall, ovule one in each of the 4 locules formed due to false septation of the ovary, placentation basal derived from axile; style gynobasic (Fig. 11.24 G), rarely terminal as in *Ajuga,* mostly bifid, rarely tetrafid as in *Cleonia,* and stigmas minute at the end of these branches. Fruit typically a group of 4 nutlets, enclosed by the persistent calyx; drupaceous in *Stenogyne.* Sometimes the pericarp is fleshy, as in *Gomphostemma,* or develops into

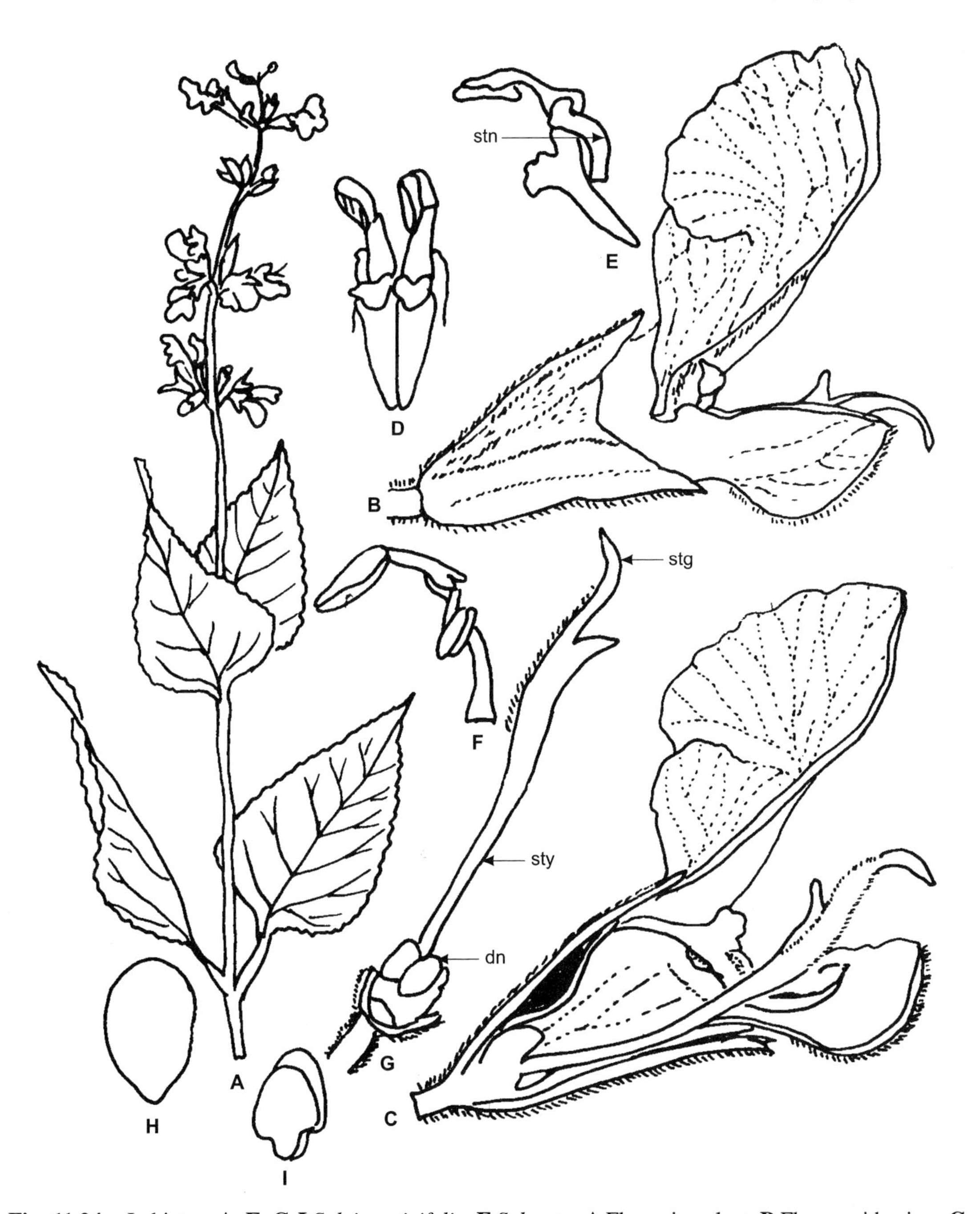

Fig. 11.24 Labiatae: **A–E**, **G-I** *Salvia urticifolia*, **F** *S. lyrata.* **A** Flowering plant. **B** Flower, side view. **C** Flower, longisection. **D** Stamens, front view **E** Side view. **F** Fertile stamen of *S. lyrata* with both the anther halves polliniferous. **G** Gynoecium, whole mount **H** Nutlet. **I** Embryo (oriented as in nutlet). (*dn* nectariferous disc, *stg* stigma. *stn* staminode. *sty* style). (Adapted from Radford 1987).

a wing-like membranous structure. In *Alvesia* of tropical Africa, the calyx enlarges and becomes bladder-like and reticulate. Seeds (Fig. 11.24 H) non- endospermous or with scanty endosperm which is absorbed by the developing embryo; embryo straight (Fig. 11.24 I).

Anatomy. Vessels small to minute, semi-ring-porous, with simple perforations. Parenchyma paratracheal, rather sparse; rays 4 to 12 cells wide, sometimes heterogeneous. Hairs of various kinds—glandular, secrete essential oils, and uniseriate, tufted or branched, nonglandular—occur on the surface of stem and leaves. Leaves dorsiventral, stomata on the lower surface, Caryophyllaceous type; crystals not very frequent.

Embryology. Pollen grains 3- to 6-colpate and 2- to 3-celled at the dispersal stage. Ovules anatropous, unitegmic, tenuinucellate. Polygonum type of embryo sac, 8-nucleate at maturity. Endosperm formation of the Cellular type.

Chromosome Number. The basic chromosome number is x = 5–11,13,17–20.

Chemical Features. Rich in volatile oils. Anthocyanin pigments of aglycone group like cyanidin, delphinidin and pelargonidin, and of acyl and glycosidic groups are present. Seed fats of various members are linolenic-rich.

Important Genera and Economic Importance. The genus *Salvia* is interesting because of its stamen structure. There are only two epipetalous stamens. The connective is thin and elongated, one anther lobe is fertile and the other rudimentary. As the insect enters the tubular corolla in search of nectar, the connective acts as a lever. Because of its movement, the fertile anther lobe brushes against the body of the insect, covering it with pollen. When the same insect visits another flower, cross-pollination takes place. *Ocimum sanctum* is a profusely branched perennial herb cultivated in many households in India (for religious purposes). Necklaces of beads from the woody twigs of this plant are worn by the followers of the Krishna cult and the Vaishnavas. *Mentha aquatica* is an aquatic plant. Leaves of *Coleus* are beautifully variegated and coloured with shades of red, violet, pink and yellow; these are often used for experimental work in laboratories.

Many genera are the source of volatile aromatic oil: *Salvia, Lavandula, Rosmarinus, Mentha* and *Pogostemon.* These oils are used in perfumery and the soap industry. Plants of *Ocimum* and *Mentha* are important medicinally. *M. piperata,* the peppermint plant, is cultivated and the oil distilled from its aerial parts is used medicinally and also in preparation of chewing gum. The leaves of *Orthosiphon aristatus* are used like tea leaves in Java. This drink is of medicinal value in kidney and bladder troubles. Some other members, like *Origanum* (marjoram), *Thymus* (thyme) and *Satureja* (savory), are important culinary herbs. The leaves of *Origanum vulgare* are used to make hair decorations in some parts of India. The ornamentals include *Salvia*, *Leonotis* (lion's head), *Dracocephalum* (dragonhead), *Nepeta* (catmint), *Scutellaria* (skull cap), *Coleus*, *Teucrium*, *Lavandula*, *Thymus* and *Pycnanthemum*. *Plectranthus vetiveroides* is a succulent herb popularly known as Black Khus root. It is in great demand for production of traditional and modern medicines in India (Sivasubramanian 2002).

Eremostachys superba grows in few and widely separated areas which are not connected by the normal dispersal ability of the plant. It is reported from Mohand, Dehra Dun in the Siwalik hills of Uttar Pradesh in India and from Peshawar in Pakistan. The plant is on the verge of extinction and may disappear within 4 to 5 years (Rao and Garg 1994).

Taxonomic Considerations. Hallier (1912), Rendle (1925) and Wettstein (1935) included Labiatae in Tubiflorae. Bentham and Hooker (1965b) and Bessey (1915) retained it in Lamiales along with the family Verbenaceae. Hutchinson (1926) also placed the two families—Verbenaceae and Labiatae—in the order Lamiales, but later (1948) Hutchinson raised Verbenaceae to Verbenales. In 1969, he pointed out that the

Labiatae is the most highly evolved family amongst the Herbaceae and Verbenaceae, amongst the Lignosae, and considered the two families unrelated. Benson (1970) placed only Verbenaceae and Labiatae in his order Lamiales, and the predominantly herbaceous family is treated as more advanced.

Cronquist (1968, 1981) and Stebbins (1974) group together Labiatae, Phrymaceae and Verbenaceae in the Order Lamiales. Dahlgren (1975a, b, 1980a, 1983a) considers Labiatae to belong to the order Lamiales, along with Verbenaceae. Takhtajan (1980, 1987) supports this assignment.

The family Labiatae has such distinctive features that it can easily be separated from others. However, the members of the two subfamilies—Ajugoideae and Prostantheroideae—resemble Verbenaceae in having terminal styles (not gynobasic). Similarly, there are some members of Verbenaceae which have a gynobasic style.

The comparative data of these two families is given in Table 11.4.

Table 11.4 Comparative Data for the Families Verbenaceae and Labiatae

Verbenaceae	*Labiatae*
Herbs, shrubs and trees, stem quadrangular	Herbs and shrubs, stem quadrangular
Inflorescence racemose or cymose	Verticillate or cymose
Flowers zygomorphic, rarely bilipped	Zygomorphic, always bilipped
Calyx persistent	Calyx persistent
Stamens 2 + 2 or 2	Stamens 2 + 2 or 2
Carpels (2), superior	Carpels (2), superior
Ovary 2- or 4-locular, placentation axile	Ovary 4-locular, placentation basal
Style simple, terminal, rarely gynobasic	Style gynobasic, rarely terminal
Fruit a drupe or berry	Fruit of 4-nutlets

From the above data, it is clear that the two families Labiatae and Verbenaceae are closely related, and should be retained in the order Lamiales.

Solanaceae

A large family of ca. 85 genera and more than 2200 species which occur chiefly in Central and South America. The largest genus *Solanum* with about 1500 species occurs over most parts of the world.

Vegetative Features. Predominantly herbaceous; shrubs, small trees, lianas and creepers are also known. Stem soft, herbaceous or woody, with hairy surface. Leaves alternate or opposite (near the inflorescence), simple or pinnatisect as in *Lycopersicon esculentum*, estipulate, often with oblique base (*Datura, Solanum nigrum*, 11.25 A).

Floral Features. Inflorescence axillary or extra-axillary cymes (11.25 A), often helicoid. Flowers solitary, axillary in *Datura, Atropa*; hermaphrodite, actinomorphic, or zygomorphic as in *Schizanthus,* hypogynous, ebracteate and ebracteolate. Calyx 5-lobed, gamosepalous but often connate only at base and free above, persistent and sometimes enlarged in fruit, as in *Physalis, Withania.* Corolla of 5 petals, gamopetalous, rotate (*Solanum*), infundibuliform (*Petunia, Datura*) or tubular (*Nicotiana tabacum*), rarely bilabiate (*Schizanthus*). Aestivation usually plicate or convolute, rarely valvate. Stamens 5, epipetalous, alternate with corolla lobes (11.25 B, F) usually unequal; all perfect and inserted at the base of the corolla. Sometimes only 4 or even 2 stamens occur and then staminodes are present. Anthers bicelled (monothecous in *Browallia*), longitudinal dehiscence or dehisce by apical pores as in *Solanum* (11.25 B, C). Hypogynous disc usually present. Gynoecium syncarpous, bicarpellary (3- to 5-carpellary in *Nicandra*), ovary superior, bilocular or 3- to 5-locular by formation of false septa, rarely unilocular with only one ovule, as in *Henoonia*[10] or apically unilocular, as in *Capsicum*, placentation axile, placenta swollen (11.25 E, F), style 1, stigma bilobed, capitate (11.25 D). Fruit a berry, sometimes enclosed in an enlarged persistent calyx, as in *Physalis* and *Withania*; or septicidal capsule, as in *Datura.* Seeds smooth or pitted, albuminous and with a straight embryo.

Anatomy. Vessels very small to medium-sized, few to numerous, perforations simple, intervascular pitting alternate; parenchyma either scanty, paratracheal or predominantly apotracheal. Rays uniseriate or up to 8 cells wide, almost homogeneous. Intraxylary phloem and crystal-sand are reported. Leaves usually dorsiventral, stomata Ranunculaceous type or sometimes Cruciferous or Caryophyllaceous type.

Embryology. Pollen grains 3- to 5- or 6-colpate, colporate or non-aperturate; shed at 2-celled stage. Ovules hemiana-, ana- or campylotropous, unitegmic, tenuinucellate. Polygonum type of embryo sac is common; Allium type occurs in *Capsicum frutescens* var. *tabasco*, *C. nigrum* and *C. pendulum* (Johri et al. 1992); 8-nucleate at maturity. Endosperm formation of the Nuclear, Cellular and intermediate type.

Chromosome Number. Basic chromosome number is x = 7 to 12.

Chemical Features. The Solanaceae are known for their tropane alkaloids and for the steroidal lactones (withanolides). The principal tropane alkaloids, hyoscine and hyoscyamine, are reported in at least 15 genera such as *Datura*, *Duboisia*, *Cyphanthera* and others (Evans 1986). Steroidal lactones are highly oxygenated C_{28} compounds and have been isolated from *Acnistus*, *Datura*, *Lycium*, *Jaborosa*, *Nicandra*, *Physalis* and *Withania.* In *W. somnifera* a number of chemotypes are known which are morphologically similar but differ in their withanolide content. Hybrids from such chemotypes often produce new withanolides not known in either parent (Eastwood et al. 1980, Nittala and Lavie 1981).

Important Genera and Economic Importance. A number of genera of this family are important. The habit of different species of *Solanum* is much varied. *S. surattense* is a spiny xerophytic prostrate herb, *S. nigrum* a mesophytic erect herb, *S. tuberosum* a herb with underground stem modified for storage of food, *S. melongena* bears edible fruits that are used as vegetable; *S. verbascifolium,* a tall unarmed shrub or small tree, has 4- to 8-inches long, elliptic-lanceolate leaves which are woolly tomentose on the lower surface, and flowers in dichotomous corymbose subterminal cymes. *S. indicum* is an erect prickly undershrub about 1 to 6 feet tall. *Lycopersicon esculentum* is the tomato plant, bearing edible fruits rich in vitamins particularly vitamins A and B. *Capsicum frutescens* and *C. annuum* are the chillies. *Physalis peruviana* is the commonly known gooseberry in which the orange-yellow berries are completely enclosed in the enlarged, papery, persistent calyx. The fruits are eaten when ripe and also made into jams. *Nicotiana* is another important genus. The two species cultivated are *N. tabacum* and *N. rustica*, the leaves yield tobacco. *Datura alba*, *D. innoxia*, *D. metel* grow wild; *D. arborea* is a small tree from the Peruvian Andes bearing solitary axillary flowers with 6- to 9-inches long white corolla. *D. suaveolens* from Brazil is also a small tree. The seeds of these plants are poisonous if eaten in large quantities, but of medicinal value when used in minute doses. *Atropa belladona* is also of medicinal importance—source of

[10]Now included in Goetziacceae (Takhtajan 1987).

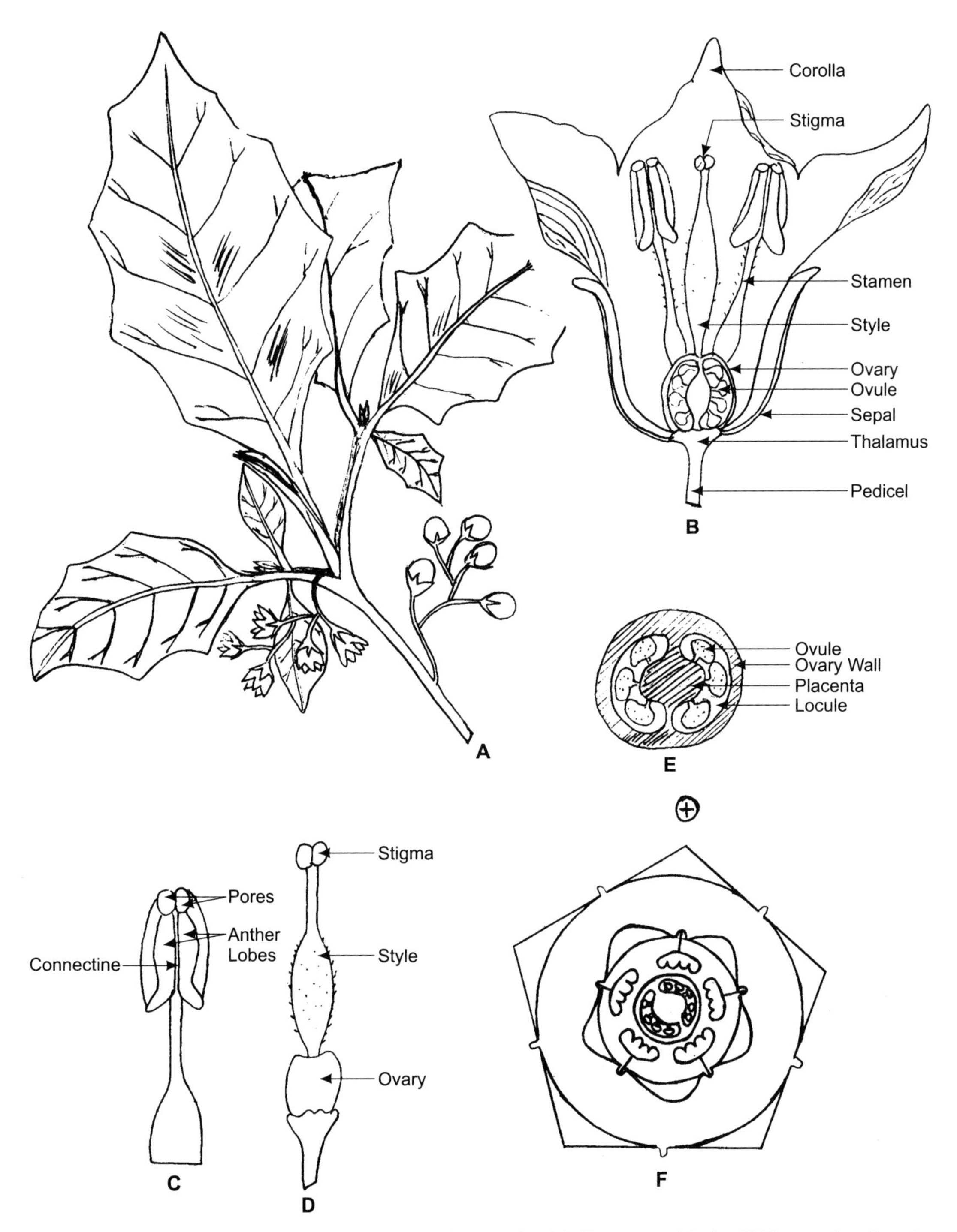

Fig. 11.25 Solanaceae: **A–F** *Solanum nigrum*. **A** Branch with flowers and fruits. **B** Flower, longisection **C** Stamen, note poricidal dehiscence. **D** Gynoecium. **E** Ovary, transverse section. **F** Floral diagram (sketched by Jaishree Chaudhary).

the alkaloid atropin used for dilating the pupil of the eye. This, too, is poisonous when administered in higher doses. Other plants of medicinal importance are *Hyoscyamus niger* and *Mandragora officinarum.*

Ornamental plants include *Petunia violacea*, *Cestrum nocturnum*, *Nicandra physaloides* and many species of *Solandra*, *Schizanthus*, *Brunfelsia* and *Browallia. Lycium europaeum* is yet another interesting plant, a spiny xerophytic shrub.

Taxonomic Considerations. The family Solanaceae has been placed variously by different authors. Bentham and Hooker (1965b) and Bessey (1915) treated it as a member of the order Polemoniales. Engler and Diels (1936), and Hallier (1912) placed it in Tubiflorae, whereas Hutchinson (1948, 1959, 1969) recognised a distinct order, Solanales, and derived it from Saxifragales. Cronquist (1968, 1981) considers it to be a member of his Polemoniales along with the families Nolanaceae, Convolvulaceae, Cuscutaceae, Polemoniaceae and others. Benson (1970) included Solanaceae and Nolanaceae in the same order, Solanales, and concluded that Solanaceae is the more advanced of the two. Varghese (1970) observed that various floral features are shared by the families Scrophulariaceae and Solanaceae. Zygomorphic flowers of Scrophulariaceae are present in *Salpioglossis*—a genus of Solanaceae. The oblique placenta as seen in Solanaceae is also present in *Scrophularia*—a member of Scrophulariaceae. However, embryologically, the two families are very distinct. In Solanaceae the endosperm formation is of the Cellular, Nuclear or Helobial type but in Scrophulariaceae it is only of the Cellular type. Embryogeny in Scrophulariaceae is of the Crucifer type and in Solanaceae it is of the Solanad type. Varghese (1970) concluded that the two families might have had a common origin but Scrophulariaceae has specialised more than Solanaceae. Stebbins (1974) included it in Polemoniales along with Nolanaceae. Dahlgren (1975a, b, 1980a, 1983a) places it in a separate order, Solanales, which also includes Nolanaceae. Takhtajan (1980) states that this family belongs to Scrophulariales but also has affinities with Convolvulaceae of the Polemoniales. According to him, these two families probably had a common origin from Loganiaceous stock. In both the families intraxylary phloem is present—in Convolvulaceae it arises in the stêm above the level of hypocotyl, and in Solanaceae it arises in the hypocotyl. At the same time, the seed structure of *Solanum* and *Lycopersicon* resembles the seed coat structure of *Strychnos* of the Loganiaceae (Corner 1976). Takhtajan (1987), however, removes Solanaceae to Solanales.

After reviewing the phylogeny of Solanaceae, it is best placed in the order Tubiflorae and close to the family Scrophulariaceae. There are certainly more resemblances of Solanaceae with Scrophulariaceae than with Convolvulaceae.

Scrophulariaceae

A large family of ca. 210 genera and nearly 3000 species, of cosmopolitan distribution and occurs in all the continents including Antarctica.

The members are mostly herbs or undershrubs (*Scoparia dulcis*), rarely trees (*Paulownia*), sometimes climbing shrubs (*Maurandya, Rhodochiton*), parasites (*Hyobanche, Harveya*), semi-parasites on roots (*Striga, Pedicularis*), or saprophytes (*Melampyrum, Castilleja*); *Hebe* spp. of New Zealand are xerophytes (resembling certain Coniferae).

Vegetative Features. Stem herbaceous (11.26.2 A) or woody. Leaves simple, estipulate, alternate, opposite (Fig. 11.26.1 A, 11.26.2 A) or whorled, often lower opposite and upper alternate; entire or pinnately lobed or incised.

In parasitic species the leaves are scale-like and devoid of chlorophyll.

Floral Features. Inflorescence variable—simple raceme or spike (indeterminate) or dichasial cyme (determinate). Solitary axillary flowers in *Linaria.* Flowers hermaphrodite (Fig. 11.26.1 B, C), zygomorphic, hypogynous, usually bracteate and bracteolate. Bracts and bracteoles brightly coloured in *Castileja.* Calyx of usually 5 sepals (or 4), gamosepalous but often deeply cleft, persistent; posterior sepal is suppressed in *Veronica* and the two anterior sepals are fused in *Calceolaria*; aestivation imbricate or valvate. Corolla of 4 or 5 (11.26.2 C) (rarely 6–8) petals, gamopetalous, aestivation imbricate or valvate. Corolla tube inconspicuous, as in *Veronica,* or prominent, as in *Digitalis,* usually bilabiate (Fig. 11.26.2 B) and personate (e.g. *Antirrhinum, Linaria, Lindenbergia* (11.26.2 B) campanulate in *Digitalis.* Corolla spurred in *Linaria,* saccate in *Antirrhinum* or the limb develops into two unequal, inflated lips (*Calceolaria*); absent from 1 species of *Synthyris.* Stamens usually 4, sometimes 5 (*Verbascun*) or 2 (*Veronica, Linaria, Gratiola*), epipetalous and fused at the base of the corolla tube (Fig. 11.26.1 C; 11.26.2 D), alternating with the petals, didynamous, anthers bicelled (Fig. 11.26.1 D; 11.26.2 E), sometimes one cell is larger than the other, dehiscence longitudinal, rarely poricidal (e.g. *Seymeria*), introrse; nectariferous disc usually present at the base. Gynoecium bicarpellary, syncarpous; superior, bilocular ovary; placentation axile, ovules numerous on enlarged placentae (Fig. 11.26.1 E, F; 11.26.2 G, H); style 1, terminal, stigma bilobed (11.26.2 F). Fruit is a dry dehiscent capsule, normally septicidal but loculicidal in *Buchnera* and poricidal in *Antirrhinum*; rarely a berry, e.g. *Halleria, Teedia, Leucocarpus* and *Dermatocalyx*, or dry indehiscent capsule as in *Hebenstretia* or 1-seeded tardily dehiscent capsule as in *Tozzia.* Seeds smooth or with rugose surface (Fig. 11.26.1 G), sometimes winged, as in *Mimulus*; endosperm fleshy (absent in *Wightia* and *Monttea*), embryo straight or slightly curved (Fig. 11.26.1 H).

Anatomy. Vessels very small and numerous, occasionally ring-porous, parenchyma usually sparse or absent, rarely abundant. Rays either absent or when present, 1 to 9 cells wide, homo- or heterogeneous. Leaves usually dorsiventral, stomata mostly Ranunculaceous type; hairs numerous on the vegetative parts and exhibit a considerable diversity of forms. Crystals not so frequent.

Embryology. The family is multipalynous. In majority members the pollen grains are tricolporate and 2-celled at the dispersal stage. Ovules ana-, hemiana- or campylotropous, unitegmic, tenuinucellate. Polygonum type of embryo sac, 8-nucleate at maturity. Endosperm formation of the Cellular type.

Chromosome Number. Basic chromosome numbers are variable: x = 6–18, 20, 21, 23–26 and 30.

Chemical Features. Various members contain Group I iridoids (Jensen et al. 1975), anthocyanin pigments like rutinoside, diglucoside and glucosylside of glycosidic type and cyanidin, pelargonidin, and delphinidin of the aglycones type. Seed fats of Scrophulariaceae members are rich in linolenic acid (Shorland 1963).

Important Genera and Economic Importance. The genus *Ixianthus* is unique in having verticillate leaves. Leaves dimorphic in *Hemiphragma*, the cauline ones are orbicular and the axillary ones linear. In the two aquatic genera *Ambulia* and *Hydrotriche*, too, leaves are dimorphic—the aerial ones entire and the submerged, highly dissected.

In *Vandellia* the flowers are totally or partially cleistogamous. In *Calceolaria* the lower corolla lip is entire, concave or slipper-shaped.

The general appearance of *Russelia equisetifolia* plants is that of a xerophyte with ridged chlorophyllous stems and scale-like leaves, resembling the plants of *Equisetum. Bacopa, Scoparia* and *Verbascum* have 5 stamens each. In semi-parasitic *Tozzia* the fruit is a single-seeded capsule. *Scrophularia nodosa* is a non-leguminous plant with root nodules. This family includes a number of ornamental plants such as *Antirrhinum, Calceolaria, Linaria, Mimulus, Nemesia, Penstemon, Russelia* and *Torenia. Striga densiflora*, a parasitic herb usually infects the roots of *Sorghum. Wightia* (2 or 3 spp.) distributed from Eastern Himalayas to southeast Asia and west Malaysia (excluding Philippines) are epiphytic shrubs; later become independent trees.

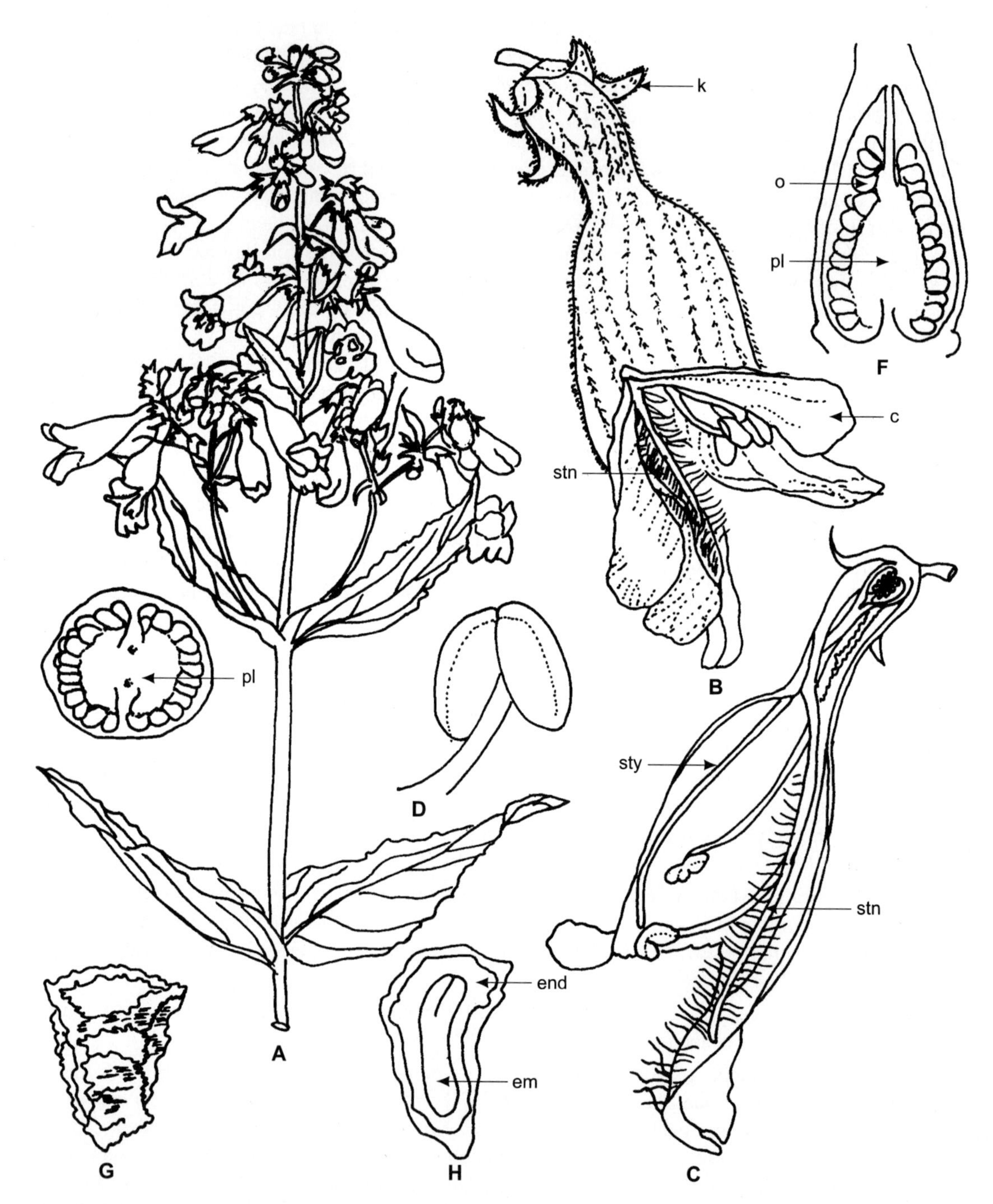

Fig. 11.26.1 Scrophulariaceae: **A-H** *Penstemon canescens.* **A** Twig with inflorescence. **B** Flower. **C** Longisection. **D** Anther. **E**. Ovary, transection. **F** Same, longisection. **G, H** Seed (**G**) and longisection (**H**). (*c* corolla, *em* embryo, *end* endosperm, *k* calyx, *o* ovule, *pl* placenta, *stn* staminode, *sty* style). (Adapted from Radford 1987).

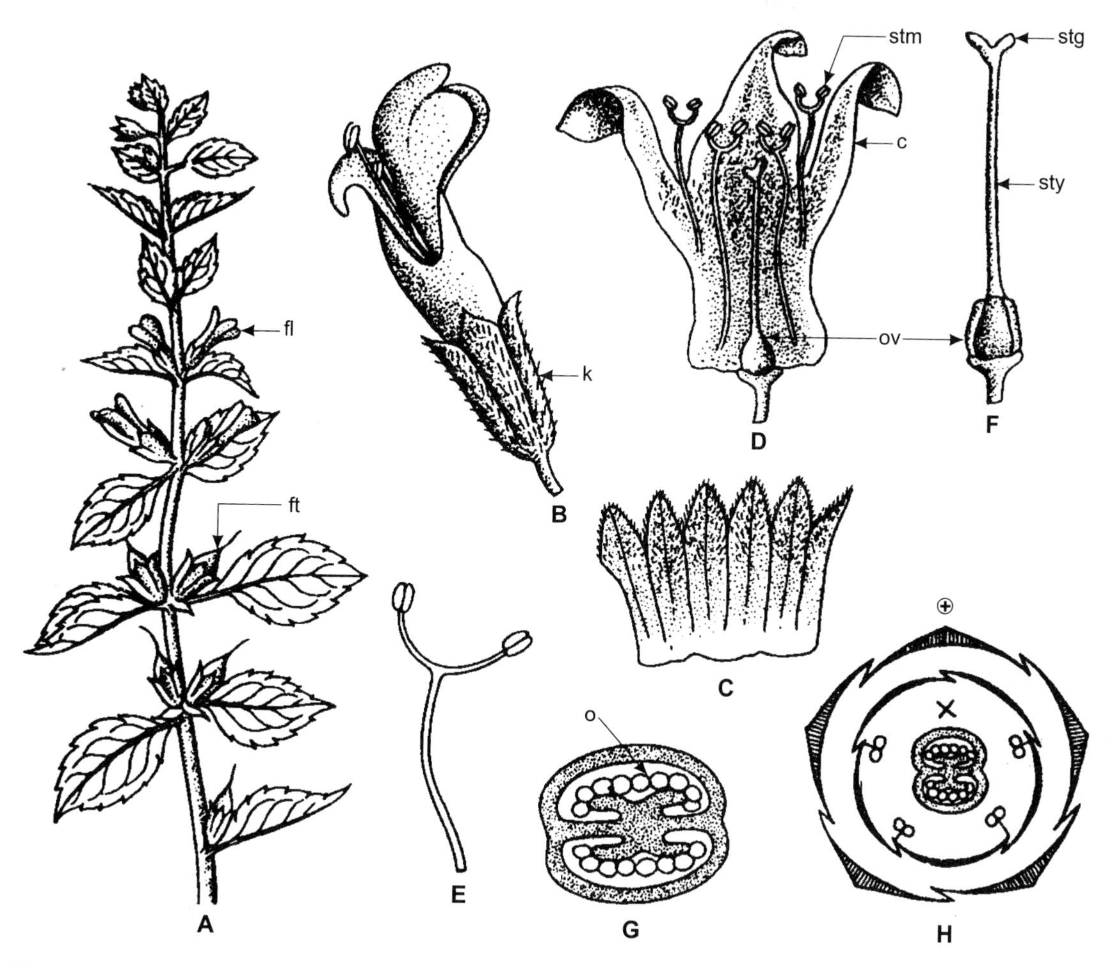

Fig. 11.26.2 Scrophulariaceae: **A–H** *Lindenbergia indica* **A** Plant with axillary flowers and fruits. **B** Flower. **C** Corolla split open. **D** Flower, longisection. **E** Stamen with elongated connective. **F** Gynoecium. **G** Ovary, transverse section. **H** Floral diagram. (*c* corolla, *fl* flower, *fr* fruit, *k* calyx, *o* ovule, *ov* ovary, *stg* stigma, *stm* stamen, *sty* style).

Digitalis is the source of the drug digitoxin obtained from dried leaves of *D. purpurea* and *D. lanata* and is used as myocardinal stimulant in congested heart failure. Seeds of *Verbascum thapsus* are to some extent, with narcotic properties, and are used to stupefy fish.

Paulwonia is a disputed genus. Its tree-like habit and winged seeds are anomalous in Scrophulariaceae, but its copious endosperm makes its position anomalous also in Bignoniaceae. It is extremely close to *Catalpa* of Bignoniaceae.

Taxonomic Considerations. The family Scrophulariaceae belongs to the order Personales according to Bentham and Hooker (1965b). Engler and Diels (1936) and Hallier (1912) retained it in Tubiflorae. Bessey (1915) and some other taxonomists placed this family in Scrophulariales, while Hutchinson (1948, 1969, 1973) included it in Personales and pointed out its derivation from Solanales through

Salpiglossidaceae. According to Cronquist (1968, 1981), it belongs to Scrophulariales and is derived from Polemoniales. Benson (1970), Stebbins (1974), and Dahlgren (1975 a, b, 1980a, 1983a) also placed it in Scrophulariales. Takhtajan (1980, 1987) too includes this family in Scrophulariales. According to him, it resembles the members of Solanaceae, especially the tribes Cestreae and Salpiglossideae, and also has close affinity with the family Buddlejaceae, particularly in embryological and chemical features. Thus, there is no controversy over the placement of Scrophulariaceae in Scrophulariales.

Acanthaceae

A large homogenous family of ca. 240 genera and 2200 species, distributed mostly in the tropical and subtropical areas of the world. The four tropical zones presumed to be the centres of distribution of the members of this family are: Indo-Malaya, Africa, Brazil and Central America.

Vegetative Features. Perennial herbs, shrubs or rarely medium-sized trees, sometimes lianas (*Mendoncia* and *Thunbergia*), some members are xerophytic herbs (*Acanthus* and *Blepharis molluginifolia*), undershrubs, e.g. *Barleria prionitis* and *Peristrophe bicalyculata* (Fig. 11.27.2 A), rarely aquatics, e.g. *Acanthus ilicifolius*, a mangrove plant. Stem herbaceous, green, often ridged. Leaves opposite-decussate, simple, estipulate, petiolate, subsessile or sessile. (Figs. 11.27.1 A; 11.27.2 B).

Floral Features. Inflorescence terminal or axillary dichasial cyme (Fig 11.27.2 A) rarely a raceme, e.g. *Beloperone*, or solitary axillary flowers, as in *Thunbergia.* Flowers bracteate and bracteolate (Fig. 11.27.1 B, C; 11.27.2 B) bracts and bracteoles often conspicuous and involucrate as in *Peristrophe* (11.27.2 C, G), sometimes spinescent as in *Acanthus mollis* and *Barleria prionitis*; zygomorphic, bisexual, pedicellate often large and showy, e.g. *Thunbergia.* Calyx of 4 or 5 sepals, deeply lobed, often with spiny or bristly margin, e.g. *Barleria,* or highly reduced as in *Thunbergia*; imbricate, rarely contorted. Corolla of 4 or 5 petals, gamopetalous, usually bilabiate (Fig. 11.27.1 C, D), (exceptions *Thunbergia* and *Ruellia*), upper lip bifid and erect, lower lip almost horizontal and three-lobed (Fig. 11.27.1 D; 11.27.2 B). Inner surface of the corolla lobes hairy and the hairs often extend up to the mouth of the corolla; in *Acanthus* the upper lip is totally missing and the stamens are protected by the calyx; aestivation imbricate or contorted. Stamens 2 to 5, epipetalous, when 4, didynamous with the fifth one suppressed or represented by a staminode; when only 2, they are attached to the anterior petals. Filaments usually long and free above (Fig. 11.27.1 E, F), and therefore the anthers are outside the floral tube (exception *Thunbergia* with short stamens within the floral tube). Form, position and number of anther lobes also vary; lobes 1 or 2, when 1-celled, there may be a rudimentary second lobe; when bicelled, the two lobes may be separated by an elongated connective as in *Peristrophe* (Fig. 11.27.2 D), *Beloperone*; sometimes the 2 lobes are unequal; anthers dorsifixed, extrorse. Gynoecium syncarpous, bicarpellary, ovary superior, bilocular, placentation axile (Fig. 11.27.1 H), 2 or more ovules per locule (Fig. 11.27.2 F), style 1, elongated, filiform (Fig. 11.27.1 G), stigma bifid, funnel-shaped in *Thunbergia.* Jaculator, a device for dispersal of seeds, may or may not be present just below the ovules. Fruit usually a loculicidal capsule (Fig. 11.27.1 I, J), often elastically dehiscent; a drupe in *Mendoncia.* Seeds (Fig. 11.27.1 J, K) mostly exalbuminous, with various types of testa such as mucilaginous, scaly, hairy or with an indurated funicle.

Anatomy. Vessels small with a radial pattern, simple perforation plates; parenchyma scanty paratracheal or vasicentric, rays 1 to 6 cells wide, uni- or multiseriate, markedly heterogeneous. Both intra- and interxylary phloem present. Leaves usually dorsiventral, rarely isobilateral, stomata of Caryophyllaceous type, occur on both the surfaces or only on the lower surface. Nonglandular hairs of unicellular or uniseriate type as well as glandular hairs are reported.

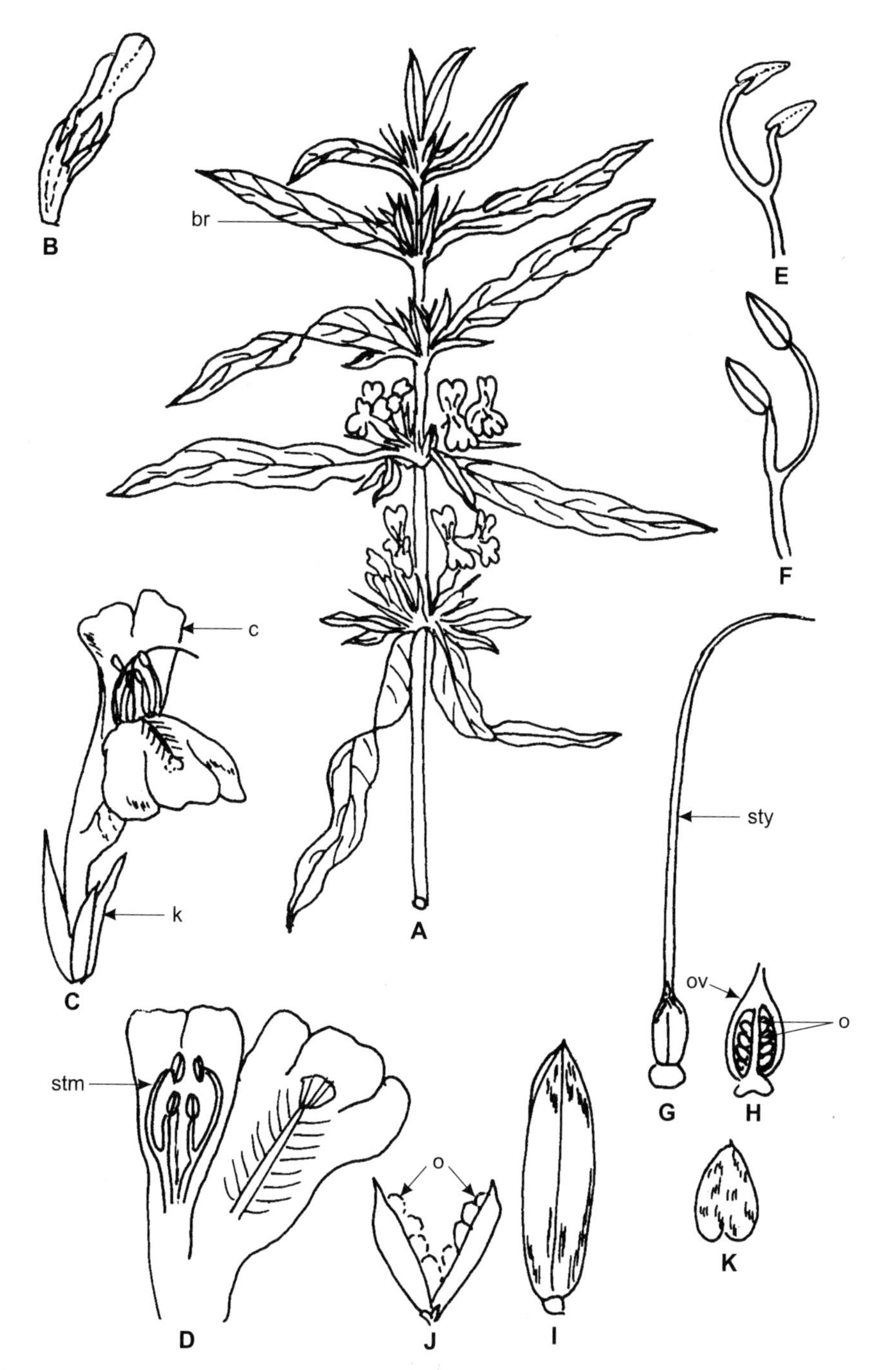

Fig. 11.27.1 Acanthaceae: **A-K** *Asteracantha longifolia.* **A** Portion of plant. **B** Bud. **C** Flower. **D** Corolla (cut open). **E, F** Stamens. **G** Pistil. **H** Ovary, longisection. **I** Capsule. **J** Dehisced capsule. **K** Seed. (*br* bract, *c* corolla, *k* calyx, *o* ovule, *stm* stamen, *sty* style). (Adapted from Hutchinson 1969)

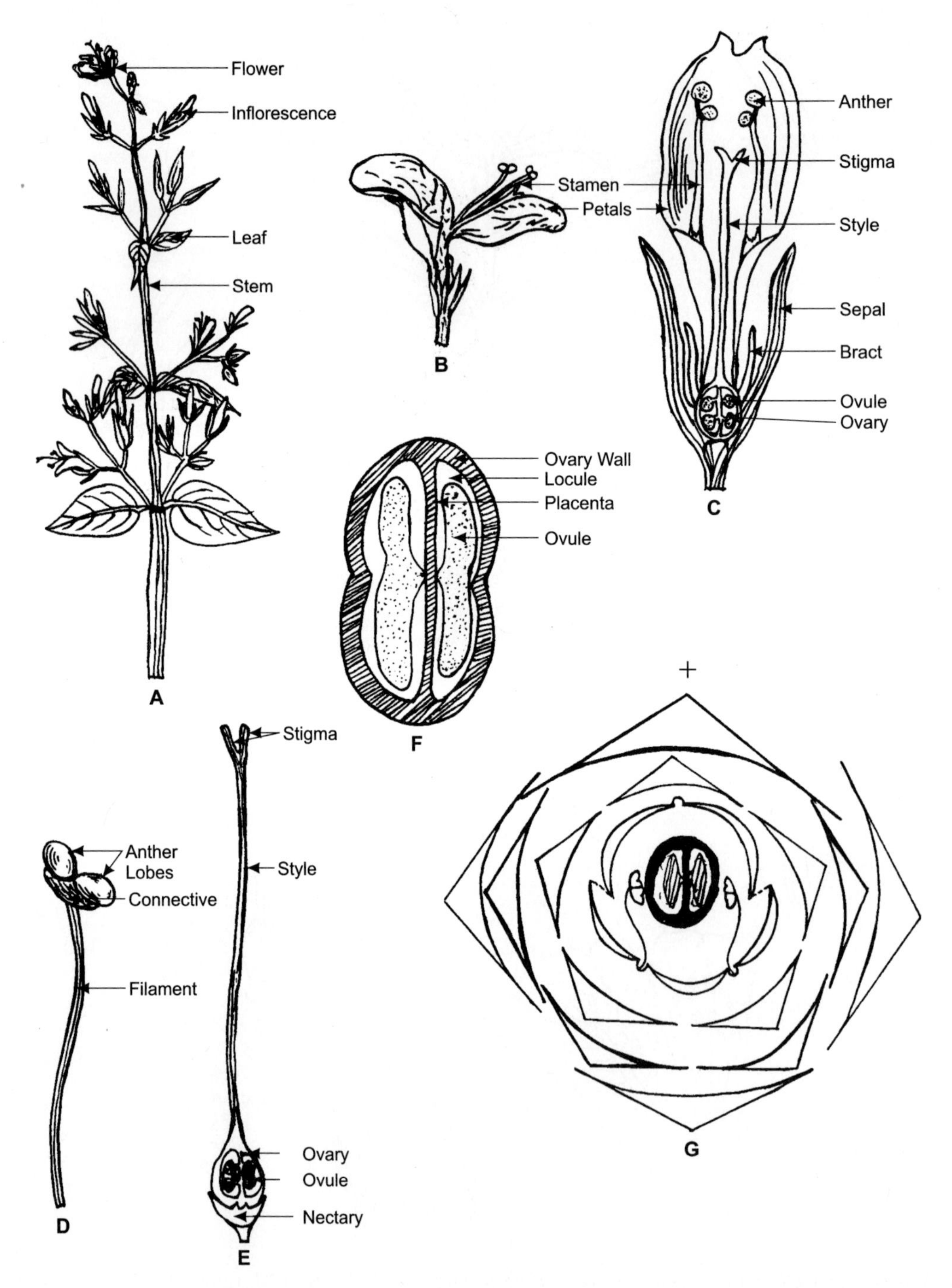

Fig. 11.27.2 Acanthaceae: **A–G** *Peristrophe bicalyculata*. **A** Flowering twig. **B** Flower. **C** Same, longisection. **D** Stamen, note: position of anther lobes. **E** Gynoecium with ovary in longisection. **F** Ovary, transverse section. **G** Floral diagram. (sketched by Uma Joshi).

Embryology. The family is multipalynous; pollen grains colpate, colporate or acolpate, 2-celled at the shedding stage. Ovules ana-, hemiana- or campylotropous, unitegmic, tenuinucellate. Polygonum type of embryo sac, 8-nucleate at maturity. Endosperm formation of the Cellular type with micropylar and chalazal haustoria.

Chromosome Number. Basic chromosome numbers are x = 7–21.

Chemical Features. The alkaloid vasicine is obtained from the dried leaves of *Adhatoda vasica.*

Important Genera and Economic Importance. *Acanthus ilicifolius* is a halophyte, growing particularly in tidal swamps. *Ruellia tuberosa*, a native of America, is common in moist places in gardens. It is an erect annual herb with bluish-pink flowers and the capsules, when mature, explode audibly. Seeds covered with hygroscopic hairs that help them to become anchored to their site of germination. *Strobilanthes dalhausianus* is a common perennial undershrub in the Western Himalayas around 2000–3000 m altitude. These plants grow in abundance, come to flower together and the seeds also mature in all the plants together. The jungle birds feed upon these seeds. Members of the type genus *Acanthus* are xerophytic; the leaves and bracts are more or less spiny, spines interpetiolar in *Barleria prionitis.* In *Mendoncia* and *Thunbergia* the calyx is reduced to an annulus. The protective function of the calyx is carried out by the large leafy bracteoles. In *Boutonia* the two bracteoles are joined together to form a tubular involucre around each axillary flower.

Barleria, *Ruellia*, *Justicia*, *Thunbergia* and *Strobilanthes* are ornamentals. The spinescent plants are sometimes grown as a hedge. *Adhatoda vasica* is medicinally important. Its active principles vasicine and adhatodic acid, obtained from the dried leaves, are components of the cough mixture, glycodin, much used as an expectorant in India. Leaves of *Andrographis paniculata* are also used medicinally against liver ailments.

Taxonomic Considerations. Acanthaceae is divided into two subfamilies depending upon the presence or absence of jaculators, i.e. the curved retinacula which support the seeds:

Subfamily Thunbergioideae—seeds without jaculators: *Nelsonia*, *Mendoncia* and *Thunbergia.*

Subfamily Acanthoideae—seeds with jaculators: *Acanthus*, *Ruellia*, *Justicia* and *Adhatoda.*

Most taxonomists presume Acanthaceae to have been derived from Scrophulariaceae or stocks ancestral to them (Lawrence 1951). Hutchinson (1969, 1973) considers it to be the most advanced taxon of the Personales. Bessey (1915) agreed with this view and treated it as the most advanced amongst the members of Scrophulariales. Cronquist (1968, 1981) also includes this family in the same order as Bessey and comments on its relationship with the family Scrophulariaceae. There has been some controversy regarding the position of the two genera *Nelsonia* and *Elytraria.* The proposal for the transfer of these genera to tribe Rhinantheae under Scrophulariaceae has been rejected by Johri and Singh (1959), Mohan Ram and Masand (1963) and P. Maheshwari (1964). Presence of jaculator (though nonfunctional) in *Elytraria* and *Nelsonia* and asymmetric development of Cellular endosperm are Acanthaceous features and not reported in Rhinantheae (see Johri et al. 1992). In addition, presence of alternate leaves, parietal placentation, endothelium, funicular obturator and albuminous seed, support their inclusion in Acanthaceae. *Elytraria* forms a link between Acanthaceae and Scrophulariaceae. This genus differs from most Acanthaceae in well-developed endosperm and its funiculus, though enlarged, does not develop into a typical jaculator. Benson (1970), Stebbins (1974) and Dahlgren (1975a, b, 1980a, 1983a) also retain it in Scrophulariales. Takhtajan (1980, 1987) reports that the Acanthaceae is closely related to the tribe Scrophularieae of the family Scrophulariaceae.

Hence, the family Acanthaceae is closely related to the Scrophulariaceae and is an advanced taxon.

Order Campanulales

An order comprising 8 families—Campanulaceae, unigeneric Sphenocleaceae and Pentaphragmataceae, Goodeniaceae, monotypic Brunoniaceae, Stylidiaceae, Calyceraceae and Compositae, the largest family of the angiosperms. These are characterised by pentamerous perianth, epipetalous stamens in a single whorl, bithecal, coherent to connate anthers, and usually unilocular ovary with a single ovule. In all the families, the ovules are anatropous, unitegmic and tenuinucellate. Iridoids are absent as a rule, except in Goodeniaceae. Some Compositae members also produce alkaloids.

The members of Campanulales are mostly perennial herbs or shrubs; sometimes may be annual; trees rare.

Compositae (= Asteraceae)

It is the largest angiosperm family including 1250 to 1300 genera and 20,000 to 25,000 species (Takhtajan 1987) distributed all over the world and in almost all habitats.

Vegetative Features. Members are much diversified, may be annual (Fig. 11.28.1 A) or perennial; xerophytes, succulents or normal mesophytes; herbs, shrubs or less commonly trees or climbers. *Espeletia hartwegiana*, and *Senecio johnstonii* from Kilimanjaro in Africa and *Vernonia arborea* are trees and *Mikania scandens* is a climber. *S. praecox* and *S. longiflorus* from southwest Africa are stem succulents. *Megalodonta beckii* is an aquatic; *S. hydrophilus* grows in wet ground or even in brackish water (Small 1919).

Normal tap root, branched and fibrous; root tubers produced in *Dahlia*, *Helianthus tuberosus* and *H. maximiliani.* Stem soft, erect or prostrate, rarely climbing, sometimes woody, usually hairy, often with milky or coloured sap (*Launaea*, *Sonchus*). Sometimes adventitious roots are borne on stem surface, e.g. *Tagetes.* Leaves radical as in *Cichorium* or cauline, alternate, rarely opposite, as in *Dahlia*, estipulate, simple or pinnatisect, e.g. *Tagetes*; smooth or hairy.

Floral Features. Inflorescence, a capitulum (Fig. 11.28.2 A) consists of a few or large number of sessile flowers arranged on the variously shaped receptacle and surrounded by one or more than one whorl of involucral bracts, that are protective in function; the receptacle may be flat disc-like, convex, concave, conical or cylindrical. Florets are either tubular or strap-shaped, i.e. ligulate. A receptacle may comprise any one type of floret or of both types:

(i) with only tubular, bisexual flowers, e.g. *Ageratum*, *Vernonia* (Fig. 11.28.1 A, B)

(ii) with only ligulate, bisexual flowers, e.g. *Launaea.*

(iii) with both types, florets—outer ligulate, neutral, and inner tubular, bisexual, e.g. *Helianthus* (Fig. 11.28.2 B, C, E).

Florets in the middle of the receptacle are disc florets and are mostly tubular. Each disc floret sessile, subtended by a scaly bract, usually hermaphrodite, actinomorphic, epigynous, pentamerous (Fig. 11.28.1 B; 11.28.2 B) except the ovary. Calyx modified into a hairy, scaly or bristly pappus, persistent and forming a parachute-like structure in mature fruit (Fig. 11.28.1 B). Corolla of 5 petals, gamopetalous, valvate. Stamens 5, epipetalous, syngenesious (Fig. 11.28.1 C); anthers basifixed, connate or coherent forming a staminal column around the style, introrse, longitudinally dehiscent (Fig. 11.28.1 B, C, F). Gynoecium syncarpous, bicarpellary, ovary unilocular, with one ovule on basal placenta (Fig. 11.28.1 E, F; 11.28.2 F) inferior; style 1, filiform, passing through the staminal column, stigma bifid, coiled (Fig. 11.28.1 D; 11.28.2 E). Fruit an achene or cypsella usually with the persistent pappus attached to it. Seeds without endosperm and with a straight embryo. Fig. 11.28.2 G represents the floral diagram of disc floret of *Helianthus.*

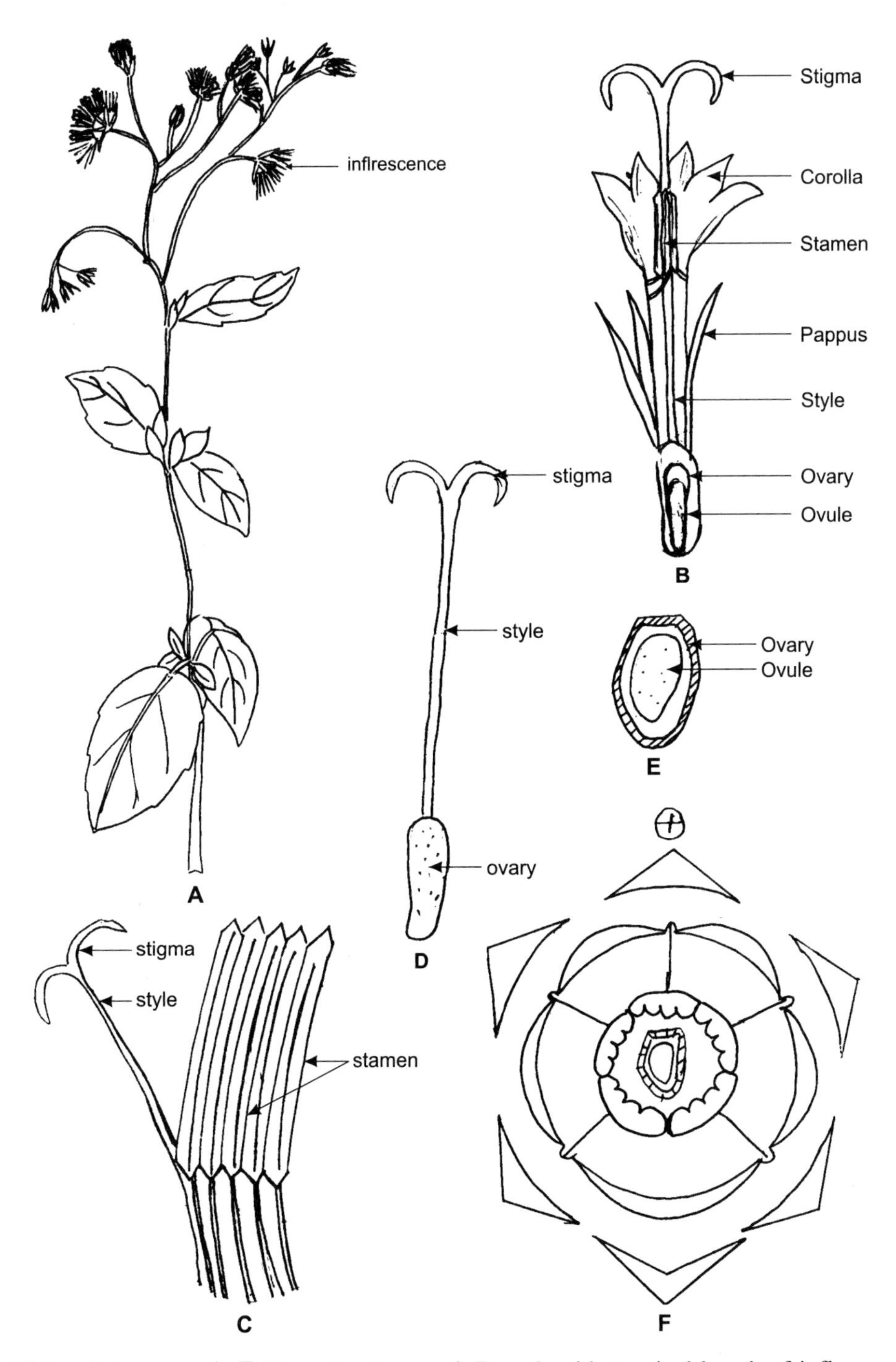

Fig. 11.28.1 Asteraceae: **A–F** *Vernonia cinerea*. **A** Branch with terminal bunch of inflorescences. **B** Flower, longisection. **C** Syngenesions stamens and style and stigma. **D** Gynoecium. **E** Ovary, transverse section. **F** Floral diagram. (sketched by Jaishree Chaudhary).

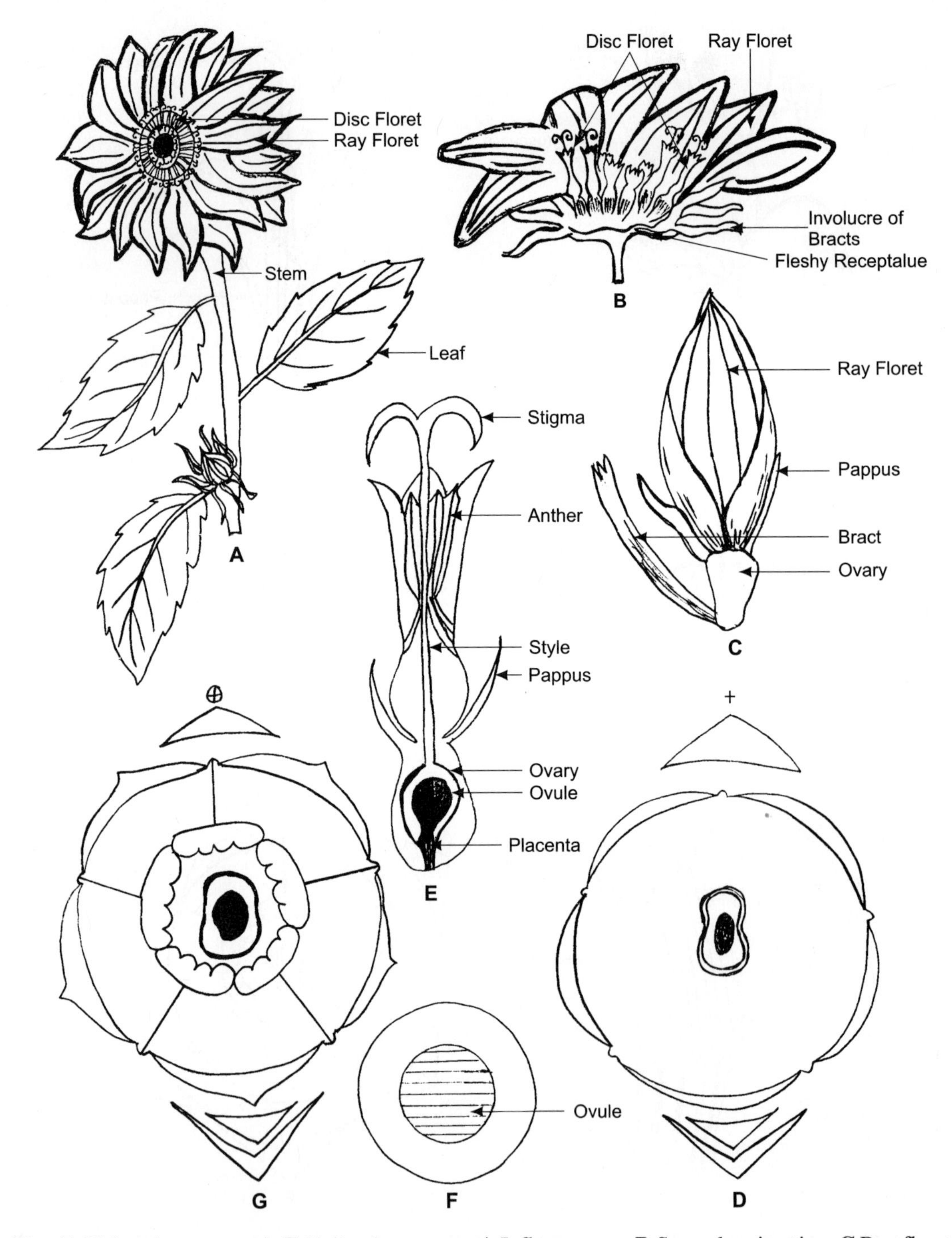

Fig. 11.28.2 Asteraceae: **A–F** *Helianthus annuus* **A** Inflorescence. **B** Same, longisection. **C** Ray floret. **D** Same, floral diagram. **E** Disc floret. **F** Ovary, transverse section. **G** Floral diagram of disc floret. (sketched by Uma Joshi).

Ray florets on the margin of the receptacle are ligulate. Calyx similar to that of tubular flowers. Corolla of 5 petals, gamopetalous, zygomorphic and ligulate (Fig. 11.28.2 E). Stamens and pistil similar to those of the tubular flowers. Mostly, the ray florets are neutral, i.e. without any stamen or pistil, sometimes pistillate (Fig. 11.28.2 C).

Anatomy. The anatomical structure of the family shows considerable diversity in correlation with their habit differences. Herbaceous stems in a transverse section usually exhibit a ring of collateral vascular bundles that are accompanied by pericyclic or bast fibers. Vessels smaller, often with radial multiples of 4 or more, sometimes with spiral thickenings; perforation plates simple and horizontal; parenchyma sparse and paratracheal. Rays 4–10 cells wide, uni- or multiseriate. Hairs both glandular and nonglandular. Presence of resin canals, secretory cavities and laticiferous vessels reported in various genera.

Leaf generally dorsiventral but variations occur such as scale leaves in *Helichrysum* and rolled leaves in *Olearia solandri.* Stomata variously distributed, mostly Ranunculaceous type, sometimes Cruciferous; absent from submerged leaves and present on both surfaces of aerial leaves in *Megalodonta beckii*, an aquatic member. Various types of anomalous secondary growth are also reported.

Embryology. Pollen grains 3- or 4-colporate, exine spinous or echinolophate; 3-celled at anthesis. Ovules anatropous, unitegmic, tenuinucellate. Polygonum type of embryo sac, 8-nucleate at maturity. Both Polygonum and Allium type of embryo sacs are seen in *Tridax trilobata*, *Sanvitalia procumbens* and *Vernonia cinerascens* (Johri et al. 1992). Fritillaria and Drusa type of embryo sac formation are also reported in many members. Pyrethrum parthenifolium type is seen in *Chrysanthemum maximum* (Johri et al. 1992).

Chromosome Number. The basic chromosome number is variable: x = 2–19 or more in various members, x = 9 being the most common number.

Chemical Features. Compositae members are rich in sesquiterpenes and polyacetylenes. Seco-iridoids are absent. An oleoresin produced by secretory ducts of *Artemisia campestris* subsp. *maritima* contains terpenoids, alkaloids, fatty acids and polyacetylenes (Ascensā and Pais 1988).

Important Genera and Economic Importance. A number of Compositae are obnoxious weeds, such as *Ageratum conyzoides*, *Blumea mollis*, *Eclipta prostrata*, *Erigeron bonariensis*, *Gnaphalium indicum*, *Launaea asplenifolia*, *L. nudicaulis*, *Parthenium hysterophorous*, *Sonchus asper*, *Sphaeranthus indicus*, *Tridax procumbens*, *Xanthium strumarium* and others.

Because of the large, showy capitula many genera are cultivated as ornamentals: *Aster*, *Calendula*, *Chrysanthemum*, *Dahlia*, *Helianthus*, *Helichrysum*, *Gerbera*, *Tagetes* and others. Compositae are also important for their food value. Many genera, e.g. *Scolymus hispanicus* in Spain, *Scorzonera hispanica* from Europe to Central Asia, and *Tragopogon porrifolius* of southern Europe are grown for their edible roots (Datta 1988). The roots of *Cichorium intybus* (commonly called chicory) are used as an adulterant of coffee in powdered form. Leaves of *Lactuca sativa* (lettuce) and *Cynara cardunculus* (cardoon) as also tubers of *Helianthus tuberosus* (Jerusalem artichoke) are edible.

The family is the source of a large number of drug plants also. Santonine used against intestinal worms is obtained from *Artemisia cina*, *A. maritima* and *A. nilagarica. Matricaria chamomilla* is the source of the medicine chamomile. Dried flower heads of *Chrysanthemum cinerariefolium* are the source of the insecticide Pyrethrum.

Oil is obtained from the seeds of *Carthamus oxycantha*, *Guizotia abyssinica* and *Helianthus annuus.* Latex of *Parthenium argentatum* of South America, *Solidago laevenworthii* and *Taraxacum kok-saghys* (in Russia) yield rubber.

Some of the tree members like *Brachylaena huillensis* and *Montanoa quadrangularis* (Colombia and Venezuela) and *Vernonia arborea* (Assam, India) furnish timber. The wood of *Tarchonanthus camphoratus* (South Africa) is used for musical instruments.

A red dye is extracted from the flowers of *Carthamus tinctorius, Tagetes erectus* and *Adenostemma tinctorium.*

A number of genera growing as weeds in pastures have poisonous effect on livestock, e.g. *Senecio, Xanthium* and *Solidago*. The pollen grains of *Ambrosia artemisifolia, A. trifida* and *Solidago* spp. are responsible for hay fever.

Taxonomic Considerations. Bentham and Hooker (1965b) placed Compositae in Asterales of Series Inferae in the Gamopetalae. Hutchinson (1969, 1973), Cronquist (1968, 1981), Dahlgren (1980a) and Takhtajan (1987) also agree to this assignment. However, Engler and Diels (1936) and Melchior (1964) include Compositae in Campanulales.

There has been diversified opinion about the phylogeny and evolution of this family. All taxonomists agree that it is the most advanced taxon of the dicots. It indicates similarity with the families Dipsacaceae and Valerianaceae, because of the similar type of inflorescences, inferior ovary, and solitary ovule, and also some embryological features like the embryo development (Asterad type in *Valeriana alitoris, Centranthus ruber*, *C. angustifolia* of Valerianaceae and *Scabiosa succisa* of Dipsacaceae), and presence of endothelium (Deshpande 1970). However, the pollen grains of Compositae are more akin to those of Brunoniaceae, Calyceraceae and Goodeniaceae and differ from those of Campanulaceae, Dipsacaceae, Stylidiaceae and Valerianaceae (G Erdtman 1952).

Many taxonomists are of the opinion that the Compositae have evolved from the Campanulaceae through the subfamily Lobelioideae, but Campanulaceae differ from the Compositae in dichasial or monochasial cyme inflorescence, pentacarpellary (Campanuloideae) or bicarpellary (Lobelioideae) ovary, axile placentation and numerous ovules. On the other hand, their pollen presentation mechanism is similar. Also, both families contain polyacetylenes (Jensen et al. 1975, Dahlgren 1980a). According to Cronquist (1968,1981): "Only the Rubiales-Dipsacales complex has the characters necessary for a near ancestor of the Compositae". Comparative morphological studies show that the ancestral prototypes of Compositae must have been woody plants. Similar pollen dispersal mechanism is seen in some Rubiales (Rubiaceae) also. In many Rubiaceae, capitulum inflorescence occurs. Phenolic compounds widespread in Compositae are also present in many Rubiales and Dipsacales but not in Campanulales.

Stebbins (1977) concluded: "The Compositae cannot be regarded as descended from or closely related to any other modern family".

From the above discussion it is clear that the Compositae is no doubt the most highly evolved family and in all probability is polyphyletic with more than one ancestral form.

The important features of this family are:

1. Mainly herbaceaous; shrubs and trees make up for only 1.5% taxa.
2. 25,000 species are approximately 10% of the dicots and occur in every conceivable spot on the earth's surface, though somewhat rare in tropical rain forests.
3. Flowers in a capitate inflorescence are a highly advantageous situation. A single insect can pollinate several flowers at a time. If, however, cross-pollination fails, the curling back of stigmas helps in self-pollination.
4. Pollen mechanism, pollen protection and nectar placement are such that any specialised insect is not required.
5. The ripening fruits are protected by the involucral bracts which bend inwards. A perfect mechanism for seed dispersal by wind is the presence of hairy pappus.

MONOCOTYLEDONS

Order Helobiae

The order Helobiae comprises 4 suborders and 9 families:

1. Suborder Alismatineae—Alismataceae and Butomaceae.
2. Suborder Hydrocharitineae—Hydrocharitaceae.
3. Suborder Scheuchzeriineae—Scheuchzeriaceae.
4. Suborder Potamogetonineae—Aponogetonaceae, Juncaginaceae, Potamogetonaceae, Zanniche-lliaceae and Najadaceae.

Most of the members are aquatic herbs, submerged or of marshy habitats. Flowers solitary or in simple or compound inflorescences, more or less enclosed in a spathe; uni- or bisexual, regular, naked or with a single or double perianth. Stamens and carpels one to numerous, carpels superior or inferior, usually free. Seeds non-endospermous, embryo large, with a strongly developed hypocotyl. Oxalate raphides and silica bodies absent, vessels absent from stems. Pollen grains 3-celled at dispersal stage.

Potamogetonaceae

A family of 8 genera (including *Zostera*, *Posidonia* and *Cymodocea*) and about 125 species, distributed widely all over the world and grow in oceanic coastal regions, brackish tidal waters as well as freshwater bodies such as ponds, pools, lakes, rivers, streams and also bogs or marshy areas.

Vegetative Features. Perennial, aquatic herbs (Fig. 11.29 A) rarely grow in marshy areas. Stem rhizomatous spreading on the ground surface under water, often jointed and nodose, the lower nodes root-bearing and upper ones foliaceous; leaves with sheathing base, sheath often apically ligulate, distichous, alternate or less often opposite; sessile or petiolate, often vaginate at the base, stipulate.

Floral Features. Flowers solitary, spicate (11.29 A, B) or cymose, bisexual (Fig. 11.29 B, C), actinomorphic, often borne in a spathe. Perianth absent or of 4 to 6, small, herbaceous or membranous, free or fused segments; stamens 1 to 6, extrorse, sessile (Fig. 11.29 E), mono- or bithecous anthers. Gynoecium of 1 to 6 carpels, free or rarely fused at base, each ovary superior, unilocular, uniovulate (Fig. 11.29 D), ovule usually pendulous from the apex. Fruit coriaceous, subwoody or membranous, 1-seeded drupelets or nutlets; seeds without endosperm, embryo axile.

Anatomy. Vessels absent in leaf and stem; in roots the vessels are with scalariform perforation plates. Sieve tube plastids of monocotyledonous type[11] occur in *Potamogeton* (Behnke 1969). Stomata of Rubiaceous type; occur only on upper surface of leaves.

Embryology. Pollen grains inaperturate, tricolpate, globose with reticulate sculpturing thread-like in *Cymodocea*; 3-celled at shedding stage. Ovules ana-, ortho- or campylotropous; bitegmic, crassinucellate; micropyle formed by inner integument only. Polygonum type of embryo sac, 8-nucleate at maturity. Endosperm formation of the Helobial type in *Potamogeton* and *Ruppia*, Endosperm formation precedes embryogenesis (Takaso and Bouman 1984).

Chromosome Number. Basic chromosome number is x = 13–15 in *Potamogeton* and *Groenlandia*; x = 8 in *Ruppia*.

---

[11]All monocotyledons investigated have the same type of sieve tube plastids: a P-type with cuneate (triangular) crystalloid bodies, generally in a considerable number per plastid (Dahlgren and Clifford 1982).

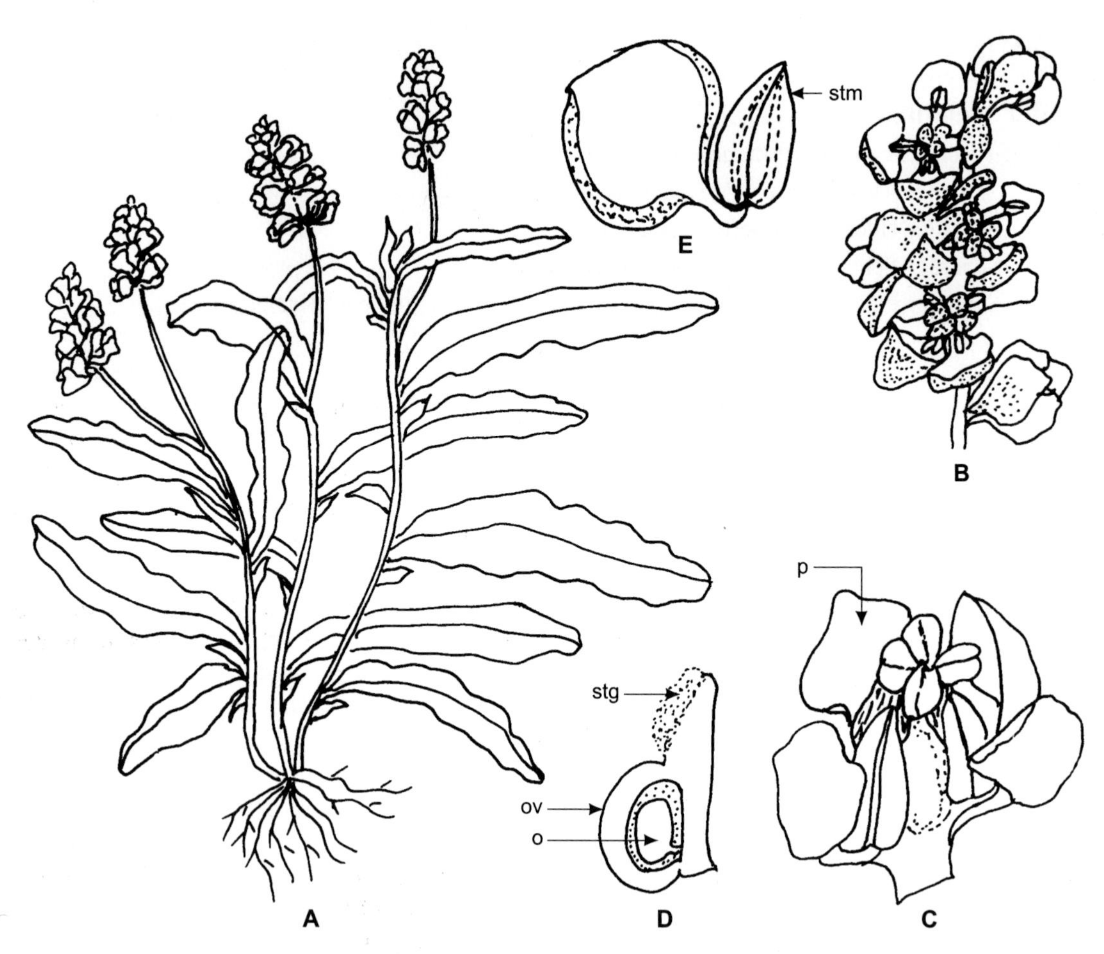

Fig. 11.29 Potamogetonaceae: **A–E** *Potamogeton crassipes* **A** Habit. **B** Inflorescence. **C** Flower. **D** Carpel, vertical section. **E** Perianth lobe with stamen attached. *o* ovule, *ov* ovary, *p* perianth, *stg* stigma, *stm* stamen. (Adapted from Lawrence 1951)

Important Genera and Economic Importance. All the known genera are aquatic weeds. *Ruppia tuberosa*—a dark green, firmly rooted plant with its rhizome completely buried—occupies hypersaline habitats in western Australia (Davis and Tomlinson 1974). *Posidonia oceanica* is a bioindicator of mercury contamination in marine environments (Maserti et al. 1988).

Taxonomic Considerations. Configuration of the family Potamogetonaceae has been rather varied. It is a heterogeneous group and has been treated variously by different authors. Bentham and Hooker (1965c) included all the genera in the Najadaceae; Engler and Diels (1936) recognised Potamogetonaceae; Benson (1970) assigned all the genera to the Zosteraceae. Lawrence (1951) recognised 8 genera of the Potamogetonaceae; Hutchinson (1969, 1973) assigns them to 4 different families—*Cymodocea*, *Diplanthera* and *Zannichellia* to Zannichelliaceae, *Potamogeton* to Potamogetonaceae, *Zostera* to Zosteraceae, and *Ruppia* to Ruppiaceae. Huber (1969) also recognises 4 families: *Ruppia*, *Groenlandia* and *Potamogeton* in Potamogetonaceae, *Zostera*, *Heterozostera* and

Phyllospadix in Zosteraceae, *Zannichellia*, *Althenia*, *Lepilaena* and *Vleisia* in Zannichelliaceae, and *Cymodocea*, *Diplanthera*, *Syringodium*, *Amphibolis*, *Halodule* and *Thalassodendron* in Cymodoceaceae. Dahlgren and Clifford (1982) also recognise these 4 families. Takhtajan (1987) raises them to the rank of order and includes 5 families: Potamogetonaceae, and Ruppiaceae in Potamogetonales, Zosteraceae in Zosterales, and Cymodoceaceae and Zannichelliaceae in Cymodoceales. The four families as recognised by Huber (1969) and Dahlgren and Clifford (1982) and also the fifth one—Ruppiaceae (Takhtajan 1987) —are so distinct from each other in their morphological (Singh 1965) as well as embryological features (Lakshmanan 1970d), and basic chromosome number, that it will be better to recognise them as distinct families under a separate order, i.e. Zosterales, as suggested by Dahlgren and Clifford (1982).

Order Liliiflorae

The order Liliiflorae comprises 5 suborders and 17 families:

1. Suborder Liliineae includes Liliaceae, Xanthorrhoeaceae, Stemonaceae, Agavaceae, Haemodoraceae, Cyanastraceae, Amaryllidaceae, Hypoxidaceae, Velloziaceae, Taccaceae and Dioscoreaceae.
2. Suborder Pontederiineae comprises the family Pontederiaceae.
3. Suborder Iridineae includes Iridaceae and Geosiridaceae.
4. Suborder Burmanniineae includes Burmanniaceae and Corsiaceae, and
5. Suborder Phylidrineae includes the family Philydraceae.

Herbs, rarely shrubs or woody, sparingly branched 'trees' without, or in several families with, secondary growth. Stems underground, mostly modified as rhizomes, corms or bulbs. Leaves usually alternate or rarely opposite or verticillate, linear to lanceolate, sessile, generally sheathing at base, lamina entire or rarely compound or digitately lobed as in Dioscoreaceae and Taccaceae; venation mostly parallel, reticulate in some.

Flowers hypo- or epigynous, actino- or zygomorphic, tepals normally petaloid. Stamens usually 6, sometimes 3, 2 or 1 as in Philydraceae, or rarely up to 9 or more as in *Vellozia* and *Pleea.* Gynoecium syncarpous, tricarpellary, placentation variable. Fruits generally capsules or berries, seeds numerous, endospermous.

Steroid saponins present.

Liliaceae

A large family of 240 genera and 4000 species widely distributed especially in the warm temperate and topical regions of the world. Except for some xerophytic representatives, Liliaceae members do not form a dominant vegetation anywhere. *Astelia solanderi* is an epiphytic member from New Zealand (Ambasht 1990).

Vegetative Features. Mostly perennial, herbs (Fig. 11.30 A, E, F) with rootstock modified to bulb, corm, tuber or rhizome; sometimes climbers (*Asparagus*, *Smilax*) and infrequently woody (*Dracaena*, *Yucca*); stem often prickly, as in *Smilax* and *Asparagus racemosus*; sometimes modified to subterranean storage organs or cladophylls (Fig. 11.30 E). Leaves radical or cauline, alternate or in rosettes, opposite only in *Scolyopus*, mostly lamellate, sometimes reduced to spines (*Asparagus*) or scale leaves (*Ruscus*), sometimes fleshy (*Aloe*) or with prickly tip and margin (*Yucca*); commonly parallel veined (parallel reticulate in *Smilax*).

Floral Features. Inflorescence a scapigerous raceme, spike or umbel (*Allium cepa*), sometimes cymose. Flowers actinomorphic, weakly zygomorphic in *Gilliesia*, *Haworthia* and *Hemerocallis*; bisexual, rarely unisexual as in *Smilax*, *Lomandra*, *Ruscus*, hypogynous, bracteate, bracts small, scarious or spathaceous; ebracteate, often highly ornamental. Perianth mostly large and showy, of 6 tepals in 2 whorls of 3 each, rarely 4 or more than 6 as in *Paris* (Fig. 11.30 A-D), generally not distinguishable into calyx and corolla, free or basally connate into a tube, imbricate or the outer whorl valvate. Stamens 6, rarely 3 as in male flower of *Ruscus*, one whorl is suppressed; filaments free or adnate to tepals (Fig. 11.30 G) often coloured as in *Gloriosa*: anthers bicelled, basifixed (Fig 11.30 B, D) or versatile as in *Lilium canadense*, extrorse or introrse, usually dehisce longitudinally. Gynoecium syncarpous, tricarpellary; ovary superior, half-inferior in *Mondo* (Dutta 1988), trilocular with axile placentation or rarely unilocular with parietal placentation; ovules numerous, biseriate; style usually 1, divided or trifid,

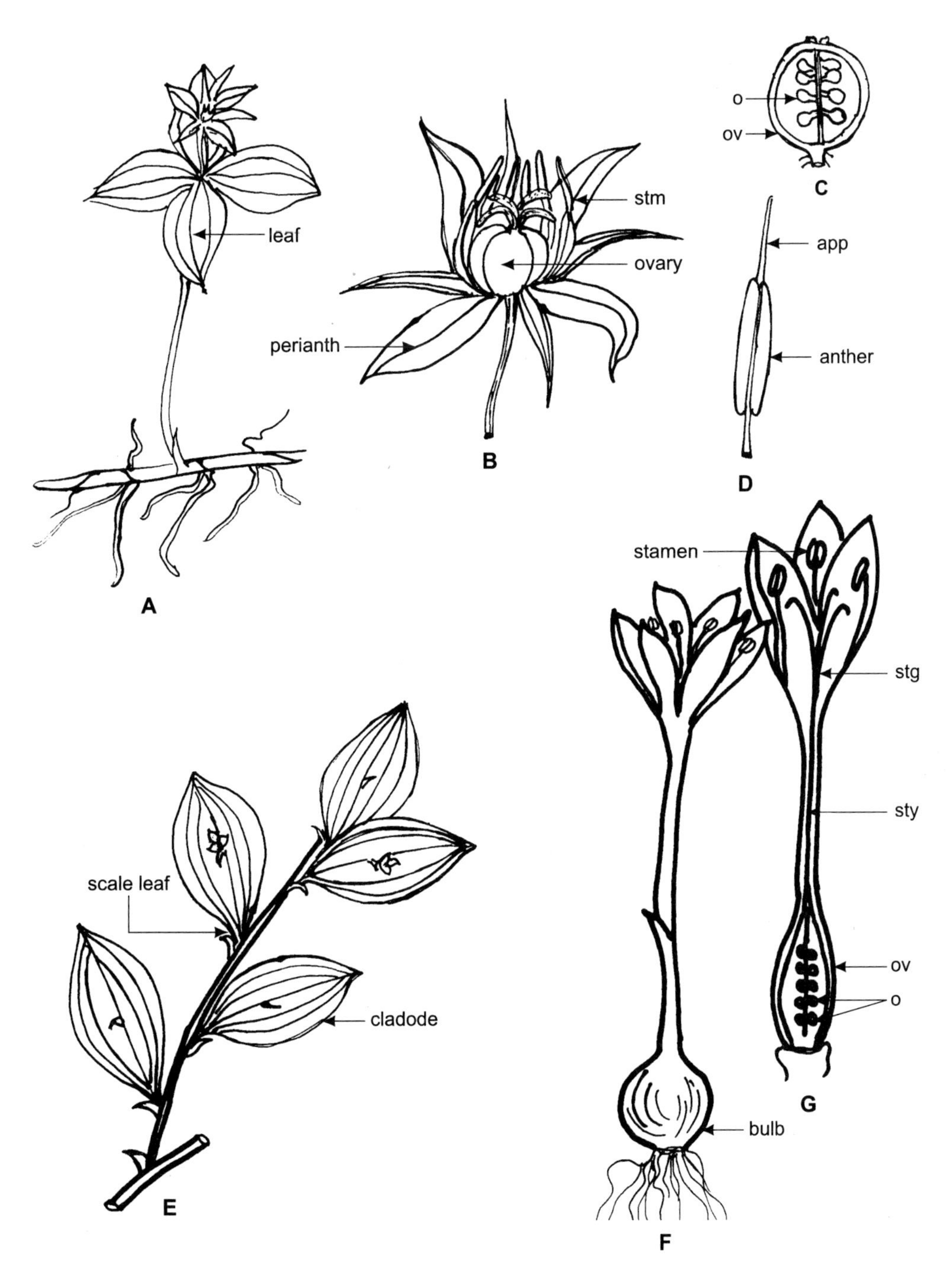

Fig. 11.30 Liliaceae: **A–D** *Paris quadrifolia*, **E** *Ruscus aculeata*, **F, G** *Colchicum autumnale.* **A** Flowering plant. **B** Flower. **C** Ovary, longisection. **D** Anther. **E** Twig of *R. aculeata.* **F** Flowering plant of *C. autumnale.* **G** Flower, longisection. (*app* appendage, *o* ovule, *ov* ovary, *stg* stigma, *stm* stamen, *sty* style). (Original).

stigmas 3 or 1 with 3 lobes. Fruit a loculi- or septicidal capsule or a berry. Seeds with small embryo and abundant endosperm.

Anatomy. Based on the observations of Cheadle and Kosakai (1971), vessels in Liliaceae occur in roots of all species and in the stems of at least some species of the tribes Asphodeleae, Herrerieae, Johnsonieae, Polygonateae, Tricyrtideae and Asparageae. Vessels in stems are almost invariably primitive except in *Tricoryne elatior*, where some vessels are with simple perforation plates. In roots they vary from highly specialized forms to very primitive forms. Raphides present in some members like *Lilium* and *Lloydia* (Dutta 1988). Leaves isobilateral, mostly with parallel venation.

Embryology. The family is multipalynous. Pollen grains trichotomocolpate, 1- or 2- or more-colpate or -porate and spiraperturate (PKK Nair 1970); 3-celled at dispersal stage (2-celled in *Urginia indica* and some others). Ovules anatropous, bitegmic, tenui- or crassinucellate; usually inner integument forms the micropyle. Embryo sac (see Johri et al. 1992) of Polygonum type (8-nucleate), Allium type (8-nucleate), Endymion type (8-nucleate), Adoxa type (8-nucleate), Fritillaria type (8-nucleate) or Drusa type (16-nucleate). Endosperm formation is of the Helobial or the Nuclear type.

Chromosome Number. Basic chromosome number is x = 12.

Chemical Features. Anthraquinone emodin occurs in the members of this family (Harborne and Turner 1984). Organic sulphides are present in various cultivated species of *Allium: A. cepa* (onion), *A. chinense* (rakkyo), *A. porrum* (leek), *A. sativum* (garlic) and *A. schoenoprasum* (chive) (Harborne and Turner 1984). The alkaloid colchicine is obtained from *Colchicum autumnale.*

Important Genera and Economic Importance. The family Liliaceae includes a large number of genera, many of these are cultivated as ornamentals: species of *Agapanthus*, *Brodiaea*, *Convallaria*, *Hemerocallis*, *Lilium*, *Scilla*, *Tulipa*, *Yucca* and many others. *Sansevieria* named after Raimond de Sangro, Prince of Sanseviero, is very abundant on the coast of Guinea and other parts of Africa. They also abound in Sri Lanka, Peninsular India and along the Bay of Bengal, extending to Java and to the coast of China. Some are important food-yielding genera, such as various species of *Allium: A. cepa*, *A. sativum*, *A. porrum*, and *A. schoenoprasum. Asparagus officinalis* yield edible young shoots; *Chlorophytum arundinaceum* produce edible roots; *Ophiopogon japonicus* yield edible tubers and *Lapageria rosea* edible fruits. Some have medicinal importance, *Aloe africana* and *A. barbadensis* yield aloin used in the drug trade; *Colchicum autumnale* yields colchicine, also used as a drug; it induces polyploidy in plants. Roots of various species of *Smilax* are the source of the drug sarsaparilla and the rhizomes of *Veratrum album* produce the drug veratrin. The red squill, used in rodent control, is obtained from the bulbs of *Urginea.* and from *Scilla* bulbs a rat poison is obtained.

The leaves of *Phormium tenax* yield a fibre, New Zealand flax, those of *Sansevieria roxburghiana* yield bow-string hemp and *Yucca filamentosa* also yield a tenacious fibre. The scented flowers of *Hyacinthus orientalis* are the source of the perfume hyacinth. *Dracaena spicata* (dragon tree) and *Yucca gloriosa* (Adam's needle) are erect woody shrubs with anomalous secondary growth. *Aloe indica* and *A. perfoliata* (Indian Aloe) are herbs with basal rosettes of thick fleshy leaves, contain mucilage which is of medicinal value. Glory lily or *Gloriosa superba* is a cultivated climber, the leaf tips are tendriller. *Asphodelus tenuifolius* is a winter weed. *Asparagus racemosus* and *Hemerocallis fulva* are garden plants.

Taxonomic Considerations. In Bentham and Hooker's (1965c) system of classification, the family Liliaceae is included in the third series Coronarieae. Engler and Diels (1936) treated it as a member of suborder Liliineae of order Liliiflorae. Cronquist (1968, 1981), Dahlgren (1975a, 1980a, 1983a), Hutchinson (1959), Takhtajan (1980, 1987) and Thorne (1968,1983) include it in the Liliales.

Hutchinson (1959) considers Liliaceae to be the stock from which many other families of monocots evolved directly or indirectly. Although formerly regarded as a primitive family, it is the most typical family of the Monocotyledons.

Cronquist (1968, 1981) includes the family Amaryllidaceae in the Liliaceae. The Liliaceae differ from the Amaryllidaceae in having superior ovary and 6 stamens. The two families are connected by the genus *Allium* (according to some Alliaceae), which has superior ovary, as in Liliaceae and umbellate inflorescence, as in Amaryllidaceae.

There is much controversy over the circumscription of this family. Hutchinson (1959, 1973) includes the tribes Agapanthieae, Allieae and Gillesieae in his Amaryllidaceae because of umbellate inflorescence. This view is substantiated by the palynological study of *Allium* by Maia (1941) and anatomical study of Monocotyledons by Cheadle (1942).

The Liliaceae have probably origniated from the Helobieae or its ancestors as some members show intermediate features, e.g. *Petrosavia* (Liliaceae) has partially-fused carpels, as in Helobieaae.

The Liliaceae is more or less a natural taxon. Further breakup into Smilacaceae, Alliaceae, Agapanthaceae and some others (Dahlgren et al. 1985) might not be necessary.

Amaryllidaceae

A family of 60-65 genera and 900 species, cosmopolitan in distribution. Majority of the members occur in the plains, plateaus and steppe areas of the tropics and subtropics (Lawrence 1951).

Vegetative Features. Perennial herbs with a rhizomatous, bulbous or cormous rootstock and aerial stem, a scape. Leaves mostly linear or lorate and basal rosettes, rarely cauline.

Floral Features. Inflorescence mostly umbellate, racemose or paniculate, sometimes reduced to single flower (*Zephyranthes*), subtended by a scarious spathe and borne on a leafless scape (Fig. 11.31 A). Flowers bisexual (Fig. 11.31 B, G), actino- or rarely zygomorphic, epigynous. Perianth of 6 tepals in 2 whorls of 3 each, sometimes polyphyllous (*Leuconium*) and in some genea (*Narcissus*) bear a cup-like corona (Fig. 11.31 F, G). Stamens 6, inserted on the perianth, in 2 whorls of 3 each, filaments short and thick or long and filamentous, usually free but in some genera filaments connected basally by a corolliform velamen or staminal corona. In *Vagaria* and *Urceolinia* the corona is represented by distinct or indistinct tooth on either side of filament base, in *Proiphys* and *Crinum* by membranes that are shortly united; anthers bicelled, basifixed or versatile, as in *Crinum*, or dorsifixed as in *Zephyranthes* (Fig. 11.31 D), dehiscence by vertical slits, rarely by terminal pores, as in *Galanthus*, usually introrse, sometimes extrorse. Gynoecium syncarpous, tricarpellary, ovary inferior, trilocular with axile placentation (Fig. 11.31 E) or unilocular with basal placentation, as in *Calostemma* or unilocular with parietal placentation, as in *Leontochir*; ovules usually many but only 1 in *Choananthus*. Style 1, stigmas 3 or 1 and 3-lobed as in *Zephyranthes* (Fig. 11.31 B, C). Fruit usually a 3-celled capsule or a berry as in *Clivia*, *Cryptostephanus*, *Haemanthus*; seeds many with small embryo in fleshy endosperm.

Anatomy. Similar to that of the members of Liliaceae.

Embryology. Pollen grains monosulcate (disulcate in *Crinum,* Dutt 1970b); 2-celled at shedding stage. Ovules bitegmic, anatropous, crassinucellate, tenuinucellate in *Zephyranthes*; the inner integument forms the micropyle. Embryo sac of Polygonum type, 8-celled at maturity. Endosperm formation of the Helobial or Nuclear type.

Chromosome Number. Basic chromosome numbers are x = 6–15, 23, 27, 29.

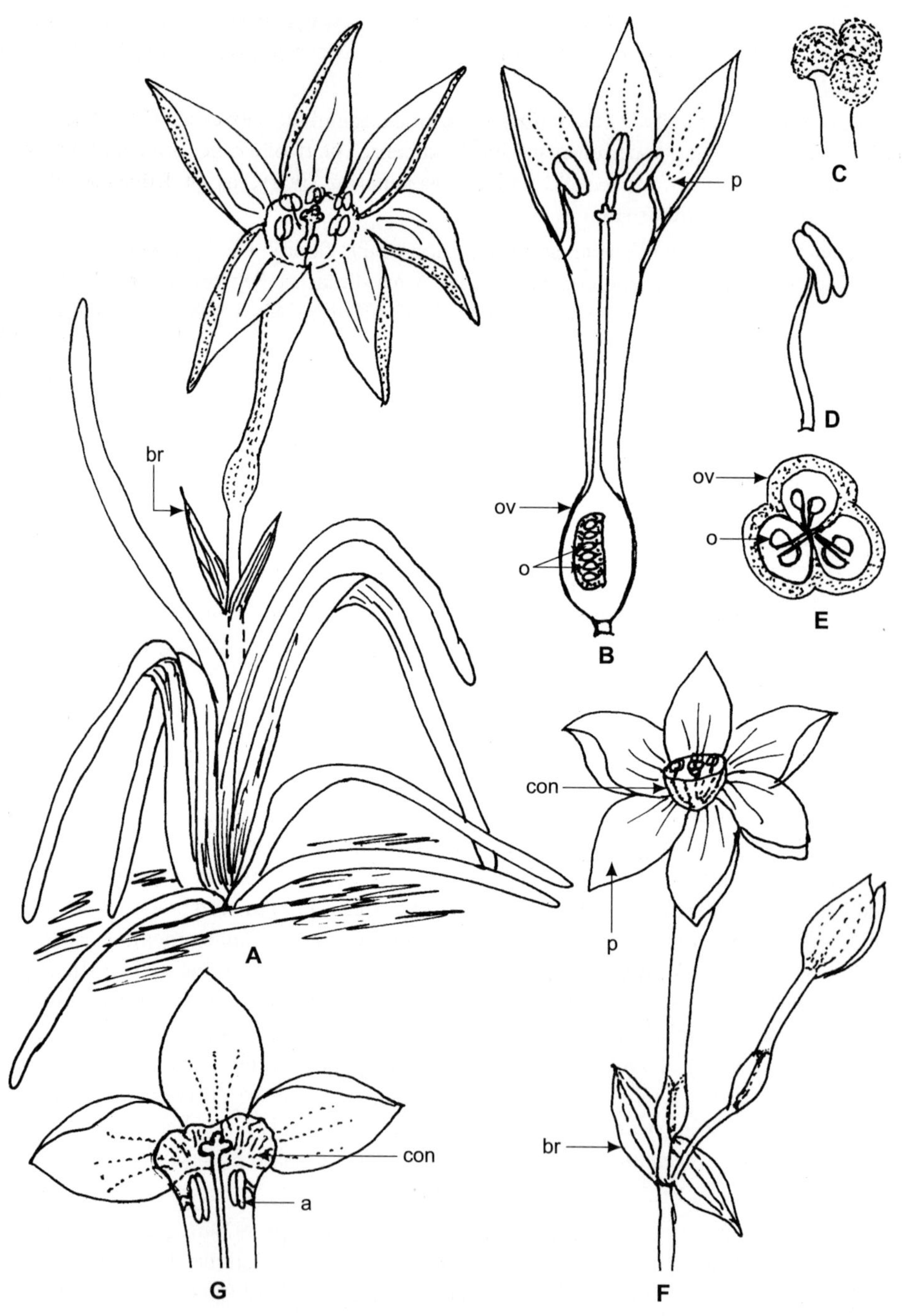

Fig. 11.31 Amaryllidaceae: **A-E** *Zephyranthes grandiflora*, **F, G** *Narcissus pseudonarcissus*. **A** Flowering plant. **B** Flower, longisection. **C** Stigma. **D** Stamen. **E** Ovary, cross section. **F, G** *N. pseudonarcissus*, part of inflorescence (**F**). and flower, longisection, carpel removed (**G**). (*a* anther, *br* bract, *con* corona, *o* ovule, *ov* ovary, *p* perianth). (Original).

Chemical Features. *Narcissus pseudonarcissus* contains the galanthamine tazettine. Amryllidaceae members are particularly rich in steroidal saponins (Dahlgren and Clifford 1982, Harborne and Turner 1984). The polysaccharide levan is the storage carbohydrate in about 15 species of this family (Pollard 1982).

Important Genera and Economic Importance. The family consists of many well known ornamental species such as *Crinum*, *Pancratium*, *Amaryllis*, *Clivia*, *Eucharis*, *Eurycles*, *Haemanthus*, *Hymenocallis*. *Lycoris*, *Narcissus*, *Nerine* and *Zephyranthes*.

Taxonomic Considerations. Bentham and Hooker (1965c) included Amaryllidaceae in the series Epigynae of the Monocotyledoneae, Engler and Diels (1936) in the order Liliiflorae, Hutchinson (1959) in the order Amaryllidales, Dahlgren (1980a, 1983a) in the order Asparagales and Takhtajan (1987) in the Liliales. Cronquist (1968) and Thorne (1968) do not recognise Amaryllidaceae as a distinct family and include its members in the Liliaceae.

The family Amaryllidaceae s.l. is a heterogeneous assemblage. Segregates of this family are Alliaceae (as suggested by Hutchinson), Agavaceae, Hypoxidaceae, Alstroemeriaceae, Taccaceae and Velloziaceae. Alliaceae is also recognised as a separate family by Takhtajan (1966, 1980, 1987) and Dahlgren (1975a, 1983a).

The Amaryllidaceae is closely related to Iridaceae (by the nature of ovary) and to the Liliaceae; the genus *Agapanthus* forms a common link between them.

Bose (1962) observes that gene mutation, chromosome repatterning, polyploidy, hybridization and apomixis have played an important role in the evolution and speciation of this family. Flory (1977) also agrees to the fact that hybridization is an important factor in the evolution of Amaryllidaceae.

Order Commelinales

The order Commelinales comprises four suborders and eight families:

1. Suborder Commelinineae—includes Commelinaceae, Mayacaceae, Xyridaceae and Rapateaceae.
2. Suborder Eriocaulineae—Eriocaulaceae.
3. Suborder Restionineae—Restionaceae and Centrolepidaceae.
4. Suborder Flagellariineae—Flagellariaceae.

Perennial or annual herbs, often of freshwater habitats (Mayacaceae, Eriocaulaceae), often succulent (Commelinaceae). Leaves simple, alternate or spirally arranged, with sheathing base. Inflorescence variable. Flowers actino- or zygomorphic, trimerous, hypogynous; with distinct sepals and petals; anthers basifixed, introrse, tetrasporangiate, longitudinally dehiscent (poricidal in Mayacaceae). Gynoecium syncarpous, tricarpellary, ovary 3- (rarely 2)- locular (unilocular in Mayacaceae), placentation axile or parietal. Fruit a capsule, rarely a berry; seeds sometimes arillate; embryo often conical, under a disc-like structure, "embryostega", beneath the seed-coat; endosperm mealy. Stomata with 4-6 subsidiary cells. Silica bodies and calcium oxalate present. Endosperm formation of the Nuclear type.

Steroid saponins occur in some members (Commelinaceae).

Commelinaceae

A moderately large family of 47 genera and 700 species, tropical and subtropical in distribution. Tribe Tradescantieae is abundant in the New World and tribe Commelinineae is represented in tropical Africa.

Vegetative Features. Perennial or annual, succulent herbs with jointed stems; roots fibrous or sometimes much thickened and tuber-like. Leaves alternate, flat, entire, sessile, linear or lanceolate to ovate, parallel-veined and with sheathing base (Fig. 11.32 A).

Floral Features. Inflorescence terminal, terminal and axillary or only axillary (*Rhoeo spathacea*), simple or compound helicoid cyme or thyrse, usually subtended by a boat-shaped or cymbiform spathe as in *Commelina* (Fig. 11.32 A), and *Tradescantia* or foliaceous, often with coloured bracts. Flowers bisexual, actinomorphic (Fig. 11.32 B, D) (subfamily Tradescantieae) or rarely zygomorphic, hypogynous. Perianth distinguished into calyx and corolla. Calyx usually green (petaloid in *Commelina* and *Rhoeo*), polysepalous, imbricate, rarely gamosepalous. Corolla of 3 petals, polypetalous, equal or unequal, rarely united basally, 1 petal often reduced in size or suppressed. Stamens usually 3 + 3 or only 3, the upper ones often staminodial or 1 whorl missing; anthers basifixed, introrse, tetrasporangiate, dehiscing longitudinally, rarely poricidal as in *Dichorisandra*; filaments free, often covered with coloured, moniliform hairs as in *Tradescantia virginiana* (Fig. 11.32 D, E). Gynoecium syncarpous, tricarpellary, ovary superior, usually trilocular (Fig. 11.32 C) (rarely bilocular), ovule 1 to a few per locule, placentation axile; style 1, stigma 1, capitate or trifid. Fruit a loculicidal capsule. Seeds with a punctiform to linear funicular scar, usually netted, muricate or ridged, sometimes arillate as in *Dichorisandra*, endospermous, endosperm mealy, embryo situated under a disc-like structure (embryostega) beneath the seed coat.

Anatomy. Vessels in stems and leaves with simple perforation plates. Unicellular or uniseriate trichomes present; stomata on both surfaces, usually surrounded by 4–6 subsidiary cells. Raphides and silica bodies present.

Embryology. Pollen grains simple, usually sulcate, 2- or rarely 3-celled at shedding stage. Ovules ortho-, ortho-campylo- or hemianatropous, bitegmic, tenui- or crassinucellate. Micropyle formed by inner

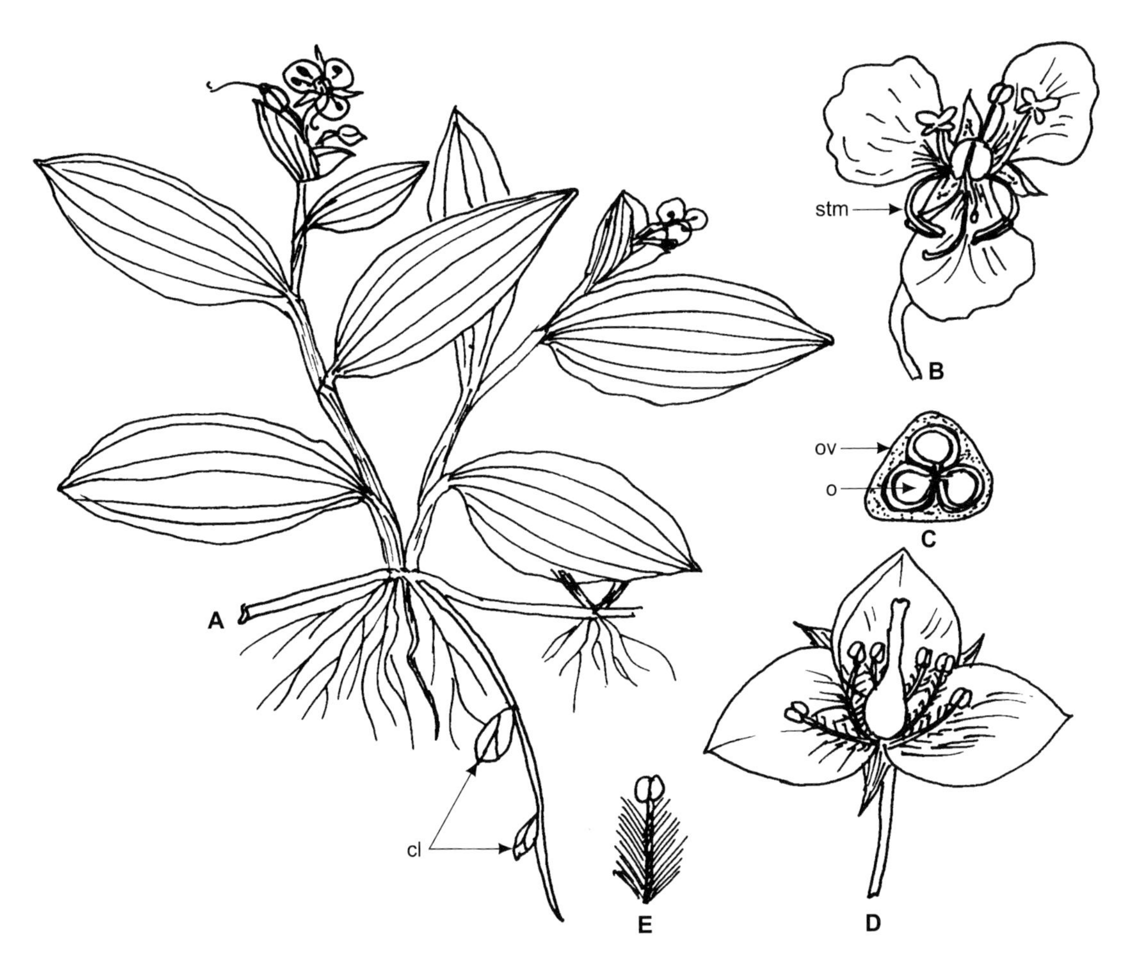

Fig. 11.32 Commelinaceae: **A-C** *Commelina forskalii*, **D, E** *Tradescantia virginiana.* **A** Habit, note cleistogamous flowers. **B** Flower. **C** Ovary, cross section. **D** Flower of *T. virginiana.* **E** Stamen. *cl* cleistogamous flower, *o* ovule, *ov* ovary, *stm* stamen. (Original)

integument or both (Johri et al. 1992). Embryo sac of Polygonum or Allium type, 8-nucleate at maturity. Endosperm formation of the Nuclear type.

Chromosome Number. Basic chromosome number is x = 5–19.

Chemical Features. Steroid saponins occur in some members of this family, such as *Cyanotis* (Dahlgren and Clifford 1982).

Important Genera and Economic Importance. Some of the important genera are *Aneilema*, *Commelina*, (commonly called dayflower), *Tradescantia*, *Cyanotis* and *Rhoeo. Tradescantia* is commonly known as spiderwort because of the soft mucilaginous substance that can be pulled from the broken ends of the stem and it hardens into a cobweb-like thread after exposure to the air (Jones and Luchsinger 1987). *Commelina benghalensis*, a common rainy season weed of the tropics, bears cleistogamous flowers on underground stems. The family is not of much economic importance except that various species of *Tradescantia*, *Commelina*, *Cyanotis*, *Zebrina* and *Rhoeo spathacea* are grown as ornamentals.

Taxonomic Considerations. There is no controversy about the position of family Commelinaceae in the Commelinales along with the families like Mayacaceae, Xyridaceae, Flagellariaceae and others. Hutchinson (1934, 1959) considered Commelinales to have been derived from the Butomales and Alismatales, and advanced over these orders in having syncarpous ovaries. Most taxonomists agree that the family is divided into 2 tribes: Tradescantieae and Commelineae.

Order Poales

A monotypic order including the only family. Poaceae (= Gramineae).

Poaceae (= Gramineae)

A very large family of 900 genera and 10,500–11,000 species (Takhtajan 1987), of cosmopolitan distribution.

Vegetative Features. Perennial or annual herbs or shrubs or trees, as in Bambusoideae. Rhizomes and stolons common; stem hollow (at internodes) or solid (at nodes) with prominent nodes; adventitious roots arising from nodes as in *Saccharum*, *Bambusa*, *Zea.* Leaves with sheathing base, usually linear or linear-lanceolate as in *Dendrocalamus*, *Setaria*, rarely petiolate as in Bambusoideae; ligulate, with parallel venation.

Floral Features. The inflorescence consists of distichous spikelets of determinate or indeterminate type, mostly in spikes (Fig. 11.33 C, D), sometimes in panicles as in *Oryza sativa* and male inflorescence of *Zea mays* (Fig. 11.33 A); female inflorescence of *Zea mays* is cob-like (Fig. 11.33 B). Mostly bi- and rarely unisexual, enclosed within a pair of sterile glumes—lemma and palea (Fig. 11.33 E). Perianth modified to two small scale-like structures, the lodicules. Stamens usually 3, rarely 3 + 3, as in *Oryza*, two species of *Bambusa* and *Thyrsostachys oliveri* (Bhanwara 1988); more rarely 1 or 2 or numerous; up to 120 in *Ochlandra.* Filaments filiform, anthers bithecous, basifixed or versatile (Fig. 11.33 F), latrorse, dehiscence longitudinal, poricidal in *Bambusa.* Gynoecium bicarpellary (Fig. 11.33 G), tricarpellary in Bambuseae and the two carpels fused to form unicarpellary condition in *Zea*; ovary unilocular with only 1 ovule, placentation basal (Fig. 11.33 H, I); style and stigmas two, stigma papillate or plumose (Fig. 11.33 G), dry. Fruit generally a caryopsis; in some Bambusoideae, a nutlet or berry (*Melocanna*). Seeds with strach-rich endosperm, embryo lateral, the cotyledon forms a haustorial tissue and a tubular structure (scutellum and coleoptile).

Anatomy. Vessels in stems and leaves with simple or both simple and scalariform perforation plates. Leaves with Rubiaceous type of stomata having dumb-bell-shaped guard cells and often with unicellular or small bicellular hairs. Oxalate raphides absent, but silica bodies of variable shape present in epidermal cells.

Embryology. Pollen grains simple, spherical, smooth, ulcerate, shed at 3-celled stage. Ovules ana-, hemiana- or campylotropous, tenui-, crassi- or pseudo-crassinucellate, bitegmic. Inner integument grows well beyond the nucellus in most tribes (except in Panicoideae); outer integument degenerates after fertilization (Bhanwara 1988). Embryo sac of Polygonum type, 8-nucleate at maturity. Endosperm formation of the Nuclear type.

Chromosome Number. Basic chromosome number variable: x = 2–23.

Chemical Features. Cyanogenic compounds and alkaloids common; luteolin, glycoflavones, flavone 5-glucosides and sulphated flavonoids are also reported. Amongst anthocyanins, cyanidin and to some extent delphinidin occur. Dehydroquinate hydrolyase isoenzyme is present in members of Gramineae (Boudet et al. 1977). Ohmoto et al. (1970) reported the occurrence of triterpenoids in various members.

Important Genera and Economic Importance. Although from the point of view of number of species, the grass family stands next to the Orchidaceae, it is the most widely distributed family of vascular plants. Its membes are known to occur in the arid and semi-arid zones (*Cenchrus*, *Duthiea*), as well as in the arctic and antarctic (*Deschampsia antarctica*); in areas where rainfall is neither very high nor very low, with moderately cold climate, forming the grassland vegetation like pampas, prairie, steppe and veldt.

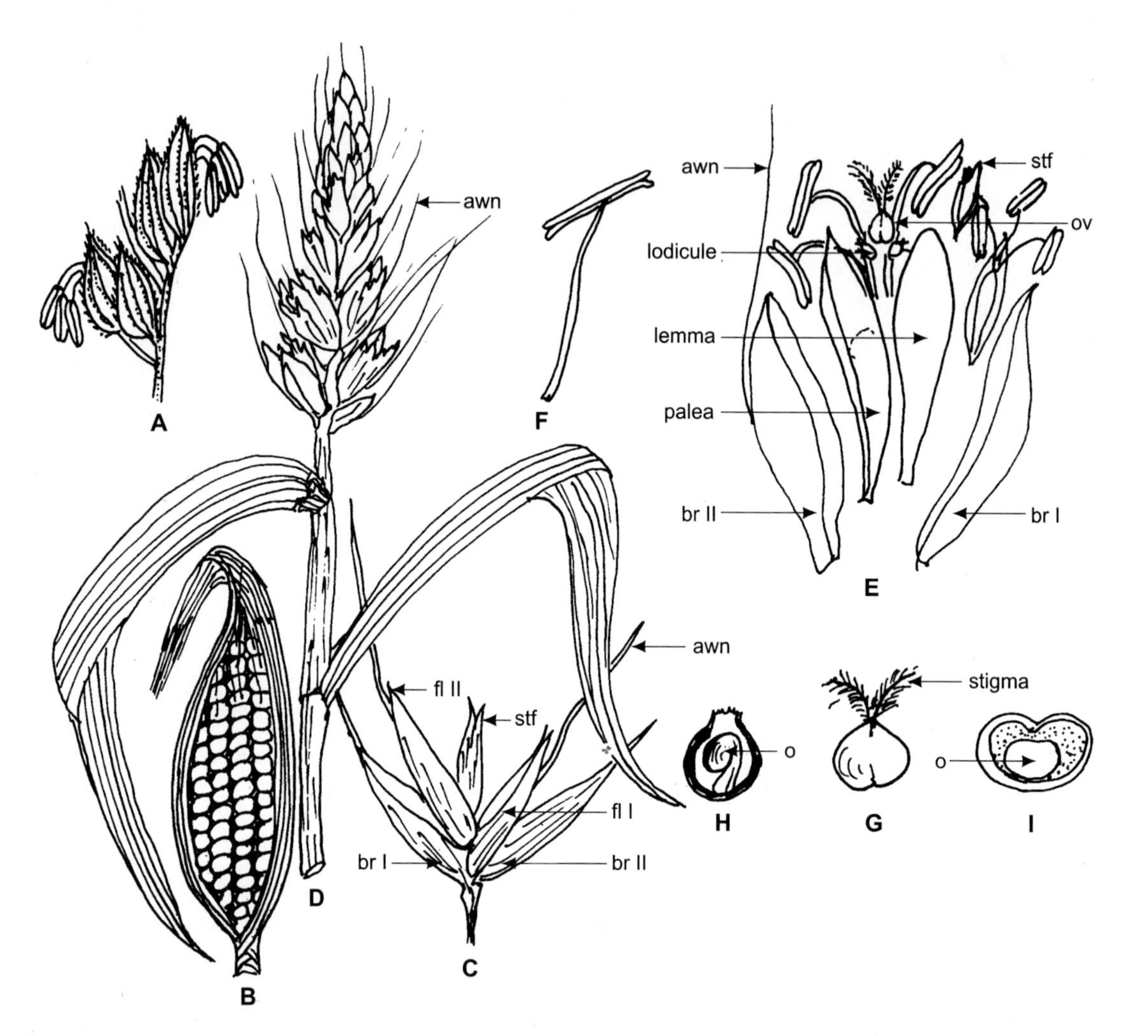

Fig. 11.33 Poaceae: **A, B** *Zea mays*, **C** *Avena sativa*, **D–I** *Triticum aestivum.* **A** Part of male inflorescence of Z. *mays.* **B** Female inflorescence of the same. **C** Dissected spikelet of *A. sativa.* **D** Inflorescence of *T. aestivum.* **E** Dissected spikelet of the same. **F** Stamen. **G** Carpel. **H, I** Vertical and cross section. (*br* bract, *fl* flower (floret), *o* ovule, *ov* ovary, *stf* sterile flower). (Original).

Bambusa pallida and *Dendrocalamus strictus* are the arborescent members, *Cynodon dactylon* is the common lawn grass.

Economically, the members of this family of vascular plants are probably the most important. The three important civilisations of the world would not have grown without the members of the grass family, The Mediterranean and the Indus Valley Civilisations grew with domestication of wheat and other related plants like rye, barley and oats. Mayacan, Aztec and Incas Civilisations grew around maize. The Chinese Civilisation domesticated rice and millet. Grasses are grown for both human and animal consumption (Robbins et al. 1957).

The economically important plants of this family can be placed under the following categories:

(a) **Cereals.** Three most important cereals are: *Triticum aestivum* (Fig. 11.33) (wheat), *Oryza sativa* (rice) and *Zea mays* (Fig. 11.33) (maize), each used as staple food the world over.

Hordeum vulgare (barley), *Avena sativa* (oats), *Secale cereale* (rye), *Eleusine coracana* (ragi), *Panicum miliaceum* (true millet), *Setaria italica* (Italian millet), *Pennisetum glaucum* (pearl millet) and many others are also used as food grains to a lesser extent.

(b) **Fodder for domestic animals.** *Agrostis alba, Bromus inermis, Cynodon dactylon, Pennisetum purpureum* and others.

(c) **Paper industry.** Many grasses are used to give good paper pulp, e.g. *Ampelodesma tenax, Eulaliopsis binata, Stipa tenacissima.*

(d) **Sugars.** *Saccharum officinarum* is the source of sugar. The juice extracted from the crushed canes is concentrated by boiling to yield molasses, which, on refining, gives white sugar.

(e) **Building material.** Various species of *Bambusa*, *Dendrocalamus*, *Guadua* and *Gynerium sagittatum* are used as building material. The hay from *Oryza sativa* and *Triticum aestivum* and the plants of *Saccharum spontaneum, Imperata cylindrica* and *I. exaltata* are used for thatching roofs.

(f) **Beverages.** A number of alcoholic beverages like sake from *Oryza sativa,* whiskey from *Hordeum vulgare, Zea mays* and *Secale cereale,* and rum from molasses, a bye-product of the sugar industry (*Saccharum officinarum*) are brewed.

(g) **Perfumery.** A number of grasses are aromatic and used in the manufacture of oils and soaps and in perfumery. *Cymbopogon citratus* (lemon grass) yields lemon-grass oil, C. *martini* (ginger grass), ginger-grass oil and C. *nardus* yields oil of citronella. Roots of *Vetiveria zizanioides* are the source of oil of vetiver.

(h) **Many grasses are grown as ornamentals.** Various species of *Cymbopogon*, many ornamental bamboos and others are grown as ornamentals. Many grasses are turf-forming and are grown in lawns and sports areas. *Ammophila arenaria* is used as a sand-binder.

Taxonomic Considerations of Poaceae

Taxonomically the Poaceae is very interesting and there are various opinions regarding its systematics. Arrangement of flowers in spikelets is an unique feature of this family, comparable to some extent only with the Cyperaceae members. The inflorescences are aggregates of spikelets arranged in spikes or racemes or panicles. The florets in each spikelet are highly reduced, enclosed within two glumes, lemma and palea, and the vestigial perianth represented by two or three lodicules (Rowlee 1898). There is controversy about the number of carpels in the florets. According to Bews (1929), Rendle (1904) and Engler and Diels (1936), the gynoecium is monocarpellary with 2 or 3 branched stigmas. Lotsy (1911), Weatherwax (1929), Arber (1934), and Randolf (1936) considered the gynoecium to be tricarpellary. According to Cronquist (1968, 1981), Jones and Luchsinger (1987), Takhtajan (1987) and Dahlgren et al. (1985), the gynoecia are mostly bicarpellary and rarely 3-carpellary. Floral anatomical studies support latter view. Belk (1939) showed that there were 3 carpels joined edge to edge, and a single ovule of the ovary was always attached to the posterior wall of the only locule. The uniovulate and unilocular ovary of the Gramineae has therefore evolved from a tricarpellary condition. The original three carpels are often suggested by the presence of 3 stigmas in some genera (Bambuseae).

Bentham and Hooker (1965c) included the Poaceae in the series Glumaceae of the Monocotyledons, which also included the Eriocaulaceae, Centrolepidaceae, Restionaceae and Cyperaceae. Rendle (1930), on the basis of wind-pollinated and monochlamydeous flowers, placed it after the Triuridales. Hutchinson (1934) treated the Gramineae as the only member of his Graminales, and derived it from Liliaceous stock. Cronquist (1968, 1981) considered the Gramineae to be allied to the Restionales or "to a broadly defined Commelinales, in which the Restionales is included". Like the members of Restionales, the members of Poales have orthotropous ovules, and position of embryo peripheral to the endosperm (also in most Commelinales). Takhtajan (1959) and Butzin (1965) suggested a relationship with Flagellariaceae, a member of the Restionales. G. Dahlgren (1989) includes this family in Poales along with Flagellariaceae, Joinvilleaceae, Restionaceae and Centrolepidaceae.

On the basis of ovular position in various families of the orders Commenilales, Restionales, Juncales and Poales, Cronquist (1968) suggests that "the Gramineae, Cyperaceae, Juncaceae and the families of the Restionales are derived from a common ancestry in the Commelinales near the Commelinaceae". Although the Restionales and the Poales are close to each other in many respects, they are separable from each other on the basis of pollen morphological data (Linder and Ferguson 1985, Kircher 1986).

Whether the two families, Cyperaceae and Poaceae, should be treated in separate orders or as members of one order is a matter of opinion. Cronquist (1968, 1981) observes that these two families are too close to be treated separately, but most others treat these in separate orders.

Order Cyperales

A monotypic order of the only family, Cyperaceae; distribution cosmopolitan. Mostly perennial, grass-like herbs, often grow in wet and marshy habitats. Stems or culms leafy or leafless, triangular. Leaves linear, in basal tufts, estipulate.

Inflorescence spikelets, variously arranged. Flowers bi- or unisexual, highly reduced. Stamens 3, ovary superior, fruit nut-like, seeds endospermous.

Cyperaceae

A very large family of 120 genera and 5,600 species, distributed throughout the world, more common in the subarctic and temperate regions of both the hemispheres. A number of genera, *Cyperus*, *Scirpus*, *Carex,* abundant in the tropics.

Vegetative Features. Perennial, or rarely annual, grass-or rush-like herbs (Fig. 11.34 A), often grow in wet, marshy or riparian habitats. Frequently rhizomatous, tufted herbs; fibrous roots arise from the base of the short or elongated and creeping rhizome. Stems and culms are leafy or leafless, mostly solid, terete, biconvex or triangular (Fig. 11.34 B), generally unbranched below the inflorescence. Leaves linear, in basal tufts or 3-ranked, with sheathing bases, estipulate; venation parallel, ligule usually absent.

Floral Features. Inflorescence mostly spikelets arranged in spikes, racemes, panicles or umbels (Fig. 11.34 A, C). Flowers bi- or unisexual (and then plants monoecious or dioecious), highly reduced in structure. Florets in the axils of glume-like, closely imbricated bracts (also called scales or glumes) (Fig. 11.34 D, F). Perianth 0 or reduced to bristles or hairs, rarely present as 3+3 bractlike scales, e.g. *Oreobolus.* Stamens 1 to 6, mostly 3, filaments thin, elongated, anthers bicelled, basifixed (Fig. 11.34 E), introrse, longitudinally dehiscent. Gynoecium syncarpous, bi- or tricarpellry, ovary superior, sometimes subtended and enveloped by a single, posterior prophyll, as in *Carex* (Fig. 11.34 F), unilocular, with a single, basal, erect ovule; style basally 1, with 2 or 3 long stylodial branches. In the *Mapania* group, there are several anthers in a flower, each subtended by a scale-like bract, Their interpretation is undecided (Dahlgren and Clifford 1982). Fruit nut-like (achene or nutlet), sometimes enclosed in a flask-shaped utricle; seeds with copious endosperm and a broad embryo.

Anatomy. Stem and leaves with vessels; perforation plates simple and/or scalariform. Oxalate raphides absent. Silica bodies present in epidermal cells, generally conical, simple or compound. Leaves isobilateral, stomata usually in parallel rows, equally distributed on both surfaces.

Embryology. Pollen grains in tetrads, with only one functional, i.e. pseudomonads; smooth, 3-celled at dispersal stage. Ovules anatropous, bitegmic, crassinucellate, inner integument forms the micropyle. Embryo sac of the Polygonum type, 8-nucleate at maturity. Endosperm formation of the Nuclear type.

Chromosome Number. Basic chromosome number is x = 5–13. Some of the Cyperaceae genera, like *Carex*, have diffuse centromeres (Sharma and Bal 1956).

Chemical Features. Flavones like tricin, luteolin, hydroxyflavonoids like 6-hydroxyluteolin, glycoflavones and flavone 5-glucoside have been reported. Chalcones and flavanones are also present. Luteolin 5-methyl ether is an interesting marker of this family. Isoenzymes of dehydroquinate (DHQ-ase) are present in Cyperaceae members (Boudet et al. 1977).

Fruit and inflorescence are coloured due to aurone aurensidin, together with the characteristic 3- deoxyanthocyanidin carexidin. Leaf flavonoids are luteolin, tricin and flavone C-glycosides (Harborne et al. 1982). Flavonoid pigmentation (in sedges) has been reported by Clifford and Harborne (1969), and presence of quinones in the genus *Cyperus* is helpful in classification (Allen et al. 1978).

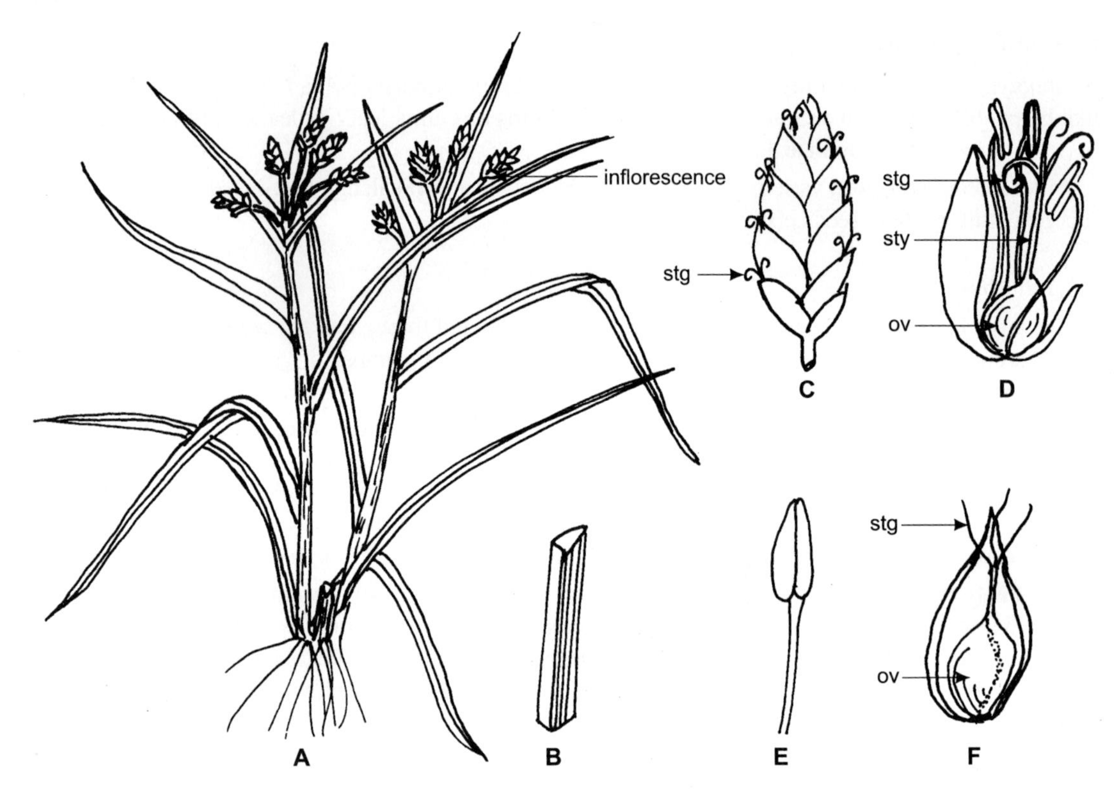

Fig. 11.34 Cyperaceae: **A–F** *Cyperus rotundas.* **A** Flowering plant. **B** Triquetrous stem. **C** Inflorescence. **D** Flower. **E** Stamen. **F** Carpel. (*br* bract, *ov* ovary, *stg* stigma, *sty* style). (Original).

Important Genera and Economic Importance. Some of the chief genera include *Cyperus* with 600 species, commonly called sedges, *Scirpus* (bulrush) with 250 species, *Eleocharis* (spike rush), with 200 species, *Fimbristylis* with 200 species, *Rhynchospora* with 250 species and *Carex* with 1100 species.

Economically, the family is not of much importance. *Cyperus papyrus*, a plant of riverine habitat, was used in making the special paper called 'papyrus' by the Egyptians, as early as 2400 B.C. The triquetrous stems were split into thin strips, which were pressed together to form a continuous structure. Both *C. papyrus* and *C. alternifolius* or the umbrella plant are grown as ornamentals. *C. rotundus* and many species of *Scirpus* are weeds and are very difficult to control in the crop fields. Dried tubers of *C. rotundus* and *C. scariosus* are used in medicine and perfumery, particularly the perfumed sticks or 'agarbatti'.

Many species of *Cyperus*, *Kyllinga* and *Carex* are used as fodder for domestic animals. *Remirea maritima* is an effective sand binder. Perianth of cotton sedge or *Eriophorum* is used as a stuffing material for cushions and pillows. *Cyperus esculentus* and *Scirpus grossus* var. *kysoor* are often grown for their edible tubers. Tubers of *Eleocharis dulcis*, the Chinese water chestnut, are also edible.

Taxonomic Considerations. The Cyperaceae, along with the Gramineae, is placed in the series Glumaceae by Bentham and Hooker (1965c) on the basis that in both the families the flowers are subtended by glumaceous bracts. Engler and Diels (1936) also placed these two families in the Glumiflorae. Hutchinson (1959,1973), Stebbins (1974), Dahlgren (1980a, 1983a), Dahlgren and Clifford

(1982), Takhtajan (1987) and G. Dahlgren (1989) recognise two separate orders for the two families. This family has a number of features common with the Juncaceae, but are advanced over Juncaceae. Possibly, the two families have a common ancester.

Taxonomic Considerations of the Order Cyperales

Cyperales, with its reduced floral structure and mostly wind pollination, are closer to the Commelinales and Juncales (sometimes included in the Commelinidae). Like these orders, Cyperales members have vessels in all vegetative organs and stomates with subsidiary cells and also 3-celled pollen grains.

The order, according to earlier authors (Bentham and Hooker 1965c, Engler and Prantl 1931, Cronquist 1968, 1981), comprised two families—Cyperaceae and Gramineae or Poaceae—and has also been termed Glumiflorae, Graminales and Poales. In recent years, some taxonomists treat these two families in two separate monotypic orders—Cyperales and Poales. The Cyperales are more closely affiliated to the Juncales, and the Poales to the Restionales (Cronquist 1968). The chemical data, however, indicate a close relationship between Juncaceae, Cyperaceae, Gramineae and Restionaceae. Keeping this in view, Cronquist (1968, 1981) points out that the chemical similarities between Gramineae and Cyperaceae are more suggestive of phyletic unity than convergence, but the occurrence of anatropous ovule and pollen tetrads in Cyperales make them distinct from the Gramineae with orthotropous ovules and single pollen grains. In addition, in Cyperales the embryo is embedded in the endosperm, and in Graminales the embryo is peripheral. There are other morphological distinctions and different methods of germination, which sharply delimit the two orders (Table 11.5).

Table 11.5 Comparison between Cyperales and Poales

Characters	*Cyperales*	*Poales*
1 Florets	Axillary	Terminal
2 Perianth	Reduced to bristles or hairs	Reduced to lodicules
3 Leaves	Tristichous, without ligule	Distichous, with ligule
4 Stem	Solid, triangular	Hollow, cylindrical
5 Pollen grains	In tetrads	As monads
6 Ovule	Anatropous	Orthotropous
7 Fruit	Nutlet or achene	Caryopsis
8 Germination of seed	Cotyledon comes out of the seed	Cotyledon does not come out of seed

The two orders Cyperales and Poales, each including one family, are distinct from each other. They also do not indicate any phyletic relationship. The order Cyperales is more closely related to the Juncales.

Order Scitamineae

An order comprising 5 families: Musaceae, Zingiberaceae, monotypic Cannaceae, Marantaceae, and Lowiaceae distributed mainly in the tropics of both the New and the Old World.

Small to very large, generally perennial herbs, rarely shrubs (*Maranta*) or "trees" (*Ravenala*) with starch rich rhizomes), rarely aquatic (*Thalia*). Vertical aerial stem often short; inflorescence-bearing stems covered with bracteate leaves. Leaves alternate, frequently distichous, with a sheathing base, usually petiolate with a large and simple or secondarily split, broad and pinnately veined lamina. Ligules present in many Zingiberaceae. Intravaginal squamules absent.

Flowers epi- or rarely perigynous, mostly with 3 + 3 petaloid, basally connate tepals; in Musaceae, 5 tepals fused to from a sheath, and one tepal free. Stamens 3 + 3 or fewer, only one in Zingiberaceae. Gynoecium syncarpous, tricarpellary, ovary mostly tri- and rarely unilocular with axile or parietal placentation. Fruit usually a loculicidal capsule, rarely a berry, nut or schizocarp; seeds arillate, endospermous with linear, capitate or curved embryo.

Vessels present in roots mostly and rarely in stems. Pollen grains in monads, 2- or 3-celled. Ovules mostly anatropous, crassinucellate; endosperm formation of the Helobial or Nuclear type.

Saponins and steroid saponins are reported in some families; some others are rich in flavonoids.

Cannaceae

A monotypic family with 55 species, native to Central America and the West Indies; a few are cultivars.

Vegetative Features. Large, perennial herbs with a tuberous rhizome and unbranched aerial stem. Leaves large, with open, eligulate sheaths, cauline, parallel veined.

Floral Features. Inflorescence terminal on leafy shoots, raceme or panicle (Fig. 11.35 A). Flowers large, showy, bisexual, irregular, epigynous (Fig. 11.35 A, B), each flower or a pair of flowers subtended by a bract and rarely with scale-like bracteoles. Perianth biseriate, distinguishable into calyx and corolla (Fig. 11.35 B). Calyx of 3 sepals, polysepalous, persistent, lanceolate or elliptic, more or less green to purple. Corolla of 3 petals, usually unequal, erect, basally connate and adnate to androecium and style to form a tube, deciduous. The staminal column consists of 4 to 6 petaloid stamens in 2 series (Fig. 11.35 C, G). The outer 3 petaloid stamens are sterile, the largest forms the labellum; inner series consists of 1 or 2 petaloid staminodes and a free petaloid fertile stamen with a monothecous anther adnate to the petaloid margin (Fig. 11.35 D). Gynoecium tricarpellary, syncarpous, ovary inferior, trilocular, with axile placentation (Fig. 11.35 G); ovules numerous per locule; style 1, petaloid, stigma 1, represented by a stigmatic crest on apical margin (Fig. 11.35 E). Fruit a warty capsule, seeds small, numerous (Fig. 11.35 C, G) subglobose, with very hard endosperm and a straight embryo.

Anatomy. Aerial stems with mucilage canals or cavities. Stomata predominantly of Rubiaceous type.

Embryology. Pollen grains mostly spherical, usually with small spinules, inaperturate (PKK Nair 1960), 3-celled at shedding stage (Davis 1966). Ovules anatropous, bitegmic, crassinucellate, micropyle formed by inner integument. Embryo sac of Polygonum type, 8-nucleate at maturity. Endosperm formation of the Nuclear type, but reduced to a thin layer in mature seed.

Chromosome Number. Basic chromosome number is x = 9.

Chemical Features. Flavonols present.

Important Genera and Economic Importance. The only genus *Canna* with its 55 species and many cultivars is grown as an ornamental. *Canna edulis* is used as a source of food for livestock; rhizomes rich

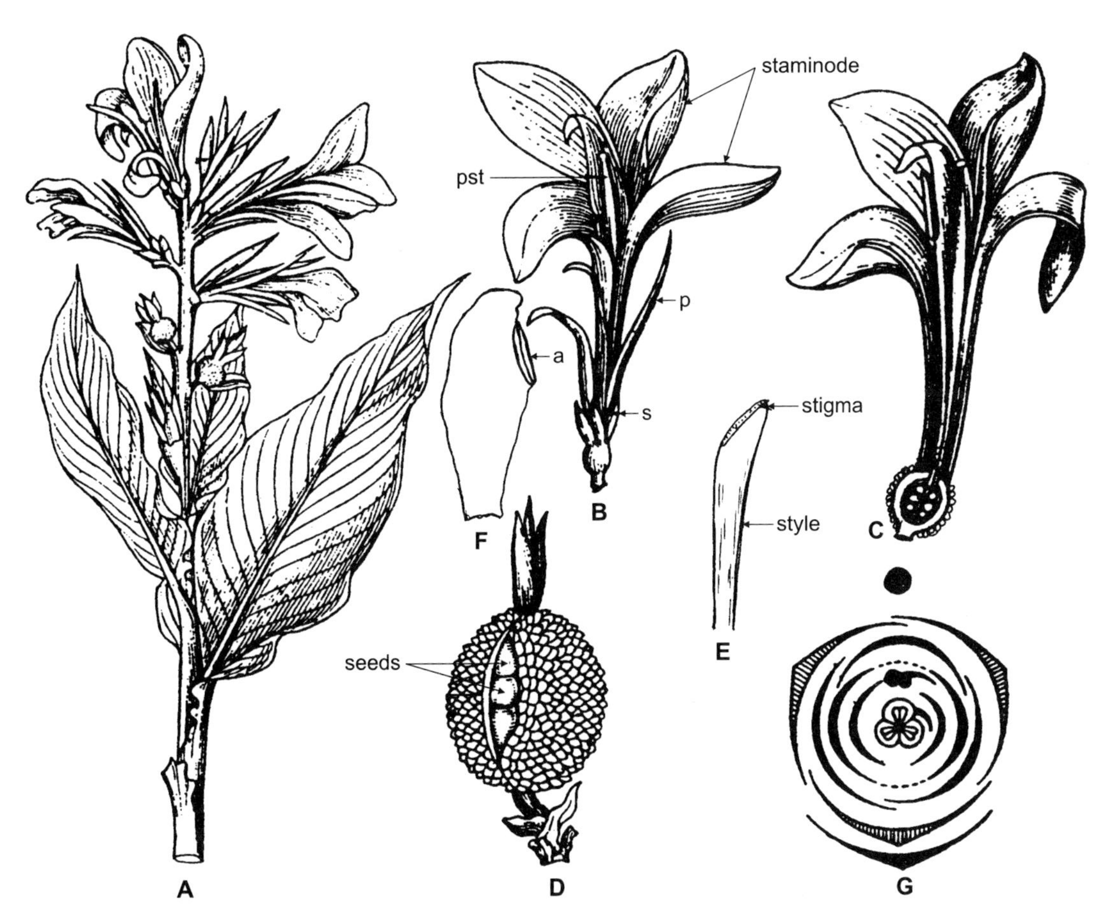

Fig. 11.35 Cannaceae: **A–F** *Canna indica* **A** Inflorescence. **B** Flower. **C** Same, longisection. **D** Fruit, dehisced to show seeds within. **E** Style and stigma. **F.** Free petaloid fertile stamen with a monothecous anther (*a*) **G** Floral diagram. (*a* anther, *p* perianth, *pst* pistil, *s* sepal).

in starch. Burning the plants is said to produce an insecticidal smoke (Rogers 1984). Extracts from C. *indica* and other species have molluscicidal activity (Mahran et al. 1977a, b).

Taxonomic Considerations. The family Cannaceae has been included in Scitamineae (or Zingiberales) by all phylogenists. It is a highly advanced family of this order.

12

Concluding Remarks: Taxonomy—Yesterday, Today and Tomorrow

The term taxonomy, when first introduced in 1813 by AF de Candolle, meant "the theory of plant classification." After prolonged usage by various botanists over the years it now includes much more. Presently, in plant science, taxonomy includes identification and nomenclature of plants in addition to their classification.

Taxonomy is basic to other sciences and at the same time dependent on them. An up-to-date account of the general principles and modern trends in taxonomy, viz. biosystematics, chemical and serological data, mathematical data have made this subject a more creative and dynamic one. Modern taxonomy, therefore, provides a sound and strong foundation for the progress of biological sciences. Various definitions for the term 'taxonomy' have been proposed, of which the best could be: "a study aimed at producing a system of classification which best reflects the totality of similarities and differences" (see Cronquist 1968).

The need for better and more objective methods of ascertaining relationships amongst various plants and plant groups was realised by the beginning of this century. The comparative studies of various plant groups from which taxonomic conclusions can be drawn, should ideally include all the characters of all the species, and from as many individual plants as possible, from the entire geographical range of the species. In this type of study, evidence from all related fields—vegetative and floral morphology, micromorphological and ultrastructural observations on anatomy, embryology, palynology, cytology and genetics, and also geographical distribution, habitat requirement, serology and chemical analysis—are used for effective diagnosis of various taxa. The base of taxonomy is historical, but the application of modern applied science has made it an ever-changing system. The study of relationships amongst plants has been revolutionised in recent years by advances in molecular genetic techniques.

The ideal situation of using "all the characters of all the individuals of a species" is yet to be attained. The availability of information is time-consuming and demands more than just an individual effort. Presently, there are numerous "gaps" in the information available. Much research in systematic botany still remains to be taken up. A large number of families have not been studied at all, or have been incompletely investigated. It will be out of context for detailed discussion of all these taxa. However, the following examples will elucidate, how important the modern molecular data is for taxonomic studies of today and tomorrow.

The Hydrostachyaceae is a monotypic family of aquatic herbs from tropical East Africa and Madagascar. The reduced or highly modified reproductive and vegetative morphology of these plants have made their placement difficult. Dahlgren (1980a) and Takhtajan (1987) treat this family as member of the Lamiales and the Scrophulariales, respectively. Thorne (1992b) places them in the Bruniales and Cronquist (1981) in the Callitrichales (because of their adaptation to aquatic habit). Absence of any diagnostic chemical compound makes it difficult to show any close link of Hydrostachyaceae to another group. Presently, however, the rbcL (Rubisco or Ribulose, 1, 5 bisphosphate carboxylase enzyme) (See Glossary), sequence data used to assess the position of this family brings it close to the Cornales s.l. (Hempel et al. 1995).

Loasaceae is yet another problematic family. Although traditionally placed in the Violales because of their parietal placentation, numerous stamens and centrifugal stamen initiation, some taxonomists have suggested different alliances on the basis of various anatomical, embryological and chemical features. Dahlgren (1975a) was the first to suggest that the Loasaceae are related to the Cornales on the basis of a chemical feature (presence of iridoids) and embryology (unitegmic, tenuinucellate ovules). Hufford (1992) also suggests the same assignment on morphological and chemical basis which is supported by rbcL sequence data provided by Hempel et al. (1995).

Similarly, use of RAPD (randomly amplified polymorphic DNA) data has proved useful in evolutionary studies and studies on relationships within and among closely related taxa. RAPD variation in *Plantago* has revealed the existence of two distinct species—*Plantago major* and *P. intermedia,* despite their morphological resemblance and absence of breeding barriers (Wolff and Morgan-Richards 1999).

The origin and relationship of the flowering plants have remained a mystery over a long period. It is an incredible task to unravel the phylogenetic relationships and prepare a family tree of the 2,50,000-3,00,000 living species belonging to 350 or more families. Qiu et al. (1999) established a family tree based upon combined gene sequences (multi gene approach) from nucleus, mitochondria and chloroplast. This study demonstrates that *Amborella*, Nymphaeales and Illiciales—Trimeniaceae—*Austrobaileya* represent the first stage of angiosperm evolution with *Amborella* being probably the most primitive angiosperm of today. According to this new phylogenetic analysis, there are two major monophyletic groups of higher dicots (or eudicots) and monocots instead of the traditional divisions—dicots and monocots.

Numerous molecular biology techniques are available for utilization in modern taxonomy. But still, it is the classical taxonomy that remains as the foundation for all taxonomic studies. Molecular studies are only complementary. These evidences can strengthen the viewpoints of classical taxonomists.

A long list of families with incomplete data has been prepared by many authors (see Dahlgren 1983b, Johri et al. 1992, Bhattacharyya and Johri 1998). The families listed by Dahlgren (1983b) are the ones, in which serological investigations should be taken up. This may reveal new connections between them.

The intregration and disintegration of taxa at all levels—species, genera, families and orders, will continue ad infinitum as long as new taxa will be discovered from explored as well as unexplored areas, and additional data become available. As a consequence, inter- and intra-familial and inter- and intra-ordinal assignments will also change. The evolutionary tendencieds will also become clearer.

Appendix

Procedures for Field and Laboratory Study

Diagnostic Features

In taxonomic studies a *character* is defined as any attribute or feature of a taxon which can be related to form, function or behaviour of that taxon. Characters are abstractions and their expression is important. For example, stout stem is a character, but stem 20"–40" in diameter is an expression. Similarly, flower colour is a character which can be expressed as blue, red, etc. Unless a character is expressed, it has no value. A particular feature or a group of features can distinguish one taxon from the related taxa. These are *diagnostic* features.

Each organism has a large number of potential characters, and anyone may be used as diagnostic feature. Some of these characters are present singly, and others in groups of two or three, are known as *correlated characters*, e.g. if in a taxon the characters such as syngenesious stamens, inferior ovary, single basal ovule and capitulum inflorescence occur together, it may be concluded that the taxon belongs to the family Asteraceae (= Compositae). Similarly, a taxon belonging to Solanaceae has gamopetalous corolla, epipetalous stamens, obliquely placed placenta and persistent calyx. Such *correlated characters* are, therefore, very useful for indentification. The two species *Cassia tora* and *C. obtusifolia* are very much alike, and have often been treated as only one species, i.e. *C. tora,* However, they can be distinguished on the basis of a few correlated characters (Table 1.1).

Correlated characters can be used very effectively in solving taxonomic problems, not only when the taxa are very closely related, but also for identification of the unknown specimens. The classification system of George Bentham and Joseph Dalton Hooker (1862-1883) is also based on correlated characters.

Table 1 Correlated Characters to distinguish *Cassia tora* from C. *obtusifolia*

	Cassia tora	*Cassia obtusifolia*
1	Leaf apex obtuse	Leaf apex mucronate
2	Extrafloral nectary between both pairs of leaflets	Extrafloral nectary between lower pair of leaflets only
3	Nectaries greenish-yellow	Nectaries deep-orange
4	Nectaries narrow at both end	Nectaries club-shaped

When a large collection of flowering plants has been made, one would like to know their names. The first step is to study their characters and to express them in a co-ordinated manner. The description of the plant and its flower is necessary to identify the undetermined taxon correctly. A critical study of the external morphology of different organs of the plants is necessary for a detailed description of a specimen.

Given below is a glossary of various morphological terms used in describing a plant specimen.

General Morphology

Aquatic plants: Grow in aquatic habitat, have slender stem and petiole, dissected leaves and poorly developed root system.

Autotrophic plants: Synthesize their own "food" from the inorganic raw materials in the presence of sunlight.

Carnivorous plants: Have variously modified leaves to trap the insects and digest them.

Epiphyte: These plants need a support and grow on the stems, branches and leaves of other plants (but are not parasitic on them). They have two sets of roots: the "clinging roots" are narrow and cling to the surface on which they grow. They also acquire nutrients from the debris collected on the bark. The other set of roots is much softer and thicker and remains suspended in the air. These are green and covered with a special tissue called velamen which absorbs moisture from the atmosphere. Most orchids are epiphytes.

Heterotrophic plants: The heterotrophs depend for their nutrition upon autotrophic plants.

Parasites: These plants grow on other plants and obtain their nutrition from the host plant through haustoria. Partial parasites are green and can prepare the carbohydrate needed for their nutrition. Total parasites are devoid of the green pigment. Root parasites are attached to the roots of the host plant; stem parasites are attached to the stems and branches.

Saprophytes: These plants grow on dead plant or animal debris or decayed products such as rotten leaves, humus etc.

Symbionts: These plants grow together without any dependence on each other.

Terrestrial: Plants grow on the surface of land, do not show any modification or adaptation due to habitat: *Euphorbia hirta.*

Xerophytic plant: Plants grow under xeric conditions and have modified green stem, flattened phylloclades or cladodes; leaves reduced to scales or spines, and stomata in stomatal pits to reduce transpiration: *Opuntia*, *Asparagus*, *Nerium.*

The Root

The root is the descending part of the plant body; positively geotropic and negatively phototropic. Root surface normally non-green, without nodes and internodes; branches or secondary roots borne at random.

Annulated tuberous roots: Appear to be formed of numerous swollen discs placed one above the other: *Cephaelis ipecacuanha* (Fig. 1 G).

Assimilatory roots: These roots are green and can carry on carbon assimilation; epiphytic roots are assimilatory.

Climbing roots: Adventitious roots which develop from nodes and help in climbing (Fig. 2 D).

Clinging roots: The epiphytes have clinging roots which enter the crevices of the support to keep the plant in place as in orchids.

Conical: Broad at the upper end gradually tapering towards the lower end, e.g. *Daucus carota* (Fig. 1 B).

Fascicled roots: Tuberous roots develop in clusters in plants like *Asparagus* (Fig. 1 F) and *Dahlia* (Fig. 1 E). These are both for perennation and propagation. Palmately branched tuberous roots occur in *Orchis* (an orchid).

Fibrous roots: Common in monocots. The radicle becomes arrested in growth and additional roots develop close to the base of the radicle. These "seminal" roots and some adventitious roots grow from the base of the plumule and form the fibrous root system of the monocots.

Floating roots: Grow at the nodes of certain aquatic plants, are spongy and help to keep the plant buoyant. They function as air floats, as in *Jussiaea repens* (Fig. 2 F).

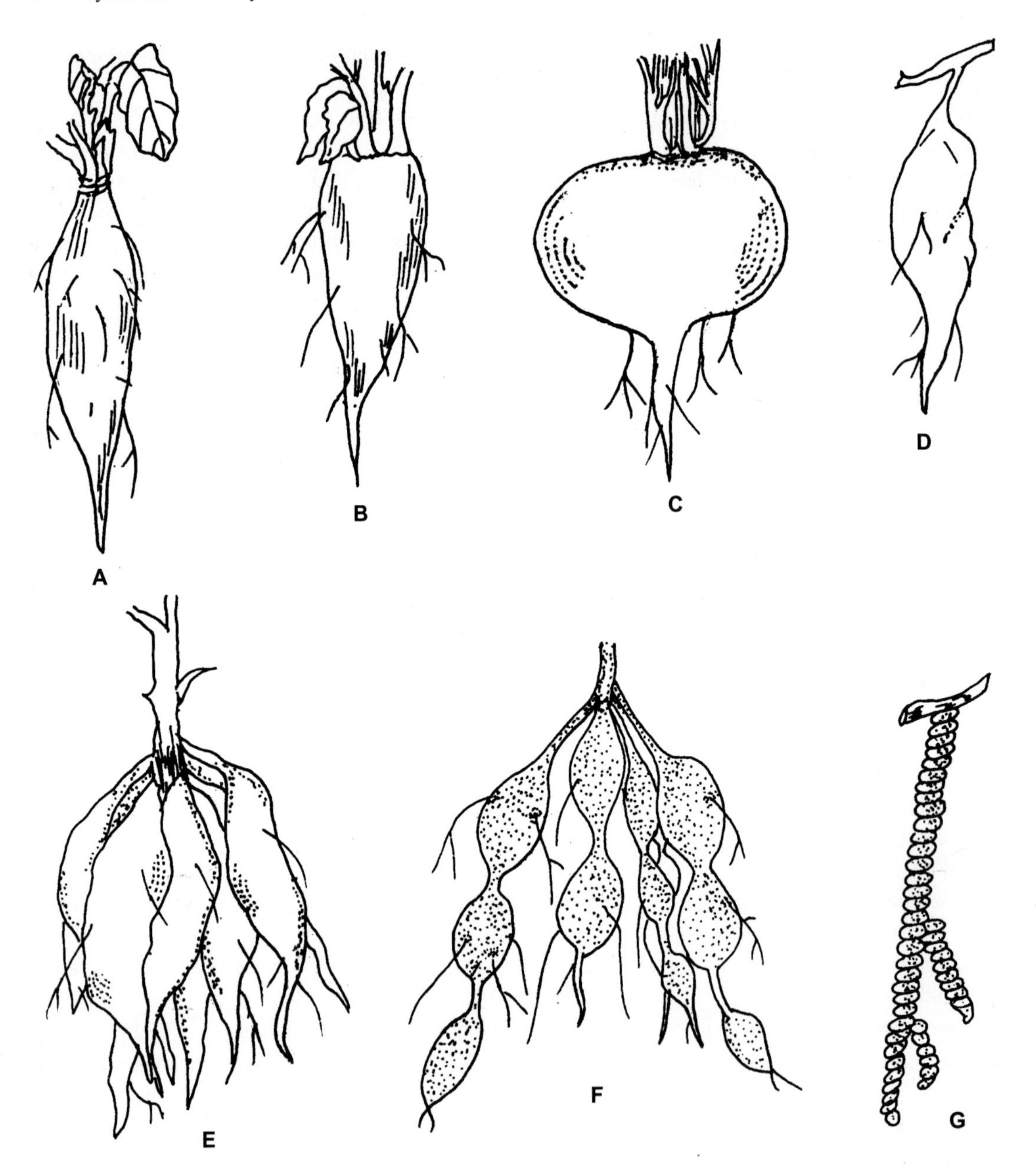

Fig. 1 **A–G.** Types of adventitious roots: **A** Fusiform root of *Raphanus sativus.* **B** Conical root of *Daucus carota.* **C** Napiform root of *Brassica campestris* var. *rapa.* **D** Tuberous root of *Ipomoea batatas.* **E** Fascicled roots of *Dahlia.* **F** Fascicled root of *Asparagus racemosus.* **G** Annulated tuberous root of *Cephaelis ipecacuanha.*

Fusiform roots: A type of storage root, swollen in the middle and gradually tapering towards both ends, e.g. *Raphanus sativus.* (Fig. 1 A)

Haustoria: These structures develop in parasites and obtain nourishment from the host plant (Fig. 2 E, G–I).

Holdfast roots or haptera: Help in attaching the plant to the surface on which it is growing. The adhesive discs at the tip of climbing roots of *Hedera helix* are an example. The thalloid branched roots of Podostemaceae attach the plants to the rock surface with such haptera.

Moniliform roots: These are tuberous roots with alternate swollen and constricted parts which gives a beaded appearance, e.g. *Asparagus racemosus* (Fig. 1 F).

Mycorrhiza: Some roots are infested with fungal mycelium and may be ecto- or endotrophic. The fungal hyphae act as root hairs do and help in absorption. Mycorrhiza is essential for seeds of orchids to germinate and growth of seedlings.

Napiform root: Storage roots swollen above and abruptly tapering towards the lower end, e.g. *Brassica campestris* var. *rapa,* (Fig. 1 C).

Nodulose tuberous roots: Only the apices of the adventitious roots are swollen, as in *Costus speciosus.*

Pneumatophores, respiratory roots: Roots of plants which grow in saline coastal marshy regions (mangroves) are negatively geotropic and grow vertically upward. These roots have small pores (lenticels) through which gaseous exchange takes place.

Prop roots: Grow adventitiously from the horizontal branches of some tropical trees such as *Ficus,* and hang vertically downward. They reach the soil, get anchored, and become thick and almost as strong as the main stem. They give support to the heavy branches and sometimes also replace the main stem (Fig. 2 B).

Root buttresses: These are broad and wing-like, radiating from the base of the stem; partly root and partly stem, and may develop in very old plants, e.g. *Adansonia digitata.*

Root tubers: These are formed at the nodes of running stems (runners); swollen in the middle and tapering at both ends, e.g. *Ipomoea batatas* (Fig. 1 D).

Root pocket: In some water plants, the root tip is protected by a different type of root cap (called root pocket) which does not regenerate.

Stilt roots: These roots provide additional support to certain shrubs and small trees which grow along the edges of water bodies (*Pandanus*; Fig. 2 C) or marshes, where anchorage of the main stem is not very strong. Adventitious roots which grow from the lower nodes of *Zea mays* or *Saccharum officinarum* also have a similar function.

Storage roots: Are swollen and fleshy due to the accumulation of stored food; may also act as perennating organs (see also Fusiform, Conical, and Napiform roots).

The Stem

The stem is the main portion of the ascending axis and develops from the plumule, bears leaves, branches, flowers and fruits. It is positively phototropic and negatively geotropic.

Acaulescent: Plants without a stem (only apparently), such as the monocots.

Adhesive climber: Plants provided with adhesive discs at the end of their climbing organs, with their help they can adhere to any surface, e.g. *Bignonia capreolata.*

Annual: Short-lived plants, complete their life cycle in one season.

Fig. 2 **A–H**. Types of adventitious roots (contd.): **A** Pneumatophores of mangrove plants. **B** Prop roots of *Ficus benghalensis.* **C** Stilt roots of *Pandanus fascicularis.* **D** Climbing roots of *Scindapsus aureus.* **E** Haustoria of *Cuscuta reflexa.* **F** Floating roots of *Jussiaea repens.* **G** Haustorial roots of *Phrygilanthus celastroides.* **H** Section through swollen junction of the host *Pittosporus tenuifolium* and parasite (stippled). **I** Section through normal stem of the host. (**G**-I after Johri and Bhatnagar 1972).

Biennial: Complete their life cycle in two seasons/years.

Bulb: An underground modified bud with highly reduced convex or conical disc-like stem and fleshy scale leaves arising from it, e.g. *Allium cepa* (Fig. 3 D). The apical bud grows into a flowering scape and the axillary buds grow into daughter bulbs. The fleshy leaves store food materials. When the scale leaves are arranged in concentric rings, as in *Allium cepa, Hyacinthus* and *Polianthes,* it is known as a *tunicated bulb.* Often the scale leaves arise from the disc loosely and appear as petals of a flower, possibly overlapping one another only at the margin, as in *Allium sativum.* These individual buds are called the *cloves*, which are enclosed by a whitish, skinny scale leaf.

Bulbil: These are modifications of axillary vegetative or floral buds, often borne on large scapiferous inflorescence, e.g. *Agave.*

Caudex or columnar stem: Usually do not branch at all. The lateral buds on the long columnar trunk remain dormant or collapse. There is a crown of leaves on the top.

Cladode: Cladode is a phylloclade of one internode only, e.g. *Asparagus racemosus* (Fig. 5 D).

Climber: Weak-stemmed, mostly herbaceous, annual or perennial plants.

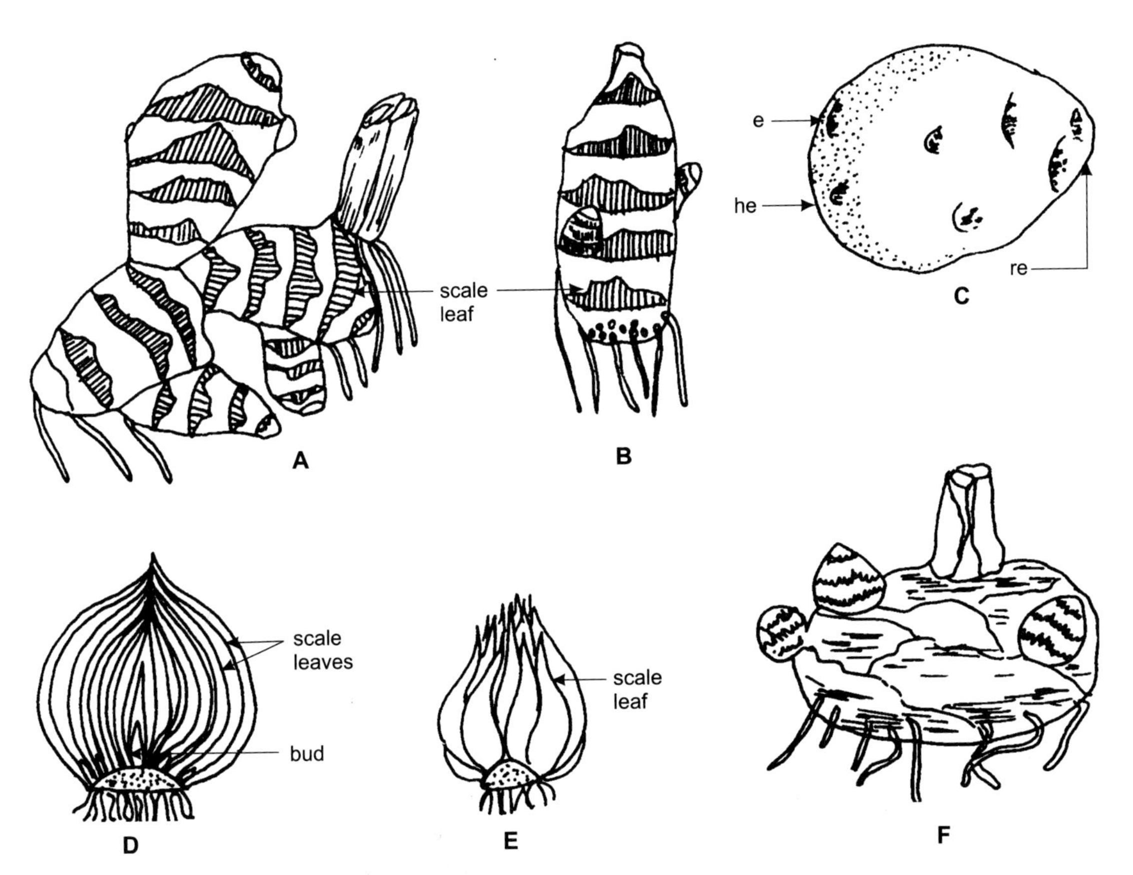

Fig. 3. **A–F**, Underground modifications of stem: **A** Rhizome of *Zingiber officinale.* **B** Rootstock of *Alocasia indica.* **C** Tuber of *Solanum tuberosum.* **D** Bulb of *Allium cepa.* **E** Bulb of *A. sativum.* **F** Corm of *Amorphophallus campanulatus. e* 'eye', *he* heel end, *re* rose end.

Corm: It is a highly condensed vertical rootstock with a large apical bud and many axillary buds. In *Amorphophallus campanulatus* (Fig. 3 F) it is a huge, condensed single internode with numerous adventitious buds and roots arising from its surface.

Creeper: Perennial, herbaceous plants; to start with the plant is erect and soon gives rise to branches which, after travelling a short distance horizontally, become rooted and give rise to another plant which behaves in a similar way and covers a large area within a short time. *Runners, stolons* and *suckers* are the examples.

Decumbent: When the branches try to straighten up after growing parallel with soil surface for some distance.

Deliquescent stem: Stem with well-developed lateral buds but a weak apical bud; such trees have a spreading habit.

Diffuse: Some trailing plants produce a number of branches in all directions; these are termed as *diffuse*.

Excurrent stem: Grows indefinitely, branches are borne at a higher level and are in acropetal order.

Herbs: Rather small, weak-stemmed and short-lived plants. May be annual, biennial or perennial.

Lianas: Perennial, woody climbers which wind round tall trees to reach the top and form the uppermost strata of the forest; mostly occur in tropical rain forests.

Offsets: These are runner-like branches of aquatic plants but are shorter and thicker, e.g. *Eichhornia crassipes* (Fig. 4 F).

Perennials: Grow for very long periods.

Procumbent: Branches lying parallel with soil surface.

Pseudobulb: Fleshy, tuberous structures metamorphosed from one or more internode of stem as seen in many aerial orchids (Fig. 5 G). These organs store large quantities of water to tide over the unfavourable season.

Phylloclade: Stem modified to a flattened or swollen, fleshy, green structure which can function as a photosynthetic organ. The leaves are metamorphosed into small scale leaves (*Ruscus*, Fig. 5 A) or spines (*Opuntia,* Fig. 5 B). Phylloclades occur commonly in xerophytic families like Cactaceae, Euphorbiaceae and one member of Polygonaceae, *Muehlenbeckia platiclados* (Fig. 5 C).

Rhizome: Dorsiventral, underground stems growing horizontally, with distinct nodes and internodes, brown scale leaves at the nodes and apical as well as axillary buds, e.g. *Zingiber officinale* (Fig. 3 A), *Curcuma longa.* Adventitious roots develop from the lower surface and the apical and axillary buds produce new shoots in favourable season. *Rootstock* is a special type of rhizome which grows vertically, e.g. *Alocasia indica* (Fig. 3 B).

Runner: These are axillary branches arising from the lower leaves which grow along the surface of soil and give rise to new daughter plant at a distance from the mother plant, e.g. *Oxalis corniculata* (Fig. 4 A), *Colocasia antiquorum* (Fig. 4 B), *Centella asiatica, Cynodon dactylon,* etc. A special type of underground runner or sobole is seen in some grasses like *Agropyron* (Fig. 4 C), *Saccharum spontaneum,* etc. These grasses are very difficult to eradicate and are good soil-binders.

Scape: The stalk-bearing inflorescence in plants with basal rosette of leaves, e.g. *Narcissus.*

Scrambler: These plants grow on other plants and retain their position with the help of small prickle-like outgrowths on their stem surface. These are common in tropical rain forests, where they form thick, impenetrable bushes because of their profuse growth, e.g. *Calamus rotung.*

Shrubs: Shorter than trees in height, and branching occurs close to the level of soil. The woody branches are not as strong as those of a tree.

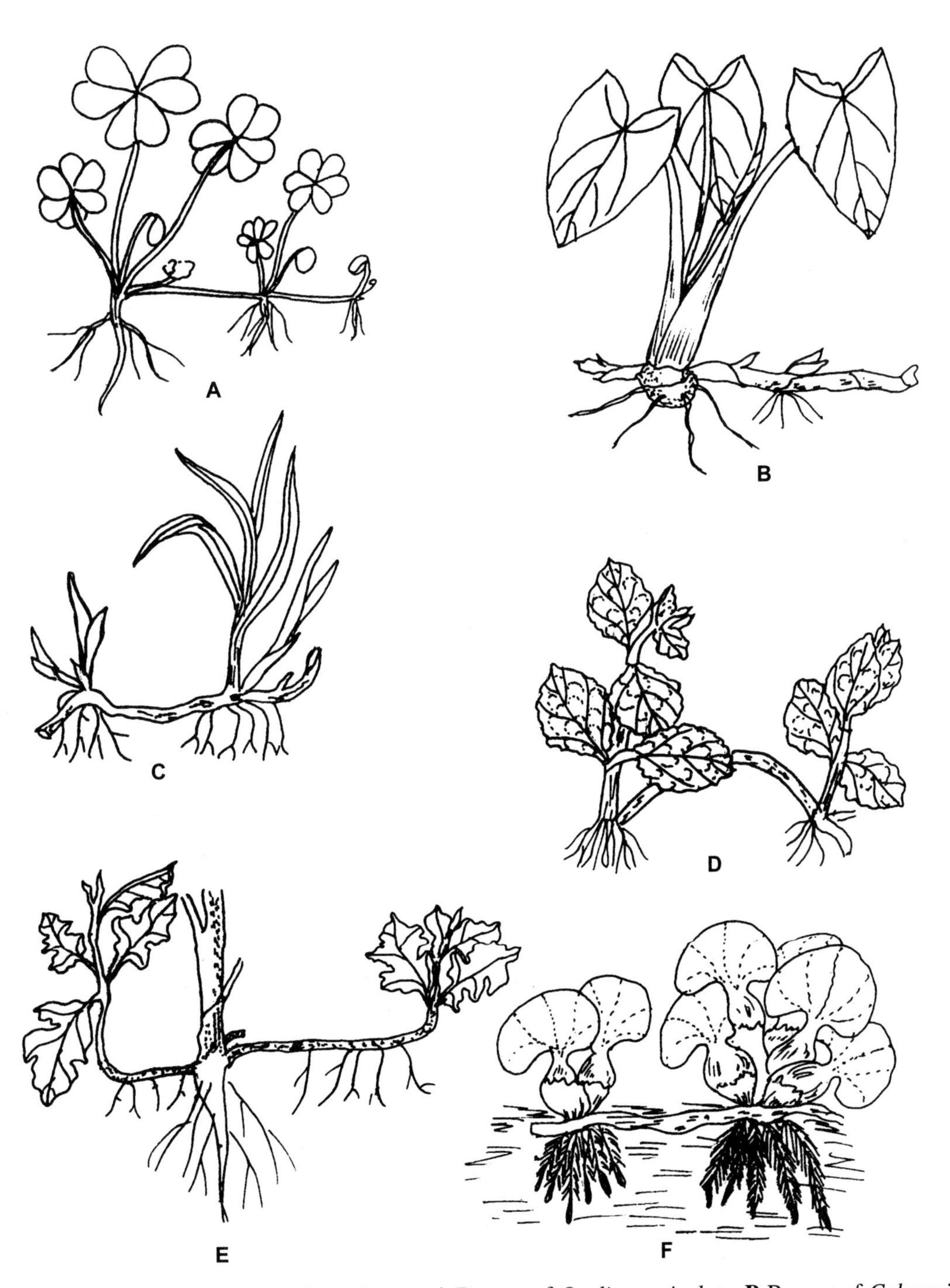

Fig. 4 **A–F** Subaerial modifications of stem: **A** Runner of *Oxalis corniculata.* **B** Runner of *Colocasia antiquorum.* **C** Sobole of *Agropyron.* **D** Stolon of *Mentha viridis.* **E** Sucker of *Chrysanthemum.* **F** Offsets of *Eichhornia crassipes.*

Stem tendril: Tendrils are climbing organs and may develop from any part of the plant. Axillary branches in *Passiflora,* inflorescence. axes of *Antigonon leptopus* (Fig. 5 F) and *Cardiospermum helicacabum* are some examples of stem tendril.

Stolon: Here the lower axillary branches grow upwards in the beginning and then arch down to meet the ground and root, where daughter plants are formed, e.g. *Mentha* (Fig. 4 D).

Suckers: These are underground runners just below the surface of the soil and form daughter plants some distance away from the mother plant after forming adventitious roots. Such daughter plants are also formed from axillary buds on the underground runner, e.g. *Chrysanthemum* (Fig. 4 E).

Tendril: Thin, wiry climbing organs which twine around any support they come in contact with. Any part of the plant body can be modified to a tendril (see also Stem tendril).

Thorn: These are axillary branches transformed into stiff pointed structures, e.g. *Punica granatum* (Fig. 5 E), *Duranta.* Thorns are different from spines and prickles, are very deep-seated and have vascular connection. Prickles are only superficial outgrowths on the epidermis and can be easily broken off, as in *Rosa.* Spines are hard, pointed structures usually modified stipules or any other part of a plant, e.g. *Acacia nilotica.* Thorns are often branched, as in *Carissa carandus.*

Tuber: Axillary or adventitious branches arising from underground part of stem swell up at the apices and form the tubers, e.g. *Solanum tuberosum.* The nodes and internodes on the tubers are not so prominent, the nodal points being marked by the presence of scale leaves at younger stages and rudimentary buds or "eyes" in the axils (Fig. 3 C). The eyes are more crowded towards the rose end (or the distal end) of the tubers. The point of attachment with the branches is the heel end.

Twiners: These are weak-stemmed plants which coil round any support. Coiling may be anti-clockwise from the top (sinistrorse climbers) or clockwise (dextrorse climbers). *Clitoria ternatea* is a sinistrorse and *Dolichos lablab* is a dextrorse climber.

The Leaf

Foliage leaf is the green flattened structure borne at nodes on stem and branches subtending a bud in its axil. Leaves develop from leaf primordia, seen as exogenic protuberances on the shoot apex.

Leaf apex: Apex of the lamina.

Acuminate: A type of acute apex whose sides are more or less concave (Fig. 9 A).

Acute: Ending in a sharp point; the sides of the tapered apex essentially straight or slightly convex (Fig. 9 B).

Apiculate: Terminated by a short sharp flexible point (Fig. 9 C).

Aristate: Tapering to a narrow, finely elongated apex (Fig. 9 D).

Caudate: Bearing a tail-like appendage (Fig. 9 E).

Cirrhose: Slender coiled apex, e.g. *Gloriosa* (Fig. 9 F).

Cleft: Divided upto ¼ or ½ distance to midpoint of blade; lobes rounded (Fig. 9 G).

Cuspidate: Acute but coriaceous and stiff (Fig. 9 H).

Emarginate: An obtuse apex that is deeply notched as in *Bauhinia* (Fig. 9 I).

Mucronate: Obtuse apex with a sharp, pointed tip as in *Catharanthus roseus* (Fig. 9 J).

Mucronulate: When the mucron in broader than long, straight (Fig. 9 K).

Obcordate: Deeply lobed at the apex; opposite of cordate (Fig. 9 L).

Obtuse: Margins straight to convex, forming a terminal angle of more than 90°, as in *Ficus benghalensis* (Fig. 9 M).

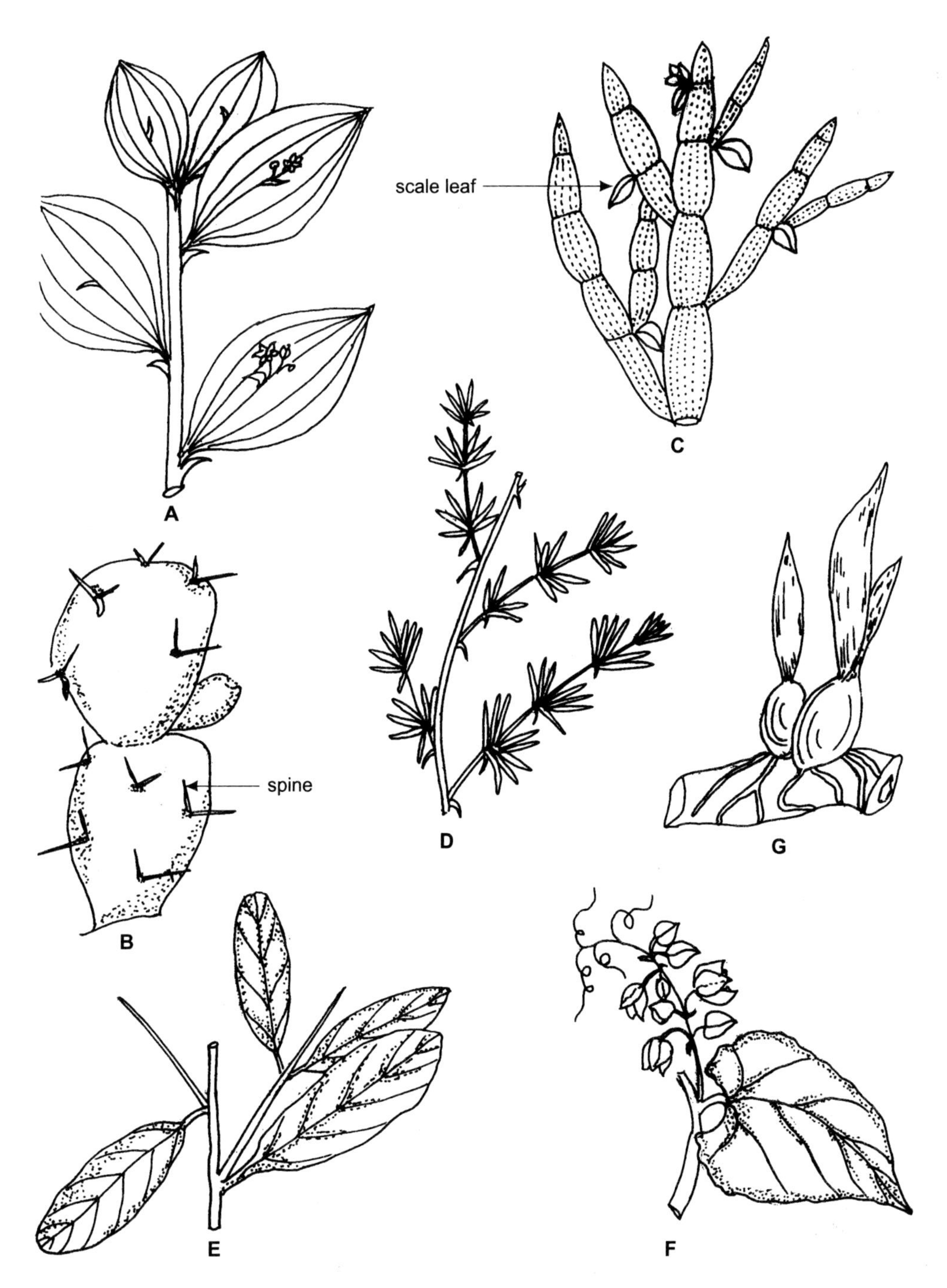

Fig. 5 **A–G** Aerial modifications of stem: **A** Phylloclade of *Ruscus.* **B** Phylloclade of *Opuntia.* **C** Phylloclade of *Muehlenbeckia platiclados.* **D** Cladode of *Asparagus racemosus.* **E** Thorn of *Punica granatum.* **F** Tendril of *Antigonon leptopus.* **G** Pseudo-bulb of orchid.

Retuse: When the obtuse apex is slightly notched, as in *Pistia* (Fig. 9 N).

Rounded: Margins and apex forming a smooth arc (Fig. 9 O).

Spinose: Margin ending into a number of spinscent end (Fig. 9 P).

Truncate: Appearing as abruptly cut off at the end (Fig. 9 Q).

Leaf Base

Basal portion of the lamina.

Attenuate: Showing a long gradual taper.

Auriculate: An ear-shaped appendage at the base, particularly in sessile leaves.

Connate: Basal portions of opposite leaves fused together, giving the appearance of a single leaf through the centre of which the stem passes as in *Swertia chirayita*, *Canscora diffusa* (Fig. 7 G).

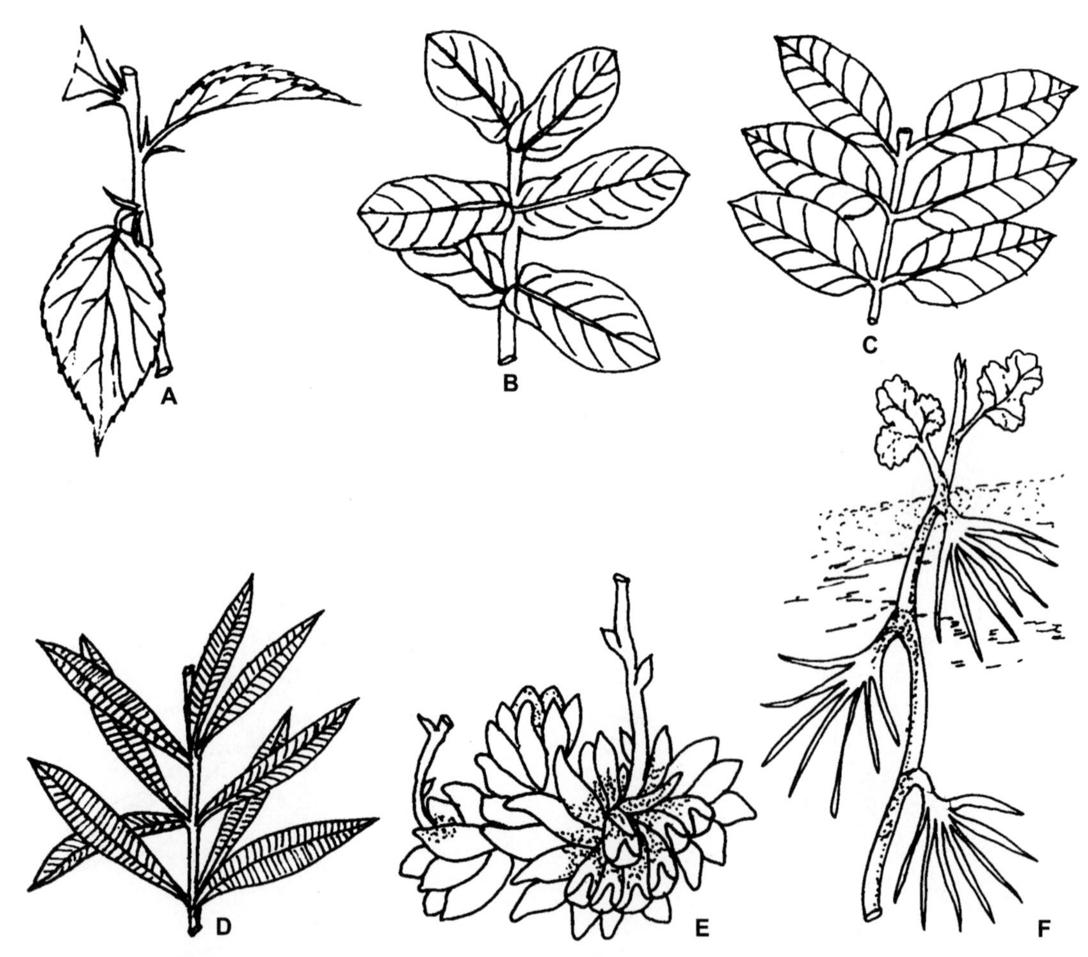

Fig. 6 **A–F** Phyllotaxy of leaf: **A-D** Cauline leaves, **E** Radical leaves, **F** Heterophylly. **A** Alternate, *Hibiscus rosa-sinensis.* **B** Opposite-decussate, *Calotropis procera.* **C** Opposite-superposed, *Quisqualis indica.* **D** Verticillate, *Nerium odorum.* **E** Radical, *Saxifraga macuabiana.* **F** Heterophylly in *Ranunculus aquatilis.*

Cordate: Rounded lobes at the base.

Cuneate: Wedge-shaped or triangular, with the narrow end at the point of attachment e.g. *Pistia stratiotes*.

Oblique: Slanting; or unequal-sided, as in Solanaceae members.

Obtuse: Blunt or rounded base.

Perfoliate: Basal lobes fused together after completely clasping the stem as in *Bupleurum* (Fig 7 G).

Truncate: Appearing as if cut off abruptly at the end.

Leaf Lamina

The flat, green part of the leaf, carrying out the most important metabolic activities like photosynthesis, respiration, and transpiration.

Abaxial: Surface of the leaf away from the stem, also called dorsal or lower surface.

Adaxial: Surface of the leaf facing the stem, also called ventral or upper surface.

Cordate: Heart-shaped; with a sinus and rounded lobes at the base and ovate in general outline.

Cuneate: Wedge-shaped; triangular, with the narrow end at point of attachment.

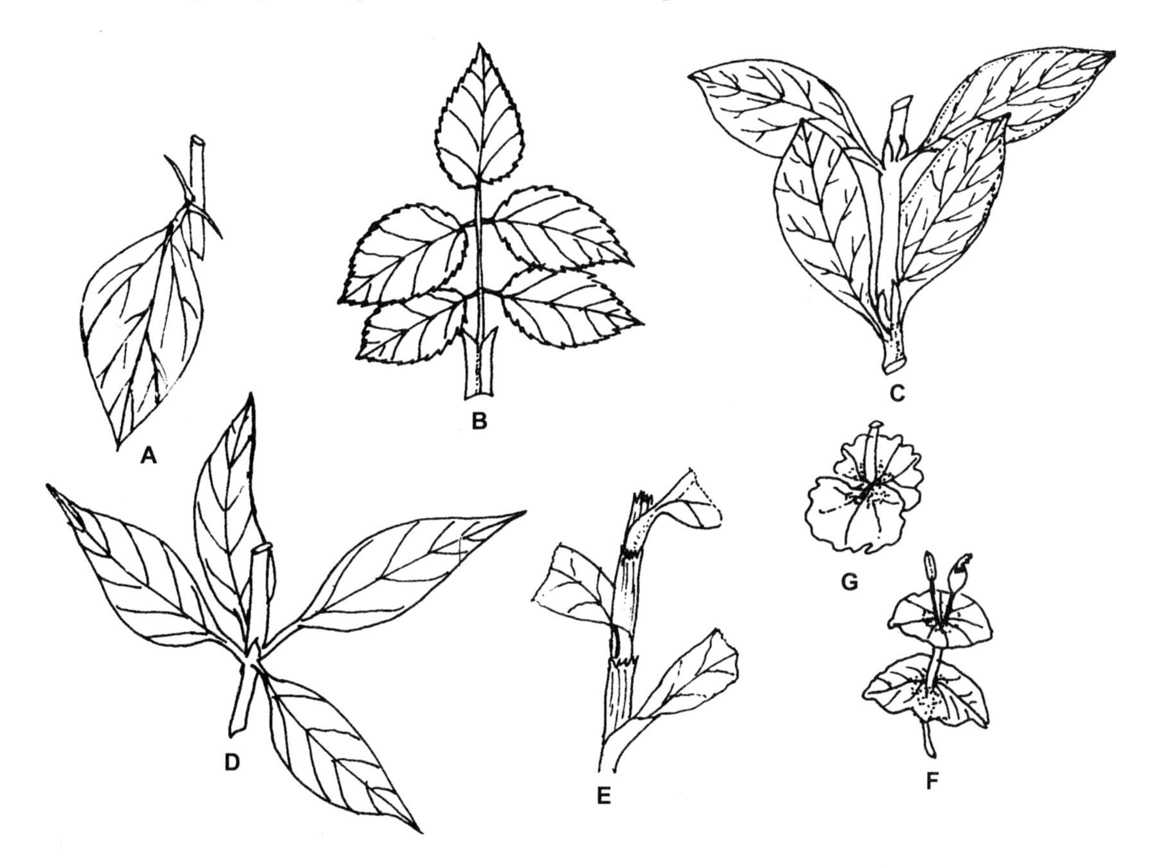

Fig. 7 **A–G** Stipules **A-E,** Leaf base **F, G:** **A** Free-lateral stipules, *Hibiscus rosa-sinensis.* **B** Adnate, *Rosa indica.* **C** Intrapetiolar, *Gardenia.* **D** Interpetiolar, *Ixora.* **E** Ochreate, *Polygonum hydropiper.* **F** Perfoliate leaf base, *Bupleurum.* **G** Connate leaf base, *Canscora diffusa.*

Dorsiventral: If the two surfaces, dorsal and ventral, differ in structure.

Elliptical: Oval in outline with narrowed to rounded ends and broadest almost in the middle.

Hastate: Triangular, with the basal lobes directed outwards, i.e. away from the petiole: *Alocasia formicata.*

Isobilateral: The two surfaces, dorsal and ventral, are similar.

Lanceolate: Lance-shaped, as in *Nerium* (Fig 6 D).

Linear: Long and slightly broad.

Lyrate: Pinnatifid, but with enlarged terminal lobe and smaller lower lobes.

Obcordate: Deeply lobed at the apex; opposite of cordate.

Oblong: More or less rectangular, as in *Musa.*

Obovate: The terminal half of the lamina is broader than the basal.

Ovate: Egg-shaped, i.e. broader towards the base and the narrower end above the middle as in *Hibiscus rosa-sinensis* (Fig 6 A).

Peltate: When the petiole is attached to the centre of the leaf lamina.

Pinnatifid: Cleft or parted in a pinnate manner.

Pinnatisect: Split up to the midrib in a pinnate manner.

Reniform: Kidney-shaped.

Rotand: Orbicular; leaf circular or disc-shaped as in *Nelumbo.*

Sagittate: Triangular, with the basal lobes pointing downward or incurved, as in *Sagittaria sagittifolia.*

Spathulate: Spoon-shaped as in *Drosera burmannii.*

Subulate: Long and narrow, tapering gradually from base to apex.

Leaf Margin

Aculeate: Prickly (Fig. 10 A).

Ciliate: With cilia or trichomes protruding from margin (Fig. 10 B).

Cleft: Indentations up to 1/4 to 1/2 distance to the midrib (Fig. 10 C).

Crenate: Shallowly ascending, round-toothed (Fig. 10 D).

Crenulate: The rounded teeth smaller than in crenate (Fig. 10 E).

Dentate: Teeth pointed and sharp, pointing outward at right angles to the midvein (Fig. 10 F).

Denticulate: Diminutive of dentate (Fig. 10 G).

Entire: With a continuous margin, may or may not be ciliate (Fig. 10 H).

Incised: Margin sharply and deeply cut at random (Fig. 10 L).

Lacerate: Margin irregularly, not sharply cut (Fig. 10 M).

Laciniate: Margin cut into ribbon-like segments (Fig. 10 N)

Parted: Cut or cleft 1/2 to 3/4 distance to the midrib (Fig. 10 R).

Palmatifid: Cut palmately almost up to the tip of the petiole (Fig. 10 P).

Pinnatifid: Cut pinnately almost up to the midrib (Fig. 10 O).

Revolute: Margin rolled towards lower side (Fig. 10 Q).

Serrate: Sharp and ascending teeth (Fig. 10 I).

Serrulate: Diminutive of serrate (Fig. 10 J).

Sinuate: Wavy in a horizontal plane; shallowly and smoothly lobed, without distinctive teeth or lobes (Fig. 10 T).

Undulate: Wavy in a vertical plane (Fig 10 S).

Leaf Types

Leaves may be simple (Fig. 8 A, B), i.e. not divided into leaflets, or compound, i.e. divided into two or more leaflets.

Bifoliate palmate or pinnate: Two leaflets articulated on the top of the petiole, e.g. *Balanites roxburghii, Hardwickia binnata* (Fig. 8 E).

Bipinnate: The rachis (or midrib) is branched once and the pinnules are borne on the rachila as in *Prosopis juliflora, Mimosa pudica.*

Decompound: In this type, the branching of the rachis is of still higher order than tripinnate, as in *Daucas carota, Foeniculum vulgare* and other members of the Umbelliferae.

Fig. 8 **A–K** Leaf types: **A** Simple leaf of Gramineae (monocot). **B** Of *Digera arvensis* (dicot). **C** Paripinnate, *Sesbania.* **D** Imparipinnate, *Murraya exotica.* **E** Bifoliate, *Hardwickia binata.* **F** Trifoliate, *Oxalis corniculata.* **G** Pinnate, trifoliate, *Dolichos lablab.* **H** Palmate, trifoliate, *Aegle marmelos.* **I** Palmately unifoliate, *Citrus.* **J, K** Palmate, quadrifoliate, *Paris* and *Marsilea.*

Imparipinnate: Rachis terminates in an unpaired or odd leaflet, e.g. *Murraya exotica* (Fig. 8 D).

Imparipinnate trifoliate: Leaflets three; an elongated rachis present between the lower pair of pinnae and the odd pinna as in *Dolichos lablab* (Fig. 8 G).

Digitate: More than four leaflets borne at the same point on the top of the petiole, as in *Cleome viscosa, C. gynandra, Salmalia malabarica.*

Palmate compound leaf: The rachis does not develop at all and the leaflets are articulated at one common point on the apex of the petiole.

Pinnate compound leaf: Leaflets (pinnae) borne on the rachis (midrib) or on its branches (rachilla) or on its branchlets.

Quadrifoliate palmate: A rare type, common in *Paris quadrifolia* (Fig. 8 J) and is with four leaflets attached at the same point on the tip of the petiole; also seen in *Marsilia quadrifolia* (Fig. 8 K).

Trifoliate palmate or ternate: Three leaflets articulated on the tip of the petiole as in *Trifolium, Oxalis* (Fig. 8 F), *Aegle marmelos* (Fig. 8 H).

Tripinnate: Third order branching of the rachis with pinnules borne on these secondary branches, e.g. *Moringa oleifera*

Unifoliate: A single leaflet borne on the tip of the petiole. Presence of the joint or the articulation proves that it is not a simple leaf, as in *Citrus* (Fig. 8 I).

Unipinnate: Leaflets or pinnae borne directly on the rachis. When rachis terminates in an unpaired or odd leaflet as in *Murraya exotica* (Fig. 8 D), it is *imparipinnate.* When the pinnae are borne in pairs as in *Sesbania sesban,* it is *paripinnate* (Fig. 8 C).

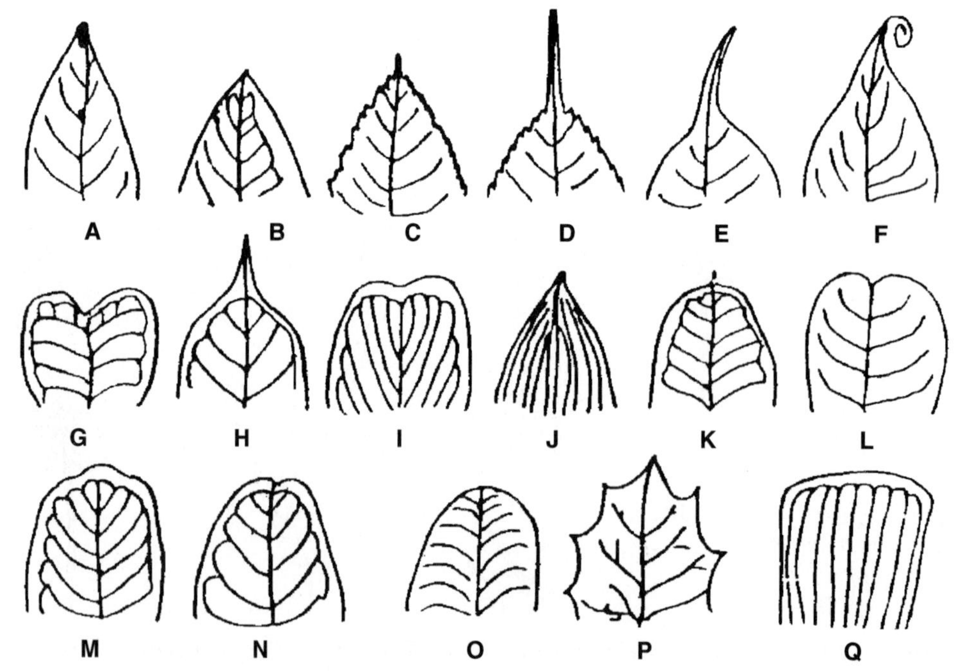

Fig. 9 A–Q Leaf apex: **A** Acuminate. **B** Acute. **C** Apiculate. **D** Aristate. **E** Caudate. **F** Cirrhose. **G** Cleft. **H** Cuspidate. **I** Emarginate. **J** Mucronate. **K** Mucronulate. **L** Obcordate. **M** Obtuse. **N** Retuse. **O** Rounded. **P** Spinose. **Q** Truncate. (Adapted from Radford 1987).

Petiole: Stalk of the leaf; leaf with petiole is *petiolate,* without it is *sessile.* The petiole is usually solid and cylindrical but may be *hollow* as in *Carica papaya* or *grooved* as in many Fabaceae or *spongy and grooved* as in *Musa, Canna; spongy and swollen* as in *Eichhornia; winged* as in *Citrus; tendriller* as in *Clematis; flattened* and *leaflike* or *phyllodial* as in *Acacia auriculiformis.* In sessile leaves of Gramineae, a membranous outgrowth, *ligule* occurs at the junction of the leaf-sheath and the lamina, on the inner surface. Sometimes the sheathing base surrounds the stem completely and then it is called *amplexicaule* as in *Polygonum orientale.* When petiole, leaf-base as well as the stem are winged, and these wings extend up to the next lower node, the condition is called *decurrent.*

Pulvinus: The swollen base of the petiole, as in many leguminous plants.

Stipules: The two lateral appendages at the base of the petiole; give protection to axillary buds. When stipules are present, the leaves are *stipulate,* when absent, they are *estipulate.* Stipules may be caducous i.e. shed very early *(Michelia champaca),* or deciduous, i.e. shed after one season (*Dillenia indica*) or persistent (*Rosa indica*). Following types of stipules are known:

Adnate: Stipules attached to petiole, e.g. *Rosa indica* (Fig. 7 B).

Free-lateral: Two, free, filiform stipules on both sides of the petiole as in *Hibiscus rosa-sinensis* (Fig. 7 A).

Foliaceous: Leafy and green, carry out photosynthesis, as in *Lathyrus aphaca.*

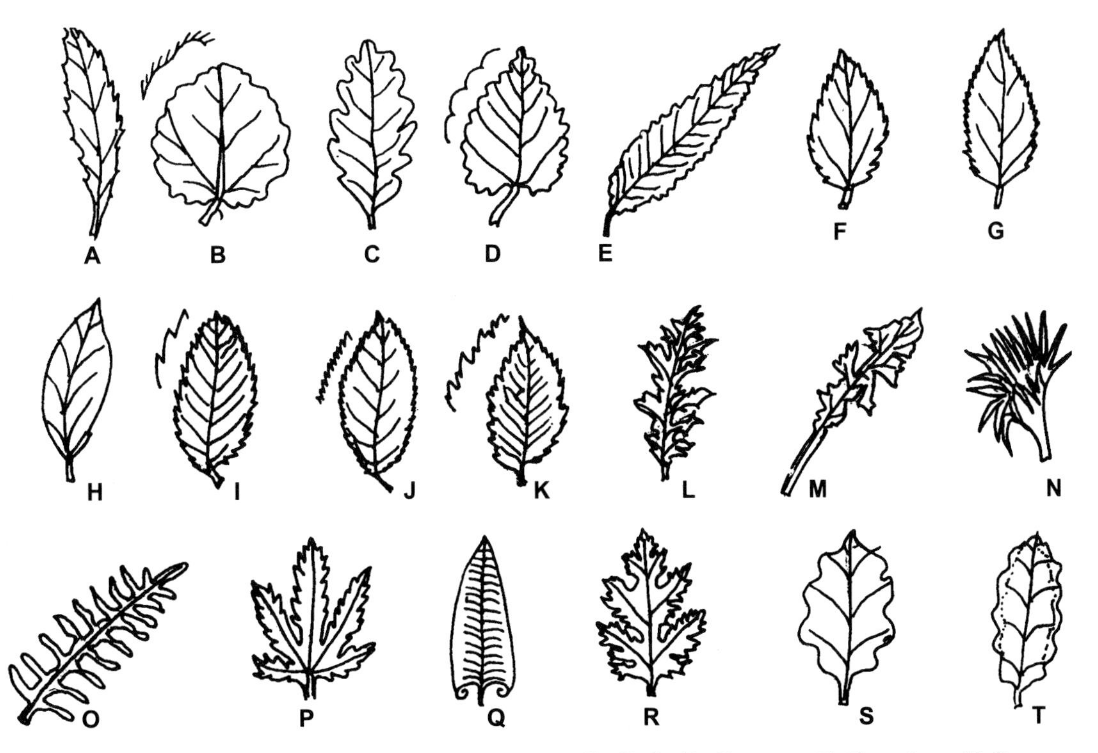

Fig. 10 **A–T** *Leaf margin*: **A** Aculeate. **B** Ciliate. **C** Cleft. **D** Crenate. **E** Crenulate. **F** Dentate. **G** Denticulate. **H** Entire. **I** Serrate. **J** Serrulate. **K** Double-serrate. **L** Incised. **M** Lacerate. **N** Laciniate. **O** Pinnatifid. **P** Palmatifid. **Q** Revolute. **R** Parted. **S** Undulate. **T** Sinuate. (Adapted from Radford 1987).

Interpetiolar: Two stipules of two opposite leaves coherent by their outer margins and appear on two sides of the stem between the petioles of two opposite leaves, e.g. *Ixora* (Fig. 7 D).

Intrapetiolar: Two stipules coherent by their inner margin, resulting in a single fused stipule in the axil of a leaf, e.g. *Gardenia* (Fig. 7 C).

Ochreate: Two stipules fusing to form a tube-like structure, covering the internode up to a certain height (Fig. 7 E), e.g. *Polygonum hydropiper.*

Tendriller: Stipules modified to tendrils as in *Smilax*; helps in climbing.

Spiny: Modified to spine serving as defensive organ, e.g. *Carissa carandus.*

Sheathing or protective: Enclosing a bud or flower.

Stipel: Stipule-like structures (scales, spines or glands) borne on the petiole of leaflets (or petiolules).

Inflorescence

Arrangement of flowers on the floral axis.

Capitulum or head inflorescence: A dense, compact inflorescence of usually sessile or only short-stalked flowers arranged on a flattened, convex or conical receptacle, in a centripetal order (Fig. 11 G). Each capitulum is surrounded by whorls of bracts called *involucral bracts*, and the individual flowers are subtended by a scaly bract each.

Catkin: A modified spike with a thin and weak axis, subtended by scaly bracts, as in many amentiferous families (Fig. 11 H).

Corymb: A more or less flat-topped indeterminate inflorescence with the outer flowers opening first, as in many Cruciferae members (Fig. 11 D).

Cyathium: A cup-shaped involucre formed by the union of bracts, enclosing a group of highly reduced staminate flowers surrounding the single, centrally placed stigmatic flower. Each flower has its individual pedicel and is subtended by a bract (Fig. 12 E-H). This is characteristic of some members of the family Euphorbiaceae (*Euphorbia, Poinsettia*).

Cymose: A determinate inflorescence in which the opening of the flowers is basipetal.

Dichasial cyme: The most common type of cymose inflorescence. Two lateral branches develop on the two sides of the terminal apical flower which is the oldest. These branches also terminate in a flower each and give out two branches which also behave in the same way (Fig. 11 O).

Helicoid cyme: A type of monochasial cyme in which the lateral branches develop on the same side forming a helix as in members of the family Boraginaceae (Fig. 11 M).

Hypanthodium: A cup-shaped receptacle with a small opening on the top and minute unisexual flowers arranged on the inner surface (Fig. 12 B–D) as in *Ficus* (Moraceae). In *Dorstenia* (Moraceae) the receptacle is saucer-shaped with slightly curved-up margin and is termed *Coenanthium* (see Ganguli et al. 1972).

Monochasial cyme: The main axis terminates in a flower and one lateral branch develops from near the base which also ends in a flower after producing another branch.

Panicle: An indeterminate branching raceme in which the branches of the primary axis are racemose and the flowers pedicellate as in *Oryza sativa* (Fig. 11 C, L).

Polychasial cyme. In this, more than two branches are formed from the base of the apical flower. These branches often end in monochasial cymes, as in *Hamelia patens* (Rubiaceae) (Fig. 12 A).

Raceme: A simple, elongated, indeterminate inflorescence with pedicelled flowers, as in *Antirrhinum*, *Brassica* (Fig. 11 A)

Racemose: An indeterminate inflorescence with older flowers near the base.

Scape: It is the flower-bearing shoot coming out from rosette of radical leaves of acaulescent, annual or perennial, sometimes bulbous plants, e.g. *Allium cepa*, *Amaryllis*; flowers may be sessile or pedicellate (Fig. 11 F).

Scorpioid cyme: A type of monochasial cyme in which the lateral branches arise alternately on opposite side of the axis; the whole structure sometimes appear as a racemose inflorescence. Within this type there may be: (i) *rhipidium,* with the lateral branches lying in the same plane as the main axis (Fig. 11 N), or (ii) *cincinnus,* with the lateral branches in angular plane, as in *Commelina.*

Spadix: A modified spike in which the rachis is thick and fleshy and the whole inflorescence is covered by one or more spathe-like bract; flowers unisexual (Fig. 11 I), e.g. *Arum.*

Spike: Similar to a raceme but the flowers are sessile, as in *Achyranthes aspera* (Fig. 11 B).

Spikelet: A secondary spike; a part of a compound inflorescence which itself is spicate as in *Triticum aestivum.* The floral unit is composed of highly reduced flowers and their subtending bracts (Fig. 11 Ja, Jb).

Strobile: A type of spike in which the individual flowers are subtended by a persistent membranous bract, as in *Humulus lupulus* (Fig. 11 K).

Thyrse: Compact, more or less compound panicle.

Umbel: A more or less flat-topped, indeterminate inflorescence in which the pedicels and peduncles arise from a common point, as in members of Umbelliferae (Fig. 11 E). Umbels may be primary or secondary.

Verticillaster: A raceme of pairs of dichasial cymes at each node, characteristic of family Labiatae. As the flowers are sessile and clustered together, they appear to form false whorls or *verticels* round the inflorescence axis. Each dichasium is subtended by a large bract (Fig. 12 I).

The Flower

The flower is the most conspicuous part of an angiospermous plant. The floral parts or the sporophylls of different types are borne on the *thalamus* or the floral axis, spirally or in a cyclic or in a spirocyclic manner. Normally, the primitive families show spiral or spirocyclic arrangement and advanced families show cyclic arrangement of the floral parts. A typical flower has four succesive floral whorls arranged on the thalamus: (i) outermost *calyx* composed of sepals, (ii) *corolla* of petals, (iii) *androecium* of stamens, and (iv) *gynoecium* of carpels. When all the four whorls are present, the flower is *complete*, if not, it is *incomplete.*

Achlamydeous: Flowers without the whorls of calyx and corolla; may be staminate, or pistillate or bisexual, subtended by bract(s).

Actinomorphic: When a flower can be cut into two equal halves through any vertical plane; also termed *regular.*

Anterior: Side of the flower facing the subtending bract, or the side opposite the stem on which it is borne.

Asymmetrical or irregular: Flowers which cannot be divided into equal halves in any plane, e.g. flower of *Canna.*

Bisexual: Flowers with both stamens and carpels (functional in each flower); also called *perfect* or *hermaphrodite.*

Bracteate: With a subtending leaf-like structure or bract at the base of the flower; *ebracteate*, when no such structure is present.

Fig. 11 **A–O** Types of inflorescence: **A** Raceme. **B** Spike. **C** Panicle. **D** Corymb. **E** Umbel. **F** Scape. **G** Capitulum. **H** Catkin. **I** Spadix. **Ja** Spike of spikelets. **Jb** Spikelet. **K** Strobile. **L** Panicle of spikelets. **M** Helicoid cyme. **N** Scorpioid cyme. **O** Dichasial cyme.

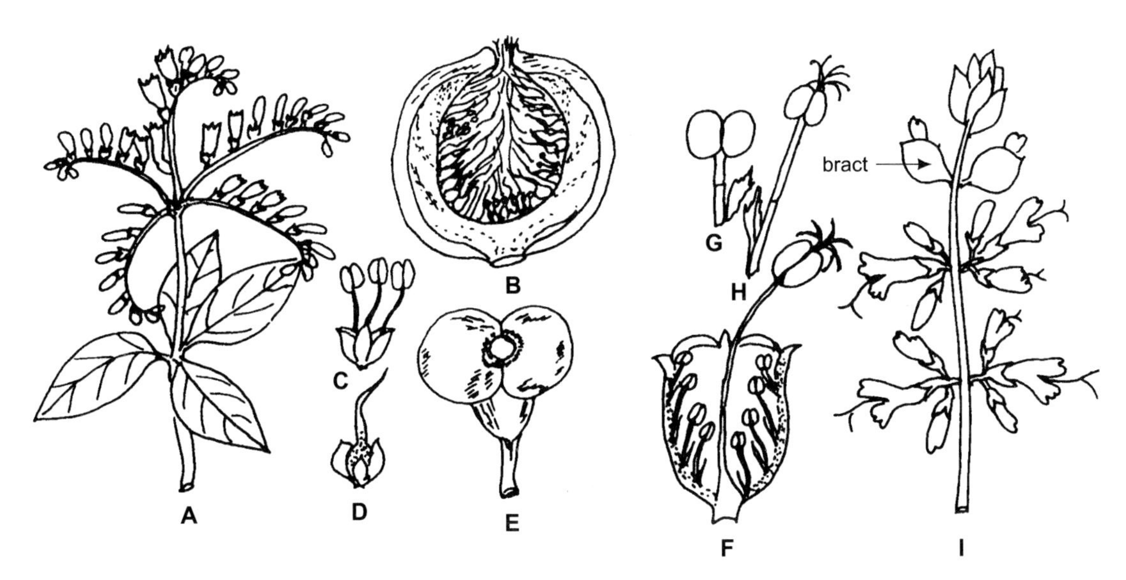

Fig. 12 **A–I** Some special types of inflorescences: **A** Polychasial cyme. **B** Hypanthodium. **C, D** Staminate (**C**) and Pistillate (**D**) flower from a hypanthodium. **E** Cyathium. **F** Vertical section. **G, H** Staminate (**G**) and Pistillate (**H**) flower from a cyathium. **I** Verticillaster inflorescence.

Bracteolate: With leaf-like structure(s) on the pedicel of the flower; *ebracteolate*, when no such structure is present on the pedicel of the flower.

Dichlamydeous: Flowers with both accessory whorls, i.e. calyx and corolla.

Dioecious: If male and female flowers are borne on separate plants, the plant is dioecious.

Epigynous: Flowers with inferior ovary.

Hypogynous: Flowers with superior ovary.

Monochlamydeous or *Haplochlamydeous:* With only one accessory whorl, i.e. calyx, corolla or perianth.

Monoecious: If male and female flowers are borne on the same plant, the plant is monoecious.

Mother axis: The axis on which the flower is borne.

Neuter or Neutral: Flowers without any functional stamen and carpel; also called *sterile.*

Perigynous: Flowers with half-inferior ovary.

Pistillate: Unisexual flowers with only carpels and sometimes with nonfunctional stamens.

Polygamodioecious: Functionally dioecious but has a few flowers of opposite sex or a few bisexual flowers on all plants at flowering time.

Polygamous: When male, female and neutral flowers are borne on the same plant.

Posterior: Side of the flower facing the stem on which it is borne.

Staminate: Unisexual flowers with only stamens, and sometimes with nonfunctional carpels.

Unisexual: Flowers with only stamens or carpels in one flower; also called *imperfect.*

Zygomorphic: The flower which can be cut into two equal halves only through one particular vertical plane.

The Calyx

The outermost (or lowermost) floral whorl; usually green, rarely petaloid, e.g. *Delphinium*; protects the flower in bud stage.

Bilabiate: Two-lipped or bilipped.

Campanulate: Bell-shaped.

Cleft: Sepals fused up to the middle.

Entire: Sepals completely united.

Gamosepalous: Sepals united to form a cup-like structure.

Partite: Sepals fused only at the base and free above.

Polysepalous: Sepals free.

Sepal: Individual part of a calyx.

Toothed: Sepals almost completely fused, only the tips are free.

Tubular: Tube-like.

Urceolate: Urn-shaped.

Modifications of Calyx

Calyptra: Calyx forms a cap-like structure, *calyptra,* which falls off at the time of anthesis, as in *Eucalyptus.*

Pappus: Calyx modified into hairy or bristly structures, as in members of Compositae.

Spines: Modified to spines, as in *Trapa bispinosa.*

Spur: One of the calyx lobes is modified into an elongated, tubular structure, i.e. spur, as in *Delphinium ajacis*, *Impatiens* spp. and *Tropaeolum majus.*

The Corolla

The second floral whorl, inner to the calyx; usually brightly coloured due to the presence of various pigments like anthocyanin, anthoxanthin and carotenoids. It protects the inner whorls and makes the flowers attractive to pollinating agents. Occasionally, petals are sepaloid. When the individual parts of corolla, the *petals* are free, the corolla is *polypetalous*, and when fused, it is *gamopetalous* or *sympetalous*

Aestivation: Arrangement of the sepals, petals or perianth lobes (tepals) in relation to one anther, in bud condition, may be of following types:

Contorted or twisted or convolute: One edge of every member overlapped by the edge of the next member in one direction, as in *Nerium, Thevetia.*

Imbricate: Floral lobes are not in one whorl; one member is completely in or overlapped from both margins and one member is completely out, i.e. overlapping the next members' margin from one side and the other three with one side in and one side out, as in *Clerodendrum.*

Quincuncial: This is a variation of imbricate type, with two members completely in and two others completely out and one member with one side overlapped and one side overlapping, as in *Psidium guajava.*

Valvate: Meeting only at the edges without overlapping, as in *Mimosa pudica.*

Vexillary: This is typical of papilionaceous corolla. The posterior largest petal is the vexillum, which overlaps the two laterals, or wings which overlap the paired anterior petals, carina or keel.

Shape (Corolla)

Bilabiate: Corolla two-lipped; gamopetalous with a posterior upper lip, of usually two petals and an anterior lower lip, of usually three petals; characteristic of the families Labiatae, Acanthaceae and Scrophulariaceae (Fig. 13 I).

Campanulate: Bell-shaped, e.g. *Cucurbita moschata* (Fig. 13 B).

Caryophyllaceous: Five free petals, with the limbs at right angles to the claws, as in the family Caryophyllaceae.

Corona: Appendages of various types such as scales or hairs developing between the corolla and stamens or on the corolla, as in some *Amaryllis*, or as outgrowths of staminal part as in *Calotropis*; function is to make the flower more attractive.

Cruciform: Four free petals arranged in the form of a Christian cross, as in family Cruciferae.

Hypocrateriform or *Salverform:* Slender corolla tube with an abruptly expanded flat limb, as in *Phlox drumondii* (Fig. 13 D).

Infundibuliform: Funnel-shaped, as in *Petunia* (Fig. 13 C).

Ligulate: Strap-shaped; a gamopetalous corolla in which the petals are united to form a very short tube, split on one side, forming a flat, ribbon-like structure, as the ray florets of many Compositae (Fig. 13 F).

Papilionaceous: See aestivation vexillary; characteristic of family Papilionaceae (Fig. 13 Ha, Hb).

Personate: A variation of bilabiate corolla, the throat is closed by a projection from the lower lip, called the palate; seen in some Scrophulariaceae members (Fig. 13 J).

Rotate: Wheel-shaped; a gamopetalous corolla with a flat and circular limb at right angles to a short or obsolete tube as in *Brunnera* (Fig. 13 Ea, Eb).

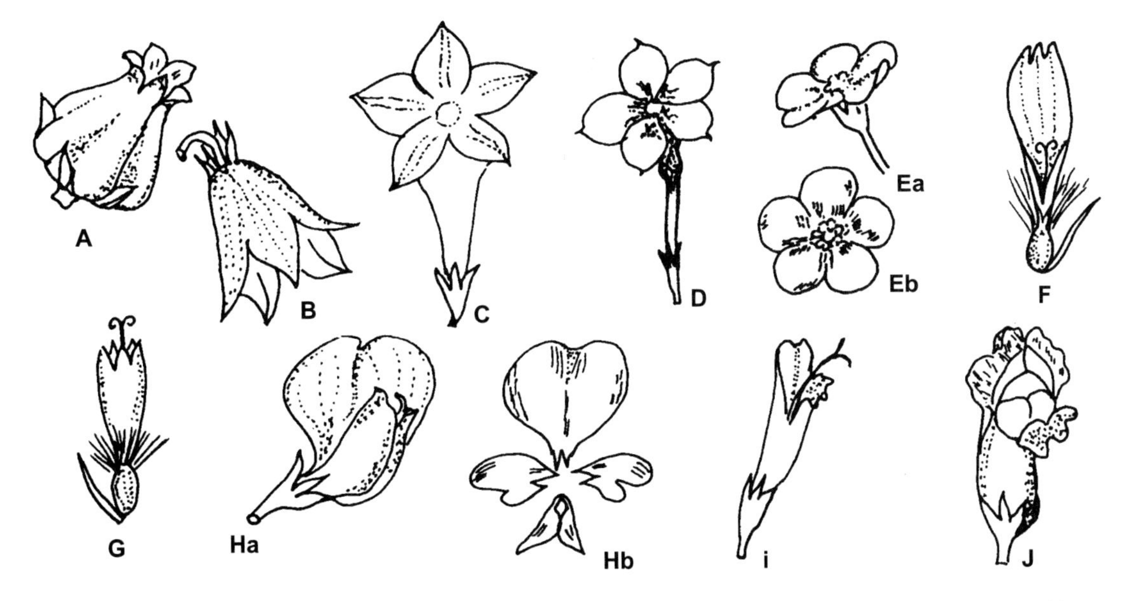

Fig. 13 **A–J** Types of corolla: **A** Urceolate. **B** Campanulate. **C** Infundibuliform. **D** Hypocrateriform. **Ea, Eb** Rotate. **F** Ligulate. **G** Tubular. **Ha, Hb** Papilionaceous. **I** Bilabiate. **J** Personate.

Saccate or *gibbous:* Corolla tube with a pouch at the base as in *Antirrhinum* (Fig. 13 J).

Spurred: Gamopetalous (*Linaria*) or polypetalous (*Aquilegia*) corolla, with one or all the petals appendaged into spurs.

Tubular: Corolla tube more or less cylindrical, limbs not spreading, e.g. the disc florets of Compositae (Fig. 13 G).

Urceolate: Urn-shaped; a gamopetalous corolla which is swollen in the middle and tapers towards both the base and the apex, as in *Pieris*, *Bryophyllum* (Fig. 13 A).

The Androecium

The third whorl of floral organs comprises the *stamens.* Each stamen consists of a stalk, termed *filament*, and a pollen-bearing *anther.*

Adnate: Filament and the connective appear to be attached to the entire length of the back of the two anther lobes, as in some Magnolialian families (Fig. 15 K).

Alternipetalous: Stamens alternate with the sepals.

Antipetalous: Stamens opposite the petals.

Antisepalous: Stamens opposite the sepals.

Apostemonous: Flower with separate stamens.

Appendiculate: Stamens with connective bearing appendage, e.g. a prolonged feathery structure in *Nerium odorum* (Fig. 15 D).

Basifixed: Filament attached to the base of the anther (Fig. 15 L).

Bithecous: Bilobed anthers.

Connective: The tissue connecting the two anther lobes.

Diadelphous: Two groups of stamens fused by their filaments (Fig. 14 B).

Didynamous: Two long and two short stamens.

Diplostemonous: Stamens in two whorls, outer opposite the sepals and inner opposite the petals.

Discrete: The connective tissue is almost nil and the anther lobes remain very close together, as in *Euphorbia* (Fig. 15 A).

Distractile: The connective is highly elongated and bears a fertile lobe on one end and the other end is sterile, e.g. *Salvia officinalis* (Fig. 15 C).

Divaricate: The connective appears to have bifurcated and the two anther lobes are separated from each other, as in *Tilia* (Fig. 15 B).

Dorsifixed: Filament attached firmly to the back of the anther (Fig. 15 M).

Epipetalous: Stamens attached to the base of the petals of a gamopetalous corolla.

Extrorse: Dehiscence of anthers longitudinal and face the outside of the flower or the petals.

Gynandril or *Gynostaminal:* Stamens and carpels fused, as in Asclepiadaceae and Orchidaceae.

Introrse: Dehiscence longitudinal and face the inside of the flower or the carpels.

Latrorse: Dehiscence longitudinal and lateral.

Longitudinal: When the thecae (anther lobes) dehisce along longitudinal sutures (Fig. 15 F).

Monadelphous: One group of stamens fused by their filaments, as in Malvaceae (Fig. 14 A).

Monothecous: Anthers with one lobe only, e.g. Malvaceae.

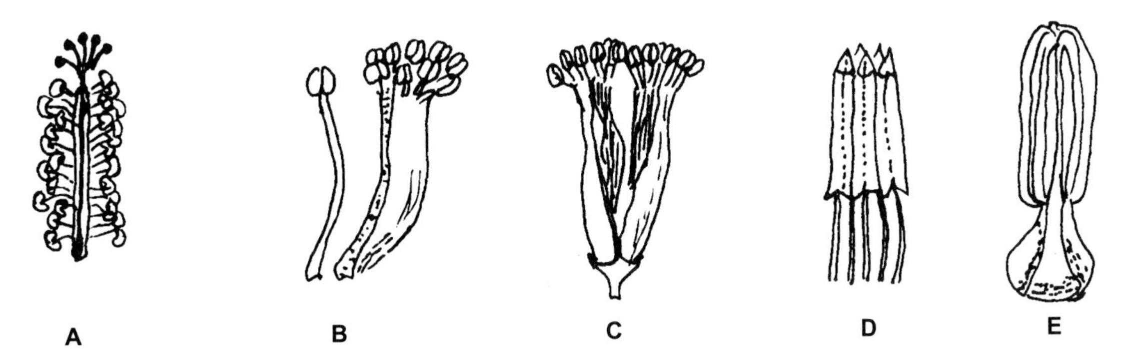

Fig. 14 **A–E** Types of androecium: **A** Monadelphous. **B** Diadelphous. **C** Polyadelphous. **D** Syngenesious. **E** Synandrous.

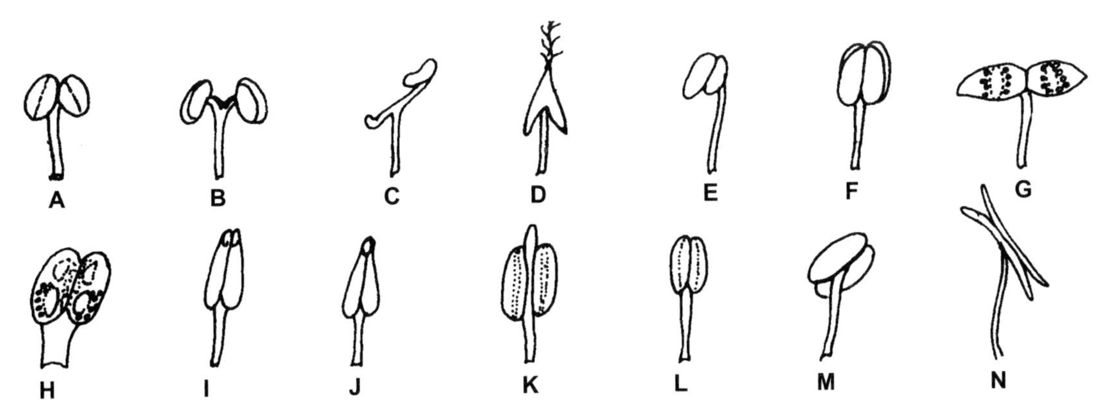

Fig. 15 **A–N** Anthers. **A-D** Types of anthers, **E-J** Types of dehiscence, **K-N** Types of attachment of anthers: **A** Discrete. **B** Divaricate. **C** Distractile. **D** Appendiculate. **E** Oblique. **F** Longitudinal. **G** Transverse. **H** Valvular. **I, J** Poricidal. **K** Adnate. **L** Basifixed. **M** Dorsifixed **N** Versatile.

Obdiplostemonous: Stamens in two whorls, outer opposite the petals and inner opposite the sepals, e.g. *Murraya exotica.*

Polyadelphous: More than two groups of stamens connate by their filaments (Fig. 14 C).

Poricidal: Dehiscence of anthers through a pore at the tip of each lobe as in *Cassia fistula, C. occidentalis* (Fig. 15 I, J).

Synandrous: Stamens in three bundles of 2 + 2 + 1; each pair of stamens has completely fused filaments and the anthers are sinuous or S-shaped, e.g. *Momordica charantia*, *Luffa cylindrica* (Cucurbitaceae) (Fig. 14 E).

Syngenesious: Stamens with free filaments but fused anthers as in Compositae (Fig. 14 D).

Tetradynamous: Four long and two short stamens.

Transverse: Dehiscence at right angles to the long axis of theca (Fig. 15 G).

Valvular: Dehiscence through pore(s) on the surface of the thecae covered by a flap of tissue as in *Berberis*, *Laurus* (Fig. 15 H).

Versatile: Filament attached merely at a point, about the middle of the connective (Fig. 15 N), to make the anthers swing, e.g. Gramineae.

The Gynoecium

The innermost or the central floral whorl comprises one or more carpels. A typical carpel has three parts: ovary, style, and stigma.

Bicarpellary: A gynoecium of two carpels.

Monocarpellary: A gynoecium of only one carpel.

Polycarpellary: A gynoecium of more than one or two carpels: may be *apocarpous* when the carpels are free, or *syncarpous* when the carpels are fused.

Syncarpous: Carpels fused; a syncarpous gynoecium may show different degrees of union. For example, in *Hibiscus rosa-sinensis*, ovaries and styles are fused but the stigmas are free. In *Dianthus,* only the ovaries are united, the styles and stigmas are free. In Apocynaceae and Asclepiadaceae, the ovaries and styles are free and the stigmas are fused to form a stigmatic disc.

The Ovule

Apical: Ovule pendulous or suspended, with the placenta near the apex of the ovary, as in *Coriandrum sativum.*

Basal: Erect, with the micropyle facing downward as in Compositae, or ascending, i.e. rising obliquely from the side of the ovary wall, as in some Ranunculaceae.

Horizontal: Ovule arising almost from the middle of the side wall, as in *Podophyllum.*

Placenta: Ovules in the locules are attached to a tissue called placenta which runs along the margin of the carpels. Placenta may also develop on a direct prolongation of the thalamus, at the base of the ovary. Depending upon the distribution of placental tissue, there may be various types of placentation (Fig. 16).

Axile: Placentae along the central axis of a syncarpous, multilocular ovary, as in members of Malvaceae, Solanaceae, Liliaceae, Commelinaceae.

Basal: Placenta at the base of an unilocular ovary, apparently placed on the tip of the thalamus at the floor of the ovary, as in Compositae.

Free Central: Placenta along the central axis of the ovary and the ovary is polycarpellary and unilocular. This condition develops in two different ways:

(a) In an ovary with axile placentation, the walls forming the locules may break down at a later stage, so that the ovary is apparently unilocular, e.g. *Dianthus.*

(b) The thalamus may extend into the ovary for a short distance and the placenta develop on this extended portion, as in members of Primulaceae.

Laminate: Placenta all over the inner surface of the ovary wall, as in *Nymphaea.* Here, the ovary is multicarpellary and multilocular.

Marginal: Placenta along the margin of a unilocular ovary, e.g. Leguminales.

Parietal: Placentae on the wall or the intruding partitions of a syncarpous but unilocular ovary, rarely bi- or trilocular, as in Cruciferae and Cucurbitaceae Bilocular condition in Cruciferae is due to the formation of a false septum called replum.

Superficial: Same as *laminate.*

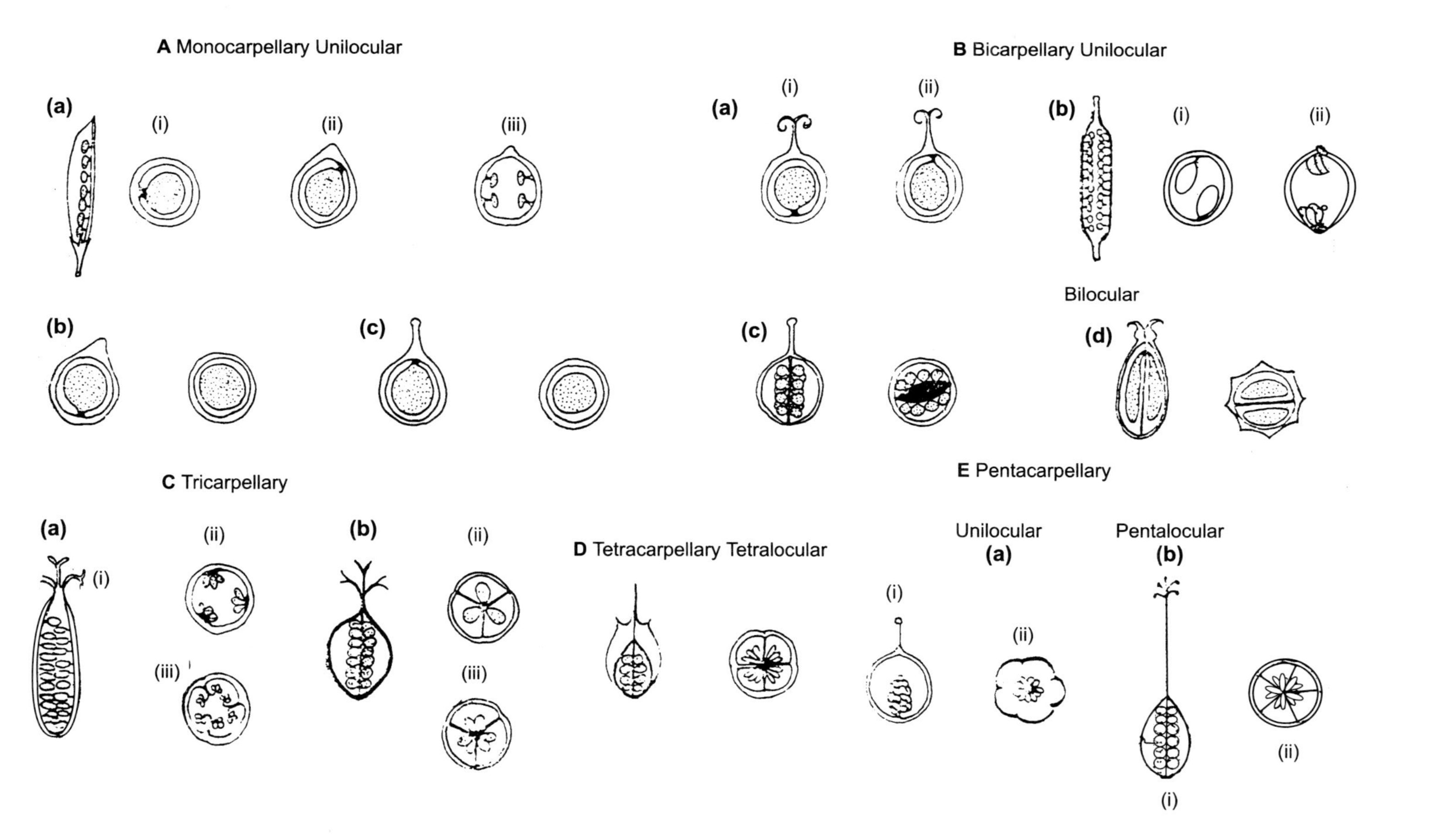

Fig. 16 **A–E** Types of placentation: **A** Monocarpellary. **a (i), (ii), (iii)** Marginal. **b** Basal. **c** Apical. **B** Bicarpellary—Unilocular **a (i)** Basal. **a (ii)** Apical. **b (i), (ii)** Parietal. Bilocular **c** Axile. **d** Apical. **C** Tricarpellary—Unilocular **a (i), (ii), (iii).** Parietal Trilocular. **b (i), (ii) (iii)** Axile. **D** Tetracarpellary—Tetralocular **a** Axile. **E** Pentacarpellary—Unilocular **a (i), (ii)** Free-central. Pentalocular **b (i), (ii)** Axile.

The stigma

The stigma is the uppermost part of the carpel, may be of the following types:

Capitate: Head-like or knob-like, e.g. *Petunia.*

Clavate: Club-shaped, e.g. *Dendrophthoe falcata, Musa paradisiaca.*

Discoid: Disc-like, e.g. *Hibiscus rosa-sinensis*.

Diffuse: Spread over a wide surface, e.g. *Elmerillia* sp.

Fimbriate: Fringed, e.g. *Najas* sp.

Lobed: Divided into parts, such as bifid (Bignoniaceae), trifid (Cucurbitaceae).

Plumose: Feather-like, as in Gramineae.

Striate: Star-like, or radiate, as in *Papaver somniferum.*

Terete: Cylindrical and elongate, e.g. *Casuarina equisetifolia.*

Truncate: Cut abruptly at right angles to the long axis, as in *Acacia nilotica.*

In *Begonia* the stigma is highly branched, in Euphorbiaceae all the three stigmas are bifid, and in *Crocus sativus.* the stigmas are funnel-shaped.

Stylopodium: The swollen base of the style as in Umbelliferae

The Style

Gynobasic: Arising from the centre of the depressed apex of the ovary, as in *Salvia*, *Ocimum.*

Lateral: Arising from the side of the ovary, as in *Mangifera indica.* In *Gloriosa superba* the style is positioned at right angles to the ovary axis.

Terminal: Arising from the tip of the ovary, common in most plants.

Glossary of Terms

A priori: Pre-assumed to be correct/predetermined.

Acropetal: Proceeding from basal or proximal end to the distal end.

Adaptation: Modified growth form in response to environmental conditions.

Adaxial: Towards the axis.

Adnate: Different organs attached with each other.

Aestivation: Arrangement of sepals and/or petals with respect to each other.

AFM: Atomic Force Microscopy

Ament: Dense, bracteate spike of unisexual and naked flowers.

Amentiferae: A group of dicotyledonous plants with inflorescence an ament or catkin.

Androgynophore: An axis which arises from the receptacle and bears both androecium and gynoecium.

Aneuploidy: A series of chromosome numbers without simple numerical relationship.

Anthesis: Flowering period when pollination takes place.

Anthocyanin: Flavonoid pigments, colour ranges from blue or violet to purple or red, occur in the central vacuole of a cell, especially in petals.

Anthostrobilus: An imaginary flower-like cone.

Anthotaxy: A term used as against phyllotaxy for the floral parts.

Anthoxanthin: A group of flavonoid pigments, closely allied to anthocyanin, colour ranges from yellow or orange to orange-red.

Antibodies: Bodies developing in the blood serum in reaction to foreign material and causing clumping of the blood.

Antigen: Bodies occurring in red blood cells. They cause clumping of blood.

Antipetalous: Opposite the petals

Antisepalous: Opposite the sepals

Antiserum: Serum from immunised blood.

Antitropous: Broader halves of leaves on alternate side of midrib.

Apogamy: Producing sporophytes from a gametophyte without fertilization.

Apomorphic: Advanced

Apotracheal: Xylem parenchyma away from the vessels.

Aptian: Geological period of Cretaceous era.

Attributes: Characters.

Barremian: Geological period during the Cretaceous.

Basal: Ancestral.

Bennettitales: A group of mesozoic, cycad-like fossil pteridosperms.

Betacyanin: A chemical class of nitrogenous, water-soluble pigments, colour ranges from blue or violet to purple or red.

Betalain: Nitrogenous, water-soluble pigments—betacyanins and betaxanthins.

Betaxanthin: Pigments allied to betacyanin, but colour ranges from yellow or orange to orange-red.

Bolster-like: Swollen like a pillow, or cushion-like.

Caruncle: A growth near the hilum of some seeds such as castor.

Cauliflory: Flowers borne on the main stem or older branches, as in *Cocoa.*

Caytoniales: Mesozoic pteridosperms.

Centrifugal: Arranged from center to periphery when applied to the sequence of maturation of stamens in a flower.

Centripetal: Arranged from periphery to center when applied to the sequence of maturation of stamens in a flower.

Cladistic: A type of analysis of characters, which attempts to summarise knowledge about the similarities among organisms in terms of a branching diagram called a cladogram.

Cladogram: A branching diagram based on cladistic analysis.

Coenospecies: These are groups of related ecospecies in which gene exchange between the assemblages is essentially absent.

Comparium: A biosystematic category corresponding to genus.

Confluent: Blending or merging together.

Connate: Attached to each other, but the organs are similar (like-organs).

Corona: A crown; any outgrowth between the androecium and corolla. It may staminal or corolline in origin.

Cretaceous: Geological period of Mesozoic era.

Cupule: Cup-like structure at the base of some fruits formed by the dry and enlarged floral envelopes.

Cycadales: A group of living gymnosperms.

Cymbiform: Boat-shaped

Cymule: Diminutive of cyme; of usually a few flowers.

Dendrogram: Representation in an elevation view or vertical axis; when based solely on phenetic data it is a *phenogram. Phylogram*: The branching pattern along the vertical axis represents ancestry in time; in *phenogram*, the representation is of the pattern of phenetic relationship and in a *cladogram*, the pattern is of ancestral relationship.

Diagnostic characters: Important distinguishing characters.

Domatia: Depressions, pockets, sacs or tufts of hairs on the principal vein, vein-axils where they occur exclusively on the abaxial surface of leaves. Predominantly seen in woody plants of humid tropical or subtropical regions (Metcalfe and Chalk 1979).

Ecospecies: A biosystematic category approximately equivalent to biological species; a group capable of hybridizing with other such groups to form hybrids showing some fertility; an assemblage of ecotypes.

Ecotype: "All the members of a species that are fitted to survive in a particular kind of environment within the total range of the species" (Clausen, Keck, and Hiesy 1945): group capable of hybridizing with other such groups to form hybrids that show complete fertility.

Embryo sac: The female gametophyte. Depending upon the number of megaspore nuclei taking part in the development, these may be monosporic, bisporic or tetrasporic.

Monosporic embryo sacs: Polygonum type—8-nucleate at maturity; Oenothera type—4-nucleate at maturity.

Bisporic embryo sacs: Allium type—8-nucleate at maturity; Podostemon type—4-nucleate at maturity.

Tetrasporic embryo sacs: Peperomia type—16-nucleate at maturity; Penaea type-16-nucleate at maturity; Drusa type—16-nucleate at maturity; Fritillaria type—8-nucleate at maturity; Plumbagella type—4-nucleate at maturity; Plumbago type—8-nucleate at maturity, tetrapolar; Adoxa type-8-nucleate at maturity; bipolar; Chrysanthemum cinerariaefolium type—10- to 12-nucleate at maturity.

Endemic: With a restricted distribution. Applied to plants and animals available in a narrow range of distribution.

Endemism: Restricted distribution of a taxon.

Endosperm: A starch- or oil-filled tissue in mature seeds. There are three types: *Cellular*—Cell walls develop immediately after mitosis of primary endosperm nucleus so that there is no initial free-nuclear stage. *Nuclear*—Endosperm which has a free-nuclear phase throughout or only during early ontogeny. Wall formation takes place only after several mitotic divisions have occurred. *Helobial*—Endosperm in which the first division of the triple-fusion-nucleus is followed by the formation of a partition wall; afterwards the micropylar chamber becomes cellular and the other remains free-nuclear. Many variations in each type.

Epiphyte: A plant growing on branches and sometimes leaves of another plant but not deriving its food or water from it.

Equitant: Overlapping of leaves in two ranks, as in members of the family Iridaceae.

Exine: The outer wall layer of a pollen grain.

Extant: Taxa living at present.

Extinct: Taxa not living at present.

Extinction: Total disappearance of a taxon from a particular site.

Farinose: Surface covered with a mealy (starch or starch-like materials) coating, as leaves of some *Primula* spp.

Fetid: With a disagreeable odour.

Fluvial: Areas that are frequently inundated.

Follicetum: An aggregate of follicles; product of a multipistillate, apocarpous gynoecium.

Follicle: Fruit derived from a single carpel that dehisces along one suture at maturity, as in *Delphinium.*

Foveolate: Pitted.

Geniculate: Bent knee-like.

Gibbous: Swollen on one side, usually at the base as in *Antirrhinum majus flower.*.

Glabrous: Surface without hairs.

Glaucous: Surface covered with very fine hairs.

Glutinous: Sticky

GLC: Gas-liquid Chromatography

Glossopteridales: A group of fossil pteridosperms.

Glumaceous: Glume-bearing or glume-like in appearance.

Gnetales: A group of living gymnosperms.

Haemolysis: Destruction of red-blood-cells (RBC).

Haustoria: The nutrition absorbing extensions of pollen, megaspore, embryo sac, endosperm and embryo suspensor. Also in roots and shoots of parasitic plants.

Heterobathomy: Occurrence of both primitive (pleseiomorphic) and advanced (apomorphic) features in a taxon at the same time due to unequal rate of evolution of different features within one lineage.

Hippocrepiform: Horseshoe-shaped.

Homonym: Same name applied to two or more taxa of the same rank.

Homotropous: Broader halves of leaves on the same side of midrib, right or left.

Hypanthium: The cup-like receptacle derived usually from the fusion of floral envelopes and androecium, and on which the calyx, corolla and stamens are apparently borne.

Hypocotyl: The axis of the embryo below the cotyledons which—on seed germination—develops into the radicle.

Hypothetical: Imaginary; without any evidence.

Indumentum: A thick pubescence.

Inflated: Bladder-like or balloon-like.

Interxylary phloem: Included phloem i.e. phloem present as patches in the secondary xylem tissue.

Intraxylary phloem: Phloem present inner to primary xylem i.e. towards the pith.

Introrse: Turned inward as an anther with the line of dehiscence towards the center of the flower.

Involucel: A secondary involucre.

Involucre: One or more whorls of small leaves or bracts borne underneath an inflorescence or a cluster or flowers.

Jurassic: Geological period of the early Mesozoic era.

Karyotype: Morphology of chromosomes at somatic metaphase.

Keel: The two anterior united petals of a papilionaceous corolla.

Labellum: Lip; part of a perianth as in the flower of orchids.

Lactiferous: Producing latex.

Laminar placentation: An arrangement of ovules on the entire inner surface of carpel.

Laminar: Thin and flat like a leaf.

Lax: Loose.

Lemma: The fertile bract or glume as in grass florets.

Lepidote: Covered with small scurfy scales (exfoliating scaly incrustations).

Leucoanthocyanins: Colourless anthocyanins.

Leveés: Natural embankment of alluvium built up by river on either side of the channel.

Ligneous: Woody.

Living fossil: Present day members with all its ancient characters retained.

Lorate: Strap-shaped (apex not acute or acuminate).

Macropodous or Macropodial: With food reserve stored in hypocotyl.

Megasporophylls: Modified leaf that bears ovules (in gymnosperms).

Mesozoic: Geological period.

Microsporophylls: Modified leaf that bears microsporangia.

Molecular data: Nuclear rRNA sequencing and chloroplast gene rbcL data.

Mosaic evolution: The phenomenon of unequal rate of evolution of different features within one lineage.

Mycorrhiza: A symbiotic association of a fungus and the root of a vascular plant.

Myrosin: An enzyme involved in the formation of mustard oil.

Nut: An indehiscent, 1-seeded hard and bony fruit.

Nutlet: A diminutive or small nut.

Obdiplostemonous: Stamens in 2 whorls, outer opposite petals, inner opposite sepals.

Ochreate: Formation of a nodal sheath by fusion of two stipules.

Ovule: Embryonic seed consisting of integument(s) and nucellus.

Ovule types: *Amphitropous*—An ovule with the body half-inverted so that the funiculus is attached near the middle.

Anatropous—The ovule with the body fully inverted so that the micropyle is basal, adjoining the funiculus.

Campylotropous—An ovule curved by uneven growth so that its axis is approximately at right angles to its funiculus (stalk).

Hemianatropous—Same as amphitropous.

Circinotropous—The ovule after attaining fully inverted position (anatropous ovule), continues the curvature until it has turned over completely and the micropylar end again points upwards as in *Opuntia.*

Orthotropous—Straight or unbent ovule with the micropyle at the opposite end from the stalk or funiculus.

Epitropous—An erect ovule with dorsal raphe, or a pendulous ovule with ventral raphe.

Crassinucellate—Ovule with the nucellus several cells thick at least at the micropylar end.

Tenuinuceliate—Ovule with the nucellus of a single layer of cells.

Paleozoic: Geological period.

Palynology: Study of pollen and spores.

Paratracheal: Xylem parenchyma in close association with vessels.

Parthenocarpy: A type of agamospermous reproduction in flowering plants.

Perisperm: Food storage tissue in seed, derived from nucellus.

Permion: Geological period.

Phenetic: Pertaining to the expressed characteristics of an individual, irrespective of its genetic nature.

Phenogram: A dendrogram of phenetic relationships.

Phenon: Groups of similar organisms recognised by numerical methods.

Phylad: A natural group of any rank, considered from the standpoint of its evolutionary history.

Phyllosporous: Ovules borne on modified leaves.

Phylogeny: The origin and evolution of taxa.

Phytochemistry: Plant chemistry.

Pleseiomorphic: Primitive.

Pollen grains: Young male gametophytes contained in the anther. Pollen grain types:

Multipalynous (syn. *eurypalynous*)—Taxa with more than one pollen morphoforms.

Unipalynous (syn. stenopalynous)—Taxa with one pollen morphoform.

Inaperturate—Without any aperture.

Pantoporate—With many apertures.

Triporate—With three apertures.

Uniporate—With one aperture.

Pollinia: Pollen grains in uniform coherent masses.

Polyclave: A multientry, order-free key.

Polygamodioecious: Functionally dioecious but with a few flowers of opposite sex or a few bisexual flowers at the time of flowering.

Population: A group of individuals of the same species.

Prophyll: Bracteole.

Pseudanthia: Cluster of small or reduced flowers, collectively appearing as a single flower.

Pseudomonomerous: An ovary apparently composed of a single carpel but phyletically derived from a compound or polycarpellary ovary.

Psychomimetics: Act as hallucinogens.

Pteridosperms: Tree ferns.

Raphe: The portion of the funiculus of an ovule that is adnate to the integument, usually represented by a ridge; present in anatropous ovules.

Raphide: Needle-shaped crystal of calcium oxalate occurring in special sac-like cells.

Raphides: Acicular or needle-shaped crystals of calcium oxalate.

Rbc L or Rubisco: Rubisco or Ribulose, 1,5 bisphosphate carboxylase enzyme is located on the stromal surface of the thylakoid membrane. It is the most abundant protein in the biosphere, particularly in the chloroplast and is involved in CO_2 fixation. This enzyme consists of 2 subunits, each encoded by a separate gene, the smaller subunit rbc S (encoded by a nuclear gene) and the larger subunit rbc L (encoded by a chloroplast gene).

Relict: Old.

Reproductive isolation: Isolation mechanism that prevents inter-breeding of closely related taxa.

Resupinated: Twisted by 180°.

Retrorse: Bent or turned over backward or downward.

Rugose: Covered with wrinkles.

Ruminate: Coarsely wrinkled or mottled. A ruminate endosperm is formed due to invaginations of outer tissues, which penetrate deeper and deeper and appear as dark wavy bands in mature seeds as in *Areca* nut, Annonaceae and Myristicaceae members.

Santonian: Geological period .

Scarious: Thin, dry and membranaous.

SEM: Scanning Electron Microscopy.

Semantide: Molecules of DNA and RNA and proteins that are information-carrying.

Serology: Study of antigen-antibody reaction by employing known serum.

Squamiform: Scale-like.

Stenopalynous: With uniform pollen morphology.

Stipel: Stipule of a leaflet.

Stomatal types:

Anomocytic or *Ranunculaceous* type—Stoma remains surrounded by a limited number of cells which cannot be distinguished from other epidermal cells.

Anisocytic or *Cruciferous* type—Stoma remains surrounded by three subsidiary cells of which one is distinctly smaller than the other two.

Diacytic or *Caryophyllaceous* type—Stoma remains enclosed by two subsidiary cells whose common wall is at right angles to the guard cells.

Paracytic or *Rubiaceous* type—Stoma is surrounded by one or more subsidiary cells on either side which lie parallel to the long axis of the stomatal pore.

Symplesiomorphy: A monophyletic group sharing primitive character states.

Sympodial branching: A type of dichotomous branching that gives rise to a false axis as one of the branches after bifurcation gets practically suppressed and the other shows normal growth.

Synapomorphy: Monophyletic groups recognised by uniquely derived characters.

Synthetic characters: Such characters are of wide occurrence among the members of a given group and help in their placement in higher taxa.

Tautonym: A binomial in which the generic name and specific epithet are exactly the same. It is a rejected name, e.g. *Malus malus.*

Taxon: A general term used for any taxonomic group irrespective of its rank.

TEM: Transmission Electron Microscopy.

Tenuinucellate: An ovule in which the nucellus is single-layered.

Tumid: Swollen.

Vasicentric: Concentrated around the vessels.

Verrucose: Warty.

Verticillate: Arranged in a whorl.

References

Akahori A (1965) Steroidal sapogenins contained in Japanese *Dioscorea* spp. Phytochem 4: 97–106

Al-Eisawi DM (1988) Resedaceae in Jordan. Bot Jb 110: 17–39

Aleykutty KM, Inamdar JA (1978) Structure, ontogeny and taxonomic significance of trichomes and stomata in some Capparidaceae. Feddes Rep 89: 19–30

Allan RD, Wells RJ, Correll RL, Macleod JK (1978) The presence of quinones in the genus *Cyperus* as an aid to classification. Phytochem 17: 263–266

Alston RE, Turner BL (1963). Natural hybridisation among four species of *Baptisia.* Am J Bot 50: 159–173

Ambasht RS (1990) A Textbook of Plant Ecology, pp 1–358. Students' Friends Publ, Varanasi (India)

Anuradha M, Pullaiah T (1999) In vitro seed culture and induction of enhanced axillary branching in *Pterocarpus santalinus* and *P. marsupium:* a method for rapid multiplication. Phytomorphology 49: 157–163

Appelqvist LA (1976) Lipids in the Cruciferace, pp 221-278 *In*: Vaughan JG, Macleod AJ, Jones BMG (eds) The Biology and Chemistry of the Cruciferae. Academic Press London, NewYork

Arber A (1925) Monocotyledons: A Morphological Study, pp 1-258. Univ Press, Cambridge (England)

Arber A (1934) The Gramineae: A Study of Cereal, Bamboo and Grass. Univ Press, Cambridge (England)

Arber EAN, Parkin J (1907) On the origin of angiosperms. Bot J Linn Soc (London) 38: 29–80

Ascensã OL, Pais MS (1988) Ultrastructure and histochemistry of secretory ducts in *Artemisia campestris* ssp. *maritima* (Compositae). Nordic J Bot 8: 283–292

Ashlock PK 1971 Monophyly and associated terms. Syst Zool 20: 63–69

Axelrod DI (1952) A theory of angiosperm evolution. Evolution 6: 29–60

Axelrod DI (1960) The evolution of flowering plants *In*: Evolution after Darwin. (ed) S.Tax Vol 1: pp 227–305 Univ of Chicago Press, Chicago

Bailey IW (1923) The cambium and its derivative tissues IV. The increase in girth of the cambium. Am J Bot 100: 499–509

Bailey IW (1944a) The comparative morphology of the Winteraceae III. Wood. J Arnold Arbor 25: 97–103

Bailey IW (1944b) The development of vessels in angiosperms and its significance in morphological research. Am J Bot 31: 421–428

Bailey IW (1949) Origin of the angiosperms: need for a broadened outlook. J Arnold Arbor 30: 64–70

Bailey IW, Swamy BGL (1951) The conduplicate carpel of dicotyledons and its initial trends of specialisation. Am J Bot 38: 373–379

Bakker RT (1978) Dinosaur feeding behaviour and the origin of flowering plants. Nature 274: 661–663

Bakker RT (1986) How dinosaurs invented flowers. Nat Hist 86: 30–38

Barthlott W, Voit G (1979) Mikromorphologie der Samenschalen und Taxonomie der Cactaceae Ein raster-elektronem-microscopischer überblick. Pl Syst Evol 1322: 205–229

Baruah A, Nath SC (1996) Stomatal diversities in *Catharanthus roseus* (L.) G. Don—Some additional information. Phytomorphology 46: 365–369

Bassle Bryson V, Vogel H (eds)(1965) Evolving Genes and Proteins. Academic Press London, New York

Bate-Smith EC (1962) The phenolic constituents of plants and their taxonomic significance I. Dicotyledons. Bot J Linn Soc 58: 95–175

Beck CB (1976) Origin and Early Evolution of Angiosperms, pp 1-341. Columbia Univ Press New York, London

Behnke H-D (1969) Die Sïebröhren-Plastiden der Monocotyledonen. Vergleichende Untersuchungen Über Feinbau und Verbreitung eines characteristischen Plastidentyps. Planta 84: 174–184

Behnke H-D (1976a) Ultrastructure of sieve-element plastids in Caryophyllales (Centrospermae): Evidence for the delimitation and classification of the order. Pl Syst Evol 126: 31–54

Behnke H-D (1976b) A tabulated survey of some characters of systematic importance in Centrospermous families. Pl Syst Evol 126: 95–98

Behnke H-D, Barthlott W (1983) New evidence from the ultrastructural and micromorphological fields in angiosperm classification. Nordic J Bot 3: 43–66

Behnke H-D, Turner BL (1971) On specific sieve-tube plastids in Caryophyllales: Further investigations with special reference to Bataceae. Taxon 20: 731–737

Belford HS, Thompson WF, Stein DB (1981) DNA hybridisation techniques for the study of plant evolution, pp 1-18. *In*: Young DA and Seigler DS (eds) Phytochemistry and Angiosperm Phylogeny. Praeger, NewYork.

Bennett RD, Ko S, Heftmann E (1966) Estrone and cholesterol from the date palm, *Phoenix dactylifera.* Phytochem 5: 231–235

Benson (1962) Plant Taxonomy: Methods and Principles. Ronald Press, New York

Benson L (1970) Plant Classification, pp 1-688. Oxford & IBH Publ, New Delhi

Bentham G, Hooker JD (1965a,b,c) Genera Plantarum, Vols 1, 2, 3 L. Reeve, London (Reprint edition).

Bessey CE (1915) The phylogenetic taxonomy of flowering plants. Ann Missouri Bot Gdn 2: 109–164

Bhanwra RK (1988) Embryology in relation to systematics of Gramineae. Ann Bot (London) 62: 215–233

Bhattacharyya B, Johri BM (1998) Flowering Plants: Taxonomy and Phylogeny, pp 1-753. Narosa Publ. New Delhi

Bhaumik PK, Mukherjee B, Juneau JP, Bhacca NS, Mukherjee R (1979) Alkaloids from leaves of *Annona squamosa.* Phytochem 18: 1584–1586

Bierner MW (1973a) Sesquiterpene lactones and the systematics of *Helenium quadridentatum* and *H. elegans* Biochem Syst 1: 95–96

Bierner MW (1973b) Chemosystematic aspects of flavonoid distribution in twenty two taxa of *Helenium.* Biochem Syst 1: 55–57

Bisset NG, Diaz MA, Ehret C, Ourisson G, Palmade M, Patil F, Pesnelle P, Streith J (1966) Chemotaxonomic studies in the Dipterocarpaceae, II Constituents of the genus *Dipterocarpus*. Phytochem 5: 865–880

Blackwelder RE (1962) Animal taxonomy and the new systematics. *In:* Glass B. (ed) Survey of Biological Progress 4: 1–57

Bordet J (1899) Sur l'agglutination et la dissolution des globules rouges par le serum d'animaux injectes de sang defibrine. Ann Inst Pasteur 13: 225–250

Bose S (1962) Cytotaxonomy of Amaryllidaceae. Bull Bot Surv India 4: 27–38

Boudet AM, Boudet A, Bouyssou H (1977) Taxonomic distribution of isoenzymes of dehydroquinate hydrolyase in the angiosperms. Phytochem 16: 919–922

Boulter D (1983) Plant protein sequence data revisited, pp 119-123. *In:* Jensen U and Fairbrothers DE (eds) Proteins and Nucleic Acids in Plant Systematics. Springer-Verlag, Berlin Heidelberg

Bremekamp CEB (1966) Remarks on the position, the delimitation and the subdivision of the Rubiaceae. Acta Bot Neerl 15: 1–53

Brenner GJ (1976) Middle Cretaceous floral provinces and early migration of angiosperms, pp 23-47. *In*: Beck CB (ed) Origin and Early Evolution of Angiosperms, pp 1–341. Columbia Univ Press, New York

Brenner GJ (1987) Paleotropical evolution of Magnoliidae in the Lower Cretaceous of Northern Gondwana. Am J Bot 74: 677–678

Brenner GJ (1990) An evolutionary model of angiosperm pollen evolution based on fossil angiosperm pollen from the Hauterivian of Israel. Am J Bot (supplement) 77: 82

Brenner GJ (1997) Evidence for the earliest stage of Angiosperm pollen evolution: A Paleoequatorial Section from Israel, pp 91–115 *In*: Taylor DW, Hickey LJ (eds) Flowering Plants: Origin, Evolution and Phylogeny, pp 1–403 Chapman & Hall Inc., New York CBS Publ. New Delhi

Brenner GH, Bickoff I (1992) Palynology and age of the Lower Cretaceous basal Kurnub Group from the coastal plain to the northern Negev of Israel. Palynology 16: 137–185

Brenner M, Niederweiser A, Pataki G (1969) Amino acids and derivatives, pp. 730-786 *In:* Stahl E(ed) Thin Layer Chromatography. George Allen and Unwin, London

Brewbaker JL (1967) The distribution and phylogenetic significance of binucleate and trinucleate pollen grains in the angiosperms. Am J Bot 54: 1069–1083

Briggs BG, Johnson LAS (1979) Evolution in the Myrtaceae: Evidence from inflorescence structure. Proc Linn Soc NSW 102: 157–272

Britton G, Liaaen-Jensen S, Pfander H (eds) (1995a) Carotenoids, Volume 1a Isolation and Analysis. Birkhauser, Basle

Britton G, Liaaen-Jensen S, Pfander H (eds) (1995b) Carotenoids, Volume1b Spectroscopty. Birkhauser, Basle

Brown R (1814) General remarks, geographical, systematical on the botany of Terra Australis. *In:* Finders M (ed) A Voyage to Terra Australis. Vol 2: 1–550

Brown WV (1958) Leaf anatomy in grass systematics. Bot Gaz 119: 170–178

Bryson V, Vogel H (eds) (1965) Evolving Genes and Proteins. Academic Press London, New York

Burger WC (1978) The Piperales and the monocots. Alternative hypothesis for the origin of monocotyledonous flowers. Bot Rev 43: 345–393

Burger WC (1981) Heresay revived: The monocot theory of angiosperm origin. Evol Theory (Chicago) 5: 189–225

Burtt BL (1965) The transfer of Cyrtandromoea from Gesneriaceae to Scrophulariaceae with notes on classification of that family. Bull Bot Surv India 7: 73–88

Canright JE (1962) Comparative morphology of pollen of Annonaceae. Pollen Spores (Paris) 4: 338–339

Carlquist S (1966) Wood anatomy of Compositae. A summary with comments on factors influencing wood evolution. Aliso 6: 25–44

Carlquist S (1975) Ecological strategies of Xylem evolution. Univ of California Press, Berkeley, USA

Carlquist S (1992) Pit membrane remnants in perforation-plates of primitive dicotyledons and their significance. Am J Bot 79: 660–672

Carlquist S (1997) Wood anatomy of primitive angiosperms: new perspectives and synthesis. *In*: Taylor DW, Hickey LJ (eds) Flowering plant Origin, Evolution and Phylogeny, pp 68–90. CBS Publ New Delhi

Carolin RC (1982) The trichomes of the Chenopodiaceae and Amaranthaceae. Bot Jb 103: 451–466

Chakravarty HL (1966) Monograph on the Cucurbitaceae of Iraq. Tech Bull No. 133. Mins Agric, Baghdad

Chakraverty RK, Mukhopadhyay (1990) A Directory of Botanical Gradens and Parks in India. Bot Surv India, Kolkata

Chase MW, Soltis DE, Olmstead RG, Morgan D, Les DH, Mishler BD, Duvall MR, Price RA, Hills HG, QiuY-L, Kron KA, Rettig JH, Conti E, Palmer JD, Manhart JR, Sytsma KJ, Michaels HJ, KressWJ, Karol KG, Clark WD, Hedren M, Gaut BS, Jansen RK, Kim K. -J, Wimpee CR, Smith JF, Furnier GR, Strauss SH, Xiang Q-Y, Plunkett GM, Soltis PS, Swensen SM, Williams SE, Gadek PA, Quinn CJ, Equiarte LE, Golenberg E, Learn J, Graham SW, Barett SCH, Dayanandan S, Albert VA (1993) Phylogenetics of seed plants: An Analysis of nucleotide sequences from the plastid gene rbcL. Ann Missouri Bot Gdn 80: 528–580

Chauhan SVS, Anuradha, Singh Jolly (2003) Stamen dimorphism in three *Cassia* species. Phytomorphology 53: 173–178

Cheadle VI (1942) The occurrence and types of vessels in the various organs of the plant in the Monocotyledoneae. Am J Bot 29: 441–450

Cheadle VI (1943) The origin and trends of specialisation of the vessel in the Monocotyledons. Am J Bot 30: 11–17

Cheadle VI, Kosakai H (1971) Vessels in Liliaceae. Phytomorphology 21: 320–333

Chrispeels MJ, Baumgartner B (1978) Serological evidence confirming the assignment of *Phaseolus aureus* and *P. mungo* to the genus *Vigna.* Phytochem 17: 125–126

Clarke EGC (1970) The forensic chemistry of alkaloids, pp 514-590. *In*: Manske HF (ed) The Alkaloids Vol X11 Academic Press, London NewYork

Clausen J, Keck DD, Heisy WM (1945) Experimental Studies on the Nature of Species-II. Plant Evolution through Amphiploidy and Autoploidy, with Examples from Madiinae. Carnegie Inst. Washington Publ. 564: vi + pp. 174.

Clifford HT, Lavarack PS (1972) The role of vegetative and reproductive attributes in the classification of the Orchidaceae. Bot J. Linn Soc 6: 97–110

Cochrane TS (1978) *Podandrogyne formosa* (Capparidaceae), a new species from Central America. Brittonia 30: 405–410

Connolly JD, Hill RA (eds) (1991) Dictionary of Terpenoids. Chapman and Hall, London

Cordell GA (1974) The biosynthesis of indole-alkaloids. Lloydia 37: 219–298

Corner EJH (1976) The Seeds of Dicotyledons. Vol 1: pp 1–311, Vol 2: pp 1–552. Univ Press, Cambridge London New York Melbourne

Cornet B (1997) A New Gnetophyte from the Late Carnian (Late Triassic) of Texas and its bearing on the origin of the angiosperm carpel and stamen, pp 32–67. *In*: Taylor DW, Hickey LJ (eds) Flowering Plant Origin Evolution and Phylogeny. CBS Publ New Delhi

Crane PR (1985) Phylogenetic analysis of seed plants and the origin of angiosperms. Ann Missouri Bot Gdn 72: 716–793

Crane PR (1987) Vegetational consequences of the angiosperm diversification, pp 107–144. *In*: E.M.Friis, WG Chaloner, PR Crane (eds) Origin of Angiosperms and their Biological Consequences, pp 1–358. Cambridge Univ Press, New York

Crane PR, Dilcher DL (1984) *Lesqueria*: an early angiosperm fruiting axis from the mid-Cretaceous. Ann Missouri Bot Gdn 71: 384–402

Crane PR, Friis EM, Pedersen KR (1995) The origin and early diversification of angiosperms. Nature 374: 27–33

Crane PR, Stockey RA (1985) Growth and reproductive biology of *Joffrea speirsii* gen. et sp. nov., A *Cercidiphyllum-* like plant from the Late Paleocene of Alberta, Canada. Can J Bot 63: 340–364

Crane PR, Friis EM, Kaj RP (1995) The origin and early diversification of angiosperms. Nature 374: 27–33

Crisci JV, Stuessy TF (1980) Determining primitive character-states for phylogenetic reconstruction. Taxon 29: 213–224

Crombie AC (1950) The notion of species in medieval philosophy and science. VI Congr Int, d'Hist Sci Amsterdam. 1: 261–269

Cronquist A (1968) The Evolution and Classification of Flowering Plants, pp 1-396. Thomas Nelson, London Edinburgh

Cronquist A (1969) On the relationship between taxonomy and evolution. Taxon 18: 177–193

Cronquist A (1977) On the taxonomic significance of secondary metabolites in angiosperms, pp179-189. *In*: Kubitzki K (ed), Plant Systematics and Evolution. Springer–Verlag, New York

Cronquist A (1981) An Integrated System of Classification of Flowering Plants, pp 1–1262. Columbia Univ Press, New York

Cronquist A (1988) The Evolution and Classification of Flowering Plants, pp 1–555, 2nd ed, The New York Bot Gdn, Bronx, New York

Culvenor CCJ (1978) Pyrrolizidine alkaloids—occurrence and systematic importance in angiosperms. Bot Notiser 131: 473–486

Curtis PJ, Meade PM (1971) Cucurbitacins from the Cruciferae. Phytochem 10: 3081–3083

Dahlgren G (1989) An updated angiosperm classification. Bot J Linn Soc 100: 197–203

Dahlgren R (1970) Current topics: Parallelism, convergence and analogy in some south African genera of Leguminosae. Bot Notiser 124: 292–304

Dahlgren R (1975a) A system of classification of the angiosperms to be used to demonstrate the distribution of characters. Bot Notiser 128: 119–147

Dahlgren R (1975b) The distribution of characters within an angiosperm system. 1. Some embryological characters. Bot Notiser 128: 181–197

Dahlgren R (1977a) A commentary on a diagrammatic presentation of the angiosperms in relation to the distribution of character-states. Pl Syst Evol Suppl 1: 253–283

Dahlgren R (1977b) A note on the taxonomy of the 'Sympetalae' and related groups. Bull Cairo Univ Herb 7-8: 83–102

Dahlgren R (1980) A revised system of classification of the angiosperms. Bot J Linn Soc 80: 91–124

Dahlgren R (1983) General aspects of angiosperm evolution and macrosystematics. Nordic J Bot 3: 119–149

Dahlgren R, Clifford HT (1982) The Monocotyledons—A Comparative Study, pp 1–378. Academic Press, London New York

Dahlgren R, Clifford HT, Yeo PF (1985) The Families of the Monocotyledons, pp 1–520. Springer, Berlin

Darwin C (1859) The Origin of Species. J Murray, London

Davis GL (1966) Systematic Embryology of the Angiosperms, pp 1-526. John Wiley New York

Davis PH, Heywood VH (1967) Principles of Angiosperm Taxonomy. pp 1-558 D Van Nostrand Princeton, New York.

De Beer GR (1954) Archaeopteryx Lithographica. A study based upon the British Musean Specimen. Brit Mus Nat Hist London

De Beer GR (1958) Embryos and Ancestors. Oxford (UK)

Dement WA, Mabry TJ (1972) Flavonoids of North American species of *Thermopsis.* Phytochem 11: 1089-1093

Deshpande PK (1970) Compositae. *In:* Proc Symp Comparative Embryology of Angiosperms. Bull Indian Natl Sci Acad No. 41: 325–333

De Wit HCD (1963) Plants of the World: The Higher Plants. 2: 1-340. Thames and Hudson, London

De Wolf GPJ (1968) Notes on making a herbarium. Arnoldia 28: 69–111

Dickson J (1936) Studies in floral anatomy. 3. An interpretation of the gynoecium in the Primulaceae. Am J Bot 23: 385–393

Dnyansagar VR (1970) Leguminosae. *In:* Proc Symp Comparative Embryology of Angiosperms. Bull Indian Natl Sci Acad No. 41: 93–103

Dobzhansky T (1950) Mendelian populations and their evolution. Amer Nat 84: 401–418

Dollo L (1893) Les lois de l'évolution. Bull Soc Belg Geol 7: 164–166

Donoghue MJ, Doyle JA (1989) Phylogenetic analysis of angiosperms and the relationships of Hamamelidae, pp 17–45. *In:* Crane PR, Blackmore S (eds) Evolution, Systematics, and Fossil History of the Hamamelidae, 2 Vols. Clarendon Press, Oxford (UK)

Douglas GE (1936) Studies in the vascular anatomy of the Primulaceae. Am J Bot 23: 199–212

Downie SR, Palmer JD (1998) Restriction site mapping of the chloroplast DNA inverted repeat: a molecular phylogeny of the Asteridae. Ann Missouri Bot Gdn 79: 266–283

Doyle JA, Donoghue MJ (1986) Seed plant phylogeny and the origin of angiosperms: An experimental cladistic approach. Bot Rev 52: 321–431

Doyle JA, Donoghue MJ (1993) Phylogenies and angiosperm diversification. Paleobiol 19: 141–167

Doyle JA, Donoghue MJ, Zimmer EA (1994) Integration of morphological and ribosomal RNA data on the origin of angiosperms. Ann Missouri Bot Gdn 81: 419–450

Doyle JA, Hickey LJ (1976) Pollen and leaves from the mid-Cretaceous Potomac Group and their bearing on early angiosperm evolution, pp 139–206. *In*: CB beck (ed) Origin and Early Evolution of Angiosperms. pp 1–341 Columbia Univ Press, New York, London

Doyle JJ, Doyle JL (1999) Nuclear protein–coding genes in phylogeny reconstruction and homology assessment: some example from Leguminosae, pp 229–254. *In*: Hollingsworth PM, Bateman RM and Gornall RJ (eds) Molecular Systematics and Plant Evolution. Taylor and Francis, London

Dreyer DL (1966) Citrus bitter principles. V. Botanical distribution and chemotaxonomy in the Rutaceae. Phytochem 5: 367–378

Dreyer DL, Trousdales EK (1978) Cucurbitacins in *Purshia tridentata.* Phytochem 17: 325–326

Drude O (1897-1898) Umbelliferae. *In:* Engler A, Prantl K (eds) Die natürlichen Pflanzenfamilien. Edn 1. 3(8): 63–250

Duke JA (1961) Preliminary revision of the genus *Drymeria.* Ann Missouri Bot Gdn 48: 173–268

Dutt BSM (1970c) Amaryllidaceae. *In:* Proc Symp Comparative Embryology of Angiosperms. Bull Indian Natl Sci Acad No. 41: 365–367

Dutta SC (1988) Systematic Botany, pp 1-645. Wiley Eastern, New Delhi

Eames AJ (1961) Morphology of the Angiosperms, pp 1-518. McGraw-Hill, New York Toronto London

Eastwood FW, Kirson I, Lavie D, Abraham A (1980) New withanolides from a cross of a South African Chemotype by Chemotype II (Israel) in *Withania somniferum.* Phytochem 19: 1503–1507

Eckardt T (1968) Zur Blütenmorphologie von *Dysphania plantaginella* F. vM. Phytomorphology 17: 165–172

Eglinton G, Gonzales AG, Hamilton RJ, Raphael RA (1962) Hydrocarbon constituents of the wax coatings of plant leaves: A taxonomic survey. Phytochem 1: 89–102

Ehrendorfer F (1964) Cytologie, taxonomie und evolution bei same pflanzen. Vistas in Botany 4:99-186

Ehrendorfer F (1976a) Chromosome numbers and differentiaion of Centrospermous families. Pl Syst Evol 126: 27–30

Ehrendorfer F (1976b) Closing remarks: Systematics and Evolution of Centrospermous families. Pl Syst Evol 126: 99–105

Emboden WA, Lewis H (1967) Terpenes as taxonomic characters in *Salvia* section *Audibertia.* Brittonia 19:152–160

Endress PK (1980a) The reproductive structures and systematic position of the Austrobaileyaceae. Bot Jb 101: 393–433

Endress PK (1980b) Floral structure and relationships of *Hortonia* (Monimiaceae). Pl Syst Evol 133: 199–221

Endress PK, Hufford LD (1989) The diversity of stamen structures and dehiscence patterns among Magnoliidae. Bot J Linn Soc (London) 100: 45–85

Endress PK, Sampson FB (1983) Flower structure and relationships of Trimeniaceae (Laurales). J Arnold Arbor 64: 447–473

Engel T, Barthlott W (1988) Epicuticular waxes of Centrosperms. Pl Syst Evol 161: 71–85

Engler A, Diels L (1936) Syllabus der Pflanzenfamilien, pp 1–419. Gebruder Borntraeger, Berlin

Engler A, Gilg E (1924) Syllabus der Pflanzenfamilien, pp 1–420. Gebruder Borntraeger, Berlin

Erber C, Leins P (1982) Zur spirale in Magnolien-Blüten. Beitr Biol Pfl 56: 225–241

Erber C, Leins P (1983) Zur Sequenz von Blüttenorganen bei einigen Magnoliiden. Bot Jb 103: 433–449

Erdtman G (1943) An Introduction to Pollen Analysis. Ronald Press, NewYork

Erdtman G (1952) Pollen Morphology and Plant Taxonomy, pp 1–539. Chronica Botanica, Waltham, Mass., USA

Erdtman G (1966) Pollen Morphology and Plant Taxonomy. Angiosperms (An Introduction to Palynology.I) Hafner Publ. Co, London

Eriksson O, Bremer B (1992) Pollination systems, dispersal modes, life forms, and diversification rates in angiosperm families. Evolution 46: 258–266

Erdtman H, Norin T (1966) The chemistry of the order Cupressales. Fortschr Chem Org Naturst 24: 206–287

Ettlinger MG, Kjaer A (1968) Sulphur compound in plants: Recent Advances in Phytochem 1: 59–144

Evans CS, Bell EA (1978) Uncommon amino acids in the seeds of 64 species of Caesalpinieae. Phytochem 17: 1127–1129

Evans WC (1986) Hybridisation and secondary metabolism in the Solanaceae, pp 179–186. *In:* D' Arcy WG (ed) Solanaceae: Biology and Systematics, pp 1-603. Columbia Univ Press, New York

Fallen ME (1985) The gynoecial development and systematic position of *Allemanda* (Apocynaceae). Am J Bot 72: 572–579

Fallen ME (1986) Floral structure in Apocynaceae: Morphological, functional and evolutionary aspects. Bot Jb 106: 245–286

Fischer NH, Oliver EJ, Fischer HD (1979) The biogenesis and chemistry of sesquiterpene lactones. Fortsch Chem Naturst 38: 1–390

Fitch WM, Margoliash E (1967) Construction of phylogenetic trees. Science 155: 279–284

Flavell RB, Rimpau J, Smith DB (1977) Repeated sequence DNA relationship in four cereal genomes. Chromosoma 63: 205–222

Flory WS (1977) Overview of chromosome evolution in the Amaryllidaceae. Nucleus 20: 70–88

Friis EM, Crane PR, Pedersen KR (1986) Floral evidence for Cretaceous chloranthoid angiosperms. Nature 320: 163–164

Friis EM, Pederson KR, Crane PR (1994) Angiosperm floral structures from the Early Cretaceous of Portugal. Pl Syst Evol Supple 8: 31–49

Frost FH (1930a) Specialisation in secondary xylem in dicotyledons, I Origin of vessels. Bot Gaz 89: 67–94

Frost FH (1930b) Specialisation in secondary xylem of dicotyledons, II Evolution of end-wall of vessel segment. Bot Gaz 91: 198–212

Frost FH (1931) Specialisation in secondary xylem in dicotyledons, III Evolution of lateral wall of vessel segment. Bot Gaz 91: 88–96

Garbari F (1980) Nasce Presso lo Studio pisano. Nel XVI secolo, la botanica moderna. *In:* "Livorno e Pisa: due citta e un territorio nella politica dei Medici"—Eds Nistri & Lischi, Pisa

Garbari F (1987) The history and present role of Pisa Botanical Gardens, pp 21–27. *In:* Nayar MP (ed) Network of Botanic Gardens. Bot Surv India. Kolkatta

Gattso SJ (1999) A secretory structure in the petiole of *Polygonum convolvulus* L. Phytomorphology 49: 405–410

Gershenzon J, Mabry TJ (1983) Secondary metabolites and the higher classification of angiosperms. Nordic J Bot 3: 5–34

Gershenzon J, Pfeil RM, Liu YL, Mabry TJ, Turner BL (1984) Sesquiterpene lactones from two newly-described species *Vernonia jonesii* and *V. pooleae* Phytochem 23: 777–780

Ghosh KC, Banerjee N (2003) Influence of plant growth regulators on in vitro micropropagation of *Rauwolfia tetraphylla* L. Phytomorphology 53: 11–19

Gibbs RD (1945) Comparative chemistry as an aid to the solution of problems in systematic botany. Trans Roy Soc Canada Sect V 39: 71–103

Goodwin TW(ed) (1976) Distribution of carotenoids, pp 225-261. *In*: Chemistry and Biochemistry of Plant Pigments 2nd edn, Vol 1 Academic Press, London

Goodspeed TH (1945) Cytotaxonomy of *Nicotiana.* Bot Rev 11: 533–592

Goodspeed TH (1947) On the evolution of genus *Nicotiana.* Proc. Nat. Acad Sci 33: 158–171

Gornall RJ, Bohn BA, Dahlgren R (1979) The distribution of flavonoids in the angiosperms. Bot Notiser 132: 1–30

Gottwald H (1997) The anatomy of secondary xylem and the classification of ancient dicotyledons, pp 111–121. *In:* Kubitzki K (ed) Plant Systematics and Evolution, Suppl I, Springer-Verlag New York

Gouth LJ (1966) A comparative study of some resins from the family Cupressaceae (Abstr) pp 175–176. 4th Intern Symp IUPAC, The chemistry of Natural Products

Grant V (1950) Genetic and taxonomic studies in *Gilia, I. Gilia capitata* Aliso 2: 239–316

Grant V (1952a) Genetic and taxonomic studies in *Gilia,* II. *Gilia capitata abrontifolia.* Aliso 2: 361–373

Grant V (1952b) Genetic and taxonomic studies in *Gilia*, III The *Gilia tricolor* complex. Aliso 2: 375–388

Grant V (1954a) Genetic and taxonomic studies in *Gilia*, IV *Gilia achilleaefolia.* Aliso 3: 1–18

Grant V (1954b) Genetic and taxonomic studies in *Gilia*. VI.Interspecific relationships in the leafy stemmed Gilias. Aliso 3: 35–49

Greuter W (1981) "XIII International Botanical Congress". Taxon 30: 904–912

Greuter W (1988) International Code of Botanical Nomenclature. Regnum Veg 118: 1–328

Gunn CR (1981) Seeds of Leguminosae. *In:* Polhill RM, Raven P (eds) Advances in Legume Systematics, Vol 2, pt 2; pp 913–925. International Legume Conference Proc. 1978 Kew, England

Gustafsson MHG, Albert VA (1999) Inferior ovaries and angiosperm diversification, pp 403–431. *In*: Hollingsworth PM, Bateman RM, Gornall RJ (eds) Molecular Systematics and Plant Evolution. Taylor & Francis London, New York

Hair JB, Beuzenberg EJ (1961) High polyploidy in a New Zealand *Poa.* Nature 189: 160

Hallier H (1905) Provisional scheme of the natural (phylogenetic) system of flowering plants. New Phytol 4: 151–162

Hallier H (1912) L'origine et le système phylétique des Angiosperms exposés à P aide de leur arbre gènéralogique. Arch Neerl Sci Exact Natu Ser 1: 146–234

Halse RR, Mishaga R (1988) Seed germination in *Sidalcea nelsoniana.* Phytologia 64: 179–184

Hamby RK, Zimmer EA (1992) Ribosomal RNA as a phylogenetic tool in plant systematics, pp 50–91. *In*: Soltis PS, Soltis DE, Doyle JJ (eds) Molecular Systematics of Plants. Chapman & Hall New York

Harborne JB (ed) (1964) Biochemistry of Phenolic Compounds. Academic Press London

Harborne JB (1968) Correlations between flavonoid pigmentation and systematics in the family Primulaceae. Phytochem 7: 1215–1230

Harborne JB (1998) Phytochemical Methods, A guide to Modern Techniques of Plant Analysis, pp 1–302, Chapman and Hall, London

Harborne JB, Williams CA (1977) Vernonieae—Chemical review, pp 523-538 *In:* Heywood VH, Harborne JB, Turner BL (eds) The Biology of the Compositae. Academic Press, London

Harborne JB, Turner BL (1984) Plant Chemosystematics, pp 1–562. Academic Press, London

Harborne JB, Williams CA Wilson KL (1982) Flavonoids in leaves and inflorescences of Australian *Cyperus* species. Phytochem 21: 2491–2507

Hardman R, Benjamin TV (1976) The co-occurrence of ecdyosomes with bufadienolides and steroidal saponins in the genus *Helleborus.* Phytochem 15: 1515–1516

Harper JL (1977) Population Biology of Plants. Academic Press, London

Hartley RD, Harris PJ (1981) Phenolic constituents of the cell walls of dicotyledons. Biochem Syst Ecol 9: 189–203

Hegnauer R (1963) The taxonomic significance of alkaloids, pp 389-427. *In:* Swain T (ed) Chemical Plant Taxonomy, pp 1–543. Academic Press, London

Hegnauer R (1964) Chemotaxonomie der Pflanzen. 3: 1–743. Birkhauser, Basel, Stuttgart

Hegnauer R (1971) Chemical patterns and relationships of the Umbelliferae, pp 267–277. *In:* Heywood VH (ed) The Biology and Chemistry of the Umbelliferae, pp 1-438. Academic Press, London

Hegnauer R (1973a) Chemotaxonomie der Pflanzen. 6:1-882. Birkhauser, Basel, Stuttgart

Hegnauer R (1973b) Zur systematischen Bedeutung des Markmals der Cyanogenese. Biochem System 1: 191–197

Hegnauer R (1977) Cyanognic compounds as systematic markers in Tracheophytes. Pl Syst Evol Suppl l: 191–209

Heilborn O (1939) Chromosome studies in Cyperaceace III-IV Hereditas 25: 224–240

Hennig W (1966) Phylogenetic systematics (Transl Davis DD, Sangerl R). Urbana

Herz W (1977) Sesquiterpene lactones of the Compositae, pp 337–358. *In:* Heywood VH, Harborne JB, Turner BL (eds) The Biology and Chemistry of the Compositae. Academic Press, London

Heywood VH (1958) The Presentation of Taxonomic Information: a short guide for contributors to Flora Europaea. Licester Univ Press

Heywood VH (1967) Plant Taxonomy. E Arnold, London

Heywood VH (1968) Plant Taxonomy today, pp 3–12. *In:* Heywood VH (ed) Modern Methods in Plant Taxonomy Academic Press, London

Heywood VH, Harborne JB, Turner BL (1977) The Biology and Chemistry of the Compositae, 2 vols. Academic Press, London, New York

Hickey LJ, Doyle JA (1977) Early Cretaceous fossil evidence for angiosperm evolution. Bot Rev 43: 3–104

Hickey LJ, Taylor DW (1977) Origin of the Angiosperm flower, pp 176-231 *In:* DW Taylor and LJ Hickey (eds) Flowering Plant Origin, Evolution and Phylogeny, pp 1–403 CBS Publ, New Delhi

Hiepko P (1965) Vergleichende-morphologische und entwicklungsgeschichtliche Untersuchungen Uber das Perianth bei den Polycarpicae. Bot Jb 84: 359–508

Hohn M, Meinschein WG (1976) Seed oil fatty acids: evolutionary significance in the Nyssaceae and Cornaceae Biochem Syst Ecol 4: 193–199

Hostettman K, Marston A (1995) Saponins. University Press, Cambridge

Hostettmann K, Wagner H (1977) Xanthone glycosides. Phytochem 16: 821–830

Huber H (1969) Die Samenmerkmale und Verwandschaftsverhatnisse der Liliifloren. Mitt Bot Staatssamml, Munich 8: 219–538

Hughes NF (1976) Paleobiology of Angiosperm Origins. Cambridge Univ Press, Cambridge

Hufford LD (1992) Rosidae and their relationship to other nonmagnoliid dicotyledons: a phylogenetic analysis using morphological and chemical data. Ann Missouri Bot Gard 79: 218–248

Hultin E, Torssell K (1965) Alkaloid screening of Swedish plants. Phytochem 4: 425–433

Hutchinson J (1926) The Families of Flowering Plants. Vol 1, Dicotyledons, pp 1–328. Macmillan, London

Hutchinson J (1934) The Families of Flowering Plants. Vol 2, Monocotyledons, pp 1-243. Macmillan, London

Hutchinson J (1948) British Flowering Plants, pp 1–374. P.R. Gawthorn Ltd., London (UK)

Hutchinson J (1959) The Families of Flowering Plants. Edn 2. Dicots, Vol 1: pp 1-510. Clarendon Press, Oxford (UK)

Hutchinson J-(1959) The Families of Flowering Plants, Edn 2. Monocots, Vol 2: pp 511–792. Clarendon Press, Oxford (UK)

Hutchinson J (1969) Evolution and Phylogeny of Flowering Plants, pp 1-717. Academic Press, London New York

Hutchinson J (1973) The Families of Flowering Plants arranged according to a New System based on their Probable Phylogeny, pp 1-968. Edn 3. Oxford Univ Press, London

Inouye H, Takeda Y, Nishimura H (1974) Two new iridoid glucosides from *Gardenia jasminoides* fruits. Phytochem 13: 2219–2224

Jagadish Chandra KS, Rachappaji S, Gowda KRD, Tharasaraswathi KJ (1999) In vitro propagation of *Pisonia alba* (L.) Spanogae (Lettuce Tree) A threatened species. Phytomorphology 49: 43–47

Jeffrey C (1980) A review of the Cucurbitaceae. Bot J Linn Soc 81: 233–247

Jensen SR, Nielsen BJ (1973) Cyanogenic glucosides in *Sambucus nigra* L. Acta Chern Scand 27: 2661–2662

Jensen SR, Nielsen BJ, Dahlgren R (1975) Iridoid compounds, their occurrence and systematic importance in the Angiosperms. Bot Notiser 128: 148–180

Jensen U, Fairbrothers DE, Boulter D (1983) Symposium statements and conclusions. pp, 395–399 *In*: Jensen U and Fairbrothers DE (eds) Proteins and Nucleic Acids in Plant Systematics. Springer-Verlag, Berlin Heidelberg

Johnson BL (1972) Seed protein profiles and the origin of the hexaploid wheats. Am J Bot 59: 952–960

Johri BM (ed) (1984) Embryology of Angiosperms. Springer-Verlag, Berlin

Johri BM, Singh H (1959). The morphology, embryology and systematic position of *Elytraria acaulis* (Linn. f.). Bot Notiser 112: 227–257

Johri BM, Ambegaokar KB, Srivastava PS (1992) Comparative Embryology of Angiosperms: Vol 1: 1–614, 2: 615–1221, Springer Berlin Heidelberg New York Tokyo

Jones DA (1972) Cyanogenic glycosides and their function pp103–124 *In*: Harborne JB (ed) Phytochemical Ecology. Academic Press, London

Jones K (1955) The role of population cytology in evolutionary studies. Heredity 9: 418

Jones SB (1977) Vernonieae-systematic review, pp 503-522 *In*: Heywood VH, Harborne JB, Turner BL (eds) The Biology and chemistry of the Compositae. Academic Press, London

Jones GN (1955) Leguminales: A new, ordinal name. Taxon 4: 188–189

Kasapligil B (1951) Morphological and ontogenetic studies of *Umbellularia californica* Nutt. and *Laurus nobilis* L. Univ Calif Publ Bot 253: 115–240

Kelsey HP, Dayton WA (1942) (eds) Standardised Plant Names, pp 1-675. Edn 2. Am Joint Com Hort Nom Harrisburg. J Horace McFarland, Co., Harrisberg, Pa.

Kelsey RG, Thomas TW, Watson TJ, Shafizadeh F (1975) Population studies in *Artemisia tridentata* spp *vaseyana:* chromosome numbers and sesquiterpene lactone races. Biochem Syst Ecol 3: 209–213

Kenrich P, Crane PR (1997) The origin and early evolution of plants on land. Nature 389: 33–39

Kjaer A (1963) The distribution of sulphur compounds. pp 453–473. *In:* Chemical Plant Taxonomy, Swain T (ed) Academic Press, London

Kjaer A (1966) The distribution of sulphur compounds, pp 187–194 *In*: Swain T (ed) Comparative Phytochemistry. Academic Press, London

Kochhar SL (1981) Economic Botany in the Tropics, pp 1–476. MacMillan India, New Delhi

Kooiman P (1972) The occurrence of iridoid glycosides in the Labiatae. Acta Bot Neerl 21: 417–427

Kostermans AJGH (1978) *Pakaraimaea dipterocarpacea* belongs to Tiliaceae. Taxon 27: 357–359

Kostermans AJGH (1985) Family status for the Monotoideae Gilg and the Pakaraimoideae Ashton, Maquire and de Zeeuw (Dipterocarpaceae). Taxon 34: 426–435

Kotresha K, Seetharam YN (2000) Epidermal micromorphology of some species of *Cassia* L. (Caesalpiniaceae). Phytomorphology 50: 229–237

Kraus R (1897) Über specifische Reactionen in Keimfrein Filtraten aus Cholera, Typhus und Pestbouillon Culturen erzeugt durch homologes serum. Weiner Klin Wechenschr 10: 136–138

Kribs DA (1935) Salient lines of structural specialisation in the wood rays of dicotyledons. Bot Gaz 96: 541–557

Kribs DA (1937) Salient lines of structural specialisation in the wood parenchyma of dicotyledons. Bull Torrey Bot Club 64: 177–186

Kubitzki K (1987) Origin and significance of trimerous flowers. Taxon 16: 21–28

Kumar V, Subramanian B (1987) Chromosome Atlas of Flowering Plants of the Indian Subcontinent. Vol 1: 1–464, Dicotyledons; Vol 2: 465–1095, Monocotyledons. Bot Surv India, Kolkata

Kuprainova LA (1974) On the evolutionary levels in the morphology of pollen and spores: Parallelism and Convergence in evolution. *In:* Nair PKK (ed) Adv Poll Sp Res 1: 31–49

Lakshmanan KK (1970) Potamogetonaceae. *In:* Proc Symp Comparative Embryology of Angiosperms. Bull lndian Natl Sci Acad No. 41: 348–351

Langlet O (1928) Einige Beobachtungen über Zytologie der Berberidaceae. Svensk Bot Tidskr 22: 169–184

Langlet O (1932) Über chromosomenverhältnisse und Systematik der Ranunculaceae. Svensk Bot Tidskr 26: 381

Lavie D, Glotter E (1971) The cucurbitanes, a group of tetracyclic triterpenes. Fortscher Chem Org Naturst 29: 307–456

Lawrence GHM (1951) Taxonomy of Vascular Plants, pp 1-823. MacMillan, New York

Leboeuf M, Cave A, Bhaumik PK, Mukherjee B, Mukherjee R (1982} The phytochemistry of the Annonaceae. Phytochem 21: 2783–2813

Lee YS, Fairbrothers DE (1978) Serological approaches to the Rubiaceae and related families. Taxon 27: 159–185

Lemesle R (1946) Les divers types de fibres a ponctuations aréolées chez les dicotyledones apocarpiques les plus archaiques et leur role dans la phylogénie. Ann Sci Nat Bot et Biol Vegetal 7: 19–40

Lersten NR, Gunn CR, Brubaker CL (1992) Comparative morphology of the lens on Legume (Fabaceae) seeds, with emphasis on species in subfamilies Caesalpinioideae and Mimosoideae. US Dep Agric Tech Bull 1791: 1–44

Les DH, Garvin DK, Wimpee CF (1991) Molecular evolutionary history of ancient aquatic angiosperms. Proc Natl Acad Sci, USA 88: 10119–10123

Lester RN, Roberts PA, Lester C (1983) Analysis of immunotaxonomic data obtained from spur identification and absorption techniques, pp 275-300. *In:* Jensen U, Fairbrothers DE (eds) Proteins and Nucleic Acids in Plant Systematics. Springer-Verlag Berlin, Heidelberg

Levitszky GA (1931) The Karyotype in systematics. Bull Appl Bot Genet Pl Breed 27: 220–240

Lewis H (1951) The origin of supernumerary chromosomes in natural populations of *Clarkia elegans.* Evolution 5: 142–157

Lewis H (1953) Chromosome phylogeny and habitat preference of *Clarkia.* Evolution 7: 102–109

Lewis H, Raven PH (1958) Rapid evolution in *Clarkia.* Evolution 12: 319–336

Le Thomas A (1981) Ultrastructural characters of the pollen grains of African Annonaceae and their significance for the phylogeny of primitive angiosperms (2nd part). Pollen Spores (Paris) 23: 5–36

Loconte H (1997) Comparison of alternate hypotheses for the origin of Angiosperms, pp 267-285. *In*: Taylor DW, Hickey LJ (eds) Flowering plant :Origin, Evolution and Phylogeny, pp 1-403. Chapman & Hall Inc New York

Loconte H, Stevenson DW (1991) Cladistics of the Magnoliidae. Cladistics 7: 267–296

Lotsy JP (1911) Vorträge über botanische Stammesgeschichte. Vol 3. Cormophyta Siphonogamia. Gustav Fischer, Jena

Luning B (1967) Studies on Orchidaceae alkaloids – IV. Screening of species for alkaloids. Phytochem 6: 857–861

Mabberley DJ (1997) The Plant Book. A portable dictionary of the vascular plants, 2nd ed. Press Syndicate of the University of Cambridge, Cambridge, UK

Mabry TJ (1976) Pigment dichotomy and DNA-RNA hybridization data for centrospermous families. Pl Syst Evol 126: 79-94

Marby TJ, Taylor A, Turner BL (1963) The betacyanins and their distribution. Phytochem 2: 61–64

Maheshwari P (1954) Embryology and systematic botany. Proc VIIIth Intl Bot Congr, Paris Sect 7–8: 254–255

Maheshwari P (1964) Embryology in relation to taxonomy. *In:* Turrril WB (ed) Vistas in Botany, Vol 4, pp 55–97.

Mahran GH, El-Hossary GA, Saleh M, Motawe HM (1977a) Isolation and identification of certain molluscicidal substances in *Canna indica* L. J Afr Med Pl 1: 107–119

Mahran GH, El-Hossary GA, Saleh M, Mohamed AM, Motawe HM (1977b) Contribution to the molluscicidal activity of *Canna* species growing in Egypt. J Afr Med Pl 1: 147–155

Maia LD'O (1941) Le grain de pollen dans l'identification et la classification des plantes. 1. Sur la position systématique du genre *Allium.* Bull Bot Soc Portugaise Sci Natu 13: 135–147

Mallick DK, Sawhney RK (2003) Seed coat ornamentation in wild and cultivated lentil taxa. Phytomorphology 53: 187-195

Markham KR, Mabry TJ, Swift WJ (1970) Distribution of flavonoids in the genus *Baptisia.* Phytochem 9: 2359–2364

Martin AC (1946) The comparative internal morphology of seeds. Am Naturalist 36: 513-660.

Maserti BE, Ferrara R, Paterno P (1988) *Posidonia oceanica* (L.) Delile as indicators of mercury contamination in seafood. Proc Intl Symp Environ Life Elements and Health, p 77. Chinese Academy of Sciences in association with Third world Academy of Sciences and WHO, Beijing, China

Mathew PM, Philip O (1983) Studies in the pollen morphology of South Indian Rubiaceae. Advances in Pollen Spore Res, Lucknow 10: 1–80

Matsubara H, Hase T (1983) Phylogenetic consideration of ferredoxin sequences in plants, particularly algae, pp 168–181. *In:* Jensen U, Fairbrothers DE (eds) Proteins and Nucleic Acids in Plant Systematics. Springer-Verlag Berlin Heidelberg

Mayr E (1969) Principles of Systematic Zoology. McGraw-Hill, New York

Mayr E (1942) Systematics and the Origin of Species. Columbia Univ Press, New York

Mears JA, Mabry TJ (1971) Alkaloids in the Leguminosae. *In:* Harborne JB (ed) Chemotaxonomy of Leguminosae. Academic Press, London

Mecklenburg HC (1966) Inflorescence hydrocarbons of some species of *Solanam* and their possible significance. Phytochem 5: 1201–1209.

Melchior H (1964) A. Engler's Syllabus der Pflanzenfamilien (Revised). Vol 2: 368-666. Borntraeger, Berlin

Metcalfe CR (1968) Current development in systematic plant anatomy, pp 45-47 *In*: Heywood VH (ed) Modern Methods in Plant Taxonomy. Academic Press, London

Metcalfe CR, Chalk L (1972) Anatomy of the Dicotyledons. Vol 1: 1–724, Vol 2: 725–1500. Clarendon Press, Oxford (UK)

Mez C (1926) Die Bedeutung der Serodiagnostik für stammesgeschichtliche Forschung. Bot Arch 16: 1–23

Mez C, Ziegenspeck H (1926) Der koenigsberger serodiagnostiche stammbaum. Bot Arch 13: 483–485

Mittal SP (1955) A contribution to the morphology of *Centella asiatica* (Linn) Urban, and some related species. J Indian Bot Soc 34: 248–261

Miyaji Y (1930) Beitrage zur Chromosomen phylogenie der Berberidaceen. Planta 11: 650–659

Mohan Ram HY, Masand P (1963) The embryology of *Nelsonia campestris* R. Br. Phytomorphology 13: 82–91

Moseley MF (1958) Morphological studies in the Nymphaeaceae, 1: The nature of stamens. Phytomorphology 8: 1–29

Muhammad AF, Sattler R (1982) Vessel structure of *Gnetum* and the origin of angiosperms. Am J Bot 69: 1004–1021

Naik VN (1977) Cytotaxonomic studies in six species of *Chlorophytum.* Bot J Linn Soc 74: 297–304

Naik VN (1984) Taxonomy of Angiosperms, pp 1-304. Tata McGraw-Hill, New Delhi

Nair LG, Seeni S (2002) Rapid clonal multiplication of *Morinda umbellata* Linn. (Rubiaceae), a medicinal liana, through cultures of nodes and shoot tips from mature plant. Phytomorphology 52: 77–81

Nair PKK (1960) Pollen grains of cultivated plants. 1. *Canna* L. J Indian Bot Soc 39: 373–381

Nair PKK (1970) Pollen Morphology of Angiosperms: A Historical and Phylogenetic Study, pp 1–160. Barnes and Noble, New York

Nayar MP (1984) Myrtaceae. *In:* Flora of India (Ser IV). Key works to the Taxonomy of Flowering Plants of India. Bot Surv India, Kolkatta 4: 123–130

Nayar MP (1987) Network of Botanical Gardens. Bot Surv India, Kolkata

Nicholas A, Baijnath H (1994) A consensus classification for the order Gentianales with additional details on the suborder Apocynineae. Bot Rev 60: 440–482

Nittala SS, Lavie D (1981) Chemistry and genetics of withanolides in *Withania somniferum* hybrids. Phytochem 20: 2741–2748

Nordby HE, Nagy S (1977) Hydrocarbons from epicuticular waxes of *Citrus* peels. Phytochem 16: 1393–1397

Nowicke JW (1970) Pollen morphology in the Nyctaginaceae. Grana 10: 79–88

Ogundipe T Oluwatoyin, Akinrinlade O Oluwadamilola (1998) Epidermal micromorphology of some species of *Albizia* Durazz (Mimosaceae). Phytomorphology 48: 325–333

Ohmoto T, Ikuse M, Natori S (1970) Triterpenoides of the Gramineae. Phytochem 9: 2137–2148

Okada H, Ueda K (1984) Cytotaxonomical studies on Asian Annonaceae. Pl Syst Evol 144: 165–177

Owen R (1848) Report on the archetype and homologies of the vertebrate skeleton. Rep. 16th Meeting Brit Assoc Adv Sci pp 169–340

Paliwal GS (1965) Development of stomata in *Basella rubra*. Phytomorphology 15: 50–53

Paliwal GS (1969) Stomatal ontogeny and phylogeny I. Monocotyledons. Acta Bot Neerl 10: 654–668

Paliwal GS (1970) Epidermal structure and ontogeny of stomata in some Bignoniaceae. Flora 159: 124–132

Paliwal GS, Kakkar L (1971) Variability in the stomatal apparatus of *Pereskia aculeat*a Mill. J Indian Bot Soc 50a: 164–171

Pant DD, Kidwai PF (1964) On the diversity in the development and organisation of stomata in *Phyla nodiflora.* Michx. Curr Sci 33: 653–654

Paris R (1963) The distribution of plant glycosides, pp 337-358. *In:* Swain T (ed) Chemical Plant Taxonomy, pp 1-543. Academic Press, London

Parkin J (1914) The evolution of the inflorescence. Bot J Linn Soc 42: 511–582

Parulekar NK (1970) Annonaceae. *In:* Proc Symp Comparative Embryology of Angiosperms. Bull Indian Natl Sci Acad No. 41: 38–41

Parveen S Nazneen, SriRamaMurthy K, Pullaiah T (2000) Leaf epidermal characters in *Crotalaria* species (Papilionoideae) from Eastern Ghats. Phytomorphology 50: 205–122

Patel VS, Skvarla JJ, Raven PH (1985) Pollen characters in relation to the delimitation of the Myrtales. Ann Missouri Bot Gdn 71: 558–969

Pax F (1927) Zur Phylogenie der Caryophyllaceae. Bot Jb 61: 223–241

Payne WW, Geissman TA, Lucas AJ, Saitoh T (1973) Chemosystematics and taxonomy of *Ambrosia chamissonis* Biochem Syst 1: 21–33

Peterson FP (1983) Pollen proteins, pp 255–272. *In:* Jensen U, Fairbrothers DE (eds) Proteins and Nucleic Acids in Plant Systematics. Springer-Verlag Berlin Heidelberg

Pettet A (1964) Studies on British Pansies –II. The status of some intermediates between *Viola tricolor* L. and *V. arvensis* Murr. Watsnoia 6: 51–69

Philipson WR (1974) Ovular morphology and the major classification of the dicotyledons. Bot J Linn Soc 68: 89–108

Philipson WR (1975) Evolutionary lines within the dicotyledons. N Z J Bot 13: 73–91

Plouvier V (1963) Distribution of aliphatic polyols and cyclitols, pp 313-336. *In:* Swain T (ed) Chemical Plant Taxonomy, pp 1-543. Academic Press, London

Pollard CJ (1982) Fructose oligosaccharides in the Monocotyledons: A possible delimitation of the order Liliales. Biochem Syst Ecol 10: 245–250

Porter CL (1967) Taxonomy of Flowering Plants, pp 1-452. W H Freeman, San Fransisco

Prance GT (1977) Floristic inventory of the tropics: Where do we stand? Ann Missouri Bot Gdn 64: 659–684

Price JR (1963) The distribution of alkaloids in the Rutaceae, pp 429–452. *In:* Swain T (ed) Chemical Plant Taxonomy, pp 1-543, Academic Press, London

Puri V (1952) Floral anatomy in relation to taxonomy. Agra Univ J Res (Sci) 1: 15–35

Puri V (1962) Floral anatomy in relation to taxonomy. Bull Bot Surv India Kolkata 4: 161–165

Qiu Y-L, Chase MW, Les DH, Parks CR (1993) Molecular phylogenetics of the Magnoliidae: cladistic analyses of nucleotide sequences of the plastid gene rbcL. Ann Missouri Bot Gdn 80: 507–606

Qiu Y-L, Lee J, Bernasconi –Quadroni F, Soltis DE, Soltis PS, Zanis M, Zimmer EA, Chen Z, Savolainen V, Chase MW (1999) The earliest angiosperms: evidence from mitochondrial, plastid and nuclear genomes. Nature 402: 404–407

Radford AE (1986) Fundamentals of Plant Science. Harper and Row Inc. NY

Ramayya N (1969) The development of trichomes in the Compositae, pp 85–113. *In:* Chowdhury KA (ed) Recent Advances in the Anatomy of Tropical Seed Plants. Hindustan Publ Corp, New Delhi

Ramayya N (1972) Classification and phylogeny of the trichomes of angiosperms, pp 91-102. *In*: Ghouse AKM, Yunus M (eds) Research Trends in Plant Taxonomy—K.A.Choudhury Commemoration Volume. Tata Mc-Graw-Hill, New Delhi

Ramayya N, Rajgopal T (1968) Foliar epidermis as a taxonomic aid in the flora of Hyderabad. Part I. Portulacaceae and Aizoaceae. J Osmania Univ 4: 147–160

Ramayya N, Rajgopal T (1971) Foliar dermotypes of Indian Aizoaceae and their use in identification. J Indian Bot Soc 50:355-362

Randolf LF (1936) Developmental morphology of the caryopsis in maize. J Agric Res 53: 881–916

Rangaswamy NS, Chakrabarty B (1966) The Leguminosae of Delhi: Some studies on their morphology and taxonomy. Bull Bot Surv India Kolkata 8: 25–41

Rao RR, Garg Arti (1994) Can *Eremostachys superba* be saved from extinction? Curr Sci 67: 80–81

Raven PH (1976) Systematic botany and plant population biology. Syst Bot 1: 284–316

Rendle AB (1904) The Classification of Flowering Plants. Vol 1: 1–412. Gymnosperms and Monocotyledons. Cambridge Univ Press, Cambridge, London

Rendle AB (1925) The Classification of Flowering Plants. Vol 2: 1–636. Dicotyledons. Cambridge Univ Press, Cambridge, London

Retallack GJ, Dilcher DL (1981) A coastal hypothesis for the dispersal and rise to dominance of flowering plants pp 27–77. *In:* Niklas KJ (ed) Paleobotany, Paleoecology and Evolution. Praeger Publ, New York

Ribereau-Gayon P (1972) Plant Phenolics. Oliver and Boyd, Edinburgh (UK)

Richardson PM (1981) Flavonoids of some controversial members of the Caryophyllales (Centrospermae). Pl Syst Evol 138: 227–233

Riveros GM, Barria O Rosa, Humana Ana Maria (1995) Self-compatibility in distylous *Hedyotis salzmannii* (Rubiaceae). Pl Syst Evol 194: 1–8

Rizk AFM (1987) The chemical constituents and economic plants of the Euphorbiaceae. Bot J Linn Soc 94: 293–326

Robbins MP, Vaughan JG (1983) Rubisco in the Brassicaceae, pp 190-204. *In*: Jensen U, Fairbrothers DE (eds) Proteins and Nucleic Acid in Plant Systematics. Springer –Verlag, Berlin Heidelberg

Robbins WW, Weier TE, Stocking CR (1957) An Introduction to Plant Science. John Wiley, New York

Rodriguez E, Towers GHN, Mitchell JC (1976) Biological activities of sesquiterpene lactones. Phytochem 15: 1573–1580

Rodriguez RL (1971) The relationships of the Umbellales. Bot J Linn Soc 64 (Suppl 1): 63–91

Rogers GK (1984) The Zingiberales (Cannaceae, Marantaceae and Zingiberaceae) in the south-eastern United States. J Arnold Arbor 65: 5–55

Rollins RC (1953) Cytogenetical approaches to the study of genera. Chronica Botanica 14: 133–139

Romeike A (1978) Tropane alkaloids-occurrence and systematic importance in angiosperms. Bot Notiser 131:85-96

Rosenthal GA (1982) Plant Non-protein Amino and Imino Acid: Biochemical and toxicological Properties. Academic Press, London

Roth I (1977) Fruits of Angiosperms. Encycl Pl Anat 10: 1–675

Ruijgrok HWL (1966) The distribution of ranunculin and cyanogenetic compounds in the Ranunculaceae, pp 175-186. *In:* Swain T (ed) Comparative Phytochemistry, pp 1-360. Academic Press, London

Rowlee WW (1898) The morphological significance of the lodicules of grasses. Bot Gaz 15: 199–203

Sahai Kanak (1999) Structural diversity in the lens of the seeds of some *Cassia* L. (Caesalpinioideae) species and its taxonomic significance. Phytomorphology 49: 203–208

Sahasrabudhe S, Stace CA (1974) Developmental and structural variation in the trichomes and stomata of some Gesneriaceae. New Botanist 1: 46–62

Sampson FB (1969) Studies on the Monimiaceae. 2. Floral morphology of *Laurelia novae-zelandiae* A. Cunn. N Z J Bot 7: 214–240

Schinz H (1934) Amaranthaceae. *In:* Engler A, Prantl K (eds) Die natürlichen Pflanzenfamilien. Edn. 16c: 7–85

Schmid R, Carlquist S, Hufford LD,Webster GL (1984) Systematic anatomy of *Oceanopapaver:* A monotypic genus of the Capparaceae from New Caledonia. Bot J Linn Soc 89: 119–152

Scora RW (1967) The essential leaf oils of the genus *Monarda* (Labiatae). Am J Bot 54: 446–452

Scora RW, Bergh BO, Hopfinger JA (1975) Leaf alkanes in *Persea* and related taxa. Biochem Syst Ecol 3: 215–218

Seaman FC (1982) Sesquiterpene lactones as taxonomic characters in the Asteraceae. Bot Rev 48: 121–595

Secor TB, Conn EC, Dunn JE, Seigler DS (1976) Detection and identification of cyanogenic glucosides in six species of *Acacia.* Phytochem 15: 1703–1706

Seibert M, Williams G, Folger G, Milne T (1986) Fuel and chemical co-production from tree crops. Biomass 9: 49–66

Seigler DS (1977) The naturally occurring cyanogenic glycosides. Progr Phytochem 4: 83–120

Seigler DS, Dunn JE, Conn EE, Holstein GL (1978) Acacipetalin from six species of *Acacia* of Mexico and Texas. Phytochem 17: 445–446

Seneviratne AS, Fowden L (1968) The amino acids of the genus *Acacia.* Phytochem 7: 1039–1045

Sengbusch PV (1983) Protein characters and their systematic value, pp 105–118. *In:* Jensen U, Fairbrothers DE (eds) Proteins and Nucleic Acid in Plant Systematics. Springer –Verlag, Berlin Heidelberg

Shamma M, Moniot JL (1978) Isoquinoline Alkaloid Research 1972-1977. Plenum, New York

Sharangpani PR, Shirke DR (1996) Scanning Electron Microscopic studies on ovarian nectaries of *Cassia occidentalis* L. Phytomorphology 46: 277–281

Sharma AK (1969) Evolution and taxonomy of monocotyledons. *In:* Darlington CD (ed) Chromosome today 2: 241–249

Sharma AK, Bal AK (1956) A cytological investigation of some members of the family Cyperaceae, Phyton (Buenos Aires) 6: 7–22

Sharma BD (1969) Pollen morphology of Tiliaceae in relation to plant taxonomy. J Palynol 5: 7–27

Sharma RK (2003) Role of herbarium in identification and authentication of crude drugs used in Indian systems of medicine. The Botanica 53: 169–176

Sherff EE (1940) The delimitation of genera from the conservative point of view. Bull Torrey Bot Club 67: 375–380

Shimizu T, Takao Shizuyo, Utami Nanda, Takao Akio (1996) Anatomy of Floral organs and its taxonomic significance in the genus *Impatiens* (Balsaminaceae), with special reference to subgenus *Acaulimpatiens.* Phytomorphology 46: 277–281

Shorland FB (1963) The distribution of fatty acid in plant lipids, pp. 253–312. *In:* Swain T (ed) Chemical Plant taxonomy, Academic Press, London

Simpson GG (1961). Principles of Animal Taxonomy. New York London

Singh G (1999) Plant Systematics—Theory and Practice. Oxford & IBH Publ. New Delhi Kolkatta

Singh J, Chauhan SVS (1999) Presence of glandular and non-glandular trichomes on anthers of *Tecoma stans* L. Phytomorphology 49: 469–472

Singh V, Jain DK (1978) Floral anatomy and systematic position of *Cyrtandromoea.* Proc Indian Acad Sci Sec B 87: 71–74

Singh U, Wadhwanl AM, Johri BM (1983) Dictionary of Economic Plants in India, pp 1–288. Second enlarged and revised Edition. Reprinted 1990, 1996. Indian Council Agricultural Research, New Delhi

Singh V (1965) Morphological and anatomical studies in Helobiae. 2. Vascular anatomy of the flower of Potamogetonaceae. Bot Gaz 126: 137–144

Sivasubramanian S, Vallinayagam S, Patric RD, Manickan VS (2002) Micropropagation of *Plectranthus vetiveroides* (Jacob) Singh and Sharma—a medicinal plant. Phytomorphology 52: 55–59

Skvarla JJ, Nowicke JW (1976) Ultrastructure of pollen exine in Centrospermous families. Pl Syst Evol 126: 55–78

Small J (1919) Origin and development of the Compositae. New Phytol 18: 201–234

Smith AC (1970) The Pacific as a key to flowering plant history. Harold L Lyon Arboretum Lecture Number 1

Smith PM (1972) Serology and species relationships in annual bromes (*Bromus* L. sect *Bromus*). Ann Bot 36: 1–30

Smith PM (1983) Proteins, mimicry and microevolution in grasses, pp 311-323. *In:* Jensen U, Fairbrothers DE (eds) Proteins and Nucleic Acids in Plant Systematics. Springer-Verlag, Berlin

Smith PM (1976) The Chemotaxonomy of Plants, pp 1–313. Edward Arnold Publ, London

Sokal RR, Crovello TJ (1970) The biological species concept: A critical evaluation. Amer Nat 104: 127–153

Snoad B (1955) Somatic instability of chromosome number in *Hymenocallis calathium*. Heredity 9: 129–134

Solbrig OT (1968) Fertility, sterility and the species problem, pp 77-96. *In:* Heywood VH (ed) Modern methods in Plant Taxonomy, Academic Press, London

Soltis PS, Soltis DE, Doyle JJ (eds) (1992) Molecular Systematics of Plants. Chapman and Hall, New York

Sporne KR (1972) Some observations on the evolution of pollen types in dicotyledons. New Phytol 71: 181–185

Sporne KR (1974) The Morphology of Angiosperms: The structure and Evolution of Flowering Plants. Hutchinson, London

Stace CA (1965) The significance of the leaf epidermis in the taxonomy of the Combretaceae.I. A general review of tribal, generic and specific characters. Bot J Linn Soc 59: 229–252

Stace CA (1981) The significance of the leaf epidermis in the taxonomy of the Combretaceae: Conclusion. Bot J Linn Soc 81: 327–339

Stace CA (1989) Plant Taxonomy and Biosystematics, pp 1-264, 2nd edn. Edward Arnold, London

Stearn WT (1957) An introduction to the "Species Plantarum" and cognate botanical works of Carl Linnaeus. Prefixed to C. Linnaeus, Species Plantarum, 1753. Ray Society Facsimile, Vol 1.

Stebbins GL (1971) Chromosomal Evolution in Higher Plants. Edward Arnold, London

Stebbins GL (1974) Flowering Plants: Evolution above the species level, pp 1–399. Arnold Press, London

Stebbins GL (1977) Development and comparative anatomy of the Compositae, pp 91-109. *In:* Heywood VH, Harborne JB, Turner BL (eds) The Biology and Chemistry of the Compositae. Vol 1: 1–619. Academic Press, London

Stix E (1960) Pollen morphologische untersuchungen in Compositen. Grana Palynologica 2: 41–104

Stoebe B, Hansmann S, Goremykin V. Kowallik KV, Martin W (1999) Proteins encoded in sequenced chloroplast genomes: an overview of gene content, phylogenetic information and endosymbiotic gene transfer to the nucleus. pp. 327–352. *In*: Hollingsworth PM, Bateman RM, Gornall RJ (eds) Molecular Systematics and Plant Evolution. Taylor and Francis, London

Strittmatter LI, Galati BG (2001) Pollen development in *Myosotis azorica* and *M. laxa* (Boraginaceae) Phytomorphology 51: 1–9

Sun G, Ji Q, Dilcher DL, Zheng S, Nixon KC, Wang X (2002) Archaefructaceae, a new basal angiosperm family. Science 296: 899–904

Sytsma KJ, Baum DA (1997) Molecular phylogenies and the diversification of the angiosperms, pp 314-340. *In*: Taylor DW, Hickey LJ (eds) Flowering Plants: Origin, Evolution and Phylogeny, pp 1-403. Chapman & Hall Inc, New York

Takaso T, Bouman F (1984) Ovule ontogeny and seed development in *Potamogeton natans* (Potamogetonaceae) with a note on the campylotropous ovule. Acta Bot Neerl 33: 519–533

Takhtajan A (1959) Die evolution der Angiospermen, pp 1–344. Gustav Fischer, Jena

Takhtajan A (1966) Systema et phylogenia Magnoliophytorum, pp 1–610. Soviet Sciences Press, Moscow Leningrad

Takhtajan A (1969) Flowering Plants: Origin and Dispersal, pp 1–310. Oliver & Boyd, Edinburgh (UK)

Takhtajan A (1980) Outline of the classification of Flowering plants (Magnoliophyta). Bot Rev 46: 225-359

Takhtajan A (1987) Systema Magnoliophytorum (in Russian), pp 1–439. Nauka Publ, Moscow

Takhtajan AL (1991) Evolutionary Trends in Flowering Plants. Columbia Univ Press, New York

Tamura M (1963) Morphology, ecology and phylogeny of the Ranunculaceae 1. Sci Rep Osaka Univ 11: 115–126

Tamura M (1972) Morphology and phyletic relationship of the Glaucidiaceae. Bot Mag Tokyo 85: 29–41

Tateoka T (1962) Starch grains of endosperm in grass systematics. Bot Mag (Tokyo) 75: 377–383

Taylor DW, Hickey LJ (1990) A new hypothesis of the morphology of the ancestral angiosperms. Am J Bot (Suppl) 77: 159 (Abstr)

Taylor DW, Hickey LJ (1992) Phylogenetic evidence for the herbaceous origin of angiosperms. Pl Syst Evol 180: 137–156

Taylor DW, Hickey LJ (1997) Introduction: The challenge of Flowering Plant History, pp 1-7. *In*: Taylor DW, Hickey LJ (eds) Flowering Plant Origin, Evolution and Phylogeny, pp1-403. Chapman & Hall Inc, New York

Taylor RJ, Campbell D (1969) Biochemical systematics and phylogenetic interpretations in the genus *Aquilegia.* Evolution 23: 153–162

Taylor WI, Farnsworth NR (eds) (1975) The *Catharanthus* Alkaloids, pp 1–323. Marcel Dekker, New York

Tetenyi P (1973) Homology of biosynthetic routes, the base in chemotaxonomy. *In:* Bendz G, Santesson J (eds) Chemistry in Botanical Classification. New York, Landon

Thien LB, Heimermann W, Holman RT (1975) Floral odours and quantitative taxonomy of *Magnolia* and *Liriodendron.* Taxon 24: 557–568

Thomas V, Mercykutty VC, Saraswathy Amma CK (1996) Seed biology of Para Rubber tree *(Hevea brasiliensis* Muell. Arg., Euphorbiaceae): A Review. Phytomorphology 46: 335–342

Thorne RF (1963) Some problems and guiding principles of angiosperm phylogeny. Am Natu 97: 287–305

Thorne RF (1968) Synopsis of a putatively phylogenetic classification of the flowering plants. Aliso 6: 57–66

Thorne RF (1973) Inclusion of the Apiaceae (Umbelliferae) in the Araliaceae. Notes R Bot Gard Edinb 32: 161–165

Thorne RF (1976) A phylogenetic classification of the Angiospermae. Evol Biol 9: 35–106

Thorne RF (1977) Some realignments in the Angiospermae. Pl Syst Evol Suppl 1: 299–319

Thorne RF (1983) Proposed new realignments in the angiosperms. Nordic J Bot 3: 85–117

Thorne RF (1992) Classification and geography of the flowering plants. Bot Rev 58: 225–348

Thorne RF (1997) The least specialised angiosperms, pp 286-313. *In*: Taylor DW, Hickey LJ (eds) Flowering Plants Origin, Evolution and Phylogeny, pp 1-403. Chapman & Hall Inc. New York

Tilman D (1988) Plant Strategies and the Dynamics and Structure of Plant Communities. Princeton Univ Press Princeton, N.J.

Tobe H (1981) Embryological studies in *Glaucidium palmatum* Sieb. et Zucc. with discussion on the taxonomy of the genus. Bot Mag (Tokyo) 94: 207–224

Tobe H, Raven PH (1983a) An embryological analysis of the Myrtales: Its definition and characteristics. Ann Missouri Bot Gdn 70: 71–94 .

Tucker SC (1960) Ontogeny of the floral apex of *Michelia fuscata.* Am J Bot 47: 266–277

Turill WB (1953) Pioneer Plant Geography: The Phytogeographical researches of Sir Joseph Dalton Hooker. Nijhoff, Hague

Turner BL (1977) Chemosystematics and its effect upon the traditionalist Ann Missouri Bot Gdn 64: 235–242

Turner BL (1981) New species and combination in *Vernonia* section *Leiboldia* and *Lepidonia* (Asteraceae). Brittonia 33: 401–412

Valentine DH, Löve Á (1958) Taxonomic and biosystematic categories. Brittonia 10: 153–166

Van Valen L (1978) Why not be a cladist? Evol Theory 3: 285–299

Varghese TM (1970) Solanaceae. *In:* Proc Symp Comparative Embryology of Angiosperms. Bull Indian Natl Sci Acad No. 41: 255–258

Venkata Rao C (1954) Embryological studies in Malvaceae. 1. Development of gametophytes. Proc Natl Inst Sci India B 20: 127–150

Verdcourt B (1970) Studies in the Leguminosae-Papilionoideae for the flora of Tropical East –Africa. Kew Bull 24: 379–447

Vickery JR (1971) The fatty acid composition of the seed oils of Proteaceae: a chemotaxonomic study. Phytochem 10: 123–130

Wagenitz G (1959) Die systematische Stellung der Rubiaceae. Bot Jb 79: 17–35

Wagenitz G (1964) Rubiaceae. *In:* Melchior H (ed) A Engler's Syllabus der Pflanzenfamilien. Vol 2: 417–424. Borntraeger, Berlin

Wagenitz G (1976) Systematics and phylogeny of the Compositae (Asteraceae). Pl Syst Evol 125: 29–46

Walker JW (1976) Comparative pollen morphology and phylogeny of the Ranalean complex, pp 241–299. *In:* Beck CB (ed) Origin and Early Evolution of Angiosperms, pp 1-341. Columbia Univ Press, New York

Walker JW, Walker AG (1984) Ultrastructure of Lower Cretaceous angiosperm pollen and origin, and early evolution of flowering plants. Ann Missouri Bot Gdn 71: 464–521

Wagner WH Jr (1961) Problems in the classification of ferns. Recent Advances in Botany, pp 841–844

Ward JV, Doyle JA, Hotton CL (1988) Probable granular magnoliid angiosperm pollen from the Early Cretaceous. Am J Bot 75; 118–119

Warmke HE (1941) Chromosome continuity and individuality. Gold Spring Herb. Symp Quant Biol. 9: 1–6

Weatherwax (1929) The morphology of the spikelet of six genera of Oryzeae. Am J Bot 16: 547–555

Webster GL (1967) The genera of Euphorbiaceae in the southeastern United States. J Arnold Arbor 48: 303–430

Weete JD (1972) Aliphatic hydrocarbons of the fungi. Phytochem 11: 1201–1205

Wernham HF (1911) Floral evolution: With particular reference to the sympetalous dicotyledons. 3. The Pentacyclidae. New Phytol 10: 145–159

Wettstein R (1935) Handbuch der Systematischen Botanik. Edn 4. pp 1–1152. Leipzing Wien

Wilber RL (1963) The Leguminous Plants of North Carolina. Tech Bull North Carolina Agric Exp Stn No, 151: 1–294

Wildman SG (1983) Polypeptide composition of Rubisco as an aid in studies of plant phylogeny, pp 182-190 *In*: Jensen U, Fairbrothers DE (eds) Proteins and Nucleic Acid in Plant Systematics. Springer-Verlag, Berlin Heidelberg

Wiley EO (1981) Phylogentics—The Theory and Practice of Phylogentic Systematics. John Wiley and Sons, New York

Willis JC (1922) Age and Area. Cambridge Univ Press, Cambridge

Willis JC (1973) A Dictionary of the Flowering Plants and Ferns. Edn VIII (revised by Airy Shaw HK), pp 1–1245. Univ press Cambridge (England)

Wink M (1985) Chemische Verteidigung der Lupinen: Zur biologischen Bedeutung der chinolizidinalkaloide. Pl Syst Evol 150: 65–81

Withner CI (1941) Stem anatomy and phylogeny of the Rhoipteleaceae. Am J Bot 28: 872–878

Wolf K, Morgan-Richards M (1999) The use of RAPD data in the analysis of population genetic structure: case studies of *Alkanna* (Boraginaceal) and *Plantago* (Plantaginaceae), pp 51-73. *In*: Hollingsworth PM, Bateman RM, Gornall RJ (eds) Molecular Systematics and Plant Evolution. The Systematic Association Special Volume, Series 51, London, New York

Wolfe JA, Doyle JA, Page VM (1976) The bases of angiosperm phylogeny: Paleobotany. Ann Missouri Bot Gdn 62: 801–824

Wollenweber E (1982) Flavones and flavonols, pp 189-259. *In:* Harborne JB, Mabry TJ (eds) The Flavonoids: Advances in Research from 1975-1981. Chapman & Hall, London

Worsdell WC (1908) A study of vascular system in certain orders of the Ranales. Ann Bot (London) 22: 561-682

Worsdell WC (1908) The affinities of *Paeonia.* J Bot (London) 46: 114–116

Yamada Y, Hagiwara, Iguchi K, Takahasi Y, Hsu H (1978) Cucurbitacins from *Anagallis arvensis* Phytochem 17: 1798

Young DA (1981) The usefulness of flavonoids in angiospermous phylogeny: Some selected examples, pp 205-232. *In:* Young DA, Siegler DS (eds) Phytochemistry and Angiosperm Phylogeny. Praeger Scientific, New York

Young DA (1981) Are the angiosperms primitively vesselless? Syst Bot 6: 313–350

Zavada MS (1983) Comparative morphology of monocot pollen and evolutionary trends of apertures and wall structures. Bot Rev 49: 331–379

Zimmer EA, Hamby RK, Arnold ML, Leblanc DA, Theriot EC (1989) Ribosomal RNA phylogenies and flowering plant evolution, pp 205-214. *In*: Fernholm B, Bremer K, Jornvall H (eds) The Hierarchy of Life. Elsevier Science Publ, New York

Plant Index